BIRKHÄUSER

Operator Theory: Advances and Applications

Vol. 198

Recent Advances in Operator Theory in Hilbert and Krein Spaces

Jussi Behrndt
Karl-Heinz Förster
Carsten Trunk
Editors

Birkhäuser
Basel · Boston · Berlin

Editors:

Jussi Behrndt
Karl-Heinz Förster
Technische Universität Berlin
Institut für Mathematik
Strasse des 17. Juni 136
10623 Berlin
Germany
e-mail: behrndt@math.tu-berlin.de
foerster@math.tu-berlin.de

Carsten Trunk
Technische Universität Ilmenau
Institut für Mathematik
Postfach 100565
98684 Ilmenau
Germany
e-mail: carsten.trunk@tu-ilmenau.de

2000 Mathematical Subject Classification: Primary: 47-06; secondary: 47B50, 47Axx, 34Bxx, 15Axx

Library of Congress Control Number: 2009937257

Bibliographic information published by Die Deutsche Bibliothek.
Die Deutsche Bibliothek lists this publication in the Deutsche Nationalbibliografie; detailed bibliographic data is available in the Internet at http://dnb.ddb.de

ISBN 978-3-0346-0179-5 Birkhäuser Verlag AG, Basel – Boston – Berlin

Basel · Boston · Berlin
P.O. Box 133, CH-4010 Basel, Switzerland
Part of Springer Science+Business Media
Printed on acid-free paper produced from chlorine-free pulp. TCF∞
Printed in Germany

ISBN 978-3-0346-0179-5
e-ISBN 978-3-0346-0180-1

9 8 7 6 5 4 3 2 1
www.birkhauser.ch

Contents

Preface

The present book is a memorial volume devoted to our friend, colleague and teacher Peter Jonas who passed away on July 18, 2007. It displays recent advances in modern operator theory in Hilbert and Krein spaces and contains a collection of original research papers written by participants of the *7th Workshop on Operator Theory in Krein Spaces and Spectral Analysis*, which was held at the Technische Universität Berlin, Germany, December 13 to 16, 2007. The articles in this volume contain new results for problems close to the area of research of Peter Jonas: Spectral and perturbation problems for operators in inner product spaces, generalized Nevanlinna functions and definitizable functions, scattering theory, extension theory for symmetric operators, fixed points, hyperbolic matrix polynomials, moment problems, indefinite spectral and Sturm-Liouville problems, and invariant subspace problems.

It is a pleasure to acknowledge the substantial financial support for the *7th Workshop on Operator Theory in Krein Spaces and Spectral Analysis* received from the

- Berlin Mathematical School (BMS)
- Gesellschaft für Angewandte Mathematik und Mechanik (GAMM)
- International Mathematical Union, Commission on Development and Exchanges
- Institute of Mathematics of the Technische Universität Berlin

The Editors

Peter Jonas (1941–2007)

In Memoriam Peter Jonas (1941–2007)

Jussi Behrndt, Karl-Heinz Förster and Carsten Trunk

Peter Jonas was born on July 18, 1941, in Memel, now Klaipeda, which was at that time the most eastern town of East Prussia. After the war, Peter Jonas moved with his mother and grandmother to Blankenfelde – a small village near Berlin, where he lived until the end of his school education.

In 1959 Peter Jonas started to study Mathematics at the Technische Universität Dresden. There he met Heinz Langer, who held at that time exercise classes in analysis, and Peter Jonas wrote his diploma thesis on stability problems of infinite-dimensional Hamiltonian systems under the supervision of Heinz Langer.

After his diploma in 1964, Peter Jonas got a position at the Karl-Weierstrass Institute of the Academy of Sciences in East Berlin where he first worked with his PhD supervisor Josef Naas on problems in differential geometry, partial differential equations and conformal mappings. At this time he married his wife Erika and his children Simon and Judith were born. After his PhD in 1969, Peter Jonas joined the mathematical physics group around Hellmut Baumgärtel, and the theory of selfadjoint and unitary operators in Krein spaces became one of his main mathematical interests.

In these early years of his research on selfadjoint and unitary operators in Krein spaces he was mainly interested in the existence of a spectral function/distribution and on stability criteria for perturbations in Krein spaces. It was in 1979 when his paper with Heinz Langer *Compact perturbations of definitizable operators* appeared, where it was proved that definitizable operators are stable under finite-dimensional perturbations. This paper is one of the milestones in perturbation theory of selfadjoint operators in Krein spaces and has inspired many colleagues and essentially initiated further research.

Peter Jonas himself continued with this topic and wrote a series of papers on compact perturbations of various classes of operators in Krein spaces applying his results to the Klein-Gordon equation. After his stay in Odessa in 1981 these activities finally culminated in the cooperation with Vadim A. Adamyan, Mark G. Krein and Heinz Langer; and led to his Habilitations thesis *Die Spurformel der Störungstheorie für einige Klassen unitärer und selbstadjungierter Operatoren im Kreinraum* (1987).

Peters deep interest in perturbation theory led him naturally to a generalization of the class of definitizable operators which appeared already in 1967 in a paper of Heinz Langer without having a name at that time: Locally definitizable operators. Peter worked on these classes of operators from about 1986 till his death in 2007. For sure one of his main contributions to operator theory in Krein spaces is the systematic and comprehensive study of unitary and selfadjoint locally definitizable operators. In several papers Peter established the existence of a local spectral function, he developed a perturbation theory and applied the general results to the Klein-Gordon equation, to problems with floating singularities and to spectral problems with eigenvalue depending boundary conditions.

Closely connected to Peters research on definitizable and locally definitizable operators was the introduction and investigation of related classes of functions: Definitizable and locally definitizable functions. Roughly speaking, these classes of functions are related to definitizable and locally definitizable operators in a similar way as generalized Nevanlinna functions are related to selfadjoint operators in Pontryagin spaces. Making use of methods from the theory of distributions Peter was able to characterize local sign type properties of these functions. In the paper *Operator representations of definitizable functions* published in 2000 he proved the existence of a minimal representating operator (or relation) in a model Krein space and he showed how the spectral information of the representing operator is encoded in the definitizable function. It is amazing to see how efficient his abstract results work in applications. This theory plays nowadays an essential role in the analysis of Titchmarsh-Weyl-coefficients associated to indefinite Sturm-Liouville operators and is applied by various groups in Germany, Canada and Eastern Europe. We wish he could have witnessed it.

Throughout his career at the Karl-Weierstrass Institute, Peter Jonas established fruitful scientific contacts with many mathematicians in the Soviet Union and other Eastern European countries. Many of these colleagues became close personal friends, among them Vadim Adamyan, Tomas Azizov, Tsembeltsogt Bajasgalan, Branko Curgus, Aurelian Gheondea, Alexander Markus, Vladimir Matsaev, Franciszek Hugon Szafraniec, Vladimir Strauss and many others. At conferences in Eastern Europe Peter met with West European colleagues, but during the Cold War it was impossible for him to visit them in their home countries.

The political changes in 1989 had a tremendous influence on Peters life. The Karl-Weierstrass Institute was closed down in 1991, Peter lost his permanent position and became a member of the so-called Wissenschaftler-Integrations-Programm; a program that was initiated to incorporate employees of scientific institutions of the former GDR into German universities. However, this program was rather inefficient and, as a result, Peters situation was vague. But it was not Peter's to complain, rather he used this situation to obtain various positions at the Technische Universität Berlin, Freie Universität Berlin and at the Universität Potsdam. After a research stay in Bellingham (USA) he finally settled down with the help of Karl-Heinz Förster at the Technische Universität where he worked until his retirement in 2006. In his last years, Peter Jonas used the possibility to meet various colleagues and friends in the USA, Israel, Austria, Venezuela, Turkey and the Netherlands. Beside his passion for mathematics, Peter was very interested in Asian culture, in particular, Buddhism; he undertook visits to countries such as Thailand, Laos, Burma and Cambodia.

The Functional Analysis group at the Technische Universität Berlin tremendously benefited from Peter. With passion he supervised PhD students (Carsten Trunk '02 and Jussi Behrndt '05) and diploma students (Antonius Gacinski '93, Carsten Trunk '95, Kai Anding '95, Michael Krause '97, Jussi Behrndt '02, Christian Kreusler '05 and Friedrich Philipp '06). Peter was an excellent and exceptional teacher: he spent whole days, sometimes nights and numerous phone calls for explaining and advising. Whenever his students had questions he always had the patience for a detailed, helpful answer. Besides this, Peter gave courses and special lectures in operator theory. In addition, he invited specialists from all over the world for research stays and to the *Operator Theory Colloquium* at the Technische Universität Berlin. Moreover, Peter initiated, together with Karl-Heinz Förster, the series of *Workshops on Operator Theory in Krein Spaces* held annually since 2001 at the Technische Universität Berlin.

Many of Peters colleagues have experienced his friendship and his hospitality during research stays or while attending conferences in Berlin. This friendship was a result of Peters life-long ties and numerous visits to his colleagues, and it was a result of his personality and his way of doing mathematics. It was his special mixture of profound and deep knowledge, his modest, calm and well-balanced style, his silent but rigorous way of doing research and his uncompromising style of writing papers which made him to this impressive person he was.

In April 2007, Peter Jonas suddenly got serious health problems and after a surgery and a short time of recovery he died at his 66th birthday on July 18, 2007. We will remember and miss him as a friend, colleague and teacher.

Jussi Behrndt and Karl-Heinz Förster
Technische Universität Berlin
Institut für Mathematik, MA 6–4
Straße des 17. Juni 136
D–10623 Berlin, Germany
e-mail: `behrndt@math.tu-berlin.de`
`foerster@math.tu-berlin.de`

Carsten Trunk
Technische Universität Ilmenau
Institut für Mathematik
Fakultät für Mathematik und Naturwissenschaften
Postfach 10 05 65
D–98684 Ilmenau, Germany
e-mail: `carsten.trunk@tu-ilmenau.de`

List of Publications of Peter Jonas

[1] Eine Bedingung für die Existenz einer Eigenspektralfunktion für gewisse Automorphismen lokalkonvexer Räume, Math. Nachr. 45 (1970), 143–160.

[2] Über Folgen von Bogenlängenfunktionen zu gleichmäßig konvergenten Kurvenfolgen, Math. Nachr. 48 (1971), 113–128.

[3] Über Folgen von Bogenlängenfunktionen zu distributionentheoretisch konvergenten Kurvenfolgen, Math. Nachr. 48 (1971), 291–308.

[4] Zur Existenz beschränkter homöomorpher Lösungen der Beltramigleichung mit Entartung am Rande, Beiträge zur komplexen Analysis und deren Anwendungen in der Differentialgeometrie, pp. 43–52. Schr. Zentralinst. Math. Mech. Akad. Wissensch. DDR, Heft 18, Akademie-Verlag, Berlin, 1974.

[5] Über die Erhaltung der Stabilität J-positiver Operatoren bei J-positiven und J-negativen Störungen, Math. Nachr. 65 (1975), 211–218.

[6] On the stability of linear Hamiltonian equations in Hilbert space, VII. int. Konf. nichtlineare Schwing. Berlin 1975, Bd. I, 1; Abh. Akad. Wiss. DDR, 3 N, 389 (1977).

[7] Zur Existenz von Eigenspektralfunktionen für J-positive Operatoren. I, Math. Nachr. 82 (1978), 241–254.

[8] Zur Existenz von Eigenspektralfunktionen für J-positive Operatoren. II, Math. Nachr. 83 (1978), 197–207.

[9] Zur Existenz von Eigenspektralfunktionen mit Singularitäten, Math. Nachr. 88 (1979), 345–361.

[10] Compact perturbations of definitizable operators, J. Operator Theory 2 no. 1 (1979), 63–77 (with H. Langer).

[11] On the local wave operators for definitizable operators in Krein space and on a paper of T. Kako, Prepr. Akad. Wiss. DDR, Zentralinst. Math. Mech. 46 (1979), 30 p.

[12] Relatively compact perturbations of uniformly J-positive operators, Prepr. Akad. Wiss. DDR, Zentralinst. Math. Mech. 15 (1980), 22 p.

[13] On the functional calculus and the spectral function for definitizable operators in Krein space, Beiträge Anal. no. 16 (1981), 121–135.

[14] Compact perturbations of definitizable operators. II, J. Operator Theory 8 no. 1 (1982), 3–18.

[15] On spectral distributions of definitizable operators in Krein space, Spectral theory (Warsaw, 1977), pp. 301–311, Banach Center Publ. 8, Warsaw, 1982.

[16] Some questions in the perturbation theory of J-nonnegative operators in Krein spaces, Math. Nachr. 114 (1983), 205–226 (with H. Langer).

[17] On a class of J-unitary operators in Krein space, Prepr. Akad. Wiss. DDR, Inst. Math. P-MATH-13/83 (1983), 25 p.

[18] Regularity criteria for critical points of definitizable operators, Oper. Theory Adv. Appl. 14 (1984), 179–195.

[19] A model for π-selfadjoint operators in π_1-spaces and a special linear pencil, Integral Equations Operator Theory 8 no. 1 (1985), 13–35 (with H. Langer).

[20] On the perturbation theory of nonnegative operators in Krein space, Topics in quantum field theory and spectral theory (Reinhardsbrunn, 1985), pp. 110–123, Rep. MATH 86-1, Akad. Wiss. DDR, Berlin, 1986.

[21] Die Spurformel der Störungstheorie für einige Klassen unitärer und selbstadjungierter Operatoren im Kreinraum, Report MATH 86-6, Akad. Wiss. DDR, Berlin, 1986.

[22] On a class of unitary operators in Krein space, Oper. Theory Adv. Appl. 17 (1986), 151–172.

[23] A characterization of spectral functions of definitizable operators, J. Operator Theory 17 no. 1 (1987), 99–119 (with A. Gheondea).

[24] On a class of selfadjoint operators in Krein space and their compact perturbations, Integral Equations Operator Theory 11 no. 3 (1988), 351–384.

[25] On the trace formula of perturbation theory. I, Math. Nachr. 137 (1988), 257–281.

[26] A note on perturbations of selfadjoint operators in Krein spaces, Oper. Theory Adv. Appl. 43 (1990), 229–235.

[27] Compact perturbations of normal operators in a Krein space, Ukrain. Mat. Zh. 42 no. 10 (1990), 1299–1306; translation in Ukrainian Math. J. 42 no. 10 (1990), 1155–1161 (with T.Ya. Azizov).

[28] On a problem of the perturbation theory of selfadjoint operators in Krein spaces, J. Operator Theory 25 no. 1 (1991), 183–211.

[29] A class of operator-valued meromorphic functions on the unit disc, Ann. Acad. Sci. Fenn. Ser. A I Math. 17 no. 2 (1992), 257–284.

[30] Models and unitary equivalence of cyclic selfadjoint operators in Pontrjagin spaces, Oper. Theory Adv. Appl. 59 (1992), 252–284 (with H. Langer and B. Textorius).

[31] On the spectral theory of operators associated with perturbed Klein-Gordon and wave type equations, J. Operator Theory 29 no. 2 (1993), 207–224.

[32] Selfadjoint extensions of a closed linear relation of defect one in a Krein space, Oper. Theory Adv. Appl. 80 (1995), 176–205 (with H. Langer).

[33] Riggings and relatively form bounded perturbations of nonnegative operators in Krein spaces, Oper. Theory Adv. Appl. 106 (1998), 259–273.

[34] On the trace formula of perturbation theory. II, Math. Nachr. 197 (1999), 29–49.

[35] Operator representations of definitizable functions, Ann. Acad. Sci. Fenn. Math. 25 no. 1 (2000), 41–72.

[36] On bounded perturbations of operators of Klein-Gordon type, Glas. Mat. Ser. III 35(55) no. 1 (2000), 59–74.

[37] Similarity between Krein space bicontractions and Hilbert space contractions, Oper. Theory Adv. Appl. 124 (2001), 109–129 (with T.Ya. Azizov, A.I. Barsukov, and A. Dijksma).

[38] On a class of analytic operator functions and their linearizations, Math. Nachr. 243 (2002), 92–133 (with C. Trunk).

[39] On locally definite operators in Krein spaces, Spectral analysis and its applications, pp. 95–127, Theta Ser. Adv. Math. 2, Theta, Bucharest, 2003.

[40] On compact perturbations of locally definitizable selfadjoint relations in Krein spaces, Integral Equations Operator Theory 52 no. 1 (2005), 17–44 (with J. Behrndt).

[41] On the infimum of quantum effects, J. Math. Phys. 46 no. 6 (2005), 062102, 11 pp (with A. Gheondea and S. Gudder).

[42] Spectral points of type π_+ and π_- of self-adjoint operators in Krein spaces, J. Funct. Anal. 226 no. 1 (2005), 114–137 (with T.Ya. Azizov and C. Trunk).

[43] Partial non-stationary perturbation determinants for a class of J-symmetric operators, Oper. Theory Adv. Appl. 162 (2006), 1–17 (with V. Adamyan and H. Langer).

[44] On operator representations of locally definitizable functions, Oper. Theory Adv. Appl. 162 (2006), 165–190.

[45] Boundary value problems with local generalized Nevanlinna functions in the boundary condition, Integral Equations Operator Theory 55 no. 4 (2006), 453–475 (with J. Behrndt).

[46] A Sturm-Liouville problem depending rationally on the eigenvalue parameter, Math. Nachr. 280 no. 15 (2007), 1709–1726 (with C. Trunk).

[47] On the spectrum of the self-adjoint extensions of a nonnegative linear relation of defect one in a Krein space, Oper. Theory Adv. Appl. 175 (2007), 121–158 (with H. Langer).

[48] On locally definitizable matrix functions, (Russian) Funktsional. Anal. i Prilozhen. 41 no. 3 (2007), 1–16; translation in Funct. Anal. Appl. 41 no. 3 (2007), 169–180 (with T.Ya. Azizov).

[49] Compact and finite rank perturbations of closed linear operators and relations in Hilbert spaces, Integral Equations Operator Theory 63 no. 2 (2009), 151–163 (with T.Ya. Azizov, J. Behrndt, and C. Trunk).

[50] Quadratic (weakly) hyperbolic matrix polynomials: direct and inverse spectral problems, Oper. Theory Adv. Appl. 198 (2009) (this volume), 11–40 (with T.Ya. Azizov, A. Dijksma, and K.H. Förster).

[51] Small perturbations of selfadjoint and unitary operators in Krein spaces, to appear in J. Operator Theory (with T.Ya. Azizov and C. Trunk).

[52] Spectral points of type π_+ and type π_- for closed linear relations in Krein spaces, submitted (with T.Ya. Azizov, J. Behrndt, and C. Trunk).

Operator Theory:
Advances and Applications, Vol. 198, 1–10

Local Scattering Problem and a Solvable Model of Quantum Network

Vadym Adamyan and Boris Pavlov

To the memory of our friend and colleague Peter Jonas

Abstract. A local scattering theory on finite spectral interval for pairs of self-adjoint operators, which are different extensions of the same densely definite symmetric operator, is developed. The present investigation was motivated by scattering problems for Schrödinger operators modeling the energy operator of electrons in real quantum networks.

Mathematics Subject Classification (2000). Primary: 47A40; Secondary: 81U20, 82099.

Keywords. Scattering theory, local scattering theory, scattering matrix, Krein formula, quantum networks.

1. Introduction

The energy operator (Hamiltonian) for carriers (electrons, holes) in a quantum network constructed from quantum dots and attached straight quantum wires is usually modeled by one-particle Schrödinger operators (Laplace operator + some potentials) on intricate domains of the three- or two-dimensional space. In fact, scattering processes in real networks are observed only on a rather small spectral interval Δ of essential spectrum centered at the Fermi-level. Therefore more simple Hamiltonians may exhibit the same scattering picture on Δ, as the original Hamiltonian of the network. In our previous papers, see [1, 3, 4, 2, 5] we tried to explain how the Hamiltonian of a real network consisting of compact connected quantum dots and a few semi-infinite quantum wires attached to them can be replaced on Δ by the solvable model in the form of second-order ordinary differential operator on a graph with segments and vertices imitating wires and quantum dots, respectively, and certain self-adjoint boundary conditions at graph vertices modeling the inner structure of corresponding quantum dots. At the same time the

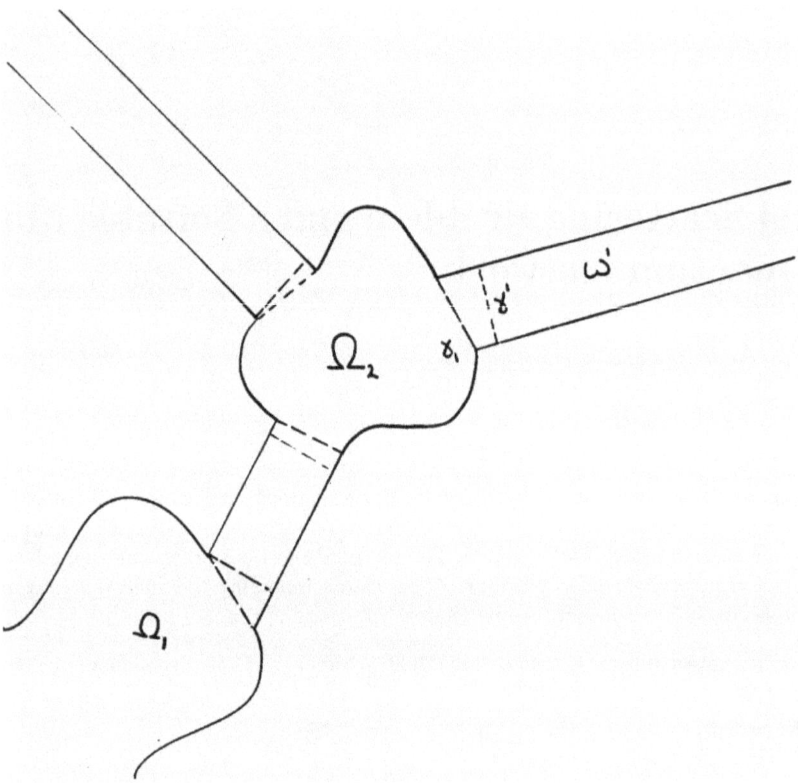

FIGURE 1. Typical fragment of quantum network

scattering theory for solvable models obtained as described above and facilitated for modeling the electron transport through quantum network in a narrow spectral interval was and is successfully developed in many papers, see, for example [6, 7, 8, 9, 10, 11, 12]. So a general approach for construction of the solvable model on a rather narrow spectral interval for quantum networks with straight wires is demanded. As it was mentioned recently in [2] a key issue for engineering of such a fitting solvable models might be a local scattering theory for a finite spectral interval Δ. The present work is devoted to elaboration of such a local theory for two self-adjoint extensions H_0 and H of the same densely defined symmetric operators B. It continues and supplements our old paper [14] and also recent papers [15, 16]. Contrary to [14, 15, 16] we do not suppose here that the resolvent difference

$$(H-z)^{-1}-(H_0-z)^{-1}, \ \mathrm{Im} z \neq 0,$$

of the extensions H, H_0 is either finite-dimensional or at least compact. However, we assume that the product of the orthogonal projection $E_0^c(\Delta)$ onto the spectral subspace of H_0 corresponding to the interval Δ of its absolutely continuous spectrum and the orthogonal projection P_B onto the defect subspace of B has finite rank. The last assumption covers completely needs of the scattering theory for quantum networks. This follows from the fact that the self-adjoint Laplace operator in an infinite cylindrical domain with Dirichlet, Neumann or other homogeneous boundary conditions on the lateral surface has absolutely continuous spectrum of finite multiplicity on each finite spectral interval Δ (for more detailed explanation, see [5]).

This paper is organized in the following way. In Section 2 we prove existence of local wave operators for the pair of extensions H, H_0 and find appropriate expressions for wave operators on the base of Krein resolvent formula. The next section is devoted to the study of local scattering operator and corresponding scattering matrix. The main result of this section is the *Krein representation* of local scattering matrix. In the last section we prove unitarity of the local scattering matrix and specify its values at points of discrete spectrum of H_0 that get into the interval Δ.

2. Local wave operators

Let H_0 and H_1 be selfadjoint operators in a Hilbert space $\mathcal{H}$. Our basic assumption are the following:

- *H_0 and H_1 are selfadjoint extensions of the same densely defined symmetric operator B in $\mathcal{H}$ with the domain $\mathcal{D}_B$;*
- *the operator H_0 has absolutely continuous spectrum on some segment $\Delta = [a, b]$, $-\infty < a < b < \infty$;*
- *the orthogonal projection P_B onto defect subspace $\mathcal{N}_B = \mathcal{H} \ominus (B + i)\mathcal{D}_B$ and the spectral projection $E_0(\lambda)$ of H_0 satisfy the condition*

$$\dim P_B E_0(\Delta)\mathcal{H} < \infty.$$

Theorem 2.1. *If H_0 and H_1 satisfy the above assumptions, then the local wave operators*

$$W_\Delta^\pm(H_1, H_0) := s - \lim_{t \to \pm\infty} e^{iH_1 t} e^{-iH_0 t} E_0(\Delta)$$

exist and can be represented in the form

$$W_\Delta^\pm(H_1, H_0) = E_0(\Delta) + s - \lim_{\varepsilon \downarrow 0} \int_a^b (H_0 + i) R_{\lambda \pm i\varepsilon}(H_0) \left[\Gamma - Q(\lambda \pm i\varepsilon)\right]^{-1} P_B dE_0(\lambda),$$

where

$$Q(z) = P_B(I + zH_0)(H_0 - z)^{-1} P_B$$

and $[\ldots]^{-1}$ stands for the inverse in $\mathcal{N}_B$.

Proof. According to the Kato-Birman theory for the pair of self-adjoint operators A_0, A_1 the local wave operators $W_\Delta^\pm(A_1, A_0)$ exist if for some joint regular point z of A_0 and A_1 (and hence for all regular points of A_0, A_1) the operator

$$\left[R_z(A_1) - R_z(A_0)\right] E_0(\Delta) P_{ac}(A_0) \left(= \left[(A_1 - z)^{-1} - (A_0 - z)^{-1}\right] E_0(\Delta) P_{ac}(A_0)\right),$$

is nuclear [17]. Here $P_{ac}(A_0)$ is the orthogonal projection onto the absolutely continuous subspace of A_0. But this is the case for the pair H_0, H_1. Indeed, for any non-real z by our assumptions and the Krein resolvent formula

$$(H_1 - z)^{-1} - (H_0 - z)^{-1} = (H_0 - z)^{-1}(H_0 + i)\left[\Gamma - Q(z)\right]^{-1} P_B(H_0 - i)(H_0 - z)^{-1},$$

where Γ is some self-adjoint operator in $\mathcal{N}_B$, the operator

$$\begin{aligned}&[R_z(H_1) - R_z(H_0)]\, E_0(\Delta)\\ &\quad = (H_0 - z)^{-1}(H_0 + i)\left[\Gamma - Q(z)\right]^{-1} \mathbf{P_B E_0(\Delta)}(H_0 - i)(H_0 - z)^{-1}\end{aligned}$$

has a finite rank.

As the strong limit defining the local wave operator $W_\Delta^+(H_1, H_0)$ exists, then applying the Krein formula we can write

$$\begin{aligned}W_\Delta^+(H_1, H_0) &= s - \lim_{\varepsilon \downarrow 0} \varepsilon \int_0^\infty e^{-\varepsilon t} e^{iH_1 t} e^{-iH_0 t} E_0(\Delta) dt\\ &= s - \lim_{\varepsilon \downarrow 0} (i\varepsilon) \int_a^b R_{\lambda - i\varepsilon}(H_1) dE_0(\lambda)\\ &= E_0(\Delta) + \lim_{\varepsilon \downarrow 0} \int_a^b (H_0 + i)(\lambda - i) R_{\lambda - i\varepsilon}(H_0)\left[\Gamma - Q(\lambda - i\varepsilon)\right]^{-1} P_B dE_0(\lambda).\end{aligned}$$

$W_\Delta(H_1, H_0)$ can be handled in the same way. □

Point out that $W_\Delta^\pm(H_1, H_0)$ are partial isometries and

$$e^{iH_1 t} W_\Delta^\pm(H_1, H_0) = W_\Delta^\pm(H_1, H_0) e^{iH_0 t}, \ -\infty < t < \infty,$$

or

$$E_1(\lambda) W_\Delta^\pm(H_1, H_0) = W_\Delta^\pm(H_1, H_0) E_0(\lambda), \ -\infty < \lambda < \infty.$$

3. Krein representation of local scattering matrix

The local scattering operator, which we define as the corresponding product of local wave operators

$$\mathfrak{S}_\Delta(H_1, H_0) := W_\Delta^+(H_1, H_0)^* W_\Delta^-(H_1, H_0)|_{E_0(\Delta)\mathcal{H}}$$

is a contraction in $E_0(\Delta)\mathcal{H}$ and

$$E_0(\lambda)\mathfrak{S}_\Delta(H_1, H_0) = \mathfrak{S}_\Delta(H_1, H_0) E_0(\lambda), \ a \le \lambda \le b.$$

Therefore for the representation of H_0 in $E_0(\Delta)\mathcal{H}$ as the multiplication operator by λ in the direct integral of Hilbert spaces $\mathfrak{h}_0(\lambda)$,

$$E_0(\Delta)\mathcal{H} = \int_a^b \oplus \mathfrak{h}_0(\lambda) d\lambda,$$

the local scattering operator $\mathfrak{S}_\Delta(H_1, H_0)$ acts as the multiplication operator by a contractive operator (matrix) function $S_\Delta(\lambda)$, which will be below referred to as the (local) scattering matrix.

We can compute $S_\Delta(\lambda)$ using the established relations

$$\begin{aligned}
\mathfrak{S}_\Delta(H_1, H_0) &= s - \lim_{t\to\infty} e^{iH_0 t} W_\Delta^-(H_1, H_0) e^{-iH_0 t} E_0(\Delta) \\
&= E_0(\Delta) + s - \lim_{t\to\infty}\Big(s - \lim_{\varepsilon\downarrow 0}\int_a^b e^{iH_0 t}(H_0 + i)(\lambda - i) \\
&\qquad \times R_{\lambda+i\varepsilon}(H_0)\left[\Gamma - Q(\lambda \pm i\varepsilon)\right]^{-1} e^{-i\lambda t} P_B dE_0(\lambda)\Big) \\
&= E_0(\Delta) + s - \lim_{t\to\infty}\Big(s - \lim_{\varepsilon\downarrow 0}\int_a^b dE_0(\mu) \\
&\qquad \times \int_a^b e^{i(\mu-\lambda)t}\frac{(\mu + i)(\lambda - i)}{\mu - \lambda - i\varepsilon}\left[\Gamma - Q(\lambda \pm i\varepsilon)\right]^{-1} P_B dE_0(\lambda)\Big).
\end{aligned}$$

Let us remind now that for vector functions f with values from $\mathcal{N}_B$, which belong to the space $\mathbb{L}^2(\mathcal{N})$ on the real axis, the operator π_+ defined by the formula

$$(\pi_+ f)(\lambda) = s - \lim_{\varepsilon\downarrow 0}\frac{1}{2\pi i}\int_{-\infty}^{\infty}\frac{1}{\mu - \lambda - i\varepsilon} f(\mu)d\mu$$

is the orthogonal projection in $\mathbb{L}^2(\mathcal{N}_B)$ on the subspace of boundary values of vector functions from the Hardy class $\mathbb{H}^2(\mathcal{N}_B)$ in the upper half-plane. Since the operators of multiplication T_t by $e^{-i\lambda t}$, $t > 0$, extend subspaces $T_t\mathbb{H}^2(\mathcal{N}_B)$ up to $\mathbb{L}^2(\mathcal{N})$ as $t \to \infty$, then for $f \in \mathbb{L}^2(\mathcal{N}_B)$ we have

$$\begin{aligned}
s - \lim_{t\to\infty}&\left(s - \lim_{\varepsilon\downarrow 0}\int_a^b e^{i(\mu-\lambda)t}\frac{1}{\mu - \lambda - i\varepsilon} f(\mu)d\mu\right) \\
&= s - \lim_{t\to\infty} 2\pi i\,(T_t\pi_+ T_{-t} f)(\lambda) \\
&= 2\pi i f(\lambda)\ a.e.
\end{aligned} \tag{3.1}$$

It follows from (3.1) that for any $f, g \in E_0(\Delta)\mathcal{H}$ such that

$$\int_a^b \left\|\left[\Gamma - Q(\lambda + i0)\right]^{-1}\frac{d}{d\lambda}P_B E_0(\lambda) f\right\|^2 d\lambda < \infty,$$

$$\int_a^b \left\|\frac{d}{d\lambda}P_B E_0(\lambda) f\right\|^2 d\lambda < \infty,$$

we can write

$$(\mathfrak{S}_\Delta(H_1,H_0)f,g) = \int_a^b d\,(E_0(\lambda)f,g)$$

$$+\,2\pi i\int_a^b(\lambda^2+1)\left([\Gamma-Q(\lambda+i0)]^{-1}\frac{d}{d\lambda}P_BE_0(\lambda)f,\frac{d}{d\lambda}P_BE_0(\lambda)g\right)d\lambda.$$

Let

$$\mathcal{N}''_B := \mathcal{N}\cap(I-E_0(\Delta))\mathcal{H},\quad \mathcal{N}'_B=\mathcal{N}_B\ominus\mathcal{N}''_B$$

and let P'_B and P''_B be the orthogonal projections onto $\mathcal{N}'_B$ and $\mathcal{N}''_B$, respectively. Evidently, $P_BE_0(\Delta)=P'_BE_0(\Delta)$ and $P_B\,(I-E_0(\Delta))=P''_B\,(I-E_0(\Delta))$. Hence, with respect to the splitting $\mathcal{N}_B=\mathcal{N}'_B\oplus\mathcal{N}''_B$ we can represent $Q(z)$, $\mathrm{Im}z\neq 0$, as the block-diagonal matrix

$$Q(z)=\begin{pmatrix} Q_1(z) & 0\\ 0 & Q_2(z)\end{pmatrix},$$

$$Q_1(z)=\int_{-\infty}^{\infty}\frac{1+tz}{t-z}dP'_BE_0(t)|_{\mathcal{N}'_B},\qquad Q_2(z)=\int_{E\setminus\Delta}\frac{1+tz}{t-z}dP''_BE_0(t)|_{\mathcal{N}''_B}.$$

Assuming further that the domain of Γ contains $\mathcal{N}'_B$ we can represent Γ for the same splitting as 2×2 selfadjoint block operator matrix

$$\Gamma=\begin{pmatrix}\Gamma_{11} & \Gamma_{12}\\ \Gamma_{21} & \Gamma_{22}\end{pmatrix}$$

with bounded entries Γ_{11}, Γ_{12} and $\Gamma_{21}=\Gamma_{12}^*$.

If the function $Q_{22}(z)-\Gamma_{22}$ is invertible in $\mathcal{N}''_B$ at least at one non-real point z, then the Nevanlinna operator function

$$G(z)=-\Gamma_{11}+\Gamma_{12}\,[\Gamma_{22}-Q_{22}(z)]^{-1}\,\Gamma_{21}$$

in $\mathcal{N}'_B$ is defined and holomorphic in Δ except of maybe a finite number of simple poles. Since $G(z)$ is the function of Nevanlinna class then $G(\lambda)$ for any $\lambda\in\Delta$ (except of those poles) is a self-adjoint operator in $\mathcal{N}'_B$. As follows if the all above assumptions hold, then

$$\begin{aligned}&E_0(\Delta)P_B\,[\Gamma-Q(\lambda+i0)]^{-1}\,P_BE_0(\Delta)\\ &\quad=E_0(\Delta)P'_B\,[G(\lambda)-Q_1(\lambda+i0)]^{-1}\,P'_BE_0(\Delta).\end{aligned}$$

Let $\mathcal{H}'$ be the minimal invariant subspace of H_0 in $\mathcal{H}$ containing $\mathcal{N}'_B$. We will assume for simplicity that the part H'_0 of H_0 on $E_0(\Delta)\mathcal{H}'$ has the Lebesgue spectrum of multiplicity $n(<\infty)$ and also that

$$\mathrm{rank}P'_BE_0(\Delta)=\mathrm{rank}P'_B=n.$$

Then there exists an isometric operator $\mathfrak{F}$, which maps $E_0(\Delta)\mathcal{H}'$ onto the space $\mathbb{L}_2([a,b];\mathbf{E}_n)$ of $\mathbf{E}_n$-vector function on $[a,b]$ and such that $\mathfrak{F}H_0'\mathfrak{F}^{-1}$ the multiplication operator by independent variable in $\mathbb{L}_2([a,b];\mathbf{E}_n)$. Then for any $f,h\in E_0(\Delta)\mathcal{H}'$ we can write

$$(f,h)_{\mathcal{H}} = \int_a^b \langle(\mathfrak{F}f)(\lambda),(\mathfrak{F}h)(\lambda)\rangle_{\mathbf{E}_n}\, d\lambda.$$

Let $g_1,\dots,g_n$ be some basis in $\mathcal{N}_B'$ (not necessarily orthonormal) and let $R(\lambda)$ denotes the $n\times n$ matrix function with elements

$$r_{ij}(\lambda) = ([Q_1(\lambda+i0)-G(\lambda)]g_j,g_i)\,.$$

We introduce the set of vector functions

$$\hat{g}_j(\lambda) = (\mathfrak{F}E_0(\Delta)g_j)\,(\lambda),\ j=1,\dots,n.$$

By our previous assumptions and notations for $\lambda\in(a,b)$ the matrix $R(\lambda)$ is the sum $R^{(1)}(\lambda)+iR^{(2)}(\lambda)$, where $R^{(1)}(\lambda),R^{(2)}(\lambda)$ are Hermitian matrices with elements

$$r_{ij}^{(1)}(\lambda) = \int_{\mathbf{E}\setminus\Delta}\frac{1+t\lambda}{t-\lambda}d\,(E_0(t)g_j,g_i) + \int_a^{b\prime}\frac{1+t\lambda}{t-\lambda}\,\langle(\hat{g}_j(t),\hat{g}_i(t)\rangle_{\mathbf{E}_n}\,dt - (G(\lambda)g_j,g_i)\,,$$

$$r_{ij}^{(2)}(\lambda) = \pi(1+\lambda^2)\,\langle(\hat{g}_j(\lambda),\hat{g}_i(\lambda)\rangle_{\mathbf{E}_n}\,,\ i,j=1,\dots,n,$$

respectively. Note that for each $\lambda\in(a,b)$ except of maybe a finite number of points, at which $G(z)$ has simple poles, the matrix $R^{(1)}$ is Hermitian and for each $\lambda\in(a,b)$ the matrix $R^{(2)}$ coincides with the Gramm matrix for the set of $\mathbf{C}_n$-vectors $\sqrt{\pi}(\lambda+i)\hat{g}_j(\lambda)$, $j=1,\dots,n$. Using the introduced matrix function $R(\lambda)$ we obtain the following *Krein representation* for the local scattering matrix:

$$S_\Delta(\lambda) = I - 2\pi i(1+\lambda^2)\sum_{\mu,\nu=1}^{n}\left([R^{(1)}(\lambda)+iR^{(2)}(\lambda)]^{-1}\right)_{\mu\nu}(\cdot\,,\,\hat{g}_\nu(\lambda))_{\mathbf{C}_n}\hat{g}_\mu(\lambda).$$

4. Unitarity of scattering matrix

Proposition 4.1. *If the above assumptions hold, then the local scattering matrix $S_\Delta(\lambda)$ is unitary.*

The unitarity of $S_\Delta(\lambda)$ is a direct consequence of the following general statement.

Proposition 4.2. *Let $n\times n$-matrix $\Upsilon=(\gamma_{\mu\nu})_1^n$ is the Gramm matrix for any set of vectors $h_1,\dots,h_n\in\mathbf{C}_n$, $\gamma_{\mu\nu}=(h_\nu,h_\mu)$, and Λ is any Hermitian $n\times n$-matrix such that the matrix $\Lambda+i\Upsilon$ is invertible. Then the matrix*

$$\Omega = I + 2i\sum_{\mu,\nu=1}^{n}\left([\Lambda-i\Upsilon)]^{-1}\right)_{\mu\nu}(\cdot\,,\,h_\nu)h_\mu$$

is unitary.

Proof. Using the fact that Υ is just the Gramm matrix for $h_1, \dots, h_n$ we see that

$$\begin{aligned}\Omega^*\Omega &= \Big[I - 2i \sum_{\mu,\nu=1}^{n} \big([\Lambda + i\Upsilon)]^{-1}\big)_{\mu\nu} (.\,, h_\nu) h_\mu\Big] \\ &\quad \times \Big[I + 2i \sum_{\mu,\nu=1}^{n} \big([\Lambda - i\Upsilon)]^{-1}\big)_{\mu\nu} (.\,, h_\nu) h_\mu\Big] \\ &= I - 2i \sum_{\mu,\nu=1}^{n} \big([\Lambda + i\Upsilon)]^{-1} - [\Lambda - i\Upsilon)]^{-1}\big)_{\mu\nu} (.\,, h_\nu) h_\mu \\ &\quad + 2i \sum_{\mu',\nu',\mu,\nu=1}^{n} \big([\Lambda + i\Upsilon)]^{-1}\big)_{\mu'\nu'} (-2i)(h_\mu, h_{\nu'}) \big([\Lambda - i\Upsilon)]^{-1}\big)_{\mu\nu} (.\,, h_\nu) h_{\mu'} \\ &= I - 2i \sum_{\mu,\nu=1}^{n} \big([\Lambda + i\Upsilon)]^{-1} - [\Lambda - i\Upsilon)]^{-1} \\ &\quad + 2i[\Lambda + i\Upsilon)]^{-1}\Upsilon[\Lambda - i\Upsilon)]^{-1}\big)_{\mu\nu} (.\,, h_\nu) h_\mu = I. \end{aligned}$$

□

Let us assume now that the Nevanlinna matrix function

$$\tilde{G}(z) := \big(r_{ij}(z) = -\left(G(z)g_j, g_i\right)_{\mathcal{N}}\big)_{ij=1}^{n}$$

has a finite number of (isolated) poles in (a, b) and that vector functions $\hat{g}_j(\lambda)$, $j = 1, \dots, n$, are continuous at those poles. Then the matrix function $R(\lambda)$ in the vicinity of a pole λ_0 has the form

$$R(\lambda) = \frac{1}{\lambda - \lambda_0} M_0 + \grave{R}^{(1)}(\lambda_0) + iR^{(2)}(\lambda_0) + o(1),$$

where M_0 is some non-negatively defined matrix. We will denote by $Q_0 = (u_{\mu\nu})_{\mu,\nu=1}^{n}$ the matrix, which is the orthogonal projection in $\mathbf{C}_n$ onto the null-space $\mathfrak{N}_0$ of M_0 and put

$$\tilde{R}^{(1)}(\lambda_0) := Q_0 \grave{R}^{(2)}(\lambda_0) Q_0, \quad \tilde{R}^{(2)}(\lambda_0) := Q_0 R^{(2)}(\lambda_0) Q_0,$$
$$\tilde{\hat{g}}_j(\lambda_0) = \sum_{i=1}^{n} u_{ij} \hat{g}_i(\lambda_0).$$

Note that $\tilde{R}^{(2)}(\lambda_0)$ is the Gramm matrix for the set $\left(\tilde{\hat{g}}_j(\lambda_0),\ j = 1, \dots, n\right)$. Applying the Schur-Frobenius factorization to the matrix function $R(\lambda)$ in the vicinity of λ_0 yields the following statement.

Proposition 4.3. *If the above assumptions hold, then*

$$S_\Delta(\lambda_0) = I - Q_0 + \tilde{S}_\Delta(\lambda_0) Q_0,$$

where

$$\tilde{S}_\Delta(\lambda_0) = I_{\mathfrak{N}_0} - 2\pi i(1+\lambda^2) \times \sum_{\mu,\nu=1}^{n} \left(\left[\tilde{R}^{(1)}(\lambda_0) + i\tilde{R}^{(2)}(\lambda_0) \right]_{\mathfrak{N}_0}^{-1} \right)_{\mu\nu} (\cdot\,, \tilde{\tilde{g}}_\nu(\lambda_0))_{\mathbf{C}_n} \tilde{\tilde{g}}_\mu(\lambda_0)$$

is a unitary matrix in $\mathfrak{N}_0$.

Corollary 4.4. *If* $\lambda_0 \in (a,b)$ *is a pole of matrix function* $G(z)$ *with the residue* M_0, *then unity is an eigenvalue of* $S_\Delta(\lambda_0)$ *of multiplicity not less than* $\operatorname{rank} M_0$.

Remark 4.5. Let us assume now that $G(z)$ is holomorphic on (a,b) and moreover that the distance Λ from the point $\frac{a+b}{2}$ to the support of the matrix measure generating $G(z)$ satisfies the condition $\frac{b-a}{\Lambda} \ll 1$. Then the λ-dependance of $G(\lambda)$ on (a,b) can be ignored and it can be replaced on (a,b) by the Hermitian matrix $G(\frac{a+b}{2})$. For such approximation the local scattering matrix $S_\Delta(\lambda)$ coincides on (a,b) with the scattering matrix for the pair of operators $H'1$, H'_0, in which H'_0 is the part of H_0 on its minimal invariant subspace $\mathcal{H}'$ in $\mathcal{H}$ containing the part of the defect subspace $\mathcal{N}'_B = \mathcal{N}_B \ominus (\mathcal{N}_B \cap (I - E_0(\Delta))\mathcal{H})$ and H'_1 is the self-adjoint operator in $\mathcal{H}'$, the resolvent of which is given by the Krein formula

$$(H'_1 - z)^{-1} - (H'_0 - z)^{-1} = (H'_0 - z)^{-1}(H'_0 + i)\left[G\left(\frac{a+b}{2}\right) - Q_1(z) \right]^{-1} P'_B(H'_0 - i)(H'_0 - z)^{-1}.$$

Acknowledgment

V. Adamyan is grateful to the Institute of Mathematics of the Technical University of Berlin for permanent attention and hospitality.

References

[1] N. Bagraev, A. Mikhailova, B.S. Pavlov, L.V. Prokhorov and A. Yafyasov. *Parameter regime of a resonance quantum switch*, Phys. Rev. B, 71:165308 (2005).

[2] A. Mikhailova, B. Pavlov and L. Prokhorov. *Intermediate Hamiltonian via Glazman's splitting and analytic perturbation for meromorphic matrix functions.* Mathematische Nachrichten, **280** 12 (2007) pp. 1376–1416.

[3] M. Harmer, B. Pavlov and A. Yafyasov. *Boundary condition at the junction*, Journal of Computational Electronics, **6**(2007) pp. 153–157.

[4] B. Pavlov and A. Yafyasov, *Standing waves and resonance transport mechanism in quantum networks*, Surface Science 601 (2007), pp. 2712–2716.

[5] V. Adamyan, B. Pavlov and A. Yafasov. *Modified Krein Formula and Analytic Perturbation Procedure for Scattering on Arbitrary Junction*, Operator Theory: Adv. and Appl. **190** (2009), pp. 3–26.

[6] N.I. Gerasimenko and B.S. Pavlov *A scattering problem on noncompact graphs*, Teoret. Mat. Fiz. **74** 1 (1988), pp. 345–359.

[7] S. Datta and B. Das Sarma *Electronic analog of the electro-optic modulator*, Appl. Phys. Lett. **56** 7 (1990) pp. 665–667.

[8] V. Adamyan *Scattering theory for microschemes* Operator Theory: Adv. and Appl. **59** (1992) pp. 1–10.

[9] S. Datta. *Electronic Transport in Mesoscopic systems*, Cambridge University Press, Cambridge, 1995.

[10] P. Kuchment *Graph models for waves in thin structures*. Waves in Periodic and Random Media, **12** 1 (2002) pp. 1–24.

[11] P. Exner and O. Post *Convergence of graph-like thin manifolds* J. Geom. Phys., **54** 1 (2005) pp. 77–115.

[12] D. Grieser *Spectra on graph neighborhoods and scattering* de.arXiv.org. math.arXiv: 0710.3405v03.

[13] B. Pavlov. *A star-graph model via operator extension*. Mathematical Proceedings of the Cambridge Philosophical Society, **142** 02 (2007) pp. 365–384.

[14] V. Adamyan and B. Pavlov *Zero-radius potential and M.G. Krein's formula for generalized resolvents*, Zap. Nauchn. Sem. LOMI **149** (1986), 7–23; translation in J. Sov. Math. **42** 2 (1988) 1537–1550.

[15] J. Behrndt, M.M. Malamud, and H. Neidhardt. Scattering theory for open quantum systems with finite rank coupling. Math. Phys. Anal. Geom., **10** 4 (2007) 313–358.

[16] J. Behrndt, M.M. Malamud, and H. Neidhardt. Scattering matrices and Weyl functions. Proc. Lond. Math. Soc. **97** 3 (2008) 568–598.

[17] M.S. Birman. *A Local Criterion for the Existence of Wave Operators*, Math. USSR Izv. **2** (1968) 879–906.

Vadym Adamyan
Department of Theoretical Physics
Odessa National I.I. Mechnikov University
Odessa 65082, Ukraine
e-mail: `vadamyan@paco.net`

Boris Pavlov
NZ Institute for Advanced Studies
Massey University, Albany Campus
Private Bag 102-904
North Shore Mail Centre
Auckland, New Zealand
e-mail: `B.Pavlov@massey.ac.nz`

Operator Theory:
Advances and Applications, Vol. 198, 11–40

Quadratic (Weakly) Hyperbolic Matrix Polynomials: Direct and Inverse Spectral Problems

T.Ya. Azizov, A. Dijksma, K.-H. Förster and P. Jonas

Abstract. Let L be a monic quadratic weakly hyperbolic or hyperbolic $n \times n$ matrix polynomial. We solve some direct spectral problems: We prove that the eigenvalues of a compression of L to an $(n-1)$-dimensional subspace of $\mathbb{C}^n$ block-interlace and that the eigenvalues of a one-dimensional perturbation of L $(-,+)$-interlace the eigenvalues of L. We also solve an inverse spectral problem: We identify two given block-interlacing sets of real numbers as the sets of eigenvalues of L and its compression.

Mathematics Subject Classification (2000). 47A56, 47A10, 47A55, 47A75, and 47B15.

Keywords. Weakly hyperbolic, hyperbolic, compression, perturbation, Nevanlinna function, eigenvalue, interlacing, root zone, direct spectral problem, inverse spectral problem.

1. Introduction

Consider the monic quadratic matrix polynomial:

$$L(\lambda) = \lambda^2 + \lambda B + C, \tag{1.1}$$

in which B and C are hermitian $n \times n$ matrices, the rank one perturbation of L:

$$L_{s,t;e}(\lambda) = L(\lambda) + (\lambda s + t)(\,\cdot\,, e)e,$$

The authors gratefully acknowledge the support from the Volkswagen-Stiftung which made it possible to work together via the Research in pairs program at the Mathematisches Forschungsinstitut Oberwolfach during the period February 25–March 10, 2001. We also thank the Netherlands Organization for Scientific Research (grant NWO 047-008-008) and the Russian Foundation for Basic Research (grant RFBR 08-01-00566-a).

Peter Jonas passed away on July 18th, 2007.

in which s and t are real numbers, e is a unit vector from $\mathbb{C}^n$ and $(\cdot,\cdot)$ is the Euclidean inner product in $\mathbb{C}^n$, and the compression of L to $\{e\}^\perp$:

$$L_{\infty;e}(\lambda) = P_e L(\lambda)|_{\operatorname{ran} P_e},$$

where $P_e = I_n - (\cdot, e)e$ is the orthogonal projection in $\mathbb{C}^n$ onto $\{e\}^\perp$. Here I_n is the $n \times n$ identity matrix; as above and in the rest of the sequel, when it is clear from the context, we write for example λ^2 instead of $\lambda^2 I_n$. To our knowledge the introduction of $L_{s,t;e}$ is new. The connection between $L_{s,t;e}(\lambda)$ and $L_{\infty;e}(\lambda)$, the second being a limit of the first as $\lambda s + t \to \infty$, is explained in Remark 4.1.

In this paper we study direct and inverse spectral problems associated with these matrix polynomials under the assumption that L is weakly hyperbolic or hyperbolic; for the definition of these and other notions we refer to Section 2 which contains the preliminaries to this paper. The *direct spectral problem* is to describe the relation between the sets of eigenvalues of L, $L_{\infty;e}$, and $L_{s,t;e}$. Denote by (i) $\{\alpha_{\pm j}\}_{j=1}^n$, (ii) $\{\beta_{\pm j}\}_{j=1}^{n-1}$, and (iii) $\{\alpha_{\pm j;s,t;e}\}_{j=1}^n$ the ordered sets of all (necessarily real) eigenvalues of L, $L_{\infty;e}$, and $L_{s,t;e}$, each eigenvalue repeated according to its multiplicity, so that, for example

$$\alpha_{-n} \le \alpha_{-n+1} \le \cdots \le \alpha_{-1} \le \alpha_1 \le \cdots \le \alpha_{n-1} \le \alpha_n.$$

We show that the eigenvalues (i) and (ii) are block-interlacing and that the eigenvalues (i) and (iii) are $(-,+)$-interlacing, see Section 4. The *inverse spectral problem* is to find conditions on two given ordered sets of real numbers under which they can be identified as the sets of eigenvalues of L and its compression or as the sets of eigenvalues of L and its perturbation. In Section 5 we consider the inverse spectral problem for L and its compression $L_{\infty;e_n}$ to $\{e_n\}^\perp$, where $e_n = \begin{pmatrix} 0 & \cdots & 0 & 1 \end{pmatrix}^\top \in \mathbb{C}^n$.

We summarize the contents of the paper. In Section 2 we recall results from the monograph [19] of A.S. Markus which are needed in the sequel. We show that the inverse of a (weakly) hyperbolic matrix polynomial of the form (1.1) can be described in terms of a rational Nevanlinna matrix function, see Lemma 2.1 (and also Lemma 3.2). This lemma will be used frequently in the proofs of the theorems in the next sections. We also prove two technical lemmas. In Section 3 we show that a weakly hyperbolic matrix polynomial of the form (1.1) can be approximated by hyperbolic matrix polynomials of the same form with simple eigenvalues, see Theorem 3.1. If the weakly hyperbolic matrix polynomial is Jacobi, that is, real symmetric and tridiagonal, then the approximating hyperbolic matrix polynomials can be chosen Jacobi also, see Theorem 3.5. In Section 4 we address the direct spectral problem. We first prove the interlacing theorems for hyperbolic matrix polynomials, see Theorem 4.2 and Theorem 4.5. Then using approximation we prove the interlacing theorems for weakly hyperbolic matrix polynomials by invoking the interlacing theorems for hyperbolic polynomials, see Theorem 4.3 and Theorem 4.6. In Section 5 we show that the inverse spectral problem mentioned above can be solved by means of Jacobi matrix polynomials, see Theorem 5.1.

In the proof of Lemma 5.2 we use similar ideas as in [1]. In that paper criteria are given under which two sets of interlacing and mutually different points are the eigenvalues of a matrix polynomial of the form (1.1) and its compression to $\{e_n\}^\perp$. The criteria are satisfied here, but we show in addition that the matrix polynomial is hyperbolic (see Remark 5.3) and that the weakly hyperbolic case can be obtained from the hyperbolic case via approximation.

For a brief survey of the study of spectral problems for operator polynomials, also called polynomial operator pencils, we refer to [19, Introduction]. Here we mention only that the quadratic matrix polynomial L arises in the theory of vibrating systems of n second-order differential equations

$$x''(t) + Bx'(t) + Cx(t) = 0, \tag{1.2}$$

see, for example, the monograph [23] by S. Timoshenko, D.H. Young, and W. Weaver, JR. The matrices B and C are related to the damping and the stiffness of the vibrating system. Substituting in (1.2) $x(t) = f e^{\lambda t}$ with $f \in \mathbb{C}^n$, we get the spectral equation $L(\lambda)f = 0$. From this equation a basis of solutions of the system (1.2) can be obtained. The inverse spectral problem applied to a vibrating system is to find necessary and sufficient conditions on its spectral data and the spectral data of the compressed system which ensure that the damping and stiffness configuration is determined by Jacobi matrices.

Interlacing theorems for J-nonnegative matrices (see Example 4.4) can be found in the paper [17] by H. Langer and B. Najman. For interlacing theorems related to the linearized pencil we refer to the paper [13] of M. Kovac-Striko and K. Veselic. We do not know any reference to a paper on quadratic direct spectral problems for rank one perturbations of the type considered here, see Theorems 4.5 and 4.6. The block interlacing Theorem 4.3 for compressions can be shown more directly by using variational principles, see R.J. Duffin [6] , H. Langer [16, Chapter II.5, Subsection 6], and the more recent paper by D. Eschwé and M. Langer [7], where further references can be found. A proof of Theorem 4.3 using Sylvester's law of inertia was given by K. Veselić [24] shortly after a lecture about this paper at the 8th Workshop Operator Theory in Krein Spaces and Inverse Problems at the Technische Universität Berlin in December 2008. In our proofs of the interlacing theorems for rank one perturbations as well as for compressions we make use of Nevanlinna functions and to our knowledge this approach is new.

The inverse spectral problem where the stiffness is assumed to be zero: $C = 0$, has been studied by H. Hochstadt [10], see also the monograph [9] by G.M.L. Gladwell and the survey paper [3] of M.T. Chu. This problem is linear in λ and concerns only one Jacobi matrix. The quadratic inverse spectral problem for two Jacobi matrices, which we study here, is first considered by Y.M. Ram and S. Elhay [22], see also Y.M. Ram [21]. In [22], and in [3, Theorem 3.8],[4, Theorem 4.7], and [15] (all three with reference to [22]), it is wrongly stated that the inverse problem has a solution if the eigenvalues of L and $L_{\infty;e_n}$ have multiplicity one and the two sets of eigenvalues are disjoint. For a counter example, see [1, Example 6.2]. Gladwell solved an inverse problem for a weakly damped system with the assumption that the eigenvalues are on the imaginary axis (and then automatically symmetric with respect to 0) and introduced the notion of block-interlacing sets.

2. Notation and preliminaries

In the sequel, unless explicitly stated otherwise, we only consider monic quadratic $n \times n$ matrix polynomials L:

$$L(\lambda) = \lambda^2 + \lambda B + C. \tag{2.1}$$

L will be called *hermitian* if the coefficients B and C are hermitian matrices.

The matrix polynomial L in (2.1) is called *weakly hyperbolic* (*hyperbolic*) if L is hermitian and for every nonzero $f \in \mathbb{C}^n$ the two zeros of the quadratic polynomial $(L(\lambda)f, f)$ are real (respectively real and distinct).

Assume L in (2.1) is weakly hyperbolic. We denote the zeros of the quadratic polynomial $(L(\lambda)f, f)$ by $p_1(f)$ and $p_2(f)$ and order them such that $p_1(f) \geq p_2(f)$. The functions p_1 and p_2 are bounded continuous functionals on the unit ball S in $\mathbb{C}^n$. Since S is compact, the *root zones* of L:

$$\Delta_i(L) := \{p_i(f) : 0 \neq f \in \mathbb{C}^n\}, \quad i = 1, 2,$$

are compact intervals $[a_i, b_i] := \Delta_i(L)$ with $a_2 \leq b_2 \leq a_1 \leq b_1$ (see [19, Theorem 31.5]). We call $[a_2, b_2]$ and $[a_1, b_1]$ the *left* and *right* root zones of L. If L is hyperbolic, then $b_2 < a_1$ (see [19, Theorem 31.3]). Conversely, if L is weakly hyperbolic and $b_2 < a_1$, then L is hyperbolic.

Recall that $\lambda \in \mathbb{C}$ is called an *eigenvalue* of an $n \times n$ matrix polynomial L if $\det L(\lambda) = 0$ and $f \in \mathbb{C}^n \setminus \{0\}$ is called an *eigenvector corresponding to the eigenvalue* λ if $L(\lambda)f = 0$. The set of eigenvalues of L is called the *spectrum of* L and denoted by $\sigma(L)$. If L is a weakly hyperbolic matrix polynomial, then $\sigma(L) \subset \mathbb{R}$ (see [19, Lemma 31.1]) and the endpoints of the root zones belong to $\sigma(L)$ (see [19, Theorem 31.9]). A matrix polynomial L is called *damped* (*strongly damped*) if it is weakly hyperbolic (hyperbolic) and $\sigma(L) \subset (-\infty, 0)$. In this case the coefficients B and C are positive matrices. They are called the *damping* and the *stiffness coefficient*, respectively. Evidently, by a suitable shift of the independent variable every weakly hyperbolic (hyperbolic) matrix polynomial can be transformed into a damped (strongly damped) matrix polynomial.

Let T_{2n} denote the collection of all ordered sets $\{\alpha_{\pm j}\}_{j=1}^n$ with $\alpha_{\pm j} \in \mathbb{R}$ and

$$\alpha_{-n} \leq \cdots \leq \alpha_{-1} \leq \alpha_1 \leq \cdots \leq \alpha_n. \tag{2.2}$$

If (2.2) holds with $\leq$ replaced by $<$ we write $\{\alpha_{\pm j}\}_{j=1}^n \in T_{2n}^s$. If the elements of the spectrum $\sigma(L)$ of a weakly hyperbolic matrix polynomial L are repeated according to their multiplicity as eigenvalues and ordered we obtain a set $\{\alpha_{\pm j}\}_{j=1}^n \in T_{2n}$. If $b_2 \neq a_1$, then there is a natural splitting between the elements with a positive and the elements with a negative index; if $b_2 = a_1$, then the splitting has to be such that there are n elements with a negative and n elements with a positive index. To be more precise about this splitting, let for some integer n_+ with $1 \le n_+ \le n$ the points

$$\alpha_{n_+} \le \alpha_{n_+ +1} \le \cdots \le \alpha_n = b_1 \tag{2.3}$$

and for some integer n_- with $1 \le n_- \le n$ the points

$$a_2 = \alpha_{-n} \le \cdots \le \alpha_{-n_- -1} \le \alpha_{-n_-} \tag{2.4}$$

be the eigenvalues of L in $\Delta_1 \setminus \{b_2\}$ and $\Delta_2 \setminus \{a_1\}$ repeated according to their multiplicity. If $b_2 \neq a_1$ then $n_+ = n_- = 1$; if $b_2 = a_1$ we set

$$b_2 = \alpha_{-n_- +1} = \cdots = \alpha_{-1} = \alpha_1 = \cdots = \alpha_{n_+ -1} = a_1, \tag{2.5}$$

compare with [14, Theorem 7.1] and [19, Theorem 31.24]. The set $\{\alpha_{\pm j}\}_{j=1}^n$ associated with $\sigma(L)$ in this way will be called the *ordered set of eigenvalues* or simply the *ordered eigenvalues* of L. If α is one of the numbers $\alpha_{\pm j}$ in (2.3) or (2.4), then for every eigenvector f of L corresponding to α we have $(L'(\alpha)f, f) > 0$ (respectively $(L'(\alpha)f, f) < 0$); if α is one of the numbers $\alpha_{\pm j}$ in (2.5) then there exist eigenvectors g and h of L corresponding to α such that

$$(L'(\alpha)g, g) \le 0 \quad \text{and} \quad (L'(\alpha)h, h) \ge 0;$$

here g and h may be equal and then of course $(L'(\alpha)g, g) = 0$. We remark that L may possess a Jordan chain of length two corresponding to the eigenvalue $b_2 = a_1$, as the simple example $L(\lambda) = (\lambda - 1)^2$ with $n = 1$ already shows.

An $n \times n$ matrix function M is called a *Nevanlinna function* if

(i) $\mathbb{C} \setminus \mathbb{R} \subset \operatorname{hol}(M)$, the domain of holomorphy of M,
(ii) M is symmetric with respect to the real axis: $M(\overline{\lambda}) = M(\lambda)^*$ for all $\lambda \in \mathbb{C}$ such that λ and $\overline{\lambda}$ belong to $\operatorname{hol}(M)$, and
(iii) M has the property that for all $f \in \mathbb{C}^n$

$$\frac{\operatorname{Im}(M(\lambda)f, f)}{\operatorname{Im}\lambda} \ge 0, \quad \lambda \in \mathbb{C} \setminus \mathbb{R}.$$

Evidently, an $n \times n$ matrix function M is a Nevanlinna function if and only if $(M(\lambda)f, f)$ is a (scalar) Nevanlinna function for every $f \in \mathbb{C}^n$. Hence results for matrix Nevanlinna functions can be derived from their counterparts for scalar Nevanlinna functions. For scalar Nevanlinna functions we refer to, for example, [5, Chapter II], where these functions are called Pick functions, and [12], where they are called R-functions ("R" for resolvent). We list some results needed in the sequel.

If M is an invertible square matrix Nevanlinna function, then also $-M^{-1}$ is a Nevanlinna function. This readily follows from the equality

$$-\frac{\operatorname{Im} M(\lambda)^{-1}}{\operatorname{Im}\lambda} = \left(M(\lambda)^{-1}\right)^* \frac{\operatorname{Im} M(\lambda)}{\operatorname{Im}\lambda} M(\lambda)^{-1}.$$

An $n \times n$ matrix function M is a Nevanlinna function if and only if it has the integral representation

$$M(\lambda) = M_1\lambda + M_0 + \int_{\mathbb{R}} \left(\frac{1}{t-\lambda} - \frac{t}{1+t^2} \right) d\Sigma(t),$$

where M_1 and M_0 are hermitian $n \times n$ matrices, $M_1 \geq 0$, and Σ is a positive $n \times n$ matrix measure such that

$$\int_{\mathbb{R}} \frac{1}{1+t^2} d(\Sigma(t)f, f) < \infty, \quad f \in \mathbb{C}^n.$$

If M is holomorphic in a neighborhood of ∞ in $\overline{\mathbb{C}}$, then M is of the form

$$M(\lambda) = M_0 + \int_I \frac{1}{t-\lambda} d\Sigma(t),$$

where I is a bounded open subset of $\mathbb{R}$. M is a rational Nevanlinna matrix function if and only if Σ in the representation of M consists of only a finite number of masses, which are nonnegative hermitian matrices, say S_j, at the points z_j, $j = 1, \ldots, k$:

$$M(\lambda) = M_1\lambda + M_0 + \sum_{j=1}^{k} \frac{S_j}{z_j - \lambda}.$$

In this formula $M_1 = 0$ if and only if M is holomorphic at ∞. In the scalar case, the function $m(\lambda)$ is a rational Nevanlinna function if and only if it is of the form

$$m(\lambda) = m_1\lambda + m_0 + \sum_{j=1}^{k} \frac{s_j}{z_j - \lambda},$$

where k is an integer ≥ 0 and m_1, m_0, s_j, and z_j are real numbers with $m_1 \geq 0$, $s_j > 0$, and $z_i \neq z_j$, $i \neq j$; see, for example, [5, p. 19].

We now prove three lemmas which we use in Sections 3 and 4. In the first lemma we describe the connection between (weakly) hyperbolic matrix polynomials and Nevanlinna matrix functions. A complex number λ_0 is called a *zero of a rational matrix function* M if there is a vector function u defined and holomorphic in a neighborhood of λ_0 such that

$$u(\lambda_0) \neq 0 \quad \text{and} \quad \lim_{\lambda \to \lambda_0} M(\lambda)u(\lambda) = 0. \tag{2.6}$$

We denote the set of zeros of M by $\operatorname{nul}(M)$.

Lemma 2.1. *Let M be a rational $n \times n$ matrix function. Then M is the inverse of a weakly hyperbolic (hyperbolic) $n \times n$ matrix polynomial L of the form (2.1) if and only if*

(i) $\operatorname{nul}(M) = \emptyset$,
(ii) $\lim_{\lambda\to\infty} \lambda^2 M(\lambda) = I_n$, *and*
(iii) *there exists a point $c \in \mathbb{R}$ (respectively $c \in \operatorname{hol}(M) \cap \mathbb{R}$) such that $-(\lambda - c)M(\lambda)$ is a Nevanlinna function.*

If these conditions hold and $[a_2, b_2]$ and $[a_1, b_1]$ are the left and right root zones of L, then c belongs to the compact interval $[b_2, a_1]$ (respectively open interval (b_2, a_1)).

Proof. (a) We begin by proving that M^{-1} exists and is a matrix polynomial if and only if $\operatorname{nul}(M) = \emptyset$: Assume $\operatorname{nul}(M) = \emptyset$. Then $\det M(\lambda) \neq 0$ and hence $M(\lambda)^{-1}$ exists for all $\lambda \in \operatorname{hol}(M)$. Assume λ_0 is a pole of M^{-1}. Then there exist a vector x_0 and a natural number p such that for $u(\lambda) = (\lambda - \lambda_0)^p M^{-1}(\lambda) x_0$ we have

$$\lim_{\lambda \to \lambda_0} u(\lambda) \neq 0,$$

hence $u(\lambda)$ is holomorphic in a neighborhood of λ_0, and

$$\lim_{\lambda \to \lambda_0} M(\lambda) u(\lambda) = \lim_{\lambda \to \lambda_0} (\lambda - \lambda_0)^p x_0 = 0,$$

that is, $\lambda_0 \in \operatorname{nul}(M) \neq \emptyset$. This contradiction implies M^{-1} has no poles and hence is a matrix polynomial. We now show the converse. Assume M^{-1} is a matrix polynomial. Suppose there is a point $\lambda_0 \in \operatorname{nul}(M)$. Then there is a vector function u defined and holomorphic in a neighborhood of λ_0 such that the relations (2.6) hold and hence

$$\lim_{\lambda \to \lambda_0} M^{-1}(\lambda) M(\lambda) u(\lambda) = u(\lambda_0) \neq 0.$$

This shows that λ_0 is a pole of M^{-1}, which contradicts the assumption. Hence $\operatorname{nul}(M) = \emptyset$.

(b) We prove the "only if" part of the lemma and assume that

$$L(\lambda) = \lambda^2 + \lambda B + C \tag{2.7}$$

is weakly hyperbolic. Then by (a) $M = L^{-1}$ satisfies (i), and, evidently, also (ii). To prove (iii) we first consider the case that L is hyperbolic. Let c be a point in the interval (b_2, a_1) between the two end zones of L. Then $(L(c)f, f) < 0$ for all vectors f in the unit ball S of $\mathbb{C}^n$. Since S is compact, we have $L(c) \ll 0$. Therefore $c \in \operatorname{hol}(L^{-1}) = \operatorname{hol}(M)$ and the function

$$\frac{L(\lambda)}{\lambda - c} = \lambda + (B + c) + \frac{L(c)}{\lambda - c} \tag{2.8}$$

is a matrix Nevanlinna function. Hence also $-(\lambda - c)M(\lambda)$ is a matrix Nevanlinna function.

If L is weakly hyperbolic and not hyperbolic, then with $c = b_2 = a_1$ we have $L(c) \leq 0$. For every $\varepsilon > 0$ the matrix function $L_\varepsilon = L - \varepsilon$ is hyperbolic and therefore $-(\lambda - c)L_\varepsilon(\lambda)^{-1}$ is a Nevanlinna function. Consequently, the limit for $\varepsilon \to 0$, that is, the function $-(\lambda - c)L(\lambda)^{-1}$, is also a Nevanlinna function.

(c) Now we prove the "if" part of the lemma. Then items (i) and (ii) in the lemma and part (a) of this proof imply that $L = M^{-1}$ is a matrix polynomial of the form (2.7). By (iii) and the fact that $c \in \mathbb{R}$ the function M is symmetric with respect to the real axis, hence L is hermitian. Again by (iii) $L(\lambda)/(\lambda - c)$ is a matrix Nevanlinna function. From (2.8) it follows that $L(c) \leq 0$. If $(L(c)f, f) \neq 0$ for some $f \in \mathbb{C}^n$, then $(L(c)f, f) < 0$ and the quadratic polynomial $(L(\lambda)f, f)$ has two different real zeros. If $(L(c)f, f) = 0$ and $f \neq 0$, then $(L(\lambda)f, f)$ has only real zeros. Hence L is weakly hyperbolic. Now assume $c \in \mathrm{hol}\,(M) \cap \mathbb{R}$. We claim that then $(L(c)f, f) < 0$ for all nonzero $f \in \mathbb{C}^n$, which, by the previous argument, implies L is hyperbolic. It remains to prove the claim. Set $A = -L(c)$. Then A is hermitian and $(Af, f) \geq 0$ for all $f \in \mathbb{C}^n$. Assume that for some $f_0 \in \mathbb{C}^n$ we have $(Af_0, f_0) = 0$. Then by the Cauchy-Schwarz inequality, $Af_0 = 0$. Since $c \in \mathrm{hol}\,(M)$, A^{-1} exists, hence $f_0 = 0$. This proves the claim.

(d) As to the last statement of the lemma, if c does not belong to the indicated interval then $(L(c)f, f) > 0$ for some $f \in \mathbb{C}^n$ and, by (2.8), $-(\lambda - c)L(\lambda)f, f)$ is not a Nevanlinna function. □

Lemma 2.2. *Let $L(\lambda) = \lambda^2 + \lambda B + C$ be a hyperbolic $n \times n$ matrix polynomial with only simple eigenvalues. Then for every unit vector $e \in \mathbb{C}^n$ and for every $\varepsilon > 0$ there exists a unit vector $e' \in \mathbb{C}^n$ such that $\|e - e'\| < \varepsilon$ and every eigenvalue of L is a pole of $(L(\lambda)^{-1}e', e')$.*

Proof. Choose c in the open interval between the root zones of L. Then $N(\lambda) = -(\lambda - c)L(\lambda)^{-1}$ is a rational Nevanlinna function which has the same poles as the function L^{-1} and

$$\lim_{\lambda \to \infty} N(\lambda) = 0.$$

Since all poles of N have multiplicity one, N can be written in the form

$$N(\lambda) = \sum_{k=1}^{2n} (\mu_k - \lambda)^{-1}(\,\cdot\,, f_k)f_k, \quad \mu_k \in \mathbb{R}, \quad 0 \neq f_k \in \mathbb{C}^n.$$

Now choose e' so that $(e', f_k) \neq 0$ for all $k = 1, \ldots, 2n$, $\|e'\| = 1$, and $\|e - e'\| < \varepsilon$. That such an e' exists can be shown by induction on k: If $(e, f_1) \neq 0$, choose $e' = e$; if $(e, f_1) = 0$, take

$$e' = \frac{e + \delta f_1}{\|e + \delta f_1\|}$$

with $\delta > 0$ so small that $\|e - e'\| < \varepsilon$. Assume that there exists an element $\widetilde{e}$ such that $(\widetilde{e}, f_j) \neq 0$, $j = 1, \ldots, k$, $\|\widetilde{e}\| = 1$, and $\|e - \widetilde{e}\| < \varepsilon$. If $(\widetilde{e}, f_{k+1}) \neq 0$ choose $e' = \widetilde{e}$; if $(\widetilde{e}, f_{k+1}) = 0$, take

$$e' = \frac{\widetilde{e} + \delta f_{k+1}}{\|\widetilde{e} + \delta f_{k+1}\|}$$

with $\delta > 0$ so small that all asserted inequalities hold. □

Lemma 2.3. *Let $L(\lambda) = \lambda^2 + \lambda B + C$ be a weakly hyperbolic $n \times n$ matrix polynomial, let $\{\alpha_{\pm j}\}_{j=1}^n \in T_{2n}$ be the eigenvalues of L, and let e be a vector in $\mathbb{C}^n$ such that the function $(L(\lambda)^{-1}e, e)$ has poles at $\alpha_{\pm n}$. Then there exist vectors $f_{\pm n} \in \ker L(\pm\alpha_n)$ such that $(f_{\pm n}, e) \neq 0$.*

Proof. We prove the lemma for α_n and $f_n \in \ker L(\alpha_n)$.

For α_{-n} and $f_{-n} \in \ker L(\alpha_{-n})$ the proof is the same. Without loss of generality we assume $\alpha_n = 0$. This implies

$$\ker C = \ker L(0) \neq \{0\},$$

and since L is weakly hyperbolic the matrices B and C are nonnegative. Without loss of generality we can assume $\ker C = \mathbb{C}^m$ for some integer m with $1 \leq m \leq n$ and

$$\mathbb{C}^n = \mathbb{C}^m \oplus \mathbb{C}^{n-m}. \tag{2.9}$$

If $m = n$, we take $f_n = e$ and the theorem is proved. Assume $m < n$ and let

$$B = \begin{pmatrix} B_1 & B_{12} \\ B_{12}^* & B_2 \end{pmatrix} \quad \text{and} \quad C = \begin{pmatrix} 0 & 0 \\ 0 & C_2 \end{pmatrix}$$

be the representations of B and C with respect to the decomposition (2.9) of $\mathbb{C}^n$. Then we can write L as the matrix function

$$L(\lambda) = \begin{pmatrix} \lambda^2 + \lambda B_1 & \lambda B_{12} \\ \lambda B_{12}^* & \lambda^2 + \lambda B_2 + C_2 \end{pmatrix}.$$

Assume the contrary: the vector e is orthogonal to $\ker C$. Then $e = \begin{pmatrix} 0 & e_2 \end{pmatrix}^\top$ and

$$(L(\lambda)^{-1}e, e) = \Big(\big(\lambda^2 + \lambda B_2 + C_2 - B_{12}^*(1 + \frac{1}{\lambda}B_1)^{-1}B_{12}\big)^{-1}e_2, e_2\Big). \tag{2.10}$$

Since B is nonnegative, we have that $\operatorname{ran} B_{12} \subset (\ker B_1)^\perp = \operatorname{ran} B_1$. Let

$$B_1' := B_1|_{\operatorname{ran} B_1}.$$

Then B_1' is invertible on $\operatorname{ran} B_1$ and we can rewrite (2.10) as:

$$(L(\lambda)^{-1}e, e) = \Big(\big(\lambda^2 + \lambda B_2 + C_2 - B_{12}^*(1 + \frac{1}{\lambda}B_1')^{-1}B_{12}\big)^{-1}e_2, e_2\Big).$$

It follows that

$$\lim_{\lambda \to 0} \big(L(\lambda)^{-1}e, e\big) = \big(C_2^{-1}e_2, e_2\big) < \infty.$$

Hence the eigenvalue $\alpha_n = 0$ is not a pole of the function $(L(\lambda)^{-1}e, e)$. We have obtained a contradiction which proves the lemma. □

3. Approximation of weakly hyperbolic matrix polynomials by hyperbolic matrix polynomials with simple eigenvalues

As for example in the proof of Lemma 2.1 a statement for weakly hyperbolic matrix polynomials can sometimes be proved first for a hyperbolic matrix polynomials and then for weakly hyperbolic matrix polynomials by approximating the weakly hyperbolic matrix polynomial by hyperbolic ones. The following theorem shows that approximating hyperbolic matrix polynomials can be chosen which have simple eigenvalues only.

Theorem 3.1. *Let $L(\lambda) = \lambda^2 + \lambda B + C$ be a weakly hyperbolic $n \times n$ matrix polynomial with left and right root zones $[a_2, b_2]$ and $[a_1, b_1]$, and let $c \in [b_2, a_1]$. Then for every $\varepsilon > 0$ there exist hermitian $n \times n$ matrices $\widetilde{B}$, $\widetilde{C}$ with $\|B - \widetilde{B}\| < \varepsilon$, $\|C - \widetilde{C}\| < \varepsilon$ such that $\widetilde{L}(\lambda) = \lambda^2 + \lambda\widetilde{B} + \widetilde{C}$ is hyperbolic, $\widetilde{L}(c) \ll 0$, and all eigenvalues of $\widetilde{L}$ are simple.*

Before proving the theorem we prove a variant of Lemma 2.1 in which condition (i) is replaced by an explicit formula for M, compare with [19, Theorem 31.19]. It only holds in the hyperbolic case; for the weakly hyperbolic case, see Remark 3.4 below.

Lemma 3.2. *Let M be a rational $n \times n$ matrix function. Then M is the inverse of a hyperbolic $n \times n$ matrix polynomial L of the form (2.1) if and only if*

(i) *M is of the form*

$$M(\lambda) = \sum_{j=1}^{s} \frac{F_j}{\lambda - \mu_j},$$

where μ_j, $j = 1, \ldots, s$, are mutually different real numbers and F_j are nonzero hermitian $n \times n$ matrices of rank r_j, $j = 1, \ldots, s$, with $r_1 + \cdots + r_s = 2n$,

(ii) *$\lim_{\lambda \to \infty} \lambda^2 M(\lambda) = I_n$, and*

(iii) *there exists a point $c \in \mathbb{R} \setminus \{\mu_1, \ldots, \mu_s\}$ such that $-(\lambda - c)M(\lambda)$ is a matrix Nevanlinna function.*

If these conditions hold, then c belongs to the open interval between the root zones of L.

Proof. (a) We prove the "if" part of the lemma. Assume that M satisfies (i)–(iii). For $j = 1, \ldots, s$ denote by G_j the compression of F_j to $\operatorname{ran} F_j$, then G_j is an hermitian $r_j \times r_j$ matrix. Relative to the decomposition $\mathbb{C}^n = \operatorname{ran} F_j \oplus \ker F_j$ (here we use that F_j is hermitian) we have that in a neighborhood of μ_j there are holomorphic matrix functions A_j, B_j, C_j, and D_j of appropriate sizes such that M has the representation

$$M(\lambda) = \begin{pmatrix} \dfrac{G_j}{\lambda - \mu_j} + A_j(\lambda) & B_j(\lambda) \\ C_j(\lambda) & D_j(\lambda) \end{pmatrix}. \tag{3.1}$$

It follows that $\det M$ has a pole at μ_j of order $\leq r_j$. Since by (ii)

$$\lim_{\lambda \to \infty} M(\lambda) = 0$$

we have by Liouville's theorem that

$$\det M(\lambda) = \frac{p(\lambda)}{(\lambda - \mu_1)^{r_1} \cdots (\lambda - \mu_s)^{r_s}},$$

where p is a real polynomial of degree $< 2n$.

By (ii) and the fact that $r_1 + \cdots + r_s = 2n$, we have $p = 1$. The definition of M and the fact that the inverse of a square matrix is the product of the reciprocal of its determinant and the transpose of the matrix of its co-factors imply that $L = M^{-1}$ has poles in $\mathbb{C}$ at most in the points μ_j, $j = 1, \ldots, s$. But from (3.1) and the just mentioned fact about the formula for the inverse of a square matrix it follows that L is holomorphic at μ_j, $j = 1, \ldots, s$. Thus L is rational and entire, hence it is a polynomial matrix function of the indicated form. From part (a) of the proof of Lemma 2.1 it follows that $\operatorname{nul}(M) = \emptyset$. To complete the proof it remains to apply Lemma 2.1.

(b) We prove the "only if" part of the lemma. Assume that $L = M^{-1}$ is hyperbolic. Then, according to Lemma 2.1, (ii) and (iii) hold with the point $c \in \operatorname{hol}(M) \cap \mathbb{R}$. It remains to prove that M has the form as in (i). Set $G(\lambda) = -(\lambda - c)M(\lambda)$, then G is a matrix Nevanlinna function which is holomorphic in some neighborhoods of c and ∞. We choose positive numbers t_1, t_2 with $t_1 < t_2$ such that G is holomorphic in $\overline{\mathbb{C}} \setminus (I_+ \cup I_-)$, where $I_+ = (c+t_1, c+t_2)$ and $I_- = (c - t_2, c - t_1)$. Let

$$G(\lambda) = \int_{I_+} \frac{d\,\Sigma_+(t)}{t-\lambda} + \int_{I_-} \frac{d\,\Sigma_-(t)}{t-\lambda}$$

be the integral representation of G. Making use of $G(c) = 0$ and the relation

$$\frac{\dfrac{1}{t-\lambda} - \dfrac{1}{t-c}}{\lambda - c} = \frac{1}{(t-\lambda)(t-c)}$$

we find that

$$-M(\lambda) = \frac{G(\lambda) - G(c)}{\lambda - c} = \int_{I_+} \frac{d\,\Sigma_+(t)}{(t-\lambda)(t-c)} + \int_{I_-} \frac{d\,\Sigma_-(t)}{(t-\lambda)(t-c)}. \tag{3.2}$$

Here $(t-c)^{-1}d\,\Sigma_+(t)$ is a positive matrix measure on I_+ with a finite number of masses $F_1^+, \dots, F_{k_+}^+$ in the points $\mu_1^+, \dots, \mu_{k_+}^+$, respectively, and $(t-c)^{-1}d\,\Sigma_-(t)$ is a negative matrix measure on I_- with a finite number of masses $F_1^-, \dots, F_{k_-}^-$ in the points $\mu_1^-, \dots, \mu_{k_-}^-$, respectively. Thus (3.2) can be written as

$$M(\lambda) = \sum_{l=1}^{k_+} (\lambda - \mu_l^+)^{-1} F_l^+ + \sum_{m=1}^{k_-} (\lambda - \mu_m^-)^{-1} F_m^-.$$

The points $\mu_j^\pm$ are the eigenvalues of L with multiplicity $\operatorname{rank}(F_j^\pm)$, $j = 1, \dots, k_\pm$, and hence these multiplicities add up to $2n$. It follows that M is of the form (i) and $\operatorname{hol}(M) = \mathbb{R} \setminus \{\mu_j^\pm : j = 1, \dots, k_\pm\}$. □

Remark 3.3. Another way to see that the equality $r_1 + \cdots + r_s = 2n$ in the proof of the "only if" part of Lemma 3.2 (i) holds is by using McMillan degrees: If $G(\lambda) = -(\lambda - c)M(\lambda)$ then

$$r_1 + \cdots + r_s = \deg M = \deg G = \deg G^{-1}.$$

On the other hand the McMillan degree of the matrix Nevanlinna function $G^{-1}(\lambda) = L(\lambda)/(\lambda - c)$ is also the dimension of the reproducing kernel Hilbert space associated with the Nevanlinna kernel

$$K_{G^{-1}}(z,\zeta) = \frac{G(z)^{-1} - (G(\zeta)^{-1})^*}{z - \overline{\zeta}} = I_n - \frac{L(c)}{(z-c)(\overline{\zeta} - c)}.$$

(For the last equality we used (2.8).) Since this Hilbert space is spanned by the vector functions given by the columns of the identity matrix I_n (constant vector functions) and the columns of $I_n/(z-c)$, we have $\deg G^{-1} = 2n$.

Remark 3.4. If M satisfies Lemma 3.2 (i), (ii), and (iii) with $c \in \mathbb{R}$ (rather than $c \in \operatorname{hol}(M) \cap \mathbb{R}$), then M^{-1} is a weakly hyperbolic matrix polynomial of the form (2.1) with eigenvalues μ_j of multiplicity r_j, $j = 1, \ldots, s$. If these conditions hold and $[a_2, b_2]$ and $[a_1, b_1]$ are the left and right root zones of L, then c belongs to the compact interval $[b_2, a_1]$. The proof is the same as the proof of the "if" part of Lemma 3.2.

Proof of Theorem 3.1. The polynomial $L_\varepsilon = L - \varepsilon/2$ is hyperbolic and $L_\varepsilon(c) \ll 0$. Then from the proof of part (b) of Lemma 3.2 it follows that $M_\varepsilon = L_\varepsilon^{-1}$ can be written as

$$M_\varepsilon(\lambda) = \sum_{l=1}^{k_+} (\lambda - \mu_l^+)^{-1} F_l^+ + \sum_{m=1}^{k_-} (\lambda - \mu_m^-)^{-1} F_m^-.$$

Here and from now on we use the same notation as in the proof of part (b) of Lemma 3.2. Using the formal equality

$$\frac{1}{\lambda - \mu} = \frac{1}{\lambda}\left(1 + \frac{\mu}{\lambda} + \frac{\mu^2}{\lambda^2} + \cdots\right)$$

we obtain that M_ε has a Taylor expansion at ∞ of the form

$$\begin{aligned} M_\varepsilon(\lambda) = \lambda^{-1} &\left(\sum_{l=1}^{k_+} F_l^+ + \sum_{m=1}^{k_-} F_m^- \right) \\ &+ \lambda^{-2} \left(\sum_{l=1}^{k_+} \mu_l^+ F_l^+ + \sum_{m=1}^{k_-} \mu_m^- F_m^- \right) + \cdots . \end{aligned}$$

From

$$\lim_{\lambda \to \infty} \lambda^2 M_\varepsilon(\lambda) = I_n$$

we obtain

$$\sum_{l=1}^{k_+} F_l^+ + \sum_{m=1}^{k_-} F_m^- = 0 \tag{3.3}$$

and

$$\sum_{l=1}^{k_+} \mu_l^+ F_l^+ + \sum_{m=1}^{k_-} \mu_m^- F_m^- = I_n. \tag{3.4}$$

We write each of the matrices F_l^+ (F_m^-) as the sum of rank one nonnegative (respectively nonpositive) hermitian matrices:

$$F_l^+ = F_{l,1}^+ + \cdots + F_{l,p_l}^+ \text{ (respectively } F_m^- = F_{m,1}^- + \cdots + F_{m,q_m}^-), \tag{3.5}$$

where $p_1 + \cdots + p_{k_+} + q_1 + \cdots + q_{k_-} = 2n$. For every $l = 1, \ldots, k_+$ we choose points $\mu_{l,1}^+, \ldots, \mu_{l,p_l}^+$ from I_+ such that $\mu_{l,p}^+ \neq \mu_{l',p'}^+$ if $l \neq l'$ or $p \neq p'$. Similarly, for every $m = 1, \ldots, k_-$ we choose points $\mu_{m,1}^-, \ldots, \mu_{m,q_m}^-$ from I_- such that $\mu_{m,q}^- \neq \mu_{m',q'}^-$ if $m \neq m'$ or $q \neq q'$.

We define

$$H(\lambda) = \sum_{l=1}^{k_+} \sum_{p=1}^{p_l} (\lambda - \mu_{l,p}^+)^{-1} F_{l,p}^+ + \sum_{m=1}^{k_-} \sum_{q=1}^{q_m} (\lambda - \mu_{m,q}^-)^{-1} F_{m,q}^-.$$

Then the Taylor expansion at ∞ of H is given by

$$\begin{aligned} H(\lambda) = \lambda^{-1} & \left(\sum_{l=1}^{k_+} \sum_{p=1}^{p_l} F_{l,p}^+ + \sum_{m=1}^{k_-} \sum_{q=1}^{q_m} F_{m,q}^- \right) \\ & + \lambda^{-2} \left(\sum_{l=1}^{k_+} \sum_{p=1}^{p_l} \mu_{l,p}^+ F_{l,p}^+ + \sum_{m=1}^{k_-} \sum_{q=1}^{q_m} \mu_{m,q}^- F_{m,q}^- \right) + \cdots . \end{aligned}$$

By (3.3) the coefficient of λ^{-1} is zero. We denote the coefficient of λ^{-2} by S, that is,

$$S = \sum_{l=1}^{k_+} \sum_{p=1}^{p_l} \mu_{l,p}^+ F_{l,p}^+ + \sum_{m=1}^{k_-} \sum_{q=1}^{q_m} \mu_{m,q}^- F_{m,q}^-.$$

Clearly, S is hermitian. By (3.4) and (3.5) there exists a $\delta > 0$ such that the relations

$$\begin{aligned} \max\{|\mu_{l,p}^+ - \mu_l^+| : \ p = 1, \dots, p_l, \ l = 1, \dots, k_+\} < \delta, \\ \max\{|\mu_{m,q}^- - \mu_m^-| : \ q = 1, \dots, q_m, \ m = 1, \dots, k_-\} < \delta \end{aligned} \tag{3.6}$$

imply $\|S - 1\| < \frac{1}{2}$. It then follows that $S^{-1/2}$ exists. We set

$$M(\lambda) = S^{-1/2} H(\lambda) S^{-1/2}.$$

Then

$$\lim_{\lambda \to \infty} \lambda^2 M(\lambda) = I_n$$

and the function

$$\begin{aligned} -(\lambda - c)M(\lambda) &= (\lambda - c) \sum_{l=1}^{k_+} \sum_{p=1}^{p_l} (\mu_{l,p}^+ - \lambda)^{-1} S^{-1/2} F_{l,p}^+ S^{-1/2} \\ & \quad + (\lambda - c) \sum_{m=1}^{k_-} \sum_{q=1}^{q_m} (\mu_{m,q}^- - \lambda)^{-1} S^{-1/2} F_{m,q}^- S^{-1/2} \\ &= \sum_{l=1}^{k_+} \sum_{p=1}^{p_l} \left(-1 - \frac{\mu_{l,p}^+ - c}{\lambda - \mu_{l,p}^+} \right) S^{-1/2} F_{l,p}^+ S^{-1/2} \\ & \quad + \sum_{m=1}^{k_-} \sum_{q=1}^{q_m} \left(1 - \frac{c - \mu_{m,q}^-}{\lambda - \mu_{m,q}^-} \right) S^{-1/2} \left(-F_{m,q}^- \right) S^{-1/2} \end{aligned}$$

is a Nevanlinna function. We conclude that M satisfies items (i)–(iii) of Lemma 3.2, hence M is the inverse of a hyperbolic $n \times n$ matrix polynomial of the form

(2.1). We denote by this polynomial by $\widetilde{L}(\lambda) = \lambda^2 + \lambda\widetilde{B} + \widetilde{C}$. The $2n$ eigenvalues of $\widetilde{L}$ are $\mu^+_{l,1}, \ldots, \mu^+_{l,p_l}$, $l = 1, \ldots, k_+$, and $\mu^-_{m,1}, \ldots, \mu^-_{m,q_m}$, $m = 1, \ldots, k_-$, and, by construction, they are distinct. Since the point c belongs to the interval $(c - t_1, c + t_1)$ and the interval is contained in the gap between the root zones of $\widetilde{L}$, we have $\widetilde{L}(c) \ll 0$.

It remains to show that the points $\mu^+_{l,p}$ and $\mu^-_{m,q}$ can be chosen so that the coefficients of $\widetilde{L}$ satisfy the inequalities

$$\|B - \widetilde{B}\| < \varepsilon \text{ and } \|C - \widetilde{C}\| < \varepsilon. \tag{3.7}$$

Let r be a real number > 1 such that the closure of $I_+ \cup I_-$ is inside the circle $r\mathbb{T}$; here $\mathbb{T}$ stands for the unit circle in $\mathbb{C}$. Then for $\lambda \in r\mathbb{T}$ we have

$$\begin{aligned}\lambda(B - \widetilde{B}) &+ (C - \varepsilon/2 - \widetilde{C})\\ &= L_\varepsilon(\lambda) - M(\lambda)^{-1}\\ &= L_\varepsilon(\lambda)\{M(\lambda) - M_\varepsilon(\lambda)\}M(\lambda)^{-1}.\end{aligned}$$

In the following estimates λ belongs to $r\mathbb{T}$ and "max" stands for the maximum over $\lambda \in r\mathbb{T}$. Choose δ in (3.6) such that for a given real number $\tau < 1$

$$\|S - 1\| < \tau, \quad \max\|M_\varepsilon(\lambda) - H(\lambda)\| < \tau. \tag{3.8}$$

The first inequality implies $1 - \tau < S < 1 + \tau$ and from this and

$$S^{-1/2} - 1 = (1 - S)S^{-1/2}(1 + S^{1/2})^{-1}$$

it follows that

$$\|S^{-1/2} - 1\|\,(\|S^{-1/2}\| + 1) \le \frac{\tau}{\sqrt{1-\tau}(1 + \sqrt{1-\tau})}\left(\frac{1}{\sqrt{1-\tau}} + 1\right) = \frac{\tau}{1-\tau}.$$

This inequality and the second inequality in (3.8) imply

$$\begin{aligned}\|H(\lambda) &- S^{-1/2}H(\lambda)S^{-1/2}\|\\ &= \|(S^{-1/2} - 1)H(\lambda)S^{-1/2} + H(\lambda)(S^{-1/2} - 1)\|\\ &\le \|S^{-1/2} - 1\|\,(\|S^{-1/2}\| + 1)\max\|H(\lambda)\|\\ &\le \frac{\tau}{1-\tau}(\max\|M_\varepsilon(\lambda)\| + \tau).\end{aligned}$$

Hence from

$$M(\lambda) - M_\varepsilon(\lambda) = (H(\lambda) - M_\varepsilon(\lambda)) + (S^{-1/2}H(\lambda)S^{-1/2} - H(\lambda))$$

it follows that

$$\|M(\lambda) - M_\varepsilon(\lambda)\| \le \frac{\tau}{1-\tau}(\max\|M_\varepsilon(\lambda)\| + \tau) + \tau =: \tau_1.$$

Since

$$\begin{aligned}
&\big|\|L_\varepsilon(\lambda)\| - \|M(\lambda)^{-1}\|\big| \\
&\qquad \le \|L_\varepsilon(\lambda) - M(\lambda)^{-1}\| \\
&\qquad = \|L_\varepsilon(\lambda)\{M(\lambda) - M_\varepsilon(\lambda)\}M(\lambda)^{-1}\| \\
&\qquad \le (\max \|L_\varepsilon(\lambda)\|)\, \tau_1\, \|M(\lambda)^{-1}\|,
\end{aligned}$$

we have

$$\|M(\lambda)^{-1}\| \le \frac{\max \|L_\varepsilon(\lambda)\|}{1 - \tau_1 \max \|L_\varepsilon(\lambda)\|}.$$

Choosing τ (and hence τ_1) sufficiently small we obtain the estimate

$$\begin{aligned}
&\|\lambda(B - \widetilde{B}) + (C - \varepsilon/2 - \widetilde{C})\| \\
&\qquad = \|L_\varepsilon(\lambda) - M(\lambda)^{-1}\| \\
&\qquad = \|L_\varepsilon(\lambda)\{M(\lambda) - M_\varepsilon(\lambda)\}M(\lambda)^{-1}\| \\
&\qquad \le \tau_1 \frac{\max \|L_\varepsilon(\lambda)\|^2}{1 - \tau_1 \max \|L_\varepsilon(\lambda)\|} \\
&\qquad < \varepsilon/2,
\end{aligned}$$

which holds for all $\lambda \in r\mathbb{T}$. The desired inequalities (3.7) follow by applying this estimate for $\lambda = r$ and $\lambda = -r$ and using the triangle inequality. □

We now prove an approximation theorem for the case that the coefficients B and C of L in (2.1) are *Jacobi* matrices, that is, real symmetric tridiagonal matrices. For a weakly hyperbolic $n \times n$ matrix polynomial L we define

$$L(\lambda; \varepsilon_1, \varepsilon_2) := L(\lambda) - \varepsilon_1 - \varepsilon_2 Z, \quad \varepsilon_1, \varepsilon_2 \in \mathbb{R},$$

where Z denotes the tridiagonal matrix $(z_{i,j})_{i,j=1}^n$ with $z_{i,j} = 1$ if $|i - j| = 1$ and $z_{i,j} = 0$ otherwise.

Theorem 3.5. *Let $L(\lambda) = \lambda^2 + \lambda B + C$ be a weakly hyperbolic $n \times n$ Jacobi matrix polynomial with left and right root zones $[a_2, b_2]$ and $[a_1, b_1]$, and let $c \in [b_2, a_1]$. Then for every $\varepsilon_1 > 0$ and $\varepsilon_2 > 0$ there exist $\varepsilon_1' \in (0, \varepsilon_1)$ and $\varepsilon_2' \in (0, \varepsilon_2)$ such that $L(\lambda; \varepsilon_1', \varepsilon_2')$ is a hyperbolic $n \times n$ Jacobi matrix polynomial with only simple eigenvalues and $L(c; \varepsilon_1', \varepsilon_2') \ll 0$.*

Proof. Since $L(c; \varepsilon_1/2, 0) \le -\varepsilon_1/2$, the matrix polynomial $L(\lambda; \varepsilon_1/2, 0)$ is hyperbolic. Let

$$e_{n,j} := \begin{pmatrix} \delta_{1,j} & \cdots & \delta_{n,j} \end{pmatrix}^\top \in \mathbb{C}^n, \quad j = 1, \dots, n.$$

We choose $\widetilde{\varepsilon}_2 \in (0, \varepsilon_2)$ sufficiently small and so that

$$L(c; \varepsilon_1/2, \widetilde{\varepsilon}_2) \le -\varepsilon_1/4 \tag{3.9}$$

and each of the linear functions

$$l_{j+1,j}(\lambda) = (L(\lambda; \varepsilon_1/2, \widetilde{\varepsilon}_2)\, e_{n,j}, e_{n,j+1}), \quad j = 1, 2, \dots, n-1,$$

which appear on the diagonal parallel to and directly below the main diagonal of the $n \times n$ matrix polynomial, is not identically equal to zero. Let $\mathcal{M}$ be the set of all $\mu \in \mathbb{R}$ such that $l_{j+1,j}(\mu) = 0$ for some $j = 1, 2, \ldots, n-1$.

If $\mathcal{M} = \emptyset$, that is, the functions $l_{j,j+1}$ are nonzero constants, then for every real λ either the number zero is not an eigenvalue or it is a simple eigenvalue of $L(\lambda; \varepsilon_1/2, \widetilde{\varepsilon}_2)$. This is a well-known property of tridiagonal matrices, see [18, Theorem 3.7.1, p. 166]. Therefore and in view of (3.9), in this case $\varepsilon_1' = \varepsilon_1/2$, $\varepsilon_2' = \widetilde{\varepsilon}_2$ have the required properties.

If $\mathcal{M} \neq \emptyset$, $\mathcal{M}$ contains no more than $n-1$ real points, say $\mu_1, \ldots, \mu_m$. Let

$$\widetilde{\varepsilon}_1 \in (0, \varepsilon_1/4) \tag{3.10}$$

be such that

$$0 \notin \sigma(L(\mu_k; \widetilde{\varepsilon}_1 + \varepsilon_1/2, \widetilde{\varepsilon}_2)), \quad k = 1, \ldots, m. \tag{3.11}$$

In this case we put $\varepsilon_1' = \widetilde{\varepsilon}_1 + \varepsilon_1/2$, $\varepsilon_2' = \widetilde{\varepsilon}_2$. By (3.9) and (3.10) we have $L(c; \varepsilon_1', \varepsilon_2') \ll 0$. By (3.11) the numbers μ_k, $k = 1, \ldots, m$, are no eigenvalues of $L(\lambda; \varepsilon_1', \varepsilon_2')$. Therefore, for every eigenvalue λ_0 of $L(\lambda; \varepsilon_1', \varepsilon_2')$ all the numbers $l_{j+1,j}(\lambda_0)$, $j = 1, \ldots, m$, are nonzero. Then it follows as above that 0 is a simple eigenvalue of $L(\lambda_0; \varepsilon_1', \varepsilon_2')$, and the theorem is proved. □

4. Rank one perturbations of (weakly) hyperbolic matrix polynomials and interlacing properties of eigenvalues

In this section we study the direct spectral problem associated with the $n \times n$ matrix polynomial

$$L(\lambda) = \lambda^2 + \lambda B + C,$$

the rank one perturbation of L of the form

$$L_{s,t;e}(\lambda) = L(\lambda) + (\lambda s + t)(\,\cdot\,, e)\, e,$$

where e is a unit vector in $\mathbb{C}^n$ and s and t are real numbers, and the compression of L to $\{e\}^\perp$ defined by

$$L_{\infty;e}(\lambda) = P_e L(\lambda)|_{\operatorname{ran} P_e},$$

where $P_e = 1 - (\,\cdot\,, e)e$ is the orthogonal projection in $\mathbb{C}^n$ onto $\{e\}^\perp$. The main assumption will be that L is (weakly) hyperbolic. Then the compression $L_{\infty;e}$ is also (weakly) hyperbolic, since $(L_{\infty;e}(\lambda)f, f) = (L(\lambda)f, f)$ for all $f \in \operatorname{ran} P_e$ and hence the zeros of the polynomials on the left- and right-hand side of the equality sign coincide. Moreover,

$$\Delta_i(L_{\infty;e}) \subset \Delta_i(L), \quad i = 1, 2.$$

Under certain assumptions on s and t, the rank one perturbation $L_{s,t;e}$ is also (weakly) hyperbolic: (a) If L is weakly hyperbolic with left and right root zones $[a_2, b_2]$ and $[a_1, b_1]$ and if

$$s < 0 \text{ and } t \leq -a_1 s \quad \text{or} \quad s \geq 0 \text{ and } t \leq -b_2 s,$$

then $L_{s,t;e}$ is weakly hyperbolic, see [19, Corollary 31.8, where the assumption $A(\lambda) \neq 0$ can be dropped as $n = 2$]. (b) If L is hyperbolic and

$$s < 0 \text{ and } t < -a_1 s \quad \text{or} \quad s \geq 0 \text{ and } t < -b_2 s,$$

then $L_{s,t;e}$ is hyperbolic, see [19, Lemma 31.15].

Remark 4.1. We show how the compression $L_{\infty;e}$ of L is related to a limit of rank one perturbations $L_{s,t;e}$ of L. For all λ such that $L(\lambda)$ is invertible and

$$1 + (\lambda s + t)(L(\lambda)^{-1}e, e) \neq 0$$

the matrix polynomial $L_{s,t;e}(\lambda)$ is invertible and its inverse is given by

$$L_{s,t;e}(\lambda)^{-1} = L(\lambda)^{-1} - \frac{\lambda s + t}{1 + (\lambda s + t)(L(\lambda)^{-1}e, e)} (L(\lambda)^{-1} \cdot, e)\, L(\lambda)^{-1}e. \tag{4.1}$$

Let Λ be the open set

$$\Lambda = \{\lambda \in \mathbb{C} : \det L(\lambda) \neq 0,\ (L(\lambda)^{-1}e, e) \neq 0\}.$$

If $\lambda \in \Lambda$ is fixed and $\lambda s + t \to \infty$, then

$$L_{s,t;e}(\lambda)^{-1} f \to \begin{cases} 0 & \text{if } f \in \operatorname{span}\{e\}, \\ L(\lambda)^{-1} f - \dfrac{(L(\lambda)^{-1} f, e)}{(L(\lambda)^{-1} e, e)} L(\lambda)^{-1} e & \text{if } f \in \operatorname{ran} P_e. \end{cases}$$

Hence, since

$$L_{\infty;e}(\lambda)^{-1} f = L(\lambda)^{-1} f - \frac{(L(\lambda)^{-1} f, e)}{(L(\lambda)^{-1} e, e)} L(\lambda)^{-1} e, \quad f \in \operatorname{ran} P_e, \tag{4.2}$$

we have that $L_{s,t;e}(\lambda)^{-1} \to L_{\infty;e}(\lambda)^{-1} P_e$ as $\lambda s + t \to \infty$. From the estimate

$$\|L_{s,t;e}(\lambda)^{-1} - L_{\infty;e}(\lambda)^{-1} P_e\| \leq \frac{\|L(\lambda)^{-1}\|^2}{|(L(\lambda)^{-1}e, e)|\, |1 + (\lambda s + t)(L(\lambda)^{-1}e, e)|}$$

and the continuity of $L(\lambda)^{-1}$ on Λ, it follows that this convergence is uniform in λ in any compact subset $K \subset \Lambda$ as $\inf_{\lambda \in K} |\lambda s + t| \to \infty$. In terms of linear relations this implies that the graph of $L_{s,t;e}(\lambda)$ converges uniformly on K in the gap metric on $\mathbb{C}^n \oplus \mathbb{C}^n$ to the linear relation

$$\left\{ \begin{pmatrix} f \\ L_{\infty;e}(\lambda) f \end{pmatrix} : f \in \operatorname{ran} P_e \right\} \oplus \left\{ \begin{pmatrix} 0 \\ we \end{pmatrix} : w \in \mathbb{C} \right\}. \tag{4.3}$$

Here the set on the left is the graph of $L_{\infty;e}(\lambda)$ in $\operatorname{ran} P_e \oplus \operatorname{ran} P_e$ considered as a subspace in $\mathbb{C}^n \oplus \mathbb{C}^n$. By the gap metric between two closed linear relations we mean the norm of the difference of the orthogonal projections onto these relations, see [2, Section 4]. To see the convergence, denote by $P_{s,t;e}(\lambda)$ the orthogonal projection in $\mathbb{C}^n \oplus \mathbb{C}^n$ onto the graph of $L_{s,t;e}(\lambda)$, by $P_{\infty;e}(\lambda)$ the orthogonal projection in $\mathbb{C}^n \oplus \mathbb{C}^n$ onto the linear relation (4.3), and by J_1 the $2n \times 2n$ signature matrix

$$J_1 = \begin{pmatrix} 0 & I_n \\ I_n & 0 \end{pmatrix}.$$

Then $J_1 P_{s,t;e}(\lambda) J_1$ and $J_1 P_{\infty;e}(\lambda) J_1$ are the orthogonal projections in $\mathbb{C}^n \oplus \mathbb{C}^n$ onto the graphs of the operators $L_{s,t;e}(\lambda)^{-1}$ and $L_{\infty;e}(\lambda)^{-1} P_e$ and we have

$$\begin{aligned}
&\|P_{s,t;e}(\lambda) - P_{\infty;e}(\lambda)\| \\
&\quad = \|J_1 (P_{s,t;e}(\lambda) - P_{\infty;e}(\lambda)) J_1\| \\
&\quad \leq \|L_{s,t;e}(\lambda)^{-1} - L_{\infty;e}(\lambda)^{-1} P_e\|.
\end{aligned}$$

The inequality holds by [11, Chapter IV, Theorem 2.14, formula (2.27)]. It readily follows that

$$\|P_{s,t;e}(\lambda) - P_{\infty;e}(\lambda)\| \to 0 \quad \text{as} \quad \inf_{\lambda \in K} |\lambda s + t| \to \infty$$

uniformly in $\lambda \in K$.

As to the direct spectral problem, in the remainder of this section we first prove two theorems about the interlacing of the eigenvalues of L and its compression $L_{\infty;e}$. In Theorem 4.2 we consider the case where L is hyperbolic and in Theorem 4.3 we consider the case where L is weakly hyperbolic. In the proof of Theorem 4.3. we approximate the weakly hyperbolic polynomial by hyperbolic polynomials and apply Theorem 4.2. Then we prove two theorems about the interlacing of the eigenvalues of L and its rank one perturbation $L_{s,t:e}$; see Theorem 4.5 where L is hyperbolic and Theorem 4.6 where L is weakly hyperbolic. Here too we apply the hyperbolic case to prove the weakly hyperbolic analog.
Theorem 4.2 and Theorem 4.3 concern block-interlacing. We say that the ordered set $\{\beta_{\pm j}\}_{j=1}^{n-1} \in T_{2n-2}$ *block-interlaces* the ordered set $\{\alpha_{\pm j}\}_{j=1}^{n} \in T_{2n}$ if

$$\alpha_{-j-1} \leq \beta_{-j} \leq \alpha_{-j} \quad \text{and} \quad \alpha_j \leq \beta_j \leq \alpha_{j+1}, \quad j = 1, \ldots, n-1,$$

and we call this block-interlacing *strict* if only the strict inequalities prevail.

Theorem 4.2. *Let $L(\lambda) = \lambda^2 + \lambda B + C$ be a hyperbolic $n \times n$ matrix polynomial and let e be a unit vector in $\mathbb{C}^n$. Assume that*

(i) *L has only simple eigenvalues: $\{\alpha_{\pm j}\}_{j=1}^{n} \in T_{2n}^s$, and*
(ii) *the function $(L(\lambda)^{-1}e, e)$ has a pole at each of these eigenvalues.*

Then the ordered eigenvalues of $L_{\infty;e} = P_e L|_{\operatorname{ran} P_e}$ strictly block-interlace the ordered eigenvalues $\{\alpha_{\pm j}\}_{j=1}^{n}$ of L.

Proof. Let $c \in (\alpha_{-1}, \alpha_1)$. Then $-(\lambda - c)(L(\lambda)^{-1}e, e)$ is a Nevanlinna function, which has a pole at each of the points $\alpha_{\pm j}$, $j = 1, \ldots, n$, and tends to 0 as $\lambda \to \infty$. Therefore, it is increasing in the open intervals between these points. It follows that the function $(L(\lambda)^{-1}e, e)$ has a zero in each of the intervals

$$(\alpha_{-n}, \alpha_{-n+1}), \ldots, (\alpha_{-2}, \alpha_{-1}), (\alpha_1, \alpha_2), \ldots, (\alpha_{n-1}, \alpha_n).$$

By (4.2) these zeros are the eigenvalues of $L_{\infty;e}$. □

Applying Theorem 4.2 we obtain the following generalization to weakly hyperbolic matrix polynomials.

Theorem 4.3. *Let $L(\lambda) = \lambda^2 + \lambda B + C$ be a weakly hyperbolic $n \times n$ matrix polynomial and let e be a unit vector in $\mathbb{C}^n$. Then the ordered eigenvalues of $L_{\infty;e} = P_e L|_{\operatorname{ran} P_e}$ block-interlace the ordered eigenvalues of L.*

Proof. Let $[a_2, b_2]$ and $[a_1, b_1]$ be the left and right root zones of L and let $b_2 \le c \le a_1$. Then by Theorem 3.1 there is a sequence of hyperbolic $n \times n$ matrix polynomials $L^{\{j\}}$ with $L^{\{j\}}(c) \ll 0$ which converges to L as $j \to \infty$. According to Lemma 2.2 there is a sequence of unit vectors $e^{\{j\}} \in \mathbb{C}^n$ converging to e such that every eigenvalue of $L^{\{j\}}$ is a pole of $(L^{\{j\}}(\lambda)^{-1} e^{\{j\}}, e^{\{j\}})$. Since $L^{\{j\}}_{\infty;e^{\{j\}}}$ converges to $L_{\infty;e}$ as $j \to \infty$, the block-interlacing follows from Theorem 4.2 □

Example 4.4. This example is about (strongly) damped $n \times n$ matrix polynomials, for the definitions we refer to Section 2, and their compressions in terms of companion matrices. Let J be the signature matrix

$$J = \begin{pmatrix} I_n & 0 \\ 0 & -I_n \end{pmatrix}.$$

A $2n \times 2n$ matrix T is called *J-nonnegative* (*J-positive*) if JT is nonnegative (positive) in $\mathbb{C}^{2n}$. Let $L(\lambda) = \lambda^2 + \lambda B + C$ be an hermitian $n \times n$ matrix polynomial and let

$$T = \begin{pmatrix} 0 & C_1 \\ -C_1^* & -B \end{pmatrix}$$

be its companion matrix, where C_1 is an $n \times n$ matrix such that $C_1^* C_1 = C$. Then the eigenvalues of T and their multiplicities coincide with the eigenvalues and their multiplicities of L. It is well known that L is damped (strongly damped) if and only if there is a $c \in \mathbb{R}$ such that $T - c$ is J-nonnegative (J-positive). If L is damped (strongly damped) then $c \in [\alpha_{-1}, \alpha_1]$ ($c \in (\alpha_{-1}, \alpha_1)$), where α_{-1} (α_1) is the largest (smallest) eigenvalue in the left (right) root zone of L, or what is the same, the right (left) endpoint of the left (right) root zone of L. By direct calculation one can verify that for a damped (strongly damped) L a companion matrix T can be constructed such that the matrix C_1 is lower triangular, that is, $C_1 = (c_{1,ij})_{i,j=1}^n$ with $c_{1,ij} = 0$ if $j < i$, $i, j = 1, \dots, n$; the diagonal entries $c_{1,ii}$ can be chosen positive and then C_1 is unique. If in this companion matrix T of L we delete the nth and the last column and the nth and the last row, then thanks to the lower triangular form of C_1, we obtain a companion matrix T_∞ of the matrix polynomial $L_{\infty;e_n}$ with $e_n = \begin{pmatrix} 0 & \cdots & 0 & 1 \end{pmatrix}^\top \in \mathbb{C}^n$. Hence the ordered eigenvalues of T_∞ block-interlace the ordered eigenvalues of T. This interlacing may not hold if the matrix C_1 is not lower triangular as the following example shows. Consider the strongly damped 2×2 matrix polynomial

$$L(\lambda) = \lambda^2 + 2\lambda + \begin{pmatrix} \sin^2 \varphi & 0 \\ 0 & \cos^2 \varphi \end{pmatrix}, \qquad \varphi \in (0, \pi/4).$$

Then $\alpha_{\pm 1} = -1 \pm \sin\varphi$, $\alpha_{\pm 2} = -1 \pm \cos\varphi$ are the ordered eigenvalues of L and the matrix

$$T = \begin{pmatrix} 0 & 0 & 0 & \cos\varphi \\ 0 & 0 & \sin\varphi & 0 \\ 0 & -\sin\varphi & -2 & 0 \\ -\cos\varphi & 0 & 0 & -2 \end{pmatrix}$$

is a companion matrix of L. If in T we delete the 2nd and 4th row and the 2nd and 4th column, we get the matrix

$$T_\infty = \begin{pmatrix} 0 & 0 \\ 0 & -2 \end{pmatrix}$$

whose eigenvalues 0 and -2 do not interlace the eigenvalues of T.

In [17] it is proved that the positive (negative) eigenvalues of a J-positive $2n \times 2n$ matrix A and of A^+ (A^-) interlace, where A^+ (A^-) is the $(2n-1)\times(2n-1)$ matrix obtained from A by cutting off its nth row and its nth column (its $2n$th row and its $2n$th column). Since $T+1$ is a J-positive matrix the example above shows that the block-interlacing rule does not work for J-positive matrices.

Theorem 4.5 and Theorem 4.6 concern $(-,+)$-interlacing. We say that the ordered set $\{\widetilde{\alpha}_{\pm j}\}_{j=1}^n \in T_{2n}$ $(-,+)$-*interlaces* the ordered set $\{\alpha_{\pm j}\}_{j=1}^n \in T_{2n}$ if

$$\widetilde{\alpha}_{-j} \le \alpha_{-j}, \quad \alpha_j \le \widetilde{\alpha}_j, \quad j = 1, \dots, n, \tag{4.4}$$

and

$$\alpha_{-j} \le \widetilde{\alpha}_{-j+1}, \quad \widetilde{\alpha}_{j-1} \le \alpha_j, \quad j = 2, \dots, n, \tag{4.5}$$

and we call this $(-,+)$-interlacing *strict* if in (4.4) and (4.5) $\le$ can be replaced by $<$. Note that $\{\widetilde{\alpha}_{\pm j}\}_{j=1}^n \in T_{2n}$ $(-,+)$-interlaces $\{\alpha_{\pm j}\}_{j=1}^n \in T_{2n}$ if and only if $\{\alpha_{\pm j}\}_{j=1}^{n-1} \in T_{2n-2}$ block-interlaces $\{\widetilde{\alpha}_{\pm j}\}_{j=1}^n$ and

$$\widetilde{\alpha}_{-n} \le \alpha_{-n} \text{ and } \alpha_n \le \widetilde{\alpha}_n.$$

Theorem 4.5. *Let $L(\lambda) = \lambda^2 + \lambda B + C$ be a hyperbolic $n \times n$ matrix polynomial, let s and t be real numbers, and let e be a unit vector in $\mathbb{C}^n$. Assume that*

(i) *L has only simple eigenvalues: $\{\alpha_{\pm j}\}_{j=1}^n \in T_{2n}^s$, and*
(ii) *the function $(L(\lambda)^{-1}e, e)$ has a pole at each of these eigenvalues.*

Then the ordered eigenvalues of

$$L_{s,t;e}(\lambda) = L(\lambda) + (\lambda s + t)(\,\cdot\,, e)\, e$$

strictly $(-,+)$-interlace the ordered eigenvalues $\{\alpha_{\pm j}\}_{j=1}^n$ of L if and only if

$$s < 0 \text{ and } t < -\alpha_{-n}s \quad \text{or} \quad s \ge 0 \text{ and } t < -\alpha_n s. \tag{4.6}$$

Proof. Assume that (4.6) holds and let $c \in (\alpha_{-1}, \alpha_1)$. We recall formula (4.1):

$$L_{s,t;e}(\lambda)^{-1} = L(\lambda)^{-1} - \frac{\lambda s + t}{1 + (\lambda s + t)(L(\lambda)^{-1}e, e)}(L(\lambda)^{-1}\,\cdot\,, e)\, L(\lambda)^{-1}e \tag{4.7}$$

and define the rational function f_c related to the quotient on the right-hand side by

$$f_c(\lambda) = -(\lambda - c)\left((\lambda s + t)^{-1} + (L(\lambda)^{-1}e, e)\right). \tag{4.8}$$

We claim that f_c is a Nevanlinna function. To prove the claim we rewrite f_c in the form

$$f_c(\lambda) = -\frac{\lambda - c}{(L(\lambda)^{-1}e, e)} + \begin{cases} -s^{-1} + \dfrac{s^{-1}(s^{-1}t + c)}{\lambda + s^{-1}t} & \text{if } s \neq 0, \\ -\lambda t^{-1} + ct^{-1} & \text{if } s = 0. \end{cases}$$

The first summand on the right-hand side is a Nevanlinna function by Lemma 2.1 and the second summand is also a Nevanlinna function, whose poles do not coincide with the poles of the first summand, because by (4.6) we have

$$\begin{array}{lll} -s^{-1}t > \alpha_n \text{ and } s^{-1}t + c < 0 & \text{if } s > 0, \\ -s^{-1}t < \alpha_{-n} \text{ and } s^{-1}t + c > 0 & \text{if } s < 0, \\ \qquad -t^{-1} > 0 & \text{if } s = 0. \end{array}$$

This proves the claim. From formula (4.8) we obtain

$$\lim_{\lambda \to \pm\infty} f_c(\lambda) = \begin{cases} -s^{-1} & \text{if } s \neq 0, \\ \pm\infty & \text{if } s = 0. \end{cases} \tag{4.9}$$

By plotting the graph of the function $f_c(\lambda)$ one sees that it has a zero at $\lambda = c$, precisely one zero in each of the $2n - 2$ open intervals

$$(\alpha_{-n}, \alpha_{-n+1}), \ldots, (\alpha_{-2}, \alpha_{-1}), (\alpha_1, \alpha_2), \ldots, (\alpha_{n-1}, \alpha_n), \tag{4.10}$$

precisely one zero in each of the two open intervals

$$\begin{cases} (-\infty, \alpha_{-n}),\ (\alpha_n, -s^{-1}t) & \text{if } s > 0, \\ (-s^{-1}t, \alpha_{-n}),\ (\alpha_n, \infty) & \text{if } s < 0, \\ (-\infty, \alpha_{-n}),\ (\alpha_n, \infty) & \text{if } s = 0, \end{cases}$$

and no other zeros. The $2n$ zeros in the open intervals, when ordered, strictly $(-,+)$-interlace $\{\alpha_{\pm j}\}_{j=1}^n$. By (4.7) they are the poles of $L_{s,t;e}^{-1}$ and hence the eigenvalues of $L_{s,t;e}$.

We now prove the converse. Assume that the ordered eigenvalues $\{\widetilde{\alpha}_{\pm j}\}_{j=1}^n$ of $L_{s,t;e}$ strictly $(-,+)$-interlace the set $\{\alpha_{\pm j}\}_{j=1}^n$. Then they lie in the union of the intervals in (4.10) and the intervals $(-\infty, \alpha_{-n})$ and (α_n, ∞), and they are the zeros of the function f_c in (4.8). We consider the three cases $s = 0$, $s > 0$, and $s < 0$ separately.

Assume $s = 0$. Then $t \neq 0$, otherwise $L_{s,t;e} = L$ contradicting the assumption about interlacing. Suppose that $t > 0$. Then

$$\lim_{\lambda \to \pm\infty} f_c(\lambda) = \mp\infty.$$

Since

$$\operatorname{Res}\{f_c;\, \alpha_{-n}\} < 0 \quad \text{and} \quad \operatorname{Res}\{f_c;\, \alpha_n\} < 0, \tag{4.11}$$

where $\operatorname{Res}\{f_c;\, \alpha\}$ stands for the residue of f_c at the point α, the number of zeros of f_c in (α_n, ∞) is 0 or even if multiplicity is counted. The same holds for the interval $(-\infty, \alpha_{-n})$. This contradicts our assumption, hence $t < 0$.

Assume $s > 0$ and suppose that $-s^{-1}t \leq \alpha_n$. Then (4.11) and (4.9) hold, and hence the number of zeros of f_c in (α_n, ∞) is 0 or ≥ 2 if multiplicities are counted. Again a contradiction, which implies that $t < -\alpha_n s$.

Finally, assume $s < 0$ and suppose that $-s^{-1}t \geq \alpha_{-n}$. Then (4.11) and (4.9) still hold and therefore the number of zeros of f_c in $(-\infty, \alpha_{-n})$ is 0 or ≥ 2 if multiplicities are counted. This contradiction implies $t < -\alpha_{-n}s$. □

The next result is a generalization of Theorem 4.5 to the weakly hyperbolic case.

Theorem 4.6. *Let $L(\lambda) = \lambda^2 + \lambda B + C$ be a weakly hyperbolic $n \times n$ matrix polynomial with eigenvalues $\{\alpha_{\pm j}\}_{j=1}^n \in T_{2n}$, let s and t be real numbers, and let e be a unit vector in $\mathbb{C}^n$. Assume that the function $(L(\lambda)^{-1}e, e)$ has a pole at each point of $\{\alpha_{\pm j}\}_{j=1}^n$. Then the ordered eigenvalues of*

$$L_{s,t;e}(\lambda) = L(\lambda) + (\lambda s + t)(\,\cdot\,, e)\, e$$

$(-,+)$-interlace the ordered eigenvalues $\{\alpha_{\pm j}\}_{j=1}^n$ of L if and only if

$$s < 0 \quad \textit{and}\ t \leq -\alpha_{-n}s \quad \textit{or} \quad s \geq 0 \quad \textit{and}\ t \leq -\alpha_n s. \tag{4.12}$$

Proof. Without loss of generality we assume that $|s| + |t| \neq 0$; otherwise $L_{s,t;e} = L$ and the theorem is trivial.

We first prove the "if part" and assume (4.12). We consider three cases.

(a) Assume L is hyperbolic. Then we can choose a point $c \in (\alpha_{-1}, \alpha_1)$ and we have $L(c) \ll 0$. By Theorem 3.1 we have that for every $\varepsilon > 0$ there exist hermitian $n \times n$ matrices B_ε and C_ε with $\|B - B_\varepsilon\| < \varepsilon$, $\|C - C_\varepsilon\| < \varepsilon$ such that the matrix polynomial $L^{\{\varepsilon\}}(\lambda) = \lambda^2 + \lambda B_\varepsilon + C_\varepsilon$ is hyperbolic, $L_\varepsilon(c) \ll 0$, and $L^{\{\varepsilon\}}$ has only simple eigenvalues $\{\alpha_{\pm j,\varepsilon}\}_{j=1}^n$. Moreover from the proof of Theorem 3.1 it follows that B_ε and C_ε can be chosen such that $\alpha_{-n,\varepsilon} > \alpha_{-n}$ and $\alpha_{n,\varepsilon} < \alpha_n$. By Lemma 2.2 there exists a unit vector $e_\varepsilon \in \mathbb{C}^n$ such that $\|e - e_\varepsilon\| < \varepsilon$ and the function $(L^{\{\varepsilon\}}(\lambda)^{-1}e_\varepsilon, e_\varepsilon)$ has a pole in each $\alpha_{\pm j,\varepsilon}$, $j = 1, \ldots, n$. Consider the matrix function

$$\left(L^{\{\varepsilon\}}\right)_{s,t;e_\varepsilon}(\lambda) = L^{\{\varepsilon\}}(\lambda) + (\lambda s + t)(\,\cdot\,, e_\varepsilon)\, e_\varepsilon.$$

Since $|s| + |t| \neq 0$ the inequalities (4.12) imply

$$s < 0 \text{ and } \ t < -\alpha_{-n,\varepsilon}s \quad \text{or} \quad s \geq 0 \text{ and } \ t < -\alpha_{n,\varepsilon}s,$$

where $\alpha_{\pm n,\varepsilon}$ is the largest/smallest eigenvalue of $\left(L^{\{\varepsilon\}}\right)_{s,t;e_\varepsilon}$. By Theorem 4.5 the ordered eigenvalues $\{\alpha_{\pm j;s,t;e_\varepsilon}\}_{j=1}^n$ of $\left(L^{\{\varepsilon\}}\right)_{s,t;e_\varepsilon}$ strongly $(-,+)$-interlace the ordered eigenvalues of L. Denote by $\{\alpha_{\pm j;s,t;e}\}_{j=1}^n$ the ordered eigenvalues of $L_{s,t;e}$. Since $\varepsilon \to 0$ implies $B_\varepsilon \to B$, $C_\varepsilon \to C$, and $e_\varepsilon \to e$ we have, as $\varepsilon \to 0$,

$$\alpha_{\pm j,\varepsilon} \to \alpha_{\pm j}, \ \alpha_{\pm j,s,t;e_\varepsilon} \to \alpha_{\pm j;s,t;e}, \quad j = 1, \ldots, n,$$

and therefore the eigenvalues of $L_{s,t;e}$ $(-,+)$-interlace the eigenvalues of L.

(b) Assume now that L is weakly hyperbolic and (4.6), that is, (4.12) with strict inequalities for t, holds. Let $\delta > 0$. Then the matrix function

$$L_\delta = L - \delta$$

is hyperbolic and for sufficiently small δ we have

$$s < 0 \text{ and } t < -\alpha_{-n,\delta} s \quad \text{or} \quad s \geq 0 \text{ and } t < -\alpha_{n,\delta} s,$$

where $\alpha_{\pm n,\delta}$ is the maximal/minimal eigenvalue of L_δ. By Lemma 2.2 there exists a unit vector $e_\delta \in \mathbb{C}^n$ such that $\|e - e_\delta\| < \delta$ and the function $(L_\delta(\lambda)^{-1} e_\delta, e_\delta)$ has a pole in each eigenvalue of L_δ. Consider the matrix function

$$\big(L_\delta\big)_{s,t;e_\delta}(\lambda) = L_\delta(\lambda) + (\lambda s + t)(\,\cdot\,, e_\delta)\, e_\delta.$$

We proved in part (a) that the eigenvalues of $\big(L_\delta\big)_{s,t;e_\delta}$ $(-,+)$-interlace the eigenvalue of L_δ. The $(-,+)$-interlacing of the ordered eigenvalues of $L_{s,t;e}$ and L is obtained by taking the limit as δ goes to 0.

(c) If in (4.12) an equality prevails, then we replace t by $t + \tau$ with $\tau < 0$, so that (4.6) holds for $L_{s,t+\tau;e}$. Then we apply (b) and take the limit as τ goes to 0 to obtain the $(-,+)$-interlacing stated in the theorem.

We now prove the "only if part." Assume that the ordered eigenvalues $\{\alpha_{\pm j;s,t;e}\}_{j=1}^n$ of $L_{s,t;e}$ $(-,+)$-interlace the ordered eigenvalues of L. We show that (4.12) holds. We shall consider the case when $s \geq 0$; the case $s < 0$ can be proved analogously. According to Lemma 2.3 there exists a vector $f \in \ker L_{s,t;e}(\alpha_{n;s,t;e})$ such that $(f, e) \neq 0$. Then, since α_n is the right endpoint of the right root zone of L and $\alpha_n \leq \alpha_{n;s,t;e}$, we have that $(L(\alpha_{n;s,t;e})f, f) \geq 0$. From

$$0 = (L_{s,t;e}(\alpha_{n;s,t;e})f, f) = (L(\alpha_{n;s,t;e})f, f) + (\alpha_{n;s,t;e}s + t)|(f, e)|^2$$

it follows that $\alpha_{n;s,t;e}s + t \leq 0$. Hence

$$\alpha_n s + t = (\alpha_n - \alpha_{n;s,t;e})s + \alpha_{n;s,t;e}s + t \leq 0,$$

that is, $t \leq -\alpha_n s$. □

5. An inverse problem

In this section we prove the following theorem. Recall that a square matrix is called a Jacobi matrix if it is real symmetric and tridiagonal.

Theorem 5.1. *Let n be an integer ≥ 2 and assume that the ordered set $\{\beta_{\pm j}\}_{j=1}^{n-1} \in T_{2n-2}$ block-interlaces the ordered set $\{\alpha_{\pm j}\}_{j=1}^n \in T_{2n}$. Then there exist $n \times n$ Jacobi matrices B and C such that*

(i) *the matrix polynomial $L(\lambda) = \lambda^2 + \lambda B + C$ is weakly hyperbolic,*
(ii) *the ordered eigenvalues of L coincide with $\{\alpha_{\pm j}\}_{j=1}^n$, and*
(iii) *the ordered eigenvalues of the compression $L_{\infty;e_n}$ of L to $\{e_n\}^\perp$ with $e_n = \begin{pmatrix}0 & \cdots & 0 & 1\end{pmatrix}^\top \in \mathbb{C}^n$ coincide with $\{\beta_{\pm j}\}_{j=1}^{n-1}$.*

If, in addition,

$$\alpha_1 - \alpha_{-1} > 0, \tag{5.1}$$

then L is hyperbolic.

The proof of the theorem is based on mathematical induction and the basic idea is as follows. Let n be an integer ≥ 2 and let

$$L(\lambda) = \lambda^2 + \lambda B + C \tag{5.2}$$

be an $n \times n$ matrix polynomial with Jacobi matrix coefficients

$$B = (b_{j,k})_{j,k=1}^n \quad \text{and} \quad C = (c_{j,k})_{j,k=1}^n .$$

Set $L_n = L$ and for $j = n, n-1, \dots, 2$, denote by L_{j-1} the compression of L_j to $\{e_j\}^\perp$ in $\mathbb{C}^j$ where $e_j = \begin{pmatrix} 0 & \cdots & 0 & 1 \end{pmatrix}^\top \in \mathbb{C}^j$:

$$L_{j-1} = (L_j)_{\infty; e_j} . \tag{5.3}$$

Then L_j is a monic quadratic $j \times j$ Jacobi matrix polynomial and

$$\begin{aligned} \det L_j(\lambda) = (\lambda^2 + \lambda b_{j,j} + c_{j,j}) \det L_{j-1}(\lambda) \\ -(\lambda b_{j-1,j} + c_{j-1,j})^2 \det L_{j-2}(\lambda), \quad j = n, n-1, \dots, 1, \end{aligned} \tag{5.4}$$

where $\det L_0 := 1$ and $\det L_{-1} := 0$. Solving the inverse spectral problem, considered in the theorem, amounts to finding Jacobi matrices B and C such that L in (5.2) is (weakly) hyperbolic and (5.4) holds when only the polynomials $\det L_n$ and $\det L_{n-1}$ with block-interlacing sets of zeros are given.

To apply mathematical induction we use the following lemma concerning formula (5.4) with $j = n$.

Lemma 5.2. *Assume $\{\alpha_{\pm j}\}_{j=1}^n \in T_{2n}^s$, and $\{\beta_{\pm j}\}_{j=1}^{n-1} \in T_{2n-2}^s$ strictly block-interlaces $\{\alpha_{\pm j}\}_{j=1}^n$, where n is an integer ≥ 2. Define the polynomials*

$$p_{2n}(\lambda) = \prod_{j=1}^{n} (\lambda - \alpha_{-j})(\lambda - \alpha_j) \text{ and } p_{2n-2}(\lambda) = \prod_{j=1}^{n-1} (\lambda - \beta_{-j})(\lambda - \beta_j).$$

Then there exist real numbers $b_{n,n}$, $c_{n,n}$, $b_{n-1,n}$, $c_{n-1,n}$ and a nonzero polynomial p_{2n-4} such that

(i) *the equality*

$$p_{2n}(\lambda) = (\lambda^2 + \lambda b_{n,n} + c_{n,n})\, p_{2n-2}(\lambda) - (\lambda b_{n-1,n} + c_{n-1,n})^2\, p_{2n-4}(\lambda)$$

holds,

(ii) *if $n = 2$ then $p_{2n-4} \equiv 1$ and if $n > 2$ then p_{2n-4} is a monic polynomial of degree $2n-4$ whose zeros are real and of order 1 and strictly block-interlace the zeros of p_{2n-2}, and*

(iii) *an equality of the form*

$$\max\{|b_{n,n}|, |c_{n,n}|, |b_{n-1,n}|, |c_{n-1,n}|\} < c$$

holds, where c *is a positive real number which only depends on* n *and any pair of numbers* c_1 *and* c_2 *satisfying*

$$\max\{|\alpha_{-n}|, |\alpha_n|\} \leq c_1, \quad \beta_1 - \beta_{-1} \geq c_2.$$

Proof. Let q be a monic polynomial of degree two:

$$q(\lambda) = \lambda^2 + q_1\lambda + q_0 \tag{5.5}$$

and r a polynomial of degree $\leq 2n-3$ such that

$$p_{2n} = q\,p_{2n-2} + r. \tag{5.6}$$

With a number $\gamma \in \mathbb{R}$ we write p_{2n} in the form

$$p_{2n} = (q+\gamma)p_{2n-2} - (\gamma\,p_{2n-2} - r) \tag{5.7}$$

and consider the polynomial $\gamma\,p_{2n-2} - r$. It has degree $2n-2$ if $\gamma \neq 0$ and degree $\leq 2n-3$ if $\gamma = 0$. By the assumption of the lemma p_{2n} has different signs at the endpoints of each of the intervals

$$(\beta_{-n+1}, \beta_{-n+2}), \ldots, (\beta_{-2}, \beta_{-1}), (\beta_1, \beta_2), \ldots, (\beta_{n-2}, \beta_{n-1}) \tag{5.8}$$

and equal signs at the endpoints of the interval (β_{-1}, β_1). Since p_{2n-2} vanishes at each of these endpoints and by (5.6), r and p_{2n} have the same value in these points. Hence $\gamma\,p_{2n-2} - r$, $\gamma \in \mathbb{R}$, has the sign properties just ascribed to the polynomial p_{2n}. It follows that $\gamma\,p_{2n-2} - r$ has at least one zero in each of $2n-4$ the intervals (5.8) and the number of zeros of $\gamma\,p_{2n-2} - r$ in the interval (β_{-1}, β_1) is either 0 or 2 counting multiplicity. If $\gamma = 0$, then since r has degree $\leq 2n-3$ the first case holds, that is, r has no zeros in $[\beta_{-1}, \beta_1]$. For $\gamma \neq 0$ the next statement holds true.

(1) If $\gamma \neq 0$ and $\gamma\,p_{2n-2} - r$ has two zeros in the interval (β_{-1}, β_1) counting multiplicity, then the other $2n-4$ zeros of $\gamma\,p_{2n-2} - r$ are real and of order 1 and they strictly block-interlace $\{\beta_{\pm j}\}_{j=1}^{n-1}$.

We now prove that there exist a real number $\gamma_0 > 0$ and a point ν in the interval (β_{-1}, β_1) such that

(2) ν is a zero of order two of the polynomial $\gamma_0\,p_{2n-2} - r$,
(3) for all $\gamma_1 < \gamma_0$ the polynomial $\gamma_1\,p_{2n-2} - r$ has no zeros in (β_{-1}, β_1), and
(4) for all $\gamma_2 > \gamma_0$ the polynomial $\gamma_2\,p_{2n-2} - r$ has precisely two different zeros of order 1 in (β_{-1}, β_1).

Evidently, it follows from (3) and (4) that

(5) γ_0 and ν in (2) are unique.

To prove (2), we consider the polynomial

$$p := r'p_{2n-2} - rp'_{2n-2}.$$

We have $p(\beta_{\pm 1}) = -r(\beta_{\pm 1})p'_{2n-2}(\beta_{\pm 1})$. As

$$\operatorname{sign} r(\beta_1) = \operatorname{sign} r(\beta_{-1}) \quad \text{and} \quad \operatorname{sign} p'_{2n-2}(\beta_1) \neq \operatorname{sign} p'_{2n-2}(\beta_{-1}),$$

we have $\operatorname{sign} p(\beta_1) \neq \operatorname{sign} p(\beta_{-1})$ and hence p has a zero $\nu \in (\beta_{-1}, \beta_1)$. It follows that ν is a zero of order two of the polynomial

$$\gamma_0\, p_{2n-2} - r \quad \text{with} \quad \gamma_0 = r(\nu) p_{2n-2}(\nu)^{-1}.$$

Since

$$\operatorname{sign} p_{2n-2}(\lambda) = (-1)^{n-1} = \operatorname{sign} p_{2n}(\beta_1) = \operatorname{sign} r(\lambda), \quad \lambda \in (\beta_{-1}, \beta_1), \tag{5.9}$$

we have that $\gamma_0 > 0$. This completes the proof of (2).

We prove (3) and (4) simultaneously. Let γ' and γ'' be two real numbers with $\gamma' < \gamma''$ such that $\gamma' p_{2n-2} - r$ has a zero $\nu' \in (\beta_{-1}, \beta_1)$. Then by (5.9) we have $\gamma' > 0$ and

$$\gamma' |p_{2n-2}(\nu')| - |r(\nu')| = 0,$$

hence

$$\gamma'' |p_{2n-2}(\nu')| - |r(\nu')| > 0.$$

On the other hand,

$$\gamma'' |p_{2n-2}(\beta_{\pm 1})| - |r(\beta_{\pm 1})| = -|r(\beta_{\pm 1})| < 0.$$

Therefore $\gamma'' p_{2n-2} - r$ has a zero $\nu''_{-1} \in (\beta_{-1}, \nu')$ and a zero $\nu''_1 \in (\nu', \beta_1)$. and the order of these zeros is one. This readily implies (4). To prove (3), we consider $\gamma_1 < \gamma_0$ and assume that $\gamma_1\, p_{2n-2} - r$ has a zero in (β_{-1}, β_1). Then $\gamma_0\, p_{2n-2} - r$ has two different zeros of order 1 in (β_{-1}, β_1). This contradicts (2) and hence (3) holds. This completes the proof of (2)–(4).

With q_0 and q_1 from (5.5) and γ_0 and ν from statement (2) we define the numbers

$$b_{n,n} = q_1, \quad c_{n,n} = q_0 + \gamma_0, \quad b_{n-1,n} = \sqrt{\gamma_0}, \quad c_{n-1,n} = -\nu\sqrt{\gamma_0}. \tag{5.10}$$

Substituting these numbers in (5.7) we obtain the formula in item (i) of the lemma. By statement (1), item (ii) of the lemma also holds.

It remains to prove item (iii). Equating the coefficients of λ^{2n-1} and of λ^{2n-2} on both sides of (5.6) we obtain the relations

$$q_1 = \sum_{j=-n+1}^{n-1}{}' \beta_j - \sum_{j=-n}^{n}{}' \alpha_j$$

and

$$q_0 = q_1 \sum_{j=-n+1}^{n-1}{}' \beta_j + \sum_{\substack{j,k=-n,\\ j<k}}^{n}{}' \alpha_j \alpha_k - \sum_{\substack{j,k=-n+1,\\ j<k}}^{n-1}{}' \beta_j \beta_k,$$

where the prime means that the indices j and k in the summation are not assigned the value 0. These relations and the inequality $\beta_{-1} < \nu < \beta_1$ imply that the numbers $|q_1|$, $|q_0|$, and $|\nu|$ are bounded by a number which only depends on n and c_1. In view of (5.10), to prove item (iii) it suffices to prove the next statement.

(6) The number γ_0 in (2) belongs to an interval $(0, d]$ where d depends only on n, c_1, and c_2.

The remainder polynomial r from (5.6) is the Lagrange interpolation polynomial corresponding to the points

$$\beta_{-n+1}, \dots, \beta_{-1}, \beta_1, \dots, \beta_{n-1}$$

and the values

$$p_{2n}(\beta_{-n+1}), \dots, p_{2n}(\beta_{-1}), p_{2n}(\beta_1), \dots, p_{2n}(\beta_{n-1}).$$

By, for example, [20, formula (2.71)] we have

$$r(\lambda) = (2\pi i)^{-1} \int_\Gamma \frac{p_{2n}(\zeta)}{p_{2n-2}(\zeta)} \frac{p_{2n-2}(\zeta) - p_{2n-2}(\lambda)}{\zeta - \lambda} \, d\zeta, \tag{5.11}$$

where Γ is the boundary of

$$\{z : \operatorname{dist}(z, \{\beta_{-n+1}, \dots, \beta_{n-1}\}) < c_2/3\}.$$

Let $\mu_0 := \frac{1}{2}(\beta_{-1} + \beta_1)$. Then by (5.11) $|r(\mu_0)|$ can be estimated from above by a number $d_1 > 0$ which depends only on n, c_1, and c_2. On the other hand, $|p_{2n-2}(\mu_0)|$ can be estimated from below by a number d_2 which depends only on n and c_2.

It follows that

$$d_1 d_2^{-1} |p_{2n-2}(\mu_0)| - |r(\mu_0)| \geq 0.$$

On the other hand

$$d_1 d_2^{-1} |p_{2n-2}(\beta_{\pm 1})| - |r(\beta_{\pm 1})| = -|r(\beta_{\pm 1})| < 0.$$

These two inequalities and (5.9) imply that the polynomial $d_1 d_2^{-1} p_{2n-2} - r$ has a zero in (β_{-1}, β_1). Therefore, by statements (2)–(4),

$$\gamma_0 \leq d_1 d_2^{-1} =: d,$$

which proves (6). □

Proof of Theorem 5.1. We consider 3 cases.

(a). We first assume that $\{\alpha_{\pm j}\}_{j=1}^n \in T_{2n}^s$ and that $\{\beta_{\pm j}\}_{j=1}^{n-1}$ strictly block-interlaces $\{\alpha_{\pm j}\}_{j=1}^n$. Define the polynomials p_{2n} and p_{2n-2} as in Lemma 5.2. Then by Lemma 5.2 (i)

$$p_{2n}(\lambda) = (\lambda^2 + \lambda b_{n,n} + c_{n,n}) p_{2n-2}(\lambda) - (\lambda b_{n-1,n} + c_{n-1,n})^2 p_{2n-4}(\lambda).$$

We apply the lemma to p_{2n-2} and p_{2n-4} and conclude that there exist real numbers $b_{n-1,n-1}$, $c_{n-1,n-1}$, $b_{n-2,n-1}$, $c_{n-2,n-1}$ and a polynomial p_{2n-6} such that

$$\begin{aligned} p_{2n-2}(\lambda) = (\lambda^2 + \lambda b_{n-1,n-1} + c_{n-1,n-1}) p_{2n-4}(\lambda) \\ -(\lambda b_{n-2,n-1} + c_{n-2,n-1})^2 p_{2n-6}(\lambda). \end{aligned}$$

We may repeat this procedure to obtain polynomials p_{2n}, p_{2n-2}, ..., p_4, p_2, and $p_0 \equiv 1$ and Jacobi matrices $B = (b_{j,k})_{j,k=1}^n$, $C = (c_{j,k})_{j,k=1}^n$ such that for $L(\lambda) = \lambda^2 + \lambda B + C$ we have $\det L = p_{2n}$ and $\det L_{\infty;e_n} = p_{2n-2}$. (In fact in the notation as in (5.3) we have $p_{2j} = \det L_j$, $j = 0, 1, \dots, n$.) The polynomial L is hyperbolic since for every point $c \in (\beta_{-1}, \beta_1)$ we have sign $p_{2j}(c) = (-1)^j$, which implies $L(c) \ll 0$. Hence for every nonzero $f \in \mathbb{C}^n$ we have

$$(L(c)f, f) < 0, \quad \lim_{\lambda \to \infty} (L(\lambda)f, f) = \infty,$$

which implies that the quadratic polynomial $(L(\lambda)f, f)$ has two distinct real zeros. The norms of the matrices B and C can be estimated from above by positive numbers which depend only on n, c_1, and c_2. This follows from Lemma 5.2 (iii).

(b). Now we assume only $\{\alpha_{\pm j}\}_{j=1}^n \in T_{2n}$ and we drop the strictness assumption on the interlacing as in (a). We assume that

$$\beta_1 - \beta_{-1} > 0. \tag{5.12}$$

We approximate the ordered sets $\{\alpha_{\pm j}\}_{j=1}^n \in T_{2n}$ and $\{\beta_{\pm j}\}_{j=1}^{n-1} \in T_{2n-2}$ by sequences of ordered sets $\{\alpha_{\pm j}^{(m)}\}_{j=1}^n \in T_{2n}^s$ and $\{\beta_{\pm j}^{(m)}\}_{j=1}^{n-1} \in T_{2n-2}^s$, $m = 1, 2, \ldots$, such that the following holds.

(i) $\{\beta_{\pm j}^{(m)}\}_{j=1}^{n-1}$ strictly block-interlaces $\{\alpha_{\pm j}^{(m)}\}_{j=1}^n$ for every m.
(ii) $\inf \{\beta_1^{(m)} - \beta_{-1}^{(m)} : m = 1, 2, \ldots\} > 0$.
(iii) There is a point $d \in \mathbb{R}$ such that $\alpha_{-1}^{(m)} < d < \alpha_1^{(m)}$ for all m and if (5.1) holds, then $\inf \{|d - \alpha_{\pm 1}^{(m)}| : m = 1, 2, \ldots\} > 0$.

Solving the inverse problems for $\{\alpha_{\pm j}^{(m)}\}_{j=1}^n$ and $\{\beta_{\pm j}^{(m)}\}_{j=1}^{n-1}$ as in part (a) of the proof we obtain a sequence of hyperbolic Jacobi $n \times n$ matrix polynomials $L^{(m)}$ the coefficients of which are uniformly bounded. Therefore, at least a subsequence of $(L^{(m)})$ and without loss of generality we may assume that the entire sequence $(L^{(m)})$ converges to a Jacobi $n \times n$ matrix polynomial L. By property (iii) L is weakly hyperbolic. If, in addition, (5.1) holds, L is hyperbolic. It follows from well-known perturbation results that the spectra of L and $L_{\infty;e_n}$ are $\{\alpha_{\pm j}\}_{j=1}^n$ and $\{\beta_{\pm j}\}_{j=1}^{n-1}$, respectively.

(c). Now we do not assume that (5.12) holds. Let

$$k_0 := \max \{k : \beta_{-k} = \beta_k\}.$$

Then we solve the problem for the systems $\{\alpha_{\pm j}\}_{j=k_0+1}^n$, $\{\beta_{\pm j}\}_{j=k_0+1}^{n-1}$ as in part (b). The orthogonal sum of this solution and the $k_0 \times k_0$ diagonal matrix polynomial with the diagonal entries $(\lambda - \beta)^2$ where $\beta := \beta_{k_0} = \beta_{-k_0}$ gives the general solution. □

Remark 5.3. Assume that the ordered set $\{\beta_{\pm j}\}_{j=1}^{n-1} \in T_{2n-2}^s$ strictly block-interlaces the ordered set $\{\alpha_{\pm j}\}_{j=1}^n \in T_{2n}^s$. Then from [1, Theorem 6.1] it follows that the construction of the sequence of polynomials $p_{2n}, p_{2n-2}, \ldots, p_0$ starting from the first two polynomials is always possible, but to prove the theorem we had to show that a special sequence exists which yields a hyperbolic matrix polynomial.

References

[1] Yu. Agranovich, T.Ya. Azizov, A. Barsukov, and A. Dijksma, *On an inverse spectral problem for a quadratic Jacobi matrix pencil*, J. of Math. Anal. Appl. **306**(1) (2005), 1–17.

[2] T.Ya. Azizov, J. Behrndt, P. Jonas, and C. Trunk, *Compact and finite rank perturbation of linear relations in Hilbert spaces*, Integral Equations Operator Theory **63** (2009), 151–163.

[3] M.T. Chu, *Inverse eigenvalue problems*, SIAM Rev. **8**(1) (1998) 1–39.

[4] M.T. Chu and G.H. Golub, *Inverse Eigenvalue Problems. Theory, Algorithms, and Applications*, Oxford University Press, 2005.

[5] W.F. Donoghue, *Monotone matrix functions and analytic continuation*, Grundlehren der mathematischen Wissenschaften **207**, Springer, Berlin, 1974.

[6] R.J. Duffin, *A minimax theory for overdamped networks*, J. Rational Mech. Anal. **4** (1955), 221–233.

[7] D. Eschwé and M. Langer, *Variational principles for eigenvalues of self-adjoint operator functions*, Integral Equations Operator Theory **49** (2004), 287–321.

[8] F. Gesztesy and B. Simon, *m-Functions and inverse spectral analysis for finite and semi-infinite Jacobi matrices*, J. d'Analyse Mathématique **73** (1997), 267–297.

[9] G.M.L. Gladwell, *Inverse problems in vibration*, Martinus Nijhoff, Dordrecht, 1986.

[10] H. Hochstadt, *On the construction of a Jacobi matrix from spectral data*, Linear Algebra Appl. **8** (1974), 435–446.

[11] T. Kato, *Perturbation theory for linear operators*, Grundlehren der mathematischen Wissenschaften **132**, Springer, Berlin, 1966.

[12] I.S Kac and M.G. Krein, *R-functions – analytic functions mapping the upper halfplane into itself*, Supplement I to the Russian edition of F.W. Atkinson, *Discrete and continuous boundary problems*, Mir, Moscow, 1968 (Russian); English transl.: Amer. Math. Soc. Transl. (2) **103** (1974), 1–18.

[13] M. Kovac-Striko and K. Veselic, *Trace minimization and definiteness of symmetric matrix pairs*, Linear Algebra Appl. **216** (1995), 139–158.

[14] M.G. Krein and H. Langer, *On some mathematical principles in the linear theory of damped oscillations of continua* (Russian); English transl.: Integral Equations Operator Theory **1** (1978), 364–399 and 539–566.

[15] Yuen-Cheng Kuo, Wen-Wei Lin, and Shu-Fang Xu, *Solution of the partially described inverse quadratic eigenvalue problem*, SIAM J. Matrix Anal. Appl., **29** (2006), 33–55.

[16] H. Langer, *Spektraltheorie linearer Operatoren in J-Räumen und einige Anwendungen auf die Schar* $L(\lambda) = \lambda^2 + \lambda B + C$, Habilitationsschrift, Dresden, 1965.

[17] H. Langer and B. Najman, *Some interlacing results for indefinite Hermitian matrices*, Linear Algebra Appl. **69** (1985), 131–154.

[18] M. Marcus and H. Minc, *A survey of matrix theory and matrix inequalities*, Allyn and Bacon, Boston, Mass., 1964.

[19] A.S. Markus, *Introduction to the spectral theory of polynomial operator pencils*, Amer. Math. Soc. Transl. **71**, Providence, RI, 1988.

[20] A.I. Markushevich, *Theory of functions of a complex variable*, vol. II, Prentice–Hall, Englewood Cliffs, N.J., 1965.

[21] Y.M. Ram, *On inverse spectral problems and zero-pole assignment*, in: Systems and Control in the twenty-first century, Birkhäuser, Basel, 1997, 281–288.

[22] Y.M. Ram and S. Elhay, *Inverse eigenvalue problem for the symmetric tridiagonal quadratic pencil with application to damped oscillatory systems*, SIAM J. Appl. Math. **56**(1) (1996), 232–244.

[23] S. Timoshenko, D.H. Young, and W. Weaver, JR., *Vibration problems in engineering*, 4-th edition, Wiley, New York, 1974.

[24] K. Veselić, *Note on interlacing for quadratic hyperbolic pencils*, Oper. Theory Adv. Appl. **198**, Birkhäuser, Basel, 2009 (this volume), 305–307.

T.Ya. Azizov
Department of Mathematics
State University of Voronezh
Universitetskaya pl. 1
Voronezh 394006, Russia
e-mail: `azizov@math.vsu.ru`

A. Dijksma
Department of Mathematics
University of Groningen
P.O. Box 407
9700 AK Groningen, The Netherlands
e-mail: `dijksma@math.rug.nl`

K.-H. Förster
Department of Mathematics
Technical University Berlin
Strasse des 17. Juni 136
D-10623 Berlin, Germany
e-mail: `foerster@math.tu-berlin.de`

P. Jonas
Department of Mathematics
Technical University Berlin
Strasse des 17. Juni 136
D-10623 Berlin, Germany

Operator Theory:
Advances and Applications, Vol. 198, 41–48

A Theorem on Existence of Invariant Subspaces for J-binoncontractive Operators

Tomas Ya. Azizov and Victor A. Khatskevich

To the memory of our friend Peter Jonas

Abstract. Let $\mathcal{H}$ be a J-space and let $V = \begin{pmatrix} V_1 & V_{12} \\ V_{21} & V_2 \end{pmatrix}$ be the matrix representation of a J-binoncontractive operator V with respect to the canonical decomposition $\mathcal{H} = \mathcal{H}^+ \oplus \mathcal{H}^-$ of $\mathcal{H}$. The main aim of this paper is to show that the assumption

$$V_{12}(V_2 - V_{21}V_1^{-1}V_{12}) \in \mathfrak{S}_\infty \tag{0.1}$$

implies the existence of a V-invariant maximal nonnegative subspace. Let us note that (0.1) is a generalization of the well-known M.G. Krein condition $V_{12} \in \mathfrak{S}_\infty$. The set of all operators satisfying (0.1) is described via Potapov–Ginsburg transform.

Mathematics Subject Classification (2000). Primary 47B50; Secondary 46C20.

Keywords. Krein space, J-binoncontractive operator, invariant subspace.

1. Introduction

Before to state the problem to be studied, let us recall some basic notions of Operator Theory in spaces with indefinite metrics. The details one can see, for example, in the monograph [4], where also priority questions are considered.

Let $\mathcal{H}$ be a Hilbert space endowed with a scalar product $(\cdot,\cdot)$, and $P^\pm$ be orthogonal projections ($P^{\pm 2} = P^\pm = P^{\pm *}$) such that $P^+ + P^- = I$. So we have the following decomposition of $\mathcal{H}$:

$$\mathcal{H} = \mathcal{H}^+ \oplus \mathcal{H}^-, \qquad \mathcal{H}^\pm = P^\pm \mathcal{H}. \tag{1.1}$$

Denote $J = P^+ - P^-$ and define a bilinear form (J-metric) in $\mathcal{H}$ as follows:

$$[x, y] = (Jx, y).$$

The research of T.Ya. Azizov was supported by the grant of RFBR 08-01-00566-a.

A space $\mathcal{H}$, endowed with J-metric is called a J-space or a Krein space with a fixed canonical decomposition and the canonical norm (the definition of a Krein space see below, p. 46). A subspace $\mathcal{L}$ is called maximal nonnegative if it is nonnegative ($[x,x] \geq 0$, $\forall x \in \mathcal{L}$) and it is not a subspace of any other nonnegative subspace $\widetilde{\mathcal{L}} \neq \mathcal{L}$. Denote by $\mathfrak{M}^+$ the set of all maximal nonnegative subspaces of $\mathcal{H}$. Let $\mathfrak{K}$ be the closed unit ball of the space of all linear bounded operators acting between $\mathcal{H}^+$ and $\mathcal{H}^-$:

$$\mathfrak{K} = \{K : \mathcal{H}^+ \to \mathcal{H}^- \mid \|K\| \leq 1\}.$$

There is a one-to-one correspondence between $\mathfrak{M}^+$ and $\mathfrak{K}$ defined by the following equality ($\mathcal{L} \in \mathfrak{M}^+$, $K \in \mathfrak{K}$):

$$\mathcal{L} = \{x^+ + Kx^+ \mid K \in \mathfrak{K},\ x^+ \in \mathcal{H}^+\}. \tag{1.2}$$

The operator K is called the angular operator of the subspace $\mathcal{L}$. A linear bounded operator V acting in $\mathcal{H}$ is called a J-noncontractive operator if $[Vx, Vx] \geq [x,x]$, $\forall x \in \mathcal{H}$, and J-binoncontractive if both V and V^* are J-noncontractive operators, where V^* is the Hilbert space adjoint to V. Equivalently, V is a J-binoncontractive operator if it is J-noncontractive and $\mathcal{L} \in \mathfrak{M}^+$ implies $\mathcal{M} := V\mathcal{L} \in \mathfrak{M}^+$. Let

$$V = \begin{bmatrix} V_1 & V_{12} \\ V_{21} & V_2 \end{bmatrix} \tag{1.3}$$

be the matrix representation of V with respect to the decomposition (1.1). The transformation $V : \mathfrak{M}^+ \to \mathfrak{M}^+$ naturally defines a linear fractional transformation $F_V : \mathfrak{K} \to \mathfrak{K}$ such that the image of the angular operator K of a subspace $\mathcal{L} \in \mathfrak{M}^+$ is the angular operator of the subspace $\mathcal{M} = V\mathcal{L} \in \mathfrak{M}^+$:

$$F_V(K) = (V_{21} + V_2 K)(V_1 + V_{12} K)^{-1}. \tag{1.4}$$

An operator $U : \mathcal{H} \to \mathcal{H}$ is called J-unitary if $U\mathcal{H} = \mathcal{H}$ and $[Ux, Uy] = [x,y]$, $\forall x, y \in \mathcal{H}$. Let $\Gamma \in \mathfrak{K}$ and $\|\Gamma\| < 1$. Then the operator

$$U(\Gamma) = \begin{bmatrix} (I - \Gamma^*\Gamma)^{-1/2} & -\Gamma^*(I - \Gamma\Gamma^*)^{-1/2} \\ \Gamma(I - \Gamma^*\Gamma)^{-1/2} & -(I - \Gamma\Gamma^*)^{-1/2} \end{bmatrix} \tag{1.5}$$

is J-unitary and also $U(\Gamma)^2 = I$.

The existence of invariant maximal nonnegative subspaces is one of central problems of the spectral theory of operators in J-spaces. The study of this problem was initiated by L.S. Pontryagin [10] after a pioneering work of S.L. Sobolev (see [11]). These investigations were extensively developed by M.G. Krein and his students and collaborators (the details concerning obtained results and priority questions one can find, for example, in the review [3] and in the monograph [4]). One of the main conditions of the existence of an invariant maximal nonnegative subspace of a J-binoncontractive operator V in a J-space or, equivalently, a fixed point of the linear fractional transformation (1.4) is the compactness of the "corner" V_{12} (M.G. Krein [9]):

$$V_{12} \in \mathfrak{S}_\infty. \tag{1.6}$$

This condition allows to use the well-known fixed point principle: a continuous transformation of a convex compact set has a fixed point. The goal of the present paper is to generalize the above Krein's condition, applying the following theorem (T. Ando [1], V.A. Khatskevich and V.S. Shulman [8]): the image $F_V(\mathfrak{K})$ of a linear fractional transformation (1.4) is convex and compact with respect to the weak operator topology. Consequently, one can consider the restriction of a transformation F_V to its image and apply the following fixed point principle:

Theorem 1.1 (I.L. Glicksberg[7]). *Let Ω be a compact convex subset of a locally convex Hausdorff topological space, and let F be a closed mapping of points $\omega \in \Omega$ into non-empty convex subsets $F(\omega) \subset \Omega$. Then the function F has at least one fixed point in Ω, i.e., there is a point $\omega_0 \in \Omega$ such that $\omega_0 \in F(\omega_0)$.*

2. Invariant subspaces

The main result of the paper is the following:

Theorem 2.1. *Let V be a J-binoncontractive operator represented by the matrix* (1.3). *If*

$$V_{12}(V_2 - V_{21}V_1^{-1}V_{12}) \in \mathfrak{S}_\infty, \tag{2.1}$$

then there exists $\mathcal{L}^+ \in \mathfrak{M}^+$ such that $V\mathcal{L}^+ = \mathcal{L}^+$.

Proof. It follows from the argument on page 42 that to prove the theorem it is sufficient (and necessary) to establish the existence of a fixed point for the linear fractional transformation (1.4), that is, there exists $K_0 \in \mathcal{K}$ such that $F_V(K_0) = K_0$. To prove the latter equality let us use the following decomposition of the operator V (see [2], [5]):

$$V = WU(\Gamma); \tag{2.2}$$

here $U(\Gamma)$ is a J-unitary operator defined by (1.5) with $\Gamma = -(V_{11}^{-1}V_{12})^*$ and $T = W$, where

$$W = \begin{pmatrix} W_1 & 0 \\ W_{21} & W_2 \end{pmatrix}, \tag{2.3}$$

$$\begin{aligned} W_1 &= (V_1 - V_{12}V_{12}^* V^{-1\,*})(I - \Gamma^*\Gamma)^{-1/2}, \\ W_{21} &= (V_{21} - V_2 V_{12}^* V_1^{-1\,*})(I - \Gamma^*\Gamma)^{-1/2}, \\ W_2 &= -(V_2 - V_{21}V^{-1}V_{12})(I - \Gamma\Gamma^*)^{-1/2}. \end{aligned} \tag{2.4}$$

Since $F_V = F_W \circ F_{U(\Gamma)}$ and $F_{U(\Gamma)}(\mathfrak{K}) = \mathfrak{K}$, then $F_V(\mathfrak{K}) = F_W(\mathfrak{K})$, i.e.,

$$F_V(\mathfrak{K}) = \{S = W_{21}W_1^{-1} + W_2 K W_1^{-1} \mid K \in \mathfrak{K}\}. \tag{2.5}$$

Evidently, $F_V(F_V(\mathfrak{K})) \subset F_V(\mathfrak{K})$. So for any $K \in \mathfrak{K}$ the following set of contractions is well defined:

$$\Phi(K) = \{Q \in \mathcal{K} \mid F_V \circ F_W(K) = F_W(Q)\}. \tag{2.6}$$

One can see that the following equality holds for Q:

$$V_{21}+V_2(W_{21}W_1^{-1}+W_2KW_1^{-1})=(W_{21}W_1^{-1}+W_2QW_1^{-1})V_1$$
$$+(W_{21}W_1^{-1}+W_2QW_1^{-1})V_{12}(W_{21}W_1^{-1}+W_2KW_1^{-1}). \qquad (2.7)$$

Hence the set (2.6) is convex and compact with respect to the weak operator topology. Since $V_{12}W_2=-V_{12}(V_2-V_{21}V^{-1}V_{12})(I-\Gamma\Gamma^*)^{-1/2}$ is compact together with the operator $V_{12}(V_2-V_{21}V^{-1}V_{12})$, then the mapping $K\mapsto\Phi(K)$ is closed with respect to the same topology. Now Theorem 1.1 implies the existence of a fixed point K_0 of Φ. Consequently, a contraction $S_0=F_W(K_0))$ is a fixed point of the linear fractional transformation F_V. As a result,

$$\mathcal{L}^+=\{x^++F_W(K_0)x^+\mid x^+\in\mathfrak{H}^+\}$$

is a maximal nonnegative V-invariant subspace. □

Remark 2.2. Let V satisfy condition (2.1) and let $\mathcal{L}_+$ be a nonnegative subspace such that $V\mathcal{L}_+=\mathcal{L}_+$. Then there exists $\mathcal{L}\in\mathfrak{M}^+$ such that $\mathcal{L}_+\subset\mathcal{L}$ and $V\mathcal{L}=\mathcal{L}$. To prove this let us define the angular operator K_+ of $\mathcal{L}_+$, $K_+:P^+\mathcal{L}_+\to\mathcal{H}^-$ like in (1.2):

$$\mathcal{L}_+=\{x^++K_+x^+\mid x^+\in P^+\mathcal{L}_+\}.$$

Then repeat the above proof substituting in it the unit operator ball $\mathfrak{K}$ by the following set

$$\mathfrak{K}_{\mathcal{L}_+}=\{K\in\mathfrak{K}\mid K_+\subset F_W(K)\}.$$

One of the important problems of Spectral Theory of Operators is to find an invariant subspace with the a priory fixed location of the spectrum of the restricted operator. For a J-binoncontractive operator V satisfying the condition (1.6) it is proved in [6] the existence of an invariant subspace $\mathcal{L}^+\in\mathfrak{M}^+$, such that $|\sigma(V|\mathcal{L}^+)|\le 1$, where $\sigma(V|\mathcal{L}^+)$ is the spectrum of the restriction of V onto $\mathcal{L}^+$. Slightly modifying the proof from [6] (see also [4, 4, 3.2.8]) one can prove the following result.

Theorem 2.3. *Under the conditions of Theorem* 2.1 *there exists an invariant subspace* $\mathcal{L}^+\in\mathfrak{M}^+$ *of the operator* V *such that* $|\sigma(V|\mathcal{L}^+)|\ge 1$.

3. Potapov–Ginsburg transform

In this section we describe the set of all J-binoncontractive operators V satisfying condition (2.1) but not necessarily satisfying the Krein condition (1.6). To obtain such a description we use the Potapov–Ginsburg transform δ:

$$\delta(V)=(P^++P^-V)(P^-+P^+V)^{-1} \qquad (3.1)$$

(for the details concerning the Potapov–Ginsburg transform see, for example, [4, § 5.1])). This transform maps one-to-one the set of all J-binoncontractive operators $\mathfrak{V}=\{V\}$ onto a set of contractions

$$\mathfrak{W}=\{W=\delta(V)\mid \|W\|\le 1,\quad 0\in\rho(P^+V|\mathcal{H}^+)\}, \qquad (3.2)$$

where $\rho(P^+V|\mathcal{H}^+)$ is the set of all regular points of the restriction of the operator P^+V onto the subspace $\mathcal{H}^+$. Note that the Potapov–Ginsburg transform is an idempotent mapping, that is $\delta(\delta(V)) = V$ for all $V \in \mathfrak{V}$. For $V \in \mathfrak{V}$ let

$$V = \begin{bmatrix} V_1 & V_{12} \\ V_{21} & V_2 \end{bmatrix} \tag{3.3}$$

be its matrix representation with respect to the decomposition (1.1). Denote by

$$W = \begin{bmatrix} W_1 & W_{12} \\ W_{21} & W_2 \end{bmatrix}$$

the corresponding matrix representation of $W = \delta(V)$. One can see that

$$W = \delta(V) = \begin{pmatrix} V_1^{-1} & -V_1^{-1}V_{12} \\ V_{21}V^{-1} & V_2 - V_{21}V_1^{-1}V_{12} \end{pmatrix}. \tag{3.4}$$

Since the operator V_1 is continuous and continuously invertible, the following Theorem 3.1 is a direct consequence of representation (3.4).

Theorem 3.1. *Let $W \in \mathfrak{W}$ and $V = \delta(W)$. Then the following statements are true:*

1° $\{V_{12} \in \mathfrak{S}_\infty\} \iff \{W_{12} \in \mathfrak{S}_\infty\}$;
2° $\{V_{12}(V_2 - V_{21}V_1^{-1}V_{12}) \in \mathfrak{S}_\infty\} \iff \{W_{12}W_2 \in \mathfrak{S}_\infty\}$;
3° $\{V_2 - V_{21}V_1^{-1}V_{12} \in \mathfrak{S}_\infty\} \iff \{W_2 \in \mathfrak{S}_\infty\}$.

Let us note that Theorem 3.1 gives a simple algorithm to construct a J-binoncontractive operator V satisfying (2.1) but not satisfying (1.6), that is with $V_{12} \notin \mathfrak{S}_\infty$. First we take a contraction $W \in \mathfrak{W}$ with $W_{12} \notin \mathfrak{S}_\infty$, but $W_{12}W_2 \in \mathfrak{S}_\infty$. Then apply the Potapov–Ginsburg transform and consider $V = \delta(W)$.

4. Operators of the class K(H)

Definition 4.1. We say that a J-binoncontractive operator V belongs to the class $(\mathbf{H})$, if the generated linear fractional transformation $F_V : \mathfrak{K} \to \mathfrak{K}$ has at least one fixed point and each fixed point of F_V can be represented as a sum $K = K_1 + K_2$, where $\|K_1\| < 1$ and K_2 is a finite-dimensional partial isometry.

Note that above definition is an equivalent formulation in terms of angular operators of the original definition from [4, § 3.5].

Definition 4.2. We say that a J-binoncontractive operator T belongs to the class $\mathbf{K}(\mathbf{H})$, if it commutes with some operator $V \in (\mathbf{H})$: there exists $V \in (\mathbf{H})$ such that $TV = VT$.

Theorem 4.3. *Let V be a J-binoncontractive operator satisfying the following condition:*

$$V_2 - V_{21}V_1^{-1}V_{12} \in \mathfrak{S}_\infty. \tag{4.1}$$

Let $\mathfrak{A} = \{A\}$ be a set of J-binoncontractive operators commuting with V: $AV = VA$, $\forall A \in \mathfrak{A}$, and containing V. Then

(i) *all operators of* $\mathfrak{A}$ *have a common nontrivial invariant subspace;*

(ii) *if, in addition,* $\mathfrak{A}$ *is a commutative family of operators then any nonnegative subspace* $\mathcal{L}_+$ *such that* $A\mathcal{L}_+ = \mathcal{L}_+$, $\forall\, A \in \mathfrak{A}$, *can be included into a maximal nonnegative subspace which is invariant under all operators* $A \in \mathfrak{A}$.

Proof. First let us check that $V \in (\mathbf{H})$. Indeed, since V is a bounded operator, condition (4.1) implies (2.1). Hence by Theorem 2.1 the operator V has an invariant maximal nonnegative subspace, or equivalently, F_V has a fixed point. Since the set of all fixed points of F_V belongs to the set (2.5), then for each such fixed point K_0 there exists $F_0 \in \mathfrak{K}$ such that

$$K_0 = W_{21}W_1^{-1} + W_2F_0W_1^{-1}, \tag{4.2}$$

where W_1, W_{21}, W_2 are entries of W from (2.2) (see (2.3)). It follows from (2.4) and (4.1) that $W_2 \in \mathfrak{S}_\infty$. Since W is a J-binoncontractive operator we have $\|W_{21}W_1^{-1}\| < 1$. Hence K_0 can be represented as a sum of a uniform contraction and a finite-dimensional partial isometry, that is $V \in (\mathbf{H})$. Now let us turn to a proof of (i) and (ii).

(i) If V has an eigenvalue λ, then $\ker(V - \lambda)$ is a common invariant subspace for all operators commuting with V. On the other hand if V has no eigenvalues, then by [4, Lemma 3.5.7] the operator V has a unique invariant subspace $\mathcal{L}^+ \in \mathfrak{M}^+$. As $VA\mathcal{L}^+ = AV\mathcal{L}^+ = A\mathcal{L}^+$ for each $A \in \mathfrak{A}$ then we have , $A\mathcal{L}^+(\mathfrak{M}^+)$ is also a maximal nonnegative subspace invariant with respect to V. Consequently $A\mathcal{L}^+ = \mathcal{L}^+$.

(ii) follows from [4, Theorem 3.5.12]. □

5. The Krein condition

Definition 5.1. A linear space $\mathcal{K}$ with an indefinite metric $[\cdot,\cdot]$ is called a Krein space if there exists the following canonical decomposition of $\mathcal{K}$:

$$\mathcal{K} = \mathcal{K}^+[\dot{+}]\mathcal{K}^-, \tag{5.1}$$

where $[x^+, x^-] = 0$ for all $x^+ \in \mathcal{K}^+$, $x^- \in \mathcal{K}^-$, $\mathcal{K}^+ \bigcap \mathcal{K}^- = \{0\}$, and $\mathcal{H}^\pm := \{\mathcal{K}^\pm; \pm[\cdot,\cdot]\}$ are Hilbert spaces.

If we introduce in $\mathcal{K}$ a scalar product $(\cdot,\cdot)$:

$$(x,y) = [x^+,y^+] - [x^-,y^-], \quad x,y \in \mathcal{K}, \quad x^\pm, y^\pm \in \mathcal{K}^\pm,$$

then $\mathcal{H} := \{\mathcal{K}, (\cdot,\cdot)\}$ becomes to be a J-space. Let us notice that in the case of a fixed canonical decomposition (5.1) of a Krein space, as it was mentioned above (see our remark after Theorem 3.1), the condition (2.1) is much more general than the Krein condition (1.6). On the other hand in the general case of a Krein space when a decomposition (5.1) is not fixed the following result is true.

Theorem 5.2. *Let $\mathcal{K}$ be a Krein space and let $V : \mathcal{K} \to \mathcal{K}$ be a J-binoncontractive operator satisfying the condition* (2.1) *with respect to a given canonical decomposition* (5.1). *Then there exists, possibly another canonical decomposition of $\mathcal{K}$, such that V satisfies the condition* (1.6).

Proof. According to [4, Remark 3.2.4] to prove the statement it is sufficient to find a uniform contraction $Q : \mathcal{H}^- \to \mathcal{H}^+$ such that

$$QV_2 + QV_{21}Q - V_{12} - V_1Q \in \mathfrak{S}_\infty.$$

Set $Q = -V_1^{-1}V_{12}$ get

$$QV_2 + QV_{21}Q - V_{12} - V_1Q = -V_1^{-1}V_{12}(V_2 - V_{21}V_1^{-1}V_{12}) \in \mathfrak{S}_\infty. \qquad \square$$

References

[1] T. Ando, *Linear operators on Krein spaces*, Sapporo, Japan, 1979.

[2] T.Ya. Azizov, *Parametric representation of operators*, Uchenye zapiski TNU, series "Mathematics. Mechanics. Computer science & Cybernetics", **19(58)**, (2006), 2, 3–11 (in Russian).

[3] T.Ya. Azizov, I.S. Iokhvidov, *Linear operators in Hilbert spaces with a G-metric*, Uspekhi matem. nauk, **26**(1971), 4, 43–92.

[4] T.Ya. Azizov, I.S. Iokhvidov, *Foundations of the theory of linear operators in spaces with an indefinite metric*, Nauka, 1986 (in Russian);
English translation: Linear operators in spaces with an indefinite metric, John Wiley & Sons, 1990.

[5] T.Ya. Azizov, V.A. Khatskevich, *Bistrict plus-operators and fractional linear operator transformations*, Ukrainskii mathem. visnik, **4**(2007), 3, 311–333 (in Russian)
English transl.: Ukrainian Mathem. Bull., **4** (2007), 3, 307–328.

[6] T.Ya. Azizov and S.A. Khoroshavin, *Invariant subspaces of operators acting in a space with an indefinite metric*, Funkzional. analis i ego pril., **14** (1980), 4, 1–7, (in Russian).

[7] I.L. Glicksberg, *A further generalization of Kakutani fixed point theorem with application to Nash equilibrium points*, Proc. Amer. Math. Soc., 1952, 3, 170–174.

[8] V.A. Khatskevich, V.S. Shulman, *Operator fractional-linear transformations: convexity and compactness of image; applications*, Studia Math., **116** (1995), 2, 191–195.

[9] M.G. Krein, *A new application of the fixed-point principle in the theory of linear operators in a space with an indefinite metric*, Doklady Akad. Nauk SSSR, **154**(1964), 5, 1026–1028.

[10] L.S. Pontryagin, *Hermitian operators in spaces with indefinite metrics*, Izvetiya AN SSSR, Series Math., **8** (1944), 243–280 (in Russian).

[11] S.L. Sobolev, *On the motion of symmetric top with a cavity filled with fluid*, J.Appl.Mech. and Tech. Phys., **3**(1960), 20–55.

Tomas Ya. Azizov
Department of Mathematics
Voronezh State University
Universitetskaya pl., 1
Voronezh, 394006, Russia
e-mail: azizov@math.vsu.ru

Victor A. Khatskevich
Department of Mathematics
ORT Braude Academic College
College Campus, P.O. Box 78
Karmiel 21982, Israel
e-mail: victor_kh@hotmail.com

Operator Theory:
Advances and Applications, Vol. 198, 49–60

Moment Problems for Real Measures on the Unit Circle

Mihály Bakonyi and Ekaterina V. Lopushanskaya

Abstract. Given a sequence $(t_j)_{j=-n}^{n}$ of complex numbers such that $t_{-j} = \bar{t}_j$, we investigate the existence of a real measure μ, such that $\mu = \mu_+ - \mu_-$, with μ_+ and μ_- positive measures supported on m respectively k distinct points on the unit circle, and the first moments of μ coincide with the given sequence.

Mathematics Subject Classification (2000). 15A48, 15A57, 15A90.

Keywords. Hermitian Toeplitz matrix, indefinite moment problem, Carathéodory type decomposition.

1. Introduction

Carathéodory proved in [1] a decomposition result for positive semidefinite Toeplitz matrices. This result, together with a theorem by Herglotz ([8]) lead to the solution of the truncated trigonometric moment problem by Carathéodory and Fejér in [2]. These results are briefly surveyed in Section 2.

Starting with [10], the problem of extending a given sequence $(t_j)_{j=-n}^{n}$ with $t_{-j} = \bar{t}_j$ to a sequence $(t_j)_{j\in\mathbb{Z}}$ such that all Toeplitz matrices $T_q = (t_{i-j})_{i,j=0}^{q}$ for $q \geq 0$ are Hermitian and have less or equal than a given number k of negative eigenvalues, has extensively been studied. In [10], such extensions which are also bounded were characterized. These results were not related so far to moment problems for real measures on the unit circle, even so, in [16] the results of [1] were extended to the case of Hermitian Toeplitz block-matrices and an announcement was made in the direction of applying in a future work such results to moment problems. The main problem we are concerned with in Section 3 is the following: given a sequence $(t_j)_{j=-n}^{n}$ of complex numbers such that $t_{-j} = \bar{t}_j$, when does there exist a real measure μ, such that $\mu = \mu_+ - \mu_-$, with μ_+ and μ_- positive measures supported on m respectively k points (all of them distinct) on the unit circle $\mathbb{T}$ of the complex plane, and the jth moment of μ coincides with t_j for $|j| \leq n$. We call such a measure μ an (m, k) *extension of* $(t_j)_{j=-n}^{n}$. If a sequence $(t_j)_{j=-n}^{n}$

admits an (m, k) extension, then the number $i_+(T_n)$ of positive eigenvalues of the Toeplitz matrix $T_n = (t_{i-j})_{i,j=0}^n$ is less or equal than m, and the number $i_-(T_n)$ of negative eigenvalues of T_n is less or equal than k. These conditions are in general not sufficient for the existence of an (m, k) extension. We are discussing necessary and sufficient conditions for the existence of an (m, k) extension in the case when $i_+(T_n) = m$ and $i_-(T_n) = k$. As it turns out, the necessary and sufficient condition is the existence of a Carathéodory type decomposition for the matrix T_n. For T_n singular, it was independently found in [6] and [3] that the existence of such decomposition is equivalent to the fact that all roots of a certain polynomial belong to $\mathbb{T}$. We present this result as Theorem 3.8 with a new proof. Later we discuss the situation when T_n is invertible. To illustrate our methods, we included several examples in Section 3.

2. Positive semidefinite Toeplitz matrices and moment problems for positive measure on the unit circle

Carathéodory proved in [1] the following result characterizing positive semidefinite Toeplitz matrices. For a Toeplitz matrix $T_n = (t_{i-j})_{i,j=0}^n$, let for $1 \leq r \leq n+1$, T_{r-1} denote the $r \times r$ principal submatrix of T_n, namely $T_{r-1} = (t_{i-j})_{i,j=0}^{r-1}$.

Theorem 2.1. *The Toeplitz matrix*

$$T_n = \begin{pmatrix} t_0 & \bar{t}_1 & \dots & \bar{t}_n \\ t_1 & t_0 & \dots & \bar{t}_{n-1} \\ \vdots & \vdots & \ddots & \vdots \\ t_n & t_{n-1} & \dots & t_0 \end{pmatrix}$$

is positive semidefinite and $1 \leq \operatorname{rank} T_n = r \leqslant n+1$, *if and only if the matrix* T_{r-1} *is invertible and there exist*

$$\alpha_j \in \mathbb{T},\ j = 1, \dots, r,\ \alpha_j \neq \alpha_k,\ \textit{for}\ j \neq k,$$

and $\rho_j > 0$, $j = 1, \dots, r$, *such that:*

$$t_k = \sum_{j=1}^{r} \rho_j \alpha_j^k, \quad k = 0, \dots, n. \tag{2.1}$$

This statement is equivalent to the fact that the extremal rays in the cone of positive semidefinite $n \times n$ Toeplitz matrices are generated by such matrices of rank one, namely each T_n as in Theorem 2.1 can be written as:

$$T_n = \rho_1 \begin{pmatrix} 1 & \bar{\alpha}_1 & \dots & \bar{\alpha}_1^n \\ \alpha_1 & 1 & \dots & \bar{\alpha}_1^{n-1} \\ \vdots & \vdots & \ddots & \vdots \\ \alpha_1^n & \alpha_1^{n-1} & \dots & 1 \end{pmatrix} + \dots + \rho_r \begin{pmatrix} 1 & \bar{\alpha}_r & \dots & \bar{\alpha}_r^n \\ \alpha_r & 1 & \dots & \bar{\alpha}_r^{n-1} \\ \vdots & \vdots & \ddots & \vdots \\ \alpha_r^n & \alpha_r^{n-1} & \dots & 1 \end{pmatrix}.$$

Carathéodory also proved that if $1 \leq r \leq n$, then $\{\alpha_1, \dots, \alpha_r\}$ are the roots of the polynomial

$$P(z) = \det \begin{pmatrix} t_0 & \bar{t}_1 & \dots & \bar{t}_r \\ t_1 & t_0 & \dots & \bar{t}_{r-1} \\ \vdots & \vdots & \vdots & \vdots \\ t_{r-1} & t_{r-2} & \dots & \bar{t}_1 \\ 1 & z & \dots & z^r \end{pmatrix},$$

which are all distinct and belong to $\mathbb{T}$. The numbers $\rho_1, \dots, \rho_r$ can be found by solving the linear system formed by the equations (2.1) for $j = 0, \dots, r-1$.

It is then straightforward that Theorem 2.1 is equivalent to the following statement. Such type factorizations of Toeplitz matrices were first considered in [6].

Proposition 2.2. *The Toeplitz matrix T_n is positive semidefinite and* rank $T_n = r$ *if and only if it can be factored as:*

$$T_n = RDR^*,$$

where R is a "pseudo-Vandermonde" matrix,

$$R = \begin{pmatrix} 1 & 1 & \dots & 1 \\ \alpha_1 & \alpha_2 & \dots & \alpha_r \\ \vdots & \vdots & \vdots & \vdots \\ \alpha_1^n & \alpha_2^n & \dots & \alpha_r^n \end{pmatrix},$$

with $\alpha_j \in \mathbb{T}$ for $j = 1, \dots, r$, and $\alpha_j \neq \alpha_p$ for $j \neq p$, and D is a diagonal matrix with positive entries.

Theorem 2.1 also implies the following characterization.

Proposition 2.3. *If the Toeplitz matrix T_n is positive semidefinite and* rank $T_n = r$, *then there exists a unitary operator U on $\mathbb{C}^r$ and vector $h \in \mathbb{C}^r$ such that:*

$$t_k = < U^k h, h >, \quad |k| \leqslant n. \tag{2.2}$$

Conversely, if U is a unitary operator on $\mathbb{C}^r$ and $h \in \mathbb{C}^r$, then defining t_k by (2.2), $T_n = (t_{i-j})_{i,j=0}^n$ is a positive semidefinite Toeplitz matrix with rank $T_n \leq r$.

Proof. For the first part, consider $U = \text{diag}\,(\alpha_1, \dots, \alpha_r)$ and $h = \begin{pmatrix} \sqrt{\rho_1} \\ \vdots \\ \sqrt{\rho_r} \end{pmatrix}$.

For the second part, let $(c_p)_{p=0}^n$ be a sequence of complex numbers. Then

$$\sum_{p,s=0}^n c_p \bar{c}_s t_{p-s} = \sum_{p,s=0}^n c_p \bar{c}_s < U^{p-s} h, h >= \| \sum_{p=0}^n c_p U^p h \|^2 \geq 0,$$

so $T_n \geq 0$. The rank condition is immediate. □

For $n \in \mathbb{Z}$, the nth moment of a finite measure μ on $\mathbb{T}$ is defined by

$$\hat{\mu}(n) = \frac{1}{2\pi} \int_{-\pi}^{\pi} e^{-int} d\mu(t).$$

A sequence $(c_n)_{n\in\mathbb{Z}}$ of complex numbers is called *positive semidefinite* if for each $n \geq 0$, the Toeplitz matrix $T_n = (c_{i-j})_{i,j=0}^n$ is positive semidefinite. The following characterization is due to Herglotz ([8]) and it represents a particular case of the so-called Bochner's Theorem (see [14]).

Theorem 2.4. *The sequence $(c_n)_{n\in\mathbb{Z}}$ of complex numbers is positive semidefinite if and only if there exists a positive measure μ on the unit circle $\mathbb{T}$, such that $c_n = \hat{\mu}(n)$ for $n \in \mathbb{Z}$.*

Theorems 2.1 and 2.4 lead then to the following result, which includes an answer to the so-called "truncated trigonometric moment problem" and was obtained by Carathéodory and Fejér in [2]. It uses in addition the fact that the nth moment of the evaluation measure (Dirac measure) μ_α at $\alpha \in \mathbb{T}$ is $\bar{\alpha}^n$.

Theorem 2.5. *Let $(t_j)_{j=-n}^n$ be given complex numbers. Then there exists a positive measure μ on $\mathbb{T}$ such that*

$$\hat{\mu}(j) = t_j, \quad |j| \leq n, \tag{2.3}$$

if and only if the Toeplitz matrix $T_n = (t_{i-j})_{i,j=0}^n$ is positive semidefinite. Moreover, if $1 \leq \operatorname{rank} T_n = r \leqslant n+1$, then there exists a positive measure μ supported on r points of the unit circle $\mathbb{T}$ which satisfies (2.3).

3. Hermitian Toeplitz matrices and moment problems for real measures on the unit circle

Consider now a sequence $(t_j)_{j=-n}^n$ of complex numbers such that $t_{-j} = \bar{t}_j$ and let $T_n = (t_{i-j})_{i,j=0}^n$, which verifies then $T_n^* = T_n$.

If the sequence $(t_j)_{j=-n}^n$ admits an (m,k) extension, it is clear that $i_+(T_n) \leq m$ and $i_-(T_n) \leq k$. The existence of an (m,k) extension of a sequence $(t_j)_{j=-n}^n$ is equivalent to the existence of a bounded extension of $(t_j)_{j=-n}^n$ to an infinite sequence $(t_j)_{j\in\mathbb{Z}}$, such that for every $q \geq n_0$, the Toeplitz matrix $T_q = (t_{i-j})_{i,j=0}^q$ has m positive and k negative eigenvalues.

The following characterization was obtained by Iohvidov and Kreĭn ([10]). The result was later extended to the case of functions on $\mathbb{R}$ in [15], for functions on amenable groups in [11] and [12], and finally for functions on locally compact groups in [13] and [14].

Theorem 3.1. *A sequence $(t_j)_{j\in\mathbb{Z}}$ is bounded and the matrices $T_q = (t_{i-j})_{i,j=0}^q$ have $k \geq 1$ negative eigenvalues for all $q \geq n_0$ if and only if the elements of the sequence $(t_j)_{j\in\mathbb{Z}}$ allow the following representation:*

$$t_l = d_l - \sum_{j=1}^{k} \rho_j \alpha_j^l, \quad l \in \mathbb{Z},$$

where $(d_l)_{l\in\mathbb{Z}}$ *is a positive semidefinite sequence,* $\rho_j > 0$, $\alpha_j \in \mathbb{T}$, $j = 1,\dots,k$, *and* $k \geq 1$ *is the smallest number with this property.*

Theorem 3.1 implies the following result.

Theorem 3.2. *If a sequence* $(t_j)_{j=-n}^n$ *admits an* (m,k) *extension then there exist positive semidefinite Toeplitz matrices* T_+ *and* T_-, *of rank* m *respectively* k, *such that:*

$$T_n = T_+ - T_-.$$

Theorem 3.2 leads to the following result inspired by work in [5].

Proposition 3.3. *Let* $(t_j)_{j=-n}^n$ *be such that* $t_{-j} = \bar{t}_j$. *If* $1 \leq i_+(T_n) = m$ *and* $1 \leq i_-(T_n) = k$, *then* $(t_j)_{j=-n}^n$ *admits a* (p,q) *extension, where* $m \leq p \leq n$ *and* $k \leq q \leq n$.

Proof. Let $(t_j)_{j=-n}^n$ be a sequence as in the statement of the proposition. The matrix $T_n = (t_{i-j})_{i,j=0}^n$ can always be written as $T_n = T_+ - T_-$, with T_+ and T_- positive semidefinite Toeplitz matrices. (For example, define $T_+ = \lambda I_n$ and $T_- = \lambda I_n - T_n$ for $\lambda > 0$ sufficiently large.) Assume $T_n = T_+ - T_-$ as above, and the trace of T_+ is minimal over all such decompositions. If both T_+ and T_- are invertible, then there exists $\lambda > 0$ such that $T_+, T_- \geq \lambda I$. Then $T_n = (T_+ - \lambda I) - (T_- - \lambda I)$, showing the trace of T_+ is not minimal. Thus one of T_+ and T_- has to be singular. Assume T_- is, and that T_+ is invertible. Then there exists $0 < \mu < 1$ such that $T_n + T_- = T_+ \geq \mu T_-$. Then $T_n = (T_+ - \mu T_-) - (T_- - \mu T_-)$, so again the trace of T_+ is not minimal. We get the same conclusion when T_- is assumed to be invertible. Thus both T_+ and T_- can be chosen to be singular. This implies the existence of a (p,q) extension for $(t_j)_{j=-n}^n$ such that $m \leq p \leq n$ and $k \leq q \leq n$. □

Example 3.4. Consider the Toeplitz matrix:

$$T_2 = \begin{pmatrix} 1 & 1 & 3 \\ 1 & 1 & 1 \\ 3 & 1 & 1 \end{pmatrix}$$

which can be written as

$$\begin{aligned} T_2 = {} & \frac{3}{2}\begin{pmatrix} 1 & 1 & 1 \\ 1 & 1 & 1 \\ 1 & 1 & 1 \end{pmatrix} - \frac{1}{2}\begin{pmatrix} 1 & i & -1 \\ -i & 1 & i \\ -1 & -i & 1 \end{pmatrix} \\ & + \frac{1}{2}\begin{pmatrix} 1 & -1 & 1 \\ -1 & 1 & -1 \\ 1 & -1 & 1 \end{pmatrix} - \frac{1}{2}\begin{pmatrix} 1 & -i & -1 \\ i & 1 & -i \\ -1 & i & 1 \end{pmatrix}. \end{aligned}$$

The above decomposition of T_2 implies that

$$\mu = \frac{3}{2}\mu_1 - \frac{1}{2}\mu_{-i} + \frac{1}{2}\mu_{-1} - \frac{1}{2}\mu_i.$$

is a $(2,2)$ extension for the sequence defining T_2. We will motivate later (see Example 3.11) why in this case there exists no $(2,1)$ extension even so $i_-(T_2) = 1$.

The most interesting question for us is: "Given a complex sequence $(t_j)_{j=-n}^n$, $t_{-j}=\bar{t}_j$, such that $i_+(T_n)=m$, and $i_-(T_n)=k$, when does there exist an (m,k) extension of $(t_j)_{j=-n}^n$?" Theorems 2.1, 2.5, and 3.2 give the following answer.

Theorem 3.5. *A sequence $(t_j)_{j=-n}^n$, $t_{-j}=\bar{t}_j$ such that $i_+(T_n)=m$, and $i_-(T_n)=k$ admits an (m,k) extension if and only if there exist $\alpha_j\in\mathbb{T}$ and $\rho_j>0$, $j=1,\dots,m+k$, with $\alpha_j\neq\alpha_p$ for $j\neq p$, such that*

$$t_l=\sum_{j=1}^{m}\rho_j\alpha_j^l-\sum_{s=m+1}^{m+k}\rho_s\alpha_s^l,\quad |l|\leqslant n. \tag{3.1}$$

It is straightforward that Theorem 3.5 is equivalent to the following factorization result similar to Proposition 2.2.

Proposition 3.6. *A sequence $(t_j)_{j=-n}^n$, $t_{-j}=\bar{t}_j$ such that $i_+(T_n)=m$, and $i_-(T_n)=k$ admits an (m,k) extension if and only if T_n can be factored as:*

$$T_n=RDR^*,$$

where R is a "pseudo-Vandermonde matrix",

$$R=\begin{pmatrix}1&1&\dots&1\\ \alpha_1&\alpha_2&\dots&\alpha_{m+k}\\ \vdots&\vdots&\vdots&\vdots\\ \alpha_1^n&\alpha_2^n&\dots&\alpha_{m+k}^n\end{pmatrix},$$

with $\alpha_j\in\mathbb{T}$ for $j=1,\dots,m+k$, and $\alpha_j\neq\alpha_p$ for $j\neq p$, and D is a diagonal matrix

$$D=\begin{pmatrix}\rho_1&\dots&0&0&\dots&0\\ \vdots&\ddots&\vdots&\vdots&&\vdots\\ 0&\dots&\rho_m&0&\dots&0\\ 0&\dots&0&-\rho_{m+1}&\dots&0\\ \vdots&&\vdots&\vdots&\ddots&\vdots\\ 0&\dots&0&0&\dots&-\rho_{m+k}\end{pmatrix},$$

with $\rho_l>0$, $l=1,\dots,m+k$.

Theorem 3.5 is also equivalent to the following characterization.

Proposition 3.7. *Let $(t_j)_{j=-n}^n$ be a sequence such that $t_{-j}=\bar{t}_j$, $i_+(T_n)=m$, and $i_-(T_n)=k$. If $(t_j)_{j=-n}^n$ admits an (m,k) extension then there exist:*

(i) *an indefinite inner product $[\cdot,\cdot]$ on $\mathbb{C}^{m+k}$ such that the maximal dimension of a subspace contained in the cone of all h with $[h,h]\leq 0$ is k,*
(ii) *a unitary operator U on the above space, and*
(iii) *a vector $h\in\mathbb{C}^{m+k}$ such that*

$$t_l=[U^lh,h],\quad |l|\leqslant n. \tag{3.2}$$

Conversely, if $T_n=(t_{i-j})_{i,j=0}^n$ where t_l is defined by (3.2) with $h\in\mathbb{C}^{m+k}$ and $[\cdot,\cdot]$ and U are verifying (i) *and* (ii) *as above, then $i_+(T_n)\leq m$ and $i_-(T_n)\leq k$.*

Given a sequence $(t_j)_{j=-n}^n$ with $t_{-j} = \bar{t}_j$, there is need for a practical method to check whether the terms of the sequence can be represented as in (3.1). We will use two different methods, depending whether T_n is singular or invertible. The methods resemble the ones used by Carathéodory in obtaining relations (2.1) in case of a positive semidefinite Toeplitz matrix.

Case 1: T_n is singular. The following result can be obtained from Theorem 3.5 and a result separately proven by Ellis and Lay ([6]) respectively Ciccarielo and Cervellino ([3]). We will provide an independent proof which is shorter, and hopefully, more intuitive.

Theorem 3.8. *A sequence $(t_j)_{j=-n}^n$, $t_{-j} = \bar{t}_j$ such that $1 \leq i_+(T_n) = m$, $1 \leq i_-(T_n) = k$, and $m + k \leq n$, admits an (m,k) extension if and only if* rank $T_n =$ rank T_{n-1} *and all roots of the polynomial*

$$P(z) = \begin{vmatrix} t_0 & \bar{t}_1 & \dots & \bar{t}_r \\ t_1 & t_0 & \dots & \bar{t}_{r-1} \\ \vdots & \vdots & \vdots & \vdots \\ t_{r-1} & t_{r-2} & \dots & \bar{t}_1 \\ 1 & z & \dots & z^r \end{vmatrix} \tag{3.3}$$

are on the unit circle, where r is the largest number $0 \leq r \leq n$ such that T_{r-1} is invertible.

Proof. Given a sequence $(t_j)_{j=-n}^n$ as in the statement of the theorem, it is known from [10] (see also [9]) that rank $T_n =$ rank T_{n-1} is a necessary and sufficient condition for the existence (of a unique) rank preserving extension $(t_j)_{j\in\mathbb{Z}}$ of $(t_j)_{j=-n}^n$, namely that for all $q \geq n$, rank $T_q = m + k$. It remains to show that the boundedness of $(t_j)_{j\in\mathbb{Z}}$ is equivalent to the condition that all roots of $P(z)$ belong to $\mathbb{T}$. Let r be the largest number such that T_{r-1} is invertible. In view of the rank condition quoted above, we observe that $r = m + k + 1$, and rank $T_q = m + k$ whenever $q \geq m + k$. Since T_{r-1} is invertible, denoting by C_j, $j = 0, \dots, r$, the jth column of T_r, there exist $\beta_0, \dots, \beta_{r-1}$ such that

$$C_r = \beta_0 C_0 + \dots + \beta_{r-1} C_{r-1}. \tag{3.4}$$

In particular,

$$t_r = \bar{\beta}_0 t_0 + \dots + \bar{\beta}_{r-1} t_{r-1}.$$

The uniqueness of the rank preserving extension together with (3.4) imply that we have

$$t_q = \bar{\beta}_0 t_{q-r} + \bar{\beta}_1 t_{q-r+1} + \dots + \bar{\beta}_{r-1} t_{q-1}, \quad q \geq r. \tag{3.5}$$

It is a well-known fact that for every sequence defined by a recursive linear relation as in (3.5), we have that

$$t_j = \gamma_1 \alpha_1^j + \dots + \gamma_r \alpha_r^j, \quad j \geq 0, \tag{3.6}$$

where $\alpha_1, \dots, \alpha_r$ are the roots of the polynomial $Q(z) = z^r - \bar{\beta}_{r-1} z^{r-1} - \dots - \bar{\beta}_0$, and $\gamma_1, \dots, \gamma_r$ are some constants uniquely determined by $t_0, \dots, t_{r-1}$. It is easy to observe that (3.4) implies that the polynomial $Q(z)$ is a scalar multiple of the polynomial $P(z)$ defined by (3.3). Using the Toeplitz structure of T_r, $P(z)$ can be normalized to the form $p(z) = \sum_{j=0}^{r} a_j z^j$, where $a_{r-j} = \bar{a}_j$, for $j = 0, \ldots, r$. If $\alpha \in \mathbb{C}$ is a root of such a polynomial $p(z)$ then evidently $\frac{1}{\bar{\alpha}}$ is one as well. Then (3.6) implies that the sequence $(t_j)_{j\in\mathbb{Z}}$ is bounded if and only if all roots of $p(z)$, and consequently of $P(z)$, are on $\mathbb{T}$. This completes the proof. □

Remark 3.9. Based on the above proof, the condition rank T_n = rank T_{n-1} in Theorem 3.8 implies that T_{m+k} is invertible. Conversely, if T_{m+k} is invertible, as rank $T_n = m + k$, we have that rank T_n = rank T_{n-1}. Thus the latter condition in Theorem 3.8 can be substituted with the one that T_{m+k} is invertible.

The following practical algorithm for deciding whether all roots of a polynomial equivalent to the one in (3.3) belong to $\mathbb{T}$ or not, was found in [3].

Theorem 3.10. *Let $z_1, \ldots, z_n$ be the roots of the polynomial*

$$p(z) = a_n z^n + a_{n-1} z^{n-1} + \cdots + a_0,$$

the coefficients of which satisfy the conditions:

$$a_m = \bar{a}_{n-m}, \quad m = 0, \ldots, n, \quad a_n \neq 0.$$

Let the numbers $\sigma_0, \ldots, \sigma_n$ depend on the polynomial via:

$$\sigma_k = \sum_{j=1}^{n} z_j^k, \quad k = 0, \ldots, n, \quad \text{and} \quad \sigma_{-k} = \bar{\sigma}_k.$$

Then the roots of $p(z)$ are all distinct and lie on the unit circle $\mathbb{T}$ if and only if: $\sigma = (\sigma_{i-j})_{i,j=0}^{n} \geqslant 0$, and rank $\sigma = n$.

Case 2: T_n is invertible. Let T_{n+1} be a singular and Hermitian one step extension of T_n. Then the entry $t_{n+1}(x)$ depends on the parameter x as described below, a and b being some constants depending on the given data:

$$t_{n+1}(x) = \begin{cases} a + bx, & x \in \mathbb{R}, \qquad \det T_{n-1} = 0 \\ a + be^{ix}, & x \in [0, 2\pi], \quad \det T_{n-1} \neq 0. \end{cases}$$

We can apply Theorem 3.10, or a different method to find out whether there exists a parameter x, such that the following polynomial (depending on x),

$$P_x(z) = \begin{vmatrix} t_0 & \bar{t}_1 & \ldots & \bar{t}_n & \bar{t}_{n+1}(x) \\ t_1 & t_0 & \ldots & \bar{t}_{n-1} & \bar{t}_n \\ \vdots & \vdots & \ddots & \vdots & \vdots \\ t_n & t_{n-1} & \ldots & t_0 & \bar{t}_1 \\ 1 & z & \ldots & z^n & z^{n+1} \end{vmatrix}$$

has all its roots on $\mathbb{T}$. Theorem 3.10 reduces then our main moment problem to: for a given $(n+2) \times (n+2)$ Hermitian matrix polynomial $p(x)$, when does there exist $x_0 \in \mathbb{T}$ (or $x_0 \in \mathbb{R}$) such that $p(x_0) \geq 0$ and rank $p(x_0) = n+1$. This issue is the subject of some future research.

Example 3.11. Consider again

$$T_2 = \begin{pmatrix} 1 & 1 & 3 \\ 1 & 1 & 1 \\ 3 & 1 & 1 \end{pmatrix},$$

for which all one step singular, Hermitian, Toeplitz extensions of T_2 are of the form

$$T_3 = \begin{pmatrix} 1 & 1 & 3 & 7+xi \\ 1 & 1 & 1 & 3 \\ 3 & 1 & 1 & 1 \\ 7-xi & 3 & 1 & 1 \end{pmatrix},$$

$x \in \mathbb{R}$. Then

$$P_x(z) = z^3 - (2 - \frac{xi}{2})z^2 - (2 + \frac{xi}{2})z + 1.$$

If its roots are $z_1 = e^{i\alpha}$, $z_2 = e^{i\beta}$, $z_3 = e^{i\gamma}$, then we have the condition $e^{i\alpha}e^{i\beta}e^{i\gamma} = -1$, which implies that $\gamma = \pi - (\alpha + \beta)$. Then

$$\cos\alpha + \cos\beta - \cos(\alpha+\beta) = 2,$$

which is easy to figure out has no solutions. Therefore, the given data does not admit a $(2,1)$ extension.

In order to apply Theorem 3.10 to the same problem, we have

$$\sigma_0(x) = 3, \qquad \sigma_1(x) = z_1 + z_2 + z_3 = 2 - \frac{xi}{2},$$

$$\sigma_2(x) = z_1^2 + z_2^2 + z_3^2 = (z_1+z_2+z_3)^2 - 2(z_1z_2 + z_2z_3 + z_1z_3) = 8 - xi - \frac{x^2}{4},$$

and

$$\sigma_3(x) = z_1^3 + z_2^3 + z_3^3 = (z_1+z_2+z_3)^3 - 3(z_1z_2 + z_2z_3 + z_1z_3)(z_1+z_2+z_3)$$
$$+ 3z_1z_2z_3 = \left(2 - \frac{xi}{2}\right)^3 + 3\left(2 - \frac{xi}{2}\right)\left(2 + \frac{xi}{2}\right) - 3.$$

Thus

$$\sigma(x) = \begin{pmatrix} 3 & 2 + \frac{xi}{2} & 8 + xi - \frac{x^2}{4} & \sigma_3(x) \\ 2 - \frac{xi}{2} & 3 & 2 + \frac{xi}{2} & 8 + xi - \frac{x^2}{4} \\ 8 - xi - \frac{x^2}{4} & 2 - \frac{xi}{2} & 3 & 2 + \frac{xi}{2} \\ \overline{\sigma_3(x)} & 8 - xi - \frac{x^2}{4} & 2 - \frac{xi}{2} & 3 \end{pmatrix}$$

and for $\sigma(x) \geq 0$ we certainly need

$$0 \leq \det \begin{pmatrix} 3 & 8 + xi - \frac{x^2}{4} \\ 8 - xi - \frac{x^2}{4} & 3 \end{pmatrix} = 9 - \left(8 - \frac{x^2}{4}\right)^2 - x^2,$$

which is an easy exercise to figure out has no real solutions. We finally remark that $\sigma(x)$ is a cubic Hermitian polynomial in a real variable.

When $k = 1$, in order to find out whether some given data admits an $(m, 1)$ extension, we can equivalently apply the following computationally simpler algorithm: determine whether there exist $\rho > 0$ and $\alpha \in \mathbb{T}$ such that

$$T_n + \rho \begin{pmatrix} 1 & \bar{\alpha} & \dots & \bar{\alpha}^n \\ \alpha & 1 & \dots & \bar{\alpha}^{n-1} \\ \vdots & \vdots & \ddots & \vdots \\ \alpha^n & \alpha^{n-1} & \dots & 1 \end{pmatrix} \geqslant 0,$$

the sum being also singular.

Example 3.12. Consider the data defining the Toeplitz matrix

$$T_2 = \begin{pmatrix} 1 & 1 & -1 \\ 1 & 1 & 1 \\ -1 & 1 & 1 \end{pmatrix}.$$

We are looking for $\rho > 0$ and $\alpha \in \mathbb{T}$ such that

$$\begin{pmatrix} 1 & 1 & -1 \\ 1 & 1 & 1 \\ -1 & 1 & 1 \end{pmatrix} + \rho \begin{pmatrix} 1 & \bar{\alpha} & \bar{\alpha}^2 \\ \alpha & 1 & \bar{\alpha} \\ \alpha^2 & \alpha & 1 \end{pmatrix} = \begin{pmatrix} 1+\rho & 1+\rho\bar{\alpha} & -1+\rho\bar{\alpha}^2 \\ 1+\rho\alpha & 1+\rho & 1+\rho\bar{\alpha} \\ -1+\rho\alpha^2 & 1+\rho\alpha & 1+\rho \end{pmatrix} \geqslant 0, \tag{3.7}$$

the matrix being also singular. Since for every $\rho > 0$ and $\alpha \in \mathbb{T}$ we have that

$$\begin{pmatrix} 1+\rho & 1+\rho\bar{\alpha} \\ 1+\rho\alpha & 1+\rho \end{pmatrix} \geq 0,$$

condition (3.7) translates into the matrix being singular, which means

$$\rho(\alpha^2 + \bar{\alpha}^2) - 2\rho(\alpha + \bar{\alpha}) - 2 = 0.$$

Using $\alpha = \cos\theta + i\sin\theta$, we obtain the equation:

$$\rho(2\cos^2\theta - 2\cos\theta - 1) = 1,$$

which has a solution for every θ such that $2\cos^2\theta - 2\cos\theta - 1 > 0$. For instance, considering $\theta = \pi$, we have that $\rho = \frac{1}{3}$. For this particular choice, (3.7) becomes:

$$\begin{pmatrix} \frac{4}{3} & \frac{2}{3} & -\frac{2}{3} \\ \frac{2}{3} & \frac{4}{3} & \frac{2}{3} \\ -\frac{2}{3} & \frac{2}{3} & \frac{4}{3} \end{pmatrix} \geqslant 0.$$

Applying the method of Carathéodory in [1] described after Theorem 2.1, we get the decomposition (2.1) of the latter matrix with $\alpha_1 = e^{i\frac{\pi}{3}}$, $\alpha_2 = e^{-i\frac{\pi}{3}}$, and $\rho_1 = \rho_2 = \frac{2}{3}$. Consequently,

$$\mu = \frac{2}{3}\mu_{e^{i\frac{\pi}{3}}} + \frac{2}{3}\mu_{e^{-i\frac{\pi}{3}}} - \frac{1}{3}\mu_{-1}$$

is a $(2, 1)$ extension of the given data.

Let T be a given Hermitian Toeplitz matrix. One can apply semidefinite programming type methods to solve the following problem, first considered in [5]:

$$\begin{cases} \min \operatorname{tr} T_+ \\ T_+ \geq T, \quad T_+ \geqslant 0 \\ T_+ \text{ Toeplitz matrix.} \end{cases}$$

The problem does not have in general a unique solution. For minimizing m and k such that the data defining T admits an (m,k) extension, one should solve the problem

$$\begin{cases} \min \operatorname{rank} T_+ \\ T_+ \geq T, \quad T_+ \geqslant 0 \\ T_+ \text{ Toeplitz matrix.} \end{cases}$$

The relationship between the solution sets of the above problems is unclear to us. We are in particular interested whether these problems always admit a common solution.

Finally, we mention that in [4] Kreĭn-space techniques were used to study indefinite extensions of Hermitian block-matrices. There is a strong indication that such techniques might be used to solve matrix-valued indefinite moment problems as well.

Acknowledgement

The authors thank Hugo J. Woerdeman (Drexel University) and Aurelian Gheondea (Bilkent University) for useful discussions during the preparation of the manuscript. Most part of the work was done while the second named author was a Fullbright Scholar at Georgia State University whom she would like to thank for its warm hospitality. Her research was also supported by the Russian Foundation for Basic Research, Grant RFBR 08-01-00566-A. Finally, we thank the anonymous referee for helping to improve the presentation of the paper and for pointing out reference [16].

References

[1] C. Carathéodory, Über den Variabilitätsbereich der Fourierschen Konstanten von positiven harmonischen Funktionen, *Rend. Circ. Matem. Palermo*, Vol. 32(1911), 193–217.

[2] C. Carathéodory and L. Fejér, Über den Zusammenhang der Extreme von harmonischen Funktionen mit ihren Koeffizienten und über den Picard-Landauschen Satz, *Rend. Circ. Mat. Palermo*, Vol. 32(1911), 218–239.

[3] S. Ciccarielo and A. Cervellino, Generalization of a theorem of Carathéodory, *Journal of Physics A: Mathematical and General*, Vol. 39(2006), 14911–14928.

[4] T. Constantinescu and A. Gheondea, On the indefinite trigonometric moment problem of I.S. Iohvidov and M.G. Kreĭn, *Math. Nachr.*, Vol. 171(1995), 79–94.

[5] G. Cybenko, Moment problems and low rank Toeplitz approximations, *Circuits Systems Signal Proc.*, Vol. 1, No. 3-4(1982), 345–366.

[6] R.L. Ellis and D.C. Lay, Factorization of finite rank Hankel and Toeplitz matrices, *Linear Algebra Appl.*, Vol. 173(1992), 19–38.

[7] U. Grenander and G. Szegő, *Toeplitz forms and their applications*, California Monographs in Mathematical Sciences, University of California Press, Berkeley, 1958.

[8] G. Herglotz, Über Potenzreihen mit positivem, reellem Teil im Einheitskreis, *Leipziger Berichte, Math.-Phys.*, Vol. 63(1911), 501–511.

[9] I.S. Iohvidov, *Hankel and Toeplitz matrices and forms*, Translated from the Russian by G. Philip A. Thijsse, Birkhäuser, Boston, 1982.

[10] I.S. Iohvidov and M.G. Kreĭn, Spectral theory of operators in spaces with an indefinite matric II., *Trudy Moskov. Mat. Obšč.*, Vol. 8(1959), 413–496 (Russian). English translation: *Amer. Math. Soc. Transl.* 2, Vol. 34(1963), 283–373.

[11] K. Sakai, On quasi-positive definite functions and unitary representations of groups in Pontrjagin spaces, *J. Math. Kyoto Univ.*, Vol. 19, No. 1(1979), 71–90.

[12] K. Sakai, On quasi-negative definite functions and certain classes of cyclic unitary representations of groups in Π_n-spaces, *Sci. Rep. Kagoshima Univ.*, Vol. 28(1979), 9–50.

[13] Z. Sasvári, Definierbare Funktionen auf Gruppen, *Dissertationes Math.*, Vol. CCLXXXI (1989), 1–83.

[14] Z. Sasvári, *Positive definite and definitizable functions*, Akademie Verlag, Berlin, 1994.

[15] V.A. Straus, On continuous Hermitian-indefinite functions, *Mat. Zametki*, Vol. 13, No. 2(1973), 303–310 (Russian).

[16] M. Tismenetsky, Matrix generalizations of a moment problem theorem. I. The Hermitian case, *SIAM J. Matrix Anal. Appl.*, Vol. 14, No. 1(1993), 92–112.

Mihály Bakonyi
Centro de Estruturas Lineares e Combinatórias
Av. Prof. Gama Pinto 2
1649-003 Lisbon, Portugal

and

Department of Mathematics
Georgia State University
P.O. Box 4110
Atlanta, GA 30302-4110, USA
e-mail: mbakonyi@cii.fc.ul.pt

Ekaterina V. Lopushanskaya
Department of Mathematics
Voronezh State University
Voronezh, Russia
e-mail: kate_lopushanskaya@yahoo.com

Operator Theory:
Advances and Applications, Vol. 198, 61–85

Finite Rank Perturbations, Scattering Matrices and Inverse Problems

Jussi Behrndt, Mark M. Malamud and Hagen Neidhardt

Dedicated to the memory of our friend Peter Jonas

Abstract. In this paper the scattering matrix of a scattering system consisting of two selfadjoint operators with finite-dimensional resolvent difference is expressed in terms of a matrix Nevanlinna function. The problem is embedded into an extension theoretic framework and the theory of boundary triplets and associated Weyl functions for (in general nondensely defined) symmetric operators is applied. The representation results are extended to dissipative scattering systems and an explicit solution of an inverse scattering problem for the Lax-Phillips scattering matrix is presented.

Mathematics Subject Classification (2000). Primary: 47A40; Secondary: 81U40, 47A55, 47B44.

Keywords. Scattering system, scattering matrix, boundary triplet, Weyl function, dissipative operator, Lax-Phillips scattering.

1. Introduction

Let A and B be selfadjoint operators in a Hilbert space $\mathfrak{H}$ and assume that the difference of the resolvents

$$(B-\lambda)^{-1}-(A-\lambda)^{-1}, \qquad \lambda\in\rho(A)\cap\rho(B), \tag{1.1}$$

is a trace class operator. Then it is well known that the wave operators $W_\pm(B,A)$ exist; they are isometries mapping the absolutely continuous subspace of A onto that of B. The scattering operator $S_{AB}:=W_+(B,A)^*W_-(B,A)$ of the scattering system $\{A,B\}$ commutes with A and is unitary on the absolutely continuous subspace of A. Therefore S_{AB} is unitarily equivalent to a multiplication operator induced by a family of unitary operators $\{S_{AB}(\lambda)\}_{\lambda\in\mathbb{R}}$ in the spectral representation of the absolutely continuous part of A. The family $\{S_{AB}(\lambda)\}_{\lambda\in\mathbb{R}}$ is called the scattering matrix of $\{A,B\}$.

One of the main objectives of this paper is to represent the scattering matrix of the scattering system $\{A, B\}$ with the help of an associated Nevanlinna function $M(\cdot)$. We restrict ourselves to the special case of finite rank perturbations in resolvent sense, i.e., it is assumed that the difference of the resolvents in (1.1) is a rank n operator, where $n < \infty$. In this case the Nevanlinna function $M(\cdot)$ will be an $n \times n$-matrix function and it will be shown in Theorem 3.1 that the scattering matrix $\{S_{AB}(\lambda)\}_{\lambda \in \mathbb{R}}$ is given by

$$S_{AB}(\lambda) = I - 2i\sqrt{\Im\mathrm{m}\,(M(\lambda + i0))}\, M(\lambda + i0)^{-1} \sqrt{\Im\mathrm{m}\,(M(\lambda + i0))} \tag{1.2}$$

for a.e. $\lambda \in \mathbb{R}$. This representation is a generalization of a recent result of the authors from [8] and an earlier different (unitarily equivalent) expression found by V.M. Adamyan and B.S. Pavlov in [5]. The formula (1.2) is obtained by embedding the scattering problem into an extension theoretic framework. More precisely, we consider the (in general nondensely defined) closed symmetric operator $S = A \cap B$ which has finite equal deficiency indices (n, n). The adjoint S^* is defined in the sense of linear relations and a so-called boundary triplet $\Pi = \{\mathbb{C}^n, \Gamma_0, \Gamma_1\}$ for S^* is chosen in such a way that the selfadjoint extensions of S corresponding to the boundary mappings Γ_0 and Γ_1 coincide with A and B, respectively. The function $M(\cdot)$ in (1.2) is the Weyl function associated with this boundary triplet – an abstract analogon of the classical Titchmarsh-Weyl m-function from singular Sturm-Liouville theory – and contains the spectral information of the operator A.

Besides selfadjoint scattering systems we also consider so-called maximal dissipative scattering systems $\{A, B\}$, that is, A is still a selfadjoint operator in the Hilbert space $\mathfrak{H}$ but B is only assumed to be maximal dissipative, i.e., $\Im\mathrm{m}\,(Bf, f) \leq 0$ and the spectrum of B is contained in $\mathbb{C}_- \cup \mathbb{R}$. As above we treat only the case of finite rank perturbations in resolvent sense. Following [7, 25, 26, 27] a minimal selfadjoint dilation L of B in the direct sum $\mathfrak{H} \oplus L^2(\mathbb{R}, \mathbb{C}^n)$ is constructed and a natural larger selfadjoint scattering system $\{K, L\}$ in $\mathfrak{H} \oplus L^2(\mathbb{R}, \mathbb{C}^n)$ is considered. From Theorem 3.1 and Theorem 3.2 we obtain a representation of the scattering matrix $\{S_{KL}(\lambda)\}_{\lambda \in \mathbb{R}}$ which is closely related to the representations found earlier in [7]. We emphasize that the lower right corner of $\{S_{KL}(\lambda)\}_{\lambda \in \mathbb{R}}$ in Proposition 4.4 can be interpreted as the Lax-Phillips scattering matrix $\{S^{LP}(\lambda)\}_{\lambda \in \mathbb{R}}$ of the Lax-Phillips scattering system $\{L, \mathcal{D}_-, \mathcal{D}_+\}$, where the incoming and outcoming subspaces $\mathcal{D}_-$ and $\mathcal{D}_+$ are $L^2(\mathbb{R}_-, \mathbb{C}^n)$ and $L^2(\mathbb{R}_+, \mathbb{C}^n)$, respectively. This also implies the well-known relation $S^{LP}(\lambda) = \Theta_B(\lambda - i0)^*$ between the Lax-Phillips scattering matrix and the characteristic function $\Theta_B(\cdot)$ of the maximal dissipative operator B found by V.M. Adamyan and D.Z. Arov in [1, 2, 3, 4].

As an application of our approach on finite rank perturbations and maximal dissipative scattering systems we prove an inverse result in Section 5. Let $W(\cdot)$ be a purely contractive analytic matrix function on $\mathbb{C}_+$. Under some mild additional assumptions it is shown in Theorem 5.1 that the limit $\{W(\lambda + i0)\}_{\lambda \in \mathbb{R}}$ can be regarded as the Lax-Phillips scattering matrix of a suitably chosen Lax-Phillips scattering system. We point out that this statement can be obtained immediately in a more abstract and general form by combining the results of B. Sz.-Nagy and

C. Foiaş in [27, Section VI] with the results of V.M. Adamyan and D.Z. Arov in [1, 2, 3, 4]. However, our approach leads to a more explicit solution of the inverse problem, in particular, we find a maximal dissipative multiplication operator B in an L^2-space and a minimal selfadjoint dilation L of B such that the Lax-Phillips scattering matrix of $\{L, \mathcal{D}_-, \mathcal{D}_+\}$ coincides with the limit of the given purely contractive analytic matrix function $W(\cdot)$; cf. Corollary 5.2.

The paper is organized as follows. In Section 2 we give a brief introduction in the theory of boundary triplets for (in general nondensely defined) closed symmetric operators. In particular, we show how a boundary triplet for the intersection $S = A \cap B$ of two selfadjoint operators A and B with a finite-dimensional resolvent difference can be chosen. Section 3 is devoted to the representation of the scattering matrix for a scattering system $\{A, B\}$ with finite rank resolvent difference and in Section 4 the results are extended to the case where the operator B is only maximal dissipative. With the help of these results we propose a solution for the inverse scattering problem in Section 5. For the convenience of the reader we add an Appendix on direct integrals, spectral representations and scattering matrices.

Notation. The Hilbert spaces in this paper are usually denoted by $\mathfrak{H}$, $\mathfrak{K}$ and $\mathcal{H}$; they are all assumed to be separable. The symbols span$\{\cdot\}$ and clospan$\{\cdot\}$ are used for the linear span and closed linear span, respectively, of a set. The algebra of everywhere defined bounded linear operators on a Hilbert space $\mathfrak{H}$ with values in a Hilbert space $\mathfrak{K}$ is denoted by $[\mathfrak{H}, \mathfrak{K}]$; we write $[\mathfrak{H}]$ if $\mathfrak{K} = \mathfrak{H}$. By $\mathcal{F}_n(\mathfrak{H})$ we denote the subset of $[\mathfrak{H}]$ that consists of linear operators with range of dimension $n \in \mathbb{N}$. The absolutely continuous part of a selfadjoint operator A in $\mathfrak{H}$ is denoted by A^{ac}, the corresponding subspace by $\mathfrak{H}^{ac}(A)$. The symbols $\rho(\cdot)$, $\sigma(\cdot)$, $\sigma_p(\cdot)$, $\sigma_c(\cdot)$, $\sigma_r(\cdot)$ stand for the resolvent set, the spectrum, the point, continuous and residual spectrum, respectively. By $E(\cdot)$ and $\Sigma(\cdot)$ we denote operator-valued measures defined on the algebra of Borel sets $\mathcal{B}(\mathbb{R})$ of the real axis $\mathbb{R}$. Usually, the symbol $E(\cdot)$ is reserved for orthogonal operator-valued measures.

2. Selfadjoint and maximal dissipative extensions of nondensely defined symmetric operators

2.1. Linear relations

Let $(\mathfrak{H}, (\cdot,\cdot))$ be a separable Hilbert space. A (closed) linear relation T in $\mathfrak{H}$ is a (closed) linear subspace of the Cartesian product space $\mathfrak{H} \times \mathfrak{H}$. The set of closed linear relations in $\mathfrak{H}$ is denoted by $\widetilde{\mathcal{C}}(\mathfrak{H})$. Linear operators in $\mathfrak{H}$ will always be identified with linear relations via their graphs. The elements of a linear relation T are pairs denoted by $\widehat{f} = \{f, f'\} \in T$, $f, f' \in \mathfrak{H}$, and the *domain*, *kernel*, *range*, and the *multi-valued part* of T are defined as

$$\begin{aligned} \mathrm{dom}\,(T) &= \{ f \in \mathfrak{H} : \{f, f'\} \in T \}, & \ker(T) &= \{ f \in \mathfrak{H} : \{f, 0\} \in T \}, \\ \mathrm{ran}\,(T) &= \{ f' \in \mathfrak{H} : \{f, f'\} \in T \}, & \mathrm{mul}\,(T) &= \{ f' \in \mathfrak{H} : \{0, f'\} \in T \}, \end{aligned}$$

respectively. Note that T is an operator if and only if $\operatorname{mul}(T) = \{0\}$. A point λ belongs to the *resolvent set* $\rho(T)$ of a closed linear relation T if $(T-\lambda)^{-1}$ is an everywhere defined bounded operator in $\mathfrak{H}$. The *spectrum* $\sigma(T)$ of T is the complement of $\rho(T)$ in $\mathbb{C}$.

A linear relation T in $\mathfrak{H}$ is called *dissipative* if $\operatorname{Im}(f', f) \leq 0$ holds for all $\{f, f'\} \in T$. A dissipative relation T is said to be *maximal dissipative* if there exists no proper dissipative extension of T in $\mathfrak{H}$. It can be shown that a dissipative relation T is maximal dissipative if and only if T is closed and $\mathbb{C}_+ \subset \rho(T)$ holds.

The *adjoint* T^* of a linear relation T in $\mathfrak{H}$ is a closed linear relation in $\mathfrak{H}$ defined by

$$T^* := \big\{ \{g, g'\} : (f', g) = (f, g') \text{ for all } \{f, f'\} \in T \big\}. \tag{2.1}$$

Observe that this definition extends the usual definition of the adjoint operator and that $\operatorname{mul}(T^*) = (\operatorname{dom}(T))^\perp$ holds. In particular, T^* is an operator if and only if T is densely defined. A linear relation T in $\mathfrak{H}$ is called *symmetric* (*selfadjoint*) if $T \subset T^*$ ($T = T^*$, respectively). It follows from the polarization identity that T is symmetric if and only if $(f', f) \in \mathbb{R}$ for all $\{f, f'\} \in T$.

A (possibly nondensely defined) symmetric operator S in $\mathfrak{H}$ is said to be *simple* if there is no nontrivial subspace in $\mathfrak{H}$ which reduces S to a selfadjoint operator. It is well known that every symmetric operator S can be written as the direct orthogonal sum $\widehat{S} \oplus S_s$ of a simple symmetric operator $\widehat{S}$ in the Hilbert space

$$\widehat{\mathfrak{H}} = \operatorname{clospan}\big\{\ker(S^* - \lambda) : \lambda \in \mathbb{C}\backslash\mathbb{R}\big\} \tag{2.2}$$

and a selfadjoint operator S_s in $\mathfrak{H} \ominus \widehat{\mathfrak{H}}$.

2.2. Boundary triplets for nondensely defined symmetric operators

Let in the following S be a (not necessarily densely defined) closed symmetric operator in the separable Hilbert space $\mathfrak{H}$ with equal deficiency indices

$$n_\pm(S) = \dim\big(\operatorname{ran}(S \pm i)^\perp\big) = \dim\big(\ker(S^* \mp i)\big) \leq \infty.$$

If $\operatorname{dom}(S)$ is not dense in $\mathfrak{H}$, then the adjoint S^* exists only in the sense of linear relations and is defined as in (2.1). Therefore, if S is not densely defined the closed extensions $S' \subset S^*$ of S in $\mathfrak{H}$ may have nontrivial multi-valued parts. However, the operator S admits also closed extensions in $\mathfrak{H}$ which are operators. We will use the concept of boundary triplets for the description of the closed extensions $S' \subset S^*$ of S in $\mathfrak{H}$; see, e.g., [12, 13, 14, 15, 18, 23]. This concept also provides a convenient criterion to check whether S' is an operator or not; cf. (2.4).

Definition 2.1. A triplet $\Pi = \{\mathcal{H}, \Gamma_0, \Gamma_1\}$ is called a *boundary triplet* for S^* if $\mathcal{H}$ is a Hilbert space and $\Gamma_0, \Gamma_1 : S^* \to \mathcal{H}$ are linear mappings such that the abstract Green's identity

$$(f', g) - (f, g') = (\Gamma_1 \hat{f}, \Gamma_0 \hat{g}) - (\Gamma_0 \hat{f}, \Gamma_1 \hat{g})$$

holds for all $\widehat{f} = \{f, f'\}$, $\widehat{g} = \{g, g'\} \in S^*$ and the mapping $\Gamma := (\Gamma_0, \Gamma_1)^\top : S^* \to \mathcal{H} \oplus \mathcal{H}$ is surjective.

We refer to [14, 15, 18, 23] for a detailed study of boundary triplets and recall only some important facts. First of all a boundary triplet $\Pi = \{\mathcal{H}, \Gamma_0, \Gamma_1\}$ for S^* exists (but is not unique) since the deficiency indices $n_\pm(S)$ of S are assumed to be equal. Then $n_\pm(S) = \dim \mathcal{H}$ holds. A standard construction of a boundary triplet will be given in the proof of Proposition 2.2.

Let $\Pi = \{\mathcal{H}, \Gamma_0, \Gamma_1\}$ be a boundary triplet for S^* and let $\mathrm{Ext}(S)$ be the set of all closed extensions $S' \subset S^*$ of S. Then $S = \ker(\Gamma)$ and the mapping

$$\Theta \mapsto S_\Theta := \Gamma^{-1}\Theta = \bigl\{\widehat{f} \in S^* : \{\Gamma_0\widehat{f}, \Gamma_1\widehat{f}\} \in \Theta\bigr\} \tag{2.3}$$

establishes a bijective correspondence between the set $\widetilde{\mathcal{C}}(\mathcal{H})$ of closed linear relations in $\mathcal{H}$ and the set of closed extensions $S_\Theta \in \mathrm{Ext}(S)$ of S. We note that the right-hand side of (2.3) can also be written as $\ker(\Gamma_1 - \Theta\Gamma_0)$ where the sum and product is interpreted in the sense of linear relations. Since $(S_\Theta)^* = S_{\Theta^*}$ holds for every $\Theta \in \widetilde{\mathcal{C}}(\mathcal{H})$ it follows that S_Θ is symmetric (selfadjoint) if and only if Θ is symmetric (selfadjoint, respectively). In particular, the extensions $A_0 := \ker(\Gamma_0)$ and $A_1 := \ker(\Gamma_1)$ are selfadjoint. The selfadjoint operator or relation A_0 will often play the role of a fixed selfadjoint extension of S in $\mathfrak{H}$. Furthermore, an extension $S_\Theta \in \mathrm{Ext}(S)$ is dissipative (maximal dissipative) if and only if Θ is dissipative (maximal dissipative, respectively). We note that S_Θ in (2.3) is an operator if and only if

$$\Theta \cap \bigl\{\{\Gamma_0\widehat{f}, \Gamma_1\widehat{f}\} : \widehat{f} = \{0, f'\} \in S^*\bigr\} = \{0\}. \tag{2.4}$$

The following proposition is a consequence of the basic properties of boundary triplets and results from [14, 15, 23]. Since it plays an important role in this paper we give a complete proof for the convenience of the reader. We also note that the statement remains true if A and B are linear relations instead of operators. Recall that $\mathcal{F}_n(\mathfrak{H})$, $n \in \mathbb{N}$, is the set of bounded finite-dimensional operators on $\mathfrak{H}$ with ranges of dimension n, i.e.,

$$\mathcal{F}_n(\mathfrak{H}) = \bigl\{T \in [\mathfrak{H}] : \dim(\mathrm{ran}\,(T)) = n\bigr\}.$$

Proposition 2.2. *Let A be a selfadjoint operator and let B be a maximal dissipative operator in $\mathfrak{H}$. Assume that*

$$(B - \lambda)^{-1} - (A - \lambda)^{-1} \in \mathcal{F}_n(\mathfrak{H})$$

holds for some (and hence for all) $\lambda \in \mathbb{C}_+$. Then the closed symmetric operator $S := A \cap B$ has finite deficiency indices (n, n) in $\mathfrak{H}$ and there exists a boundary triplet $\Pi = \{\mathbb{C}^n, \Gamma_0, \Gamma_1\}$ for S^* and a dissipative $n \times n$-matrix D such that $A = \ker(\Gamma_0)$ and $B = \ker(\Gamma_1 - D\Gamma_0)$ holds.

Proof. Let $\lambda_0 \in \rho(A) \cap \rho(B)$ and let $n \in \mathbb{N}$, $\{e_1, \dots, e_n\}$ and $\{f_1, \dots, f_n\}$ be linearly independent vectors such that

$$(B - \lambda_0)^{-1} - (A - \lambda_0)^{-1} = \sum_{i=1}^{n} (\cdot, e_i) f_i. \tag{2.5}$$

The operator $S = A \cap B$, that is,

$$Sf = Af = Bf, \quad \mathrm{dom}\, S = \bigl\{f \in \mathrm{dom}\, A \cap \mathrm{dom}\, B : Af = Bf\bigr\},$$

is a (in general non-densely defined) symmetric operator in $\mathfrak{H}$ and it is easy to check that

$$(S-\lambda_0)^{-1} = (A-\lambda_0)^{-1} \cap (B-\lambda_0)^{-1} \tag{2.6}$$

holds. The intersection in (2.6) is understood in the sense of linear relations. Hence (2.5) and (2.6) imply $\dim(\operatorname{ran}(A-\lambda_0)^{-1})/\operatorname{ran}(S-\lambda_0)^{-1})) = n$. Therefore $\dim(A/S) = n$ and S has deficiency indices (n,n). Note that $(S-\lambda_0)^{-1}$ is defined on the subspace $\mathfrak{H} \ominus \operatorname{span}\{e_1,\dots,e_n\}$ which has codimension n in $\mathfrak{H}$.

It is not difficult to verify that S^* coincides with the direct sum of the graph of A and the linear subspace

$$\widehat{\mathcal{N}}_{\lambda_0} = \bigl\{\{f_{\lambda_0}, \lambda_0 f_{\lambda_0}\} : f_{\lambda_0} \in \mathcal{N}_{\lambda_0} = \ker(S^*-\lambda_0)\bigr\}$$

$\in \mathfrak{H} \times \mathfrak{H}$. Let us decompose the elements $\widehat{f} \in S^*$ accordingly, i.e.,

$$\widehat{f} = \{f, f'\} = \bigl\{f_A + f_{\lambda_0}, Af_A + \lambda_0 f_{\lambda_0}\bigr\}, \quad f_A \in \operatorname{dom} A,\ f_{\lambda_0} \in \mathcal{N}_{\lambda_0}, \tag{2.7}$$

and denote by P_0 the orthogonal projection onto the closed subspace $\mathcal{N}_{\lambda_0}$. Then $\Pi = \{\mathcal{N}_{\lambda_0}, \Gamma_0, \Gamma_1\}$, where

$$\Gamma_0 \widehat{f} := f_{\lambda_0} \quad \text{and} \quad \Gamma_1 \widehat{f} := P_0\bigl((A-\bar{\lambda}_0)f_A + \lambda_0 f_{\lambda_0}\bigr),$$

$\widehat{f} \in S^*$, is a boundary triplet with $A = A_0 := \ker(\Gamma_0)$. In fact, for $\widehat{f}$ as in (2.7) and $\widehat{g} = \{g, g'\} = \{g_A + g_{\lambda_0}, Ag_A + \lambda_0 g_{\lambda_0}\}$ we obtain from $(Af_A, g_A) = (f_A, Ag_A)$ that

$$\begin{aligned}(f', g) - (f, g') &= \bigl((A-\bar{\lambda}_0)f_A + \lambda_0 f_{\lambda_0}, g_{\lambda_0}\bigr) - \bigl(f_{\lambda_0}, (A-\bar{\lambda}_0)g_A + \lambda_0 g_{\lambda_0}\bigr)\\ &= (\Gamma_1 \widehat{f}, \Gamma_0 \widehat{g}) - (\Gamma_0 \widehat{f}, \Gamma_1 \widehat{g})\end{aligned}$$

holds. The surjectivity of the mapping $\Gamma = (\Gamma_0, \Gamma_1)^\top : S^* \to \mathcal{N}_{\lambda_0} \oplus \mathcal{N}_{\lambda_0}$ follows from $\bar{\lambda}_0 \in \rho(A)$ since for $x, x' \in \mathcal{N}_{\lambda_0}$ we can choose $f_A \in \operatorname{dom} A$ such that $(A-\bar{\lambda}_0)f_A = x' - \lambda_0 x$ holds. Then obviously $\widehat{f} := \{f_A + x, Af_A + \lambda_0 x\}$ satisfies $\Gamma \widehat{f} = (x, x')^\top$. Moreover, from the definition of Γ_0 we immediately obtain that the extension $A_0 = \ker(\Gamma_0)$ coincides with the operator A. As the deficiency indices of S are (n,n) we can identify $\mathcal{N}_{\lambda_0}$ with $\mathbb{C}^n$.

Since B is a maximal dissipative extension of the symmetric operator S, $B \in \operatorname{Ext}(S)$. Hence $B \subset \operatorname{dom}(\Gamma) = S^*$ and the linear relation

$$D := \Gamma B = \bigl\{\{\Gamma_0 \widehat{f}, \Gamma_1 \widehat{f}\} : \widehat{f} = \{f, Bf\} \in B\bigr\}$$

is maximal dissipative in $\mathbb{C}^n$ and B coincides with the maximal dissipative extension S_D via (2.3). We claim that D is a matrix, i.e., $\operatorname{mul}(D) = \{0\}$. In fact, assume that D is multi-valued, that is, there exists $\widehat{f} = \{f, Bf\} \in B$ such that $\{0, \Gamma_1 \widehat{f}\} \in D$ with $\Gamma_1 \widehat{f} \neq 0$. In particular, $\Gamma_0 \widehat{f} = 0$, i.e., $\widehat{f} \in A_0 = A$ and therefore $\widehat{f} \in A \cap B = S = \ker(\Gamma_0, \Gamma_1)^\top$, however, this is a contradiction. Thus D is a dissipative $n \times n$-matrix and it follows from (2.3) that $B = \ker(\Gamma_1 - D\Gamma_0)$ holds. □

2.3. Weyl functions and Krein's formula

Again let S be a (in general nondensely defined) closed symmetric operator in $\mathfrak{H}$ with equal deficiency indices as in the previous section. If $\lambda \in \mathbb{C}$ is a point of regular type of S, i.e., $(S-\lambda)^{-1}$ is a bounded operator, we denote the *defect subspace* of S at λ by $\mathcal{N}_\lambda = \ker(S^* - \lambda)$ and we agree to write

$$\widehat{\mathcal{N}}_\lambda = \{\{f, \lambda f\} : f \in \mathcal{N}_\lambda\} \subset S^*.$$

Let $\Pi = \{\mathcal{H}, \Gamma_0, \Gamma_1\}$ be a boundary triplet for S^* and let $A_0 = \ker(\Gamma_0)$ be the fixed selfadjoint extension of S. Recall that for every $\lambda \in \rho(A_0)$ the relation S^* is the direct sum of the selfadjoint relation A_0 and $\widehat{\mathcal{N}}_\lambda$ and denote by π_1 the orthogonal projection onto the first component of $\mathfrak{H} \oplus \mathfrak{H}$. The operator-valued functions

$$\gamma(\cdot) : \rho(A_0) \to [\mathcal{H}, \mathfrak{H}], \quad \lambda \mapsto \gamma(\lambda) = \pi_1 \big(\Gamma_0 \restriction \widehat{\mathcal{N}}_\lambda\big)^{-1}$$

and

$$M(\cdot) : \rho(A_0) \to [\mathcal{H}], \quad \lambda \mapsto M(\lambda) = \Gamma_1 \big(\Gamma_0 \restriction \widehat{\mathcal{N}}_\lambda\big)^{-1}$$

are called the γ*-field* and the *Weyl function*, respectively, corresponding to the boundary triplet $\Pi = \{\mathcal{H}, \Gamma_0, \Gamma_1\}$; see, e.g., [13, 14, 15, 23]. It can be shown that both $\gamma(\cdot)$ and $M(\cdot)$ are holomorphic on $\rho(A_0)$ and that the identities

$$\gamma(\mu) = \big(I + (\mu - \lambda)(A_0 - \mu)^{-1}\big)\gamma(\lambda), \qquad \lambda, \mu \in \rho(A_0), \tag{2.8}$$

and

$$M(\lambda) - M(\mu)^* = (\lambda - \bar{\mu})\gamma(\mu)^*\gamma(\lambda), \qquad \lambda, \mu \in \rho(A_0), \tag{2.9}$$

are valid; see [14, 23]. The identity (2.9) yields that $M(\cdot)$ is a $[\mathcal{H}]$-valued *Nevanlinna function*, that is, $M(\cdot)$ is holomorphic on $\mathbb{C}\backslash\mathbb{R}$, $\Im\mathrm{m}\,(M(\lambda))$ is a nonnegative operator for all $\lambda \in \mathbb{C}_+$ and $M(\lambda) = M(\bar{\lambda})^*$ holds for all $\lambda \in \mathbb{C}\backslash\mathbb{R}$. Moreover, it follows from (2.9) that $0 \in \rho(\Im\mathrm{m}\,(M(\lambda)))$ for all $\lambda \in \mathbb{C}\backslash\mathbb{R}$ and, in particular,

$$\frac{\Im\mathrm{m}\,(M(\lambda))}{\Im\mathrm{m}\,(\lambda)} = \gamma(\lambda)^*\gamma(\lambda), \qquad \lambda \in \mathbb{C}\backslash\mathbb{R}. \tag{2.10}$$

The following inverse result is essentially a consequence of [21], see also [15, 23].

Theorem 2.3. *Let $M : \mathbb{C}\backslash\mathbb{R} \to [\mathcal{H}]$ be a Nevanlinna function such that $0 \in \rho(\Im\mathrm{m}\,(M(\lambda)))$ for some (and hence for all) $\lambda \in \mathbb{C}\backslash\mathbb{R}$ and assume that the condition*

$$\lim_{\eta \to +\infty} \frac{1}{\eta}(M(i\eta)h, h) = 0 \tag{2.11}$$

holds for all $h \in \mathcal{H}$. Then there exists a separable Hilbert space $\mathfrak{H}$, a closed simple symmetric operator S in $\mathfrak{H}$ and a boundary triplet $\Pi = \{\mathcal{H}, \Gamma_0, \Gamma_1\}$ for the adjoint relation S^* such that $A_0 = \ker(\Gamma_0)$ is a selfadjoint operator and the Weyl function of Π coincides with $M(\cdot)$ on $\mathbb{C}\backslash\mathbb{R}$. The symmetric operator S is densely defined if and only if the conditions (2.11) *and*

$$\lim_{\eta \to +\infty} \eta\, \Im\mathrm{m}\,(M(i\eta)h, h) = \infty, \qquad h \in \mathcal{H},\ h \neq 0,$$

are satisfied.

The spectrum and the resolvent set of the closed extensions in $\mathrm{Ext}(S)$ can be described with the help of the Weyl function. More precisely, if $S_\Theta \in \mathrm{Ext}(S)$ is the extension corresponding to $\Theta \in \widetilde{\mathcal{C}}(\mathcal{H})$ via (2.3), then a point $\lambda \in \rho(A_0)$ belongs to $\rho(S_\Theta)$ ($\sigma_i(S_\Theta)$, $i = p, c, r$) if and only if $0 \in \rho(\Theta - M(\lambda))$ ($0 \in \sigma_i(\Theta - M(\lambda))$, $i = p, c, r$, respectively). Moreover, for $\lambda \in \rho(A_0) \cap \rho(S_\Theta)$ the well-known resolvent formula

$$(S_\Theta - \lambda)^{-1} = (A_0 - \lambda)^{-1} + \gamma(\lambda)\bigl(\Theta - M(\lambda)\bigr)^{-1}\gamma(\bar\lambda)^* \tag{2.12}$$

holds, see [14, 23]. Formula (2.12) and Proposition 2.2 imply the following statement which will be used in Section 4.

Corollary 2.4. *Let A be a selfadjoint operator and let B be a maximal dissipative operator in $\mathfrak{H}$ such that*

$$(B - \lambda)^{-1} - (A - \lambda)^{-1} \in \mathcal{F}_n(\mathfrak{H})$$

holds for some (and hence for all) $\lambda \in \mathbb{C}_+$. Let $\Pi = \{\mathbb{C}^n, \Gamma_0, \Gamma_1\}$ be the boundary triplet from Proposition 2.2 such that $A = \ker(\Gamma_0)$ and $B = \ker(\Gamma_1 - D\Gamma_0)$ holds with some dissipative $n \times n$-matrix D and denote the γ-field and the Weyl function of Π by $\gamma(\cdot)$ and $M(\cdot)$, respectively. Then

$$(B - \lambda)^{-1} - (A - \lambda)^{-1} = \gamma(\lambda)\bigl(D - M(\lambda)\bigr)^{-1}\gamma(\bar\lambda)^* \tag{2.13}$$

holds for all $\lambda \in \rho(B) \cap \rho(A)$.

If the maximal dissipative operator B in Proposition 2.2 and Corollary 2.4 is even selfadjoint the representation of the resolvent difference in (2.13) can be further simplified.

Corollary 2.5. *Let A and B be selfadjoint operators in $\mathfrak{H}$ such that*

$$(B - \lambda)^{-1} - (A - \lambda)^{-1} \in \mathcal{F}_n(\mathfrak{H})$$

holds for some (and hence for all) $\lambda \in \mathbb{C}\backslash\mathbb{R}$. Then the closed symmetric operator $S = A \cap B$ has finite deficiency indices (n, n) in $\mathfrak{H}$ and there exists a boundary triplet $\Pi = \{\mathbb{C}^n, \Gamma_0, \Gamma_1\}$ for S^* such that $A = \ker(\Gamma_0)$ and $B = \ker(\Gamma_1)$ holds. Moreover, if $\gamma(\cdot)$ and $M(\cdot)$ denote the γ-field and Weyl function of Π, then

$$(B - \lambda)^{-1} - (A - \lambda)^{-1} = -\gamma(\lambda)M(\lambda)^{-1}\gamma(\bar\lambda)^*$$

holds for all $\lambda \in \rho(B) \cap \rho(A)$.

Proof. According to Proposition 2.2 there is a boundary triplet $\Pi' = \{\mathbb{C}^n, \Gamma_0', \Gamma_1'\}$ for S^* such that $A = \ker(\Gamma_0')$ and $B = \ker(\Gamma_1' - D\Gamma_0')$. Here the dissipative matrix D is even symmetric since B is selfadjoint. A simple calculation shows that $\Pi = \{\mathbb{C}^n, \Gamma_0, \Gamma_1\}$, where

$$\Gamma_0 := \Gamma_0' \quad \text{and} \quad \Gamma_1 := \Gamma_1 - D\Gamma_0,$$

is also a boundary triplet for S^*. If $M(\cdot)$ is the Weyl function corresponding to the boundary triplet Π', then $\lambda \mapsto M(\lambda) - D$ is the Weyl function corresponding to the boundary triplet Π. This together with Proposition 2.2 and Corollary 2.4 implies the statement. □

3. A representation of the scattering matrix

In this section we consider scattering systems $\{A, B\}$ consisting of two selfadjoint operators A and B in a separable Hilbert space $\mathfrak{H}$ and we assume that the difference of the resolvents of A and B is a finite rank operator, that is, for some $n \in \mathbb{N}$ we have

$$(B-\lambda)^{-1} - (A-\lambda)^{-1} \in \mathcal{F}_n(\mathfrak{H}) \tag{3.1}$$

for one (and hence for all) $\lambda \in \rho(A) \cap \rho(B)$. Then the *wave operators*

$$W_\pm(B, A) := \text{s-}\lim_{t\to\pm\infty} e^{itB}e^{-itA}P^{ac}(A)$$

exist and are complete, where $P^{ac}(A)$ denotes the orthogonal projection onto the absolutely continuous subspace $\mathfrak{H}^{ac}(A)$ of A. Completeness means that the ranges of $W_\pm(B, A)$ coincide with the absolutely continuous subspace $\mathfrak{H}^{ac}(B)$ of B; cf. [6, 19, 28, 29]. The *scattering operator* S_{AB} of the *scattering system* $\{A, B\}$ is defined by

$$S_{AB} := W_+(B, A)^* W_-(B, A).$$

Since the scattering operator commutes with A and is unitary in $\mathfrak{H}^{ac}(A)$ it follows that S_{AB} is unitarily equivalent to a multiplication operator induced by a family $\{S_{AB}(\lambda)\}_{\lambda\in\mathbb{R}}$ of unitary operators in a spectral representation of $A^{ac} := A \restriction \mathrm{dom}\,(A) \cap \mathfrak{H}^{ac}(A)$. The aim of this section is to generalize a representation result of this so-called *scattering matrix* $\{S_{AB}(\lambda)\}_{\lambda\in\mathbb{R}}$ from [8].

According to (3.1) and Corollary 2.5 the (possibly nondensely defined) closed symmetric operator $S = A \cap B$ has deficiency indices (n, n) and there exists a boundary triplet $\Pi = \{\mathbb{C}^n, \Gamma_0, \Gamma_1\}$ for S^* such that $A = \ker(\Gamma_0)$ and $B = \ker(\Gamma_1)$. The Weyl function $M(\cdot)$ corresponding to the boundary triplet Π is a $[\mathbb{C}^n]$-valued Nevanlinna function. Therefore the limit

$$M(\lambda) := M(\lambda + i0) = \lim_{\varepsilon\to+0} M(\lambda + i\varepsilon) \tag{3.2}$$

from the upper half-plane $\mathbb{C}_+$ exists for a.e. $\lambda \in \mathbb{R}$; see [16, 17]. As $\Im\mathrm{m}\,(M(\lambda))$ is uniformly positive (uniformly negative) for all $\lambda \in \mathbb{C}_+$ ($\lambda \in \mathbb{C}_-$, respectively) the inverses $M(\lambda)^{-1}$ exist for all $\lambda \in \mathbb{C}\backslash\mathbb{R}$ and $-M(\cdot)^{-1}$ is also a $[\mathbb{C}^n]$-valued Nevanlinna function. Hence it follows that the limit $\lim_{\varepsilon\to+0} M(\lambda + i\varepsilon)^{-1}$ exists for a.e. $\lambda \in \mathbb{R}$ and coincides with the inverse of $M(\lambda)$ in (3.2) for a.e. $\lambda \in \mathbb{R}$.

In the following theorem we find a representation of the scattering matrix $\{S_{AB}(\lambda)\}_{\lambda\in\mathbb{R}}$ of the scattering system $\{A, B\}$ in the direct integral $L^2(\mathbb{R}, d\lambda, \mathcal{H}_\lambda)$, where

$$\mathcal{H}_\lambda := \mathrm{ran}\,(\Im\mathrm{m}\,(M(\lambda + i0)) \quad \text{for a.e.} \quad \lambda \in \mathbb{R}, \tag{3.3}$$

cf. Appendix A. We will formulate and prove our result first for the case of a simple symmetric operator $S = A \cap B$ and discuss the general case afterwards in Theorem 3.2. For the special case that the simple symmetric operator $S = A \cap B$ is densely defined Theorem 3.1 reduces to [8, Theorem 3.8]. We remark that the proof of Theorem 3.1 differs from the proof of [8, Theorem 3.8]. Here we make use of the abstract representation result Theorem A.2.

Theorem 3.1. *Let A and B be selfadjoint operators in $\mathfrak{H}$ such that* (3.1) *is satisfied, suppose that the symmetric operator $S = A \cap B$ is simple and let $\Pi = \{\mathbb{C}^n, \Gamma_0, \Gamma_1\}$ be a boundary triplet for S^* such that $A = \ker(\Gamma_0)$ and $B = \ker(\Gamma_1)$; cf. Corollary* 2.5. *Let $M(\cdot)$ be the corresponding Weyl function and define the spaces $\mathcal{H}_\lambda$ for a.e. $\lambda \in \mathbb{R}$ as in* (3.3).

Then $L^2(\mathbb{R}, d\lambda, \mathcal{H}_\lambda)$ forms a spectral representation of A^{ac} such that the scattering matrix $\{S_{AB}(\lambda)\}_{\lambda\in\mathbb{R}}$ of the scattering system $\{A, B\}$ admits the representation

$$S_{AB}(\lambda) = I_{\mathcal{H}_\lambda} - 2i\sqrt{\Im\mathrm{m}\,(M(\lambda))}\, M(\lambda)^{-1}\sqrt{\Im\mathrm{m}\,(M(\lambda))} \in [\mathcal{H}_\lambda] \tag{3.4}$$

for a.e. $\lambda \in \mathbb{R}$, where $M(\lambda) = M(\lambda + i0)$.

Proof. In order to verify the representation (3.4) of the scattering matrix $\{S_{AB}(\lambda)\}_{\lambda\in\mathbb{R}}$ we will make use of Theorem A.2. For this let us first rewrite the difference of the resolvents $(B - i)^{-1}$ and $(A - i)^{-1}$ as in (A.3). According to Corollary 2.5 we have

$$(B - i)^{-1} - (A - i)^{-1} = -\gamma(i)M(i)^{-1}\gamma(-i)^*. \tag{3.5}$$

Using (2.8) we find

$$(B - i)^{-1} - (A - i)^{-1} = -(A + i)(A - i)^{-1}\gamma(-i)M(i)^{-1}\gamma(-i)^*.$$

and hence the representation (A.3) follows if we set

$$\phi(t) := \frac{t+i}{t-i}, \quad t \in \mathbb{R}, \qquad C = \gamma(-i) \quad \text{and} \quad G = -M(i)^{-1}. \tag{3.6}$$

Moreover, since S is simple it follows from (2.2) that

$$\mathfrak{H} = \mathrm{clospan}\big\{\ker(S^* - \lambda) : \lambda \in \mathbb{C}\backslash\mathbb{R}\big\}$$

holds. As $\mathrm{ran}\, C = \mathrm{ran}\,\gamma(-i) = \ker(S^* + i)$ one concludes in the same way as in the proof of [8, Lemma 3.4] that the condition

$$\mathfrak{H}^{ac}(A) = \mathrm{clospan}\big\{E_A^{ac}(\delta)\mathrm{ran}\,(C) : \delta \in \mathcal{B}(\mathbb{R})\big\}$$

is satisfied.

Next we express the $[\mathbb{C}^n]$-valued function

$$\lambda \mapsto K(\lambda) = \frac{d}{d\lambda}\, C^* E_A((-\infty, \lambda))C$$

and its square root $\lambda \mapsto \sqrt{K(\lambda)}$ in terms of the Weyl function $M(\cdot)$ for a.e. $\lambda \in \mathbb{R}$. We have

$$\begin{aligned} K(\lambda) &= \lim_{\varepsilon\to+0} \frac{1}{2\pi i}\gamma(-i)^* \left((A - \lambda - i\varepsilon)^{-1} - (A - \lambda + i\varepsilon)^{-1}\right)\gamma(-i) \\ &= \lim_{\varepsilon\to+0} \frac{\varepsilon}{\pi}\gamma(-i)^*(A - \lambda - i\varepsilon)^{-1}(A - \lambda + i\varepsilon)^{-1}\gamma(-i) \end{aligned} \tag{3.7}$$

and on the other hand by (2.10)

$$\Im\mathrm{m}\,(M(\lambda + i\varepsilon)) = \varepsilon\gamma(\lambda + i\varepsilon)^*\gamma(\lambda + i\varepsilon).$$

Inserting $\gamma(\lambda + i\varepsilon) = (I + (\lambda + i\varepsilon + i)(A - \lambda - i\varepsilon)^{-1}\gamma(-i))$ (cf. (2.8)) we obtain

$$\Im\mathrm{m}\,(M(\lambda + i\varepsilon)) = \varepsilon\gamma(-i)^*(I + A^2)(A - \lambda - i\varepsilon)^{-1}(A - \lambda + i\varepsilon)^{-1}\gamma(-i) \quad (3.8)$$

and by comparing (3.7) and (3.8) we find

$$\Im\mathrm{m}\,(M(\lambda)) = \lim_{\varepsilon\to 0+} \Im\mathrm{m}\,(M(\lambda + i\varepsilon)) = \pi(1 + \lambda^2)K(\lambda) \quad (3.9)$$

for a.e. $\lambda \in \mathbb{R}$. In particular, the finite-dimensional subspaces $\mathrm{ran}\,(K(\lambda))$ in Theorem A.2 coincide with the spaces $\mathcal{H}_\lambda = \mathrm{ran}\,(\Im\mathrm{m}\,(M(\lambda)))$ for a.e. $\lambda \in \mathbb{R}$ and therefore $L^2(\mathbb{R}, d\lambda, \mathcal{H}_\lambda)$ is a spectral representation of A^{ac} and the scattering matrix $\{S_{AB}(\lambda)\}_{\lambda\in\mathbb{R}}$ admits the representation (A.4). Inserting the square root $\sqrt{K(\lambda)}$ from (3.9) into (A.4) we find

$$S_{AB}(\lambda) = I_{\mathcal{H}_\lambda} + 2i(1 + \lambda^2)\sqrt{\Im\mathrm{m}\,(M(\lambda))}Z(\lambda)\sqrt{\Im\mathrm{m}\,(M(\lambda))} \quad (3.10)$$

and it remains to compute

$$Z(\lambda) = \frac{1}{\lambda + i}Q^*Q + \frac{\phi(\lambda)}{(\lambda + i)^2}G + \lim_{\varepsilon\to 0+} Q^*(B - \lambda - i\varepsilon)^{-1}Q, \quad (3.11)$$

where $Q = \phi(A)CG = -\gamma(i)M(i)^{-1}$, cf. (A.5), (3.6) and (3.5).

It follows from [8, Lemma 3.2] that

$$Q^*(B - \lambda - i0)^{-1}Q = \frac{1}{1 + \lambda^2}\left(M(i)^{-1} - M(\lambda)^{-1}\right) + \frac{1}{\lambda + i}\Im\mathrm{m}\,(M(i)^{-1}) \quad (3.12)$$

holds for a.e. $\lambda \in \mathbb{R}$ and from (2.10) we obtain

$$\begin{aligned} Q^*Q &= (M(i)^{-1})^*\gamma(i)^*\gamma(i)M(i)^{-1} = (M(i)^{-1})^*\Im\mathrm{m}\,(M(i))M(i)^{-1} \\ &= -\Im\mathrm{m}\,(M(i)^{-1}). \end{aligned} \quad (3.13)$$

Therefore we conclude from (3.13) and (3.6) that (3.11) takes the form

$$Z(\lambda) = -\frac{1}{\lambda + i}\Im\mathrm{m}\,(M(i)^{-1}) - \frac{1}{1 + \lambda^2}M(i)^{-1} + Q^*(B - \lambda - i0)^{-1}Q$$

and by inserting (3.12) we find $Z(\lambda) = -(1 + \lambda^2)^{-1}M(\lambda)^{-1}$. Hence (3.10) turns into the representation (3.4) of the scattering matrix $\{S_{AB}(\lambda)\}$. □

In general it may happen that the operator $S = A \cap B$ is not simple, that is, there is a nontrivial decomposition of the Hilbert space $\mathfrak{H} = \widehat{\mathfrak{H}} \oplus \mathfrak{K}$ such that

$$S = \widehat{S} \oplus H, \quad (3.14)$$

where $\widehat{S}$ is simple symmetric operator in $\widehat{\mathfrak{H}}$ and H is a selfadjoint operator in $\mathfrak{K}$, cf. Section 2.1. Then there exist selfadjoint extensions $\widehat{A}$ and $\widehat{B}$ of $\widehat{S}$ in $\widehat{\mathfrak{H}}$ such that

$$A = \widehat{A} \oplus H \qquad \text{and} \qquad B = \widehat{B} \oplus H. \quad (3.15)$$

The next result extends the representation of the scattering matrix in Theorem 3.1 to the case of a non-simple S.

Theorem 3.2. *Let A and B be selfadjoint operators in $\mathfrak{H}$ such that* (3.1) *is satisfied, let $S = A \cap B$ be decomposed as in* (3.14) *and let $\Pi = \{\mathbb{C}^n, \Gamma_0, \Gamma_1\}$ be a boundary triplet for S^* such that $A = \ker(\Gamma_0)$ and $B = \ker(\Gamma_1)$; cf. Corollary* 2.5. *Furthermore, let $L^2(\mathbb{R}, d\lambda, \mathcal{K}_\lambda)$ be a spectral representation of the absolutely continuous part H^{ac} of the selfadjoint operator H in the Hilbert space $\mathfrak{K}$.*

Then $L^2(\mathbb{R}, d\lambda, \mathcal{H}_\lambda \oplus \mathcal{K}_\lambda)$ is a spectral representation of A^{ac} and the scattering matrix $\{S_{AB}(\lambda)\}_{\lambda\in\mathbb{R}}$ is given by

$$S_{AB}(\lambda) = \begin{pmatrix} S_{\widehat{A},\widehat{B}}(\lambda) & 0 \\ 0 & I_{\mathcal{K}_\lambda} \end{pmatrix} \in [\mathcal{H}_\lambda \oplus \mathcal{K}_\lambda]$$

for a.e. $\lambda \in \mathbb{R}$, where $\mathcal{H}_\lambda = \operatorname{ran}(\Im\mathrm{m}\,(M(\lambda + i0)))$, $M(\cdot)$ is the Weyl function corresponding to the boundary triplet Π and

$$S_{\widehat{A}\widehat{B}}(\lambda) = I_{\mathcal{H}_\lambda} - 2i\sqrt{\Im\mathrm{m}\,(M(\lambda))}M(\lambda)^{-1}\sqrt{\Im\mathrm{m}\,(M(\lambda))} \in [\mathcal{H}_\lambda]$$

is the scattering matrix of the scattering system $\{\widehat{A}, \widehat{B}\}$ from (3.15).

Proof. It follows from the decomposition (3.15) that the absolutely continuous subspaces $\mathfrak{H}^{ac}(A)$ and $\mathfrak{H}^{ac}(B)$ can be written as the orthogonal sums

$$\mathfrak{H}^{ac}(A) = \widehat{\mathfrak{H}}^{ac}(\widehat{A}) \oplus \mathfrak{K}^{ac}(H) \quad \text{and} \quad \mathfrak{H}^{ac}(B) = \widehat{\mathfrak{H}}^{ac}(\widehat{B}) \oplus \mathfrak{K}^{ac}(H)$$

of the absolutely continuous subspaces of $\widehat{A}$ and $\widehat{B}$, and the absolutely continuous subspace $\mathfrak{K}^{ac}(H)$ of the selfadjoint operator H in $\mathfrak{K}$. Therefore the wave operators of $W_\pm(B, A)$ of the scattering system $\{A, B\}$ can be written with the wave operators $W_\pm(\widehat{B}, \widehat{A})$ of the scattering system $\{\widehat{A}, \widehat{B}\}$ in the form

$$W_\pm(B, A) = W_\pm(\widehat{B}, \widehat{A}) \oplus I_{\mathfrak{K}^{ac}(H)}.$$

This implies the corresponding decomposition of the scattering operator S_{AB} in $S = S_{\widehat{A}\widehat{B}} \oplus I_{\mathfrak{K}^{ac}(H)}$ and hence the scattering matrix $\{S_{AB}(\lambda)\}_{\lambda\in\mathbb{R}}$ of the scattering system $\{A, B\}$ coincides with the orthogonal sum of the scattering matrix $\{S_{\widehat{A}\widehat{B}}(\lambda)\}_{\lambda\in\mathbb{R}}$ of the scattering system $\{\widehat{A}, \widehat{B}\}$ and the identity operator in the spectral representation $L^2(\mathbb{R}, d\lambda, \mathcal{K}_\lambda)$ of H^{ac}.

It is not difficult to see that $\widehat{\Pi} = \{\mathbb{C}^n, \widehat{\Gamma}_0, \widehat{\Gamma}_1\}$, where $\widehat{\Gamma}_0$ and $\widehat{\Gamma}_1$ denote the restrictions of the boundary mappings Γ_0 and Γ_1 from $S^* = (\widehat{S})^* \oplus H$ onto $(\widehat{S})^*$, is a boundary triplet for $(\widehat{S})^*$ such that $\widehat{A} = \ker(\widehat{\Gamma}_0)$ and $\widehat{B} = \ker(\widehat{\Gamma}_1)$. Moreover, the Weyl function corresponding to $\widehat{\Pi}$ coincides with the Weyl function $M(\cdot)$ corresponding to Π. Hence, by Theorem 3.1 the scattering matrix $\{S_{\widehat{A}\widehat{B}}(\lambda)\}_{\lambda\in\mathbb{R}}$ is given by (3.4). □

4. Dissipative and Lax-Phillips scattering systems

In this section we consider a scattering systems $\{A, B\}$ consisting of a selfadjoint operator A and a maximal dissipative operator B in the Hilbert space $\mathfrak{H}$. As above it is assumed that

$$(B-\lambda)^{-1} - (A-\lambda)^{-1} \in \mathcal{F}_n(\mathfrak{H}), \qquad \lambda \in \rho(A) \cap \rho(B), \tag{4.1}$$

holds for some $n \in \mathbb{N}$. Then the closed symmetric operator $S = A \cap B$ is in general not densely defined and its deficiency indices are (n, n). By Corollary 2.4 there exists a boundary triplet $\Pi = \{\mathbb{C}^n, \Gamma_0, \Gamma_1\}$ for S^* and a dissipative $n \times n$-matrix D such that $A = \ker(\Gamma_0)$, $B = \ker(\Gamma_1 - D\Gamma_0)$ and

$$(B-\lambda)^{-1} - (A-\lambda)^{-1} = \gamma(\lambda)\big(D - M(\lambda)\big)^{-1}\gamma(\bar{\lambda})^*$$

holds. For our later purposes in Section 5 it is sufficient to investigate the special case $\ker(\Im\mathrm{m}\,(D)) = \{0\}$, the general case can be treated in the same way as in [7, 9].

For the investigation of the dissipative scattering system $\{A, B\}$ it is useful to construct a so-called minimal selfadjoint dilation L of the maximal dissipative operator B. For the explicit construction of L we will use the following lemma which also shows how the constant function $\mathbb{C}_+ \ni \lambda \mapsto -i\Im\mathrm{m}\,(D)$, $\lambda \in \mathbb{C}_+$, can be realized as a Weyl function. A detailed proof of Lemma 4.1 can be found in [7].

Lemma 4.1. *Let T be the first-order differential operator in the Hilbert space $L^2(\mathbb{R}, \mathbb{C}^n)$ defined by*

$$(Tg)(x) = -ig'(x), \qquad \mathrm{dom}\,(T) = \big\{g \in W_2^1(\mathbb{R}, \mathbb{C}^n) : g(0) = 0\big\}.$$

Then the following holds.

(i) *T is a densely defined closed simple symmetric operator with deficiency indices (n, n).*

(ii) *The adjoint operator is*

$$(T^*g)(x) = -ig'(x), \qquad \mathrm{dom}\,(T^*) = W_2^1(\mathbb{R}_-, \mathbb{C}^n) \oplus W_2^1(\mathbb{R}_+, \mathbb{C}^n).$$

(iii) *The triplet $\Pi_T = \{\mathbb{C}^n, \Upsilon_0, \Upsilon_1\}$, where*

$$\Upsilon_0\hat{g} := \frac{1}{\sqrt{2}}(-\Im\mathrm{m}\,(D))^{-\frac{1}{2}}\big(g(0+) - g(0-)\big),$$

$$\Upsilon_1\hat{g} := \frac{i}{\sqrt{2}}(-\Im\mathrm{m}\,(D))^{\frac{1}{2}}\big(g(0+) + g(0-)\big), \quad \hat{g} = \{g, T^*g\},$$

is a boundary triplet for T^* and $T_0 = \ker(\Upsilon_0)$ is the selfadjoint first-order differential operator in $L^2(\mathbb{R}, \mathbb{C}^n)$ defined on $W_2^1(\mathbb{R}, \mathbb{C}^n)$.

(iv) *The Weyl function $\tau(\cdot)$ corresponding to the boundary triplet in* (iii) *is given by*

$$\tau(\lambda) = \begin{cases} -i\Im\mathrm{m}\,(D), & \lambda \in \mathbb{C}_+, \\ i\Im\mathrm{m}\,(D), & \lambda \in \mathbb{C}_-. \end{cases}$$

Let $S = A \cap B$ and let T be the first-order differential operator from Lemma 4.1. It is clear that

$$\begin{pmatrix} S & 0 \\ 0 & T \end{pmatrix} \tag{4.2}$$

is a closed symmetric operator in the Hilbert space $\mathfrak{H} \oplus L^2(\mathbb{R}, \mathbb{C}^n)$ with deficiency indices $(2n, 2n)$ and the adjoint of (4.2) is the orthogonal sum of the relation S^* and the operator T^* from Lemma 4.1. The next theorem, which is a variant of [7, Theorem 3.2], shows how a minimal selfadjoint dilation of the dissipative operator $B = \ker(\Gamma_1 - D\Gamma_0)$ can be constructed. For the particular case of Sturm-Liouville operators with dissipative boundary conditions this construction goes back to B.S. Pavlov; cf. [25, 26].

Theorem 4.2. *Let A be a selfadjoint operator and let B be a maximal dissipative operator in $\mathfrak{H}$ such that (4.1) holds. Let $\Pi = \{\mathbb{C}^n, \Gamma_0, \Gamma_1\}$ be a boundary triplet for S^*, $S = A \cap B$, and let D be a dissipative $n \times n$-matrix with* $\ker(\Im\mathrm{m}\,(D)) = \{0\}$ *such that $A = \ker(\Gamma_0)$ and $B = \ker(\Gamma_1 - D\Gamma_0)$; cf. Proposition 2.2. If $\Pi_T = \{\mathbb{C}^n, \Upsilon_0, \Upsilon_1\}$ is the boundary triplet of T^* introduced in Lemma 4.1, then the operator*

$$L\begin{pmatrix} f \\ g \end{pmatrix} = \begin{pmatrix} f' \\ T^* g \end{pmatrix}, \qquad \hat f = \{f, f'\} \in S^*, \ \ \hat g = \{g, T^* g\},$$

with domain

$$\mathrm{dom}\,(L) = \left\{ \begin{pmatrix} f \\ g \end{pmatrix} \in \mathrm{dom}\,(S^*) \oplus \mathrm{dom}\,(T^*) : \begin{array}{c} \Gamma_0 \hat f - \Upsilon_0 \hat g = 0 \\ (\Gamma_1 - \Re\mathrm{e}\,(D)\Gamma_0)\hat f = -\Upsilon_1 \hat g \end{array} \right\}$$

is a minimal selfadjoint dilation of the maximal dissipative operator B, that is, for all $\lambda \in \mathbb{C}_+$

$$P_{\mathfrak{H}} \left(L - \lambda\right)^{-1} \restriction_{\mathfrak{H}} = (B - \lambda)^{-1}$$

holds and the condition $\mathfrak{H} \oplus L^2(\mathbb{R}, \mathbb{C}^n) = \mathrm{clospan}\{(L - \lambda)^{-1}\mathfrak{H} : \lambda \in \mathbb{C}\backslash\mathbb{R}\}$ is satisfied.

Proof. Besides the assertion that L is an operator the proof of Theorem 4.2 is essentially the same as the proof of [7, Theorem 3.2]. The fact that the restriction L of the relation $S^* \oplus T^*$ is an operator can be seen as follows: Suppose that $\hat f \oplus \hat g \in L$, where $\{0, f'\} \in S^*$, $\{0, g'\} \in T^*$. Since T^* is an operator we have $g' = 0$ and this implies $\hat g = 0$. Therefore we obtain from the boundary conditions in $\mathrm{dom}\,(L)$ that

$$\Gamma_0 \hat f = \Upsilon_0 \hat g = 0$$

holds. Hence $\hat f = \{f, f'\}$ belongs to $A = \ker(\Gamma_0)$ which is an operator. Therefore $f' = 0$ and L is an operator. □

Let L be the minimal selfadjoint dilation of the maximal dissipative operator B from Theorem 4.2 and define a selfadjoint operator K in $\mathfrak{H} \oplus L^2(\mathbb{R}, \mathbb{C}^n)$ by

$$K := \begin{pmatrix} A & 0 \\ 0 & T_0 \end{pmatrix}, \tag{4.3}$$

where $T_0 = \ker(\Upsilon_0)$ is the selfadjoint first-order differential operator from Lemma 4.1. In the following theorem we consider the scattering system $\{K, L\}$ in the Hilbert space $\mathfrak{H} \oplus L^2(\mathbb{R}, \mathbb{C}^n)$. The operator $R := K \cap L$ is symmetric and may have a nontrivial selfadjoint part H which acts in the Hilbert space

$$(\mathfrak{H} \oplus L^2(\mathbb{R}, \mathbb{C}^n)) \ominus \operatorname{clospan}\{\ker(R^* - \lambda) : \lambda \in \mathbb{C}\backslash\mathbb{R}\}.$$

Hence the operators K and L admit the decompositions

$$K = \widehat{K} \oplus H \quad \text{and} \quad L = \widehat{L} \oplus H,$$

with selfadjoint operators $\widehat{K}$ and $\widehat{L}$ in $\operatorname{clospan}\{\ker(R^* - \lambda) : \lambda \in \mathbb{C}\backslash\mathbb{R}\}$ and we have $R = \widehat{R} \oplus H$, where $\widehat{R} = \widehat{K} \cap \widehat{L}$. In particular, $\widehat{K}$ and $\widehat{L}$ are both selfadjoint extensions of the closed simple symmetric operator $\widehat{R}$. We remark that the symmetric operator R is an n-dimensional extension of the orthogonal sum in (4.2); this follows easily from the next theorem. In the following we assume that $L^2(\mathbb{R}, d\lambda, \mathcal{K}_\lambda)$ is a spectral representation of the absolutely continuous part H^{ac} of H.

Theorem 4.3. *Let A be a selfadjoint operator and let B be a maximal dissipative operator in $\mathfrak{H}$ such that* (4.1) *holds. Let $\Pi = \{\mathbb{C}^n, \Gamma_0, \Gamma_1\}$ be a boundary triplet for S^*, $S = A \cap B$, and let D be a dissipative $n \times n$-matrix with $\ker(\Im\mathrm{m}\,(D)) = \{0\}$ such that $A = \ker(\Gamma_0)$ and $B = \ker(\Gamma_1 - D\Gamma_0)$; cf. Proposition* 2.2. *If L is the minimal self-adjoint dilation of B in Theorem* 4.2 *and K is given by* (4.3), *then*

$$(K - \lambda) - (L - \lambda)^{-1} \in \mathcal{F}_n, \qquad \lambda \in \mathbb{C}\backslash\mathbb{R}. \tag{4.4}$$

Moreover, if $L^2(\mathbb{R}, d\lambda, \mathcal{K}_\lambda)$ is a spectral representation of H^{ac}, where H is the maximal self-adjoint part of $R = K \cap L$, then $L^2(\mathbb{R}, d\lambda, \mathbb{C}^n \oplus \mathcal{K}_\lambda)$ is a spectral representation of K and the scattering matrix $\{S_{KL}(\lambda)\}_{\lambda \in \mathbb{R}}$ of the scattering system $\{K, L\}$ admits the representation

$$S_{KL}(\lambda) = \begin{pmatrix} S_{\widehat{K}\widehat{L}}(\lambda) & 0 \\ 0 & I_{\mathcal{K}_\lambda} \end{pmatrix} \in [\mathbb{C}^n \oplus \mathcal{K}_\lambda]$$

for a.e. $\lambda \in \mathbb{R}$, where

$$S_{\widehat{K}\widehat{L}}(\lambda) = I_{\mathbb{C}^n} - 2i\sqrt{\Im\mathrm{m}\,(M(\lambda) - D)}\,\big(M(\lambda) - D\big)^{-1}\sqrt{\Im\mathrm{m}\,(M(\lambda) - D)}$$

is the scattering matrix of the scattering system $\{\widehat{K}, \widehat{L}\}$, $M(\cdot)$ is the Weyl function of the boundary triplet Π and $M(\lambda) = M(\lambda + i0)$.

Proof. We are going to apply Theorem 3.2 to the scattering system $\{K, L\}$. For this we consider the symmetric operator $R = K \cap L$ and note that the operator K is given by $\ker(\Gamma_0) \oplus \ker(\Upsilon_0)$. Hence the boundary condition $\Gamma_0 \hat{f} - \Upsilon_0 \hat{g} = 0$ in $\operatorname{dom}(L)$ is automatically fulfilled and this implies that the intersection $R = K \cap L$ is given by

$$R\begin{pmatrix} f \\ g \end{pmatrix} = K\begin{pmatrix} f \\ g \end{pmatrix} = L\begin{pmatrix} f \\ g \end{pmatrix}$$

$$\operatorname{dom}(R) = \left\{\begin{pmatrix} f \\ g \end{pmatrix} \in \operatorname{dom} K : (\Gamma_1 - \Re\mathrm{e}\,(D)\Gamma_0)\hat{f} = -\Upsilon_1 \hat{g}\right\},$$

where $\hat{f} = \{f, Af\}$ and $\hat{g} = \{g, T_0 g\}$. It is not difficult to verify that the adjoint operator R^* has the form

$$R^* = \left\{\hat{f} \oplus \hat{g} \in S^* \oplus T^* : \Gamma_0 \hat{f} = \Upsilon_0 \hat{g}\right\}$$

and $\widetilde{\Pi} = \{\mathbb{C}^n, \widetilde{\Gamma}_0, \widetilde{\Gamma}_1\}$, where

$$\widetilde{\Gamma}_0(\hat{f} \oplus \hat{g}) = \Gamma_0 \hat{f} \quad \text{and} \quad \widetilde{\Gamma}_1(\hat{f} \oplus \hat{g}) = (\Gamma_1 - \Re\mathrm{e}\,(D)\Gamma_0)\hat{f} + \Upsilon_1 \hat{g},$$

is a boundary triplet for R^*. Observe that $K = \ker(\widetilde{\Gamma}_0)$ and $L = \ker(\widetilde{\Gamma}_1)$. This also implies that the difference of the resolvents of K and L in (4.4) is a rank n operator; cf. Corollary 2.5.

Let us compute the Weyl function $\widetilde{M}$ corresponding to the boundary triplet $\widetilde{\Pi}$. For $\lambda \in \mathbb{C}_+$ and $\hat{f} \oplus \hat{g} \in \widehat{\mathcal{N}}_{\lambda, R^*}$ we have $\hat{f} \in \widehat{\mathcal{N}}_{\lambda, S^*}$, $\hat{g} \in \widehat{\mathcal{N}}_{\lambda, T^*}$ and $\Gamma_0 \hat{f} = \Upsilon_0 \hat{g}$. Hence the definition of the Weyl function and Lemma 4.1 imply

$$\begin{aligned}
\widetilde{M}(\lambda)\widetilde{\Gamma}_0(\hat{f} \oplus \hat{g}) &= \widetilde{\Gamma}_1(\hat{f} \oplus \hat{g}) = \Gamma_1 \hat{f} - \Re\mathrm{e}\,(D)\Gamma_0 \hat{f} + \Upsilon_1 \hat{g} \\
&= M(\lambda)\Gamma_0 \hat{f} - \Re\mathrm{e}\,(D)\Gamma_0 \hat{f} - i\Im\mathrm{m}\,(D)\Upsilon_0 \hat{g} \\
&= (M(\lambda) - D)\Gamma_0 \hat{f} = (M(\lambda) - D)\widetilde{\Gamma}_0(\hat{f} \oplus \hat{g})
\end{aligned}$$

and therefore $\widetilde{M}(\lambda) = M(\lambda) - D$ for $\lambda \in \mathbb{C}_+$. As D is a dissipative matrix and $\ker(\Im\mathrm{m}\,(D)) = \{0\}$ by assumption it follows that

$$\Im\mathrm{m}\,(\widetilde{M}(\lambda + i0)) = \Im\mathrm{m}\,(M(\lambda + i0)) - \Im\mathrm{m}\,(D)$$

is uniformly positive and hence $\mathrm{ran}\,(\Im\mathrm{m}\,(\widetilde{M}(\lambda + i0))) = \mathbb{C}^n$. Now Theorem 3.2 applied to the boundary triplet $\widetilde{\Pi}$ and the corresponding Weyl function $\widetilde{M}$ yields the statement of Theorem 4.3. □

In view of our later purposes it is useful to express the scattering matrix $\{S_{KL}(\lambda)\}_{\lambda \in \mathbb{R}}$ in Theorem 4.3 in a slightly different form. The following proposition extends [7, Theorem 3.6] to the case where $S = A \cap B$ is not necessarily densely defined. The proof is almost the same and will not be repeated.

Proposition 4.4. *Let the assumptions of Theorem 4.3 be satisfied, assume, in addition, that $S = A \cap B$ is simple and let $L^2(\mathbb{R}, d\lambda, \mathcal{H}_\lambda)$, $\mathcal{H}_\lambda = \mathrm{ran}\,(\Im\mathrm{m}\,(M(\lambda)))$, $M(\lambda) := M(\lambda + i0)$, be a spectral representation of A^{ac}.*

Then $L^2(\mathbb{R}, d\lambda, \mathcal{H}_\lambda \oplus \mathbb{C}^n)$ is a spectral representation of $K^{ac} = A^{ac} \oplus T_0$ such that the scattering matrix $\{S_{KL}(\lambda)\}_{\lambda \in \mathbb{R}}$ of the scattering system $\{K, L\}$ can be expressed by

$$S_{KL}(\lambda) = \begin{pmatrix} I_{\mathcal{H}_\lambda} & 0 \\ 0 & I_{\mathbb{C}^n} \end{pmatrix} + 2i \begin{pmatrix} S_{11}(\lambda) & S_{12}(\lambda) \\ S_{21}(\lambda) & S_{22}(\lambda) \end{pmatrix} \in [\mathcal{H}_\lambda \oplus \mathbb{C}^n]$$

for a.e. $\lambda \in \mathbb{R}$, *where*

$$\begin{aligned}
S_{11}(\lambda) &= \sqrt{\Im\mathrm{m}\,(M(\lambda))}\,\big(D - M(\lambda)\big)^{-1}\sqrt{\Im\mathrm{m}\,(M(\lambda))},\\
S_{12}(\lambda) &= \sqrt{\Im\mathrm{m}\,(M(\lambda))}\,\big(D - M(\lambda)\big)^{-1}\sqrt{-\Im\mathrm{m}\,(D)},\\
S_{21}(\lambda) &= \sqrt{-\Im\mathrm{m}\,(D)}\,\big(D - M(\lambda)\big)^{-1}\sqrt{\Im\mathrm{m}\,(M(\lambda))},\\
S_{22}(\lambda) &= \sqrt{-\Im\mathrm{m}\,(D)}\,\big(D - M(\lambda)\big)^{-1}\sqrt{-\Im\mathrm{m}\,(D)}.
\end{aligned}$$

Remark 4.5. *If* $S = A \cap B$ *is simple we find by combining Theorem* 4.3 *with Proposition* 4.4 *that* $\dim(\mathcal{K}_\lambda) = \dim(\mathcal{H}_\lambda)$ *holds for a.e.* $\lambda \in \mathbb{R}$, *i.e., the spectral multiplicity of* H^{ac}, *where* H *is the maximal self-adjoint part of* $R = K \cap L$ *is equal to the spectral multiplicity of* A^{ac}.

In the following we are going to interpret the right lower corner $I + 2iS_{22}$ of the scattering matrix $\{S_{KL}(\lambda)\}_{\lambda\in\mathbb{R}}$ in Proposition 4.4 as the scattering matrix corresponding to a Lax-Phillips scattering system; see, e.g., [6, 22] for further details. For this purpose we decompose $L^2(\mathbb{R}, \mathbb{C}^n)$ into the orthogonal sum of the subspaces

$$\mathcal{D}_- := L^2(\mathbb{R}_-, \mathbb{C}^n) \quad \text{and} \quad \mathcal{D}_+ := L^2(\mathbb{R}_+, \mathbb{C}^n),$$

and denote the natural embeddings of $\mathcal{D}_\pm$ into $\mathfrak{H} \oplus L^2(\mathbb{R}, \mathbb{C}^n)$ by $J_\pm$. The subspaces $\mathcal{D}_+$ and $\mathcal{D}_-$ are called *outgoing* and *incoming subspaces*, respectively, for the selfadjoint dilation L in $\mathfrak{H} \oplus L^2(\mathbb{R}, \mathbb{C}^n)$, i.e.,

$$e^{-itL}\mathcal{D}_\pm \subseteq \mathcal{D}_\pm, \quad t \in \mathbb{R}_\pm, \quad \text{and} \quad \bigcap_{t\in\mathbb{R}} e^{-itL}\mathcal{D}_\pm = \{0\}$$

hold. The system $\{L, \mathcal{D}_-, \mathcal{D}_+\}$ is called *Lax-Phillips scattering system* and the corresponding *Lax-Phillips wave operators* are defined by

$$\Omega_\pm := \text{s-}\lim_{t\to\pm\infty} e^{itL} J_\pm e^{-itT_0} : L^2(\mathbb{R}, \mathbb{C}^n) \to \mathfrak{H} \oplus L^2(\mathbb{R}, \mathbb{C}^n);$$

cf. [6]. Since $\text{s-lim}_{t\to\pm\infty} J_\mp e^{-itT_0} = 0$ the restrictions of the wave operators $W_\pm(L, K)$ of the scattering system $\{K, L\}$ onto $L^2(\mathbb{R}, \mathbb{C}^n)$ coincide with the Lax-Phillips wave operators $\Omega_\pm$ and hence the *Lax-Phillips scattering operator* $S^{LP} := \Omega_+^*\Omega_-$ is given by $S^{LP} = P_{L^2} S_{KL}\, \iota_{L^2}$, where S_{KL} is the scattering operator of the scattering system $\{K, L\}$, P_{L^2} is the orthogonal projection from $\mathfrak{H} \oplus L^2(\mathbb{R}, \mathbb{C}^n)$ onto $L^2(\mathbb{R}, \mathbb{C}^n)$ and ι_{L^2} denotes the canonical embedding. Hence the Lax-Phillips scattering operator S^{LP} is a contraction in $L^2(\mathbb{R}, \mathbb{C}^n)$ and commutes with the selfadjoint differential operator T_0. Therefore S^{LP} is unitarily equivalent to a multiplication operator induced by a family $\{S^{LP}(\lambda)\}_{\lambda\in\mathbb{R}}$ of contractive operators in $L^2(\mathbb{R}, \mathbb{C}^n)$; this family is called the *Lax-Phillips scattering matrix*.

The above considerations together with Proposition 4.4 immediately imply the following corollary on the representation of the Lax-Phillips scattering matrix; cf. [7, Corollary 3.10].

Corollary 4.6. *Let the assumptions of Proposition* 4.4 *be satisfied. If* $\{L, \mathcal{D}_-, \mathcal{D}_+\}$ *is the Lax-Phillips scattering system from above, then the Lax-Phillips scattering matrix* $\{S^{LP}(\lambda)\}_{\lambda\in\mathbb{R}}$ *admits the representation*

$$S^{LP}(\lambda) = I_{\mathbb{C}^n} + 2i\sqrt{-\Im\mathrm{m}\,(D)}\bigl(D - M(\lambda)\bigr)^{-1}\sqrt{-\Im\mathrm{m}\,(D)}$$

for a.e. $\lambda \in \mathbb{R}$, *where* $M(\lambda) = M(\lambda + i0)$.

We mention that Corollary 4.6 also implies a well-known result of Adamyan and Arov in [1, 2, 3, 4] on the relation of the Lax-Phillips scattering matrix with the characteristic function of the maximal dissipative operator B; see [7] for further details.

5. An inverse scattering problem

Let $W : \mathbb{C}_+ \to [\mathbb{C}^n]$ be a contractive analytic matrix function defined on the upper half-plane $\mathbb{C}_+$. Then the limit

$$W(\lambda) = W(\lambda + i0) = \lim_{y\to+0} W(\lambda + iy)$$

exists for a.e. $\lambda \in \mathbb{R}$. In the following theorem we show that under some mild additional conditions the limit of the function W can be regarded as the scattering matrix of a Lax-Phillips scattering system $\{L, \mathcal{D}_-, \mathcal{D}_+\}$, where L is the minimal selfadjoint dilation of some maximal dissipative operator in a Hilbert space $\mathfrak{H}$ as in the previous section.

Theorem 5.1. *Let* $W : \mathbb{C}_+ \to [\mathbb{C}^n]$ *be a contractive analytic function such that the conditions*

$$\ker(I - W(\eta)^*W(\eta)) = \{0\}, \qquad \eta \in \mathbb{C}_+, \tag{5.1}$$

and

$$\lim_{y\to+\infty} \frac{1}{y}(I - W(iy))^{-1} = 0 \tag{5.2}$$

are satisfied. Then the following holds:

(i) *There exist a separable Hilbert space* $\mathfrak{H}$, *a (in general nondensely defined) simple symmetric operator* S *with deficiency indices* (n, n) *in* $\mathfrak{H}$, *a boundary triplet* $\Pi = \{\mathbb{C}^n, \Gamma_0, \Gamma_1\}$ *for* S^* *with Weyl function* $M(\cdot)$ *and a dissipative matrix* $D \in [\mathbb{C}^n]$ *with* $\ker(\Im\mathrm{m}\,(D)) = \{0\}$ *such that* $W(\cdot)$ *admits the representation*

$$W(\mu) = I + 2i\sqrt{-\Im\mathrm{m}\,(D)}\bigl(D - M(\mu)\bigr)^{-1}\sqrt{-\Im\mathrm{m}\,(D)} \tag{5.3}$$

for all $\mu \in \mathbb{C}_+$ *and a.e.* $\mu \in \mathbb{R}$, *where* $W(\mu) = W(\mu + i0)$ *and* $M(\mu) = M(\mu + i0)$.

(ii) *The function* $\mathbb{R} \ni \mu \mapsto W(\mu)$ *is the Lax-Phillips scattering matrix of the Lax-Phillips scattering system* $\{L, \mathcal{D}_-, \mathcal{D}_+\}$, *where* L *is the minimal selfadjoint dilation of the maximal dissipative extension* $B = \ker(\Gamma_1 - D\Gamma_0)$ *of* S *in Theorem* 4.2 *and the incoming and outgoing subspaces are* $\mathcal{D}_\pm = L^2(\mathbb{R}_\pm, \mathbb{C}^n)$.

(iii) *If, in addition, the condition*

$$\lim_{y\to+\infty} y^{1/2}\left\|\sqrt{I-W(iy)^*W(iy)}\,(I-W(iy))^{-1}h\right\| = \infty \tag{5.4}$$

holds for all $h \in \mathcal{H}$, $h \neq 0$, *then the symmetric operator* S *is densely defined.*

(iv) *If* $W(\cdot)$ *is an inner function, i.e.,* $W(\mu + i0)$ *is unitary for a.e.* $\mu \in \mathbb{R}$, *then the spectrum of the selfadjoint operator* $A = \ker(\Gamma_0)$ *is purely singular and the absolutely continuous part* A^{ac} *of* A *is trivial.*

Proof. (i) Observe that condition (5.1) together with $\|W(\eta)\| \leq 1$ implies that $\ker(I - W(\eta)) = \{0\}$ holds for all $\eta \in \mathbb{C}_+$. Indeed, for $x \in \ker(I - W(\eta))$ we have $\|W(\eta)x\| = \|x\|$ and hence

$$((I - W(\eta)^*W(\eta))x, x) = 0$$

which yields $x = 0$. We define a function $M : \mathbb{C}_+ \to [\mathbb{C}^n]$ by

$$\mathbb{C}_+ \ni \eta \mapsto M(\eta) := i(I + W(\eta))(I - W(\eta))^{-1} \tag{5.5}$$

and we extend M to the lower half-plane by $\mathbb{C}_- \ni \eta \mapsto M(\eta) := M(\bar{\eta})^*$. Then M is analytic and a straightforward computation shows

$$\Im\mathrm{m}\,(M(\eta)) = (I - W(\eta)^*)^{-1}(I - W(\eta)^*W(\eta))(I - W(\eta))^{-1} \geq 0. \tag{5.6}$$

for $\eta \in \mathbb{C}_+$. Hence M is a Nevanlinna function and condition (5.1) implies

$$\ker(\Im\mathrm{m}\,(M(\eta))) = \{0\}.$$

From condition (5.2) we obtain

$$\text{s-}\lim_{y\to+\infty} \frac{1}{y}M(iy) = \text{s-}\lim_{y\to+\infty} \frac{i}{y}(I + W(iy))(I - W(iy))^{-1} = 0.$$

By Theorem 2.3 there exists a separable Hilbert space $\mathfrak{H}$, a (in general nondensely defined) simple symmetric operator S with deficiency indices (n, n) and a boundary triplet $\Pi = \{\mathbb{C}^n, \Gamma_0, \Gamma_1\}$ for S^* such that M is the corresponding Weyl function. For $\eta \in \mathbb{C}_+$ we have

$$W(\eta) = I - 2i(M(\eta) + i)^{-1}, \quad \eta \in \mathbb{C}_+. \tag{5.7}$$

Setting $D := -iI$ we have $\sqrt{-\Im\mathrm{m}\,(D)} = I$ and hence the representation (5.3) follows from (5.7).

(ii) From Corollary 4.6 one immediately gets that W can be regarded as the Lax-Phillips scattering matrix of the Lax-Phillips scattering system $\{L, \mathcal{D}_-, \mathcal{D}_+\}$.

(iii) Making use of (5.6) one easily verifies that the condition (5.4) yields

$$\lim_{y\to\infty} y\,\Im\mathrm{m}\,(M(iy)h, h) = \infty, \qquad h \in \mathcal{H} \setminus \{0\}.$$

Hence the operator S is densely defined by Theorem 2.3.

(iv) We consider the analytic function $w(\eta) := \det(I - W(\eta))$, $\eta \in \mathbb{C}_+$. Since the limit $W(\lambda + i0) := \lim_{y\to+0} W(\lambda + iy)$ exists for a.e. $\lambda \in \mathbb{R}$ the limit $w(\lambda + i0) := \lim_{y\to+0} w(\lambda + iy)$ exist for a.e. $\lambda \in \mathbb{R}$, too. If the Lebesgue measure of the set $\{\lambda \in \mathbb{R} : w(\lambda + i0) = 0\}$ is different from zero, then $w(\eta) = 0$ for all

$\eta \in \mathbb{C}_+$ by the Lusin-Privalov theorem [20, Section III] but this is impossible by assumption (5.1), cf. proof of (i). Hence, the set $\{\lambda \in \mathbb{R} : w(\lambda + i0) = 0\}$ has Lebesgue measure zero. Therefore, the operator $(I - W(\lambda + i0))^{-1}$ exists for a.e. $\lambda \in \mathbb{R}$. Using (5.6) we find that $\lim_{y \to +0} \Im m\,(M(\lambda + iy)) = 0$ for a.e. $\lambda \in \mathbb{R}$. By [11, Theorem 4.3(iii)] we get that the selfadjoint operator $A = \ker(\Gamma_0)$ has no absolutely continuous spectrum, i.e., the absolutely continuous part of A is trivial. □

We remark, that the representation (5.3) can also be obtained from [15, Proposition 7.5]. In fact, in the special case considered here some parts of the proof of Theorem 5.1 (i) coincide with the proof of [15, Proposition 7.5].

The Lax-Phillips scattering system and the selfadjoint dilation L in Theorem 5.1 can be made more explicit. Let $W : \mathbb{C}_+ \to [\mathbb{C}^n]$ be as in the assumptions of Theorem 5.1 and define the function M by

$$M(\eta) = i(I + W(\eta))(I - W(\eta))^{-1}, \qquad \eta \in \mathbb{C}_+$$

and $M(\eta) = M(\bar{\eta})^*$ as in (5.5). Then M is $[\mathbb{C}^n]$-valued Nevanlinna function and hence M admits an integral representation of the form

$$M(\eta) = \alpha + \int_{\mathbb{R}} \left(\frac{1}{t - \eta} - \frac{t}{1 + t^2} \right) d\Sigma(t), \qquad \eta \in \mathbb{C} \backslash \mathbb{R}, \tag{5.8}$$

where α is a symmetric matrix and $t \mapsto \Sigma(t)$ is a $[\mathbb{C}^n]$-valued nondecreasing symmetric matrix function on $\mathbb{R}$ such that $\int (1 + t^2)^{-1} d\Sigma(t) \in [\mathbb{C}^n]$. We note that due to condition (ii) in Theorem 5.1 the linear term in the integral representation (5.8) is absent. Let $L^2_\Sigma(\mathbb{R}, \mathbb{C}^n)$ be the Hilbert space of $\mathbb{C}^n$-valued functions as in [10, 18, 24]. It was shown in [24] that the mapping

$$f \mapsto \int_{\mathbb{R}} d\Sigma(t) f(t)$$

defined originally on the space $C_0(\mathbb{R}, \mathbb{C}^n)$ of (strongly) continuous functions with compact support admits a continuous extension to an operator from the domain $\mathrm{dom}\,(Q_\Sigma)$ of the multiplication operator $Q_\Sigma : f \longrightarrow tf$ in $L2_\Sigma(\mathbb{R}, \mathbb{C}^n)$ into $\mathbb{C}^n$. According to [15, 24] the adjoint of the (in general nondensely defined) closed symmetric operator

$$\begin{aligned} &(Sf)(t) = tf(t), \\ &\mathrm{dom}\,(S) = \left\{ f \in L^2_\Sigma(\mathbb{R}, \mathbb{C}^n) : tf(t) \in L^2_\Sigma(\mathbb{R}, \mathbb{C}^n), \int_{\mathbb{R}} d\Sigma(t) f(t) = 0 \right\}, \end{aligned}$$

is given by the linear relation

$$S^* = \left\{ \left\{ f(t) + \frac{t}{1 + t^2} h, tf(t) - \frac{1}{1 + t^2} h \right\} : f(t), tf(t) \in L^2_\Sigma(\mathbb{R}, \mathbb{C}^n),\, h \in \mathbb{C}^n \right\},$$

and that $\{\mathbb{C}^n, \Gamma_0, \Gamma_1\}$, where

$$\Gamma_0 \hat{f} := h \qquad \text{and} \qquad \Gamma_1 \hat{f} := \alpha h + \int_{\mathbb{R}} d\Sigma(t) f(t),$$

$\hat{f} = \{f(t) + t(1+t^2)^{-1}h, tf(t) - (1+t^2)^{-1}h\} \in S^*$, is a boundary triplet for S^* with corresponding Weyl function $M(\cdot)$. Note that here $A_0 = \ker(\Gamma_0)$ is the usual maximal multiplication operator in $L^2_\Sigma(\mathbb{R}, \mathbb{C}^n)$.

Corollary 5.2. *Let $W : \mathbb{C}_+ \to [\mathbb{C}^n]$ be a contractive analytic function which satisfies the conditions* (5.1) *and* (5.2) *in Theorem* 5.1. *Then there exists a symmetric matrix $\alpha \in [\mathbb{C}^n]$ and a $[\mathbb{C}^n]$-valued nondecreasing symmetric matrix function $\Sigma(\cdot)$ on $\mathbb{R}$ such that*

$$W(\mu) = I - 2i\left(\alpha + i + \int_{\mathbb{R}} \left(\frac{1}{t-\mu} - \frac{t}{1+t^2}\right) d\Sigma(t)\right)^{-1}$$

holds for all $\mu \in \mathbb{C}_+$ and $\int (1+t^2)^{-1} d\Sigma(t) \in [\mathbb{C}^n]$. The function $W(\cdot)$ coincides with the Lax-Phillips scattering matrix of the system $\{L, \mathcal{D}_-, \mathcal{D}_+\}$, where $\mathcal{D}_\pm = L^2(\mathbb{R}_\pm, \mathbb{C}^n)$ and

$$L\begin{pmatrix} f(t) + \frac{t}{1+t^2}h \\ g \end{pmatrix} = \begin{pmatrix} tf(t) - \frac{1}{1+t^2}h \\ ig' \end{pmatrix}, \qquad \begin{array}{l} f(t), tf(t) \in L^2_\Sigma(\mathbb{R}, \mathbb{C}^n),\ h \in \mathbb{C}^n, \\ g \in W^1_2(\mathbb{R}_-, \mathbb{C}^n) \oplus W^1_2(\mathbb{R}_+, \mathbb{C}^n), \end{array}$$

$$\mathrm{dom}\,(L) = \left\{ \begin{pmatrix} f(t) + \frac{t}{1+t^2}h \\ g \end{pmatrix} : \begin{array}{c} \frac{1}{\sqrt{2}}(g(0+) - g(0-)) = h \\ \frac{i}{\sqrt{2}}(g(0+) + g(0-)) = \alpha h + \int d\Sigma(t) f(t) \end{array} \right\}$$

is the minimal selfadjoint dilation in $L^2_\Sigma(\mathbb{R}, \mathbb{C}^n) \oplus L^2(\mathbb{R}, \mathbb{C}^n)$ of the maximal dissipative multiplication operator $B = \ker(\Gamma_1 + i\Gamma_0)$ in $L^2_\Sigma(\mathbb{R}, \mathbb{C}^n)$.

Appendix A. Spectral representations and scattering matrices

Let A be a selfadjoint operator in the separable Hilbert space $\mathfrak{H}$ and let $E(\cdot)$ be the corresponding spectral measure defined on the σ-algebra $\mathcal{B}(\mathbb{R})$ of Borel subsets of $\mathbb{R}$. The absolutely continuous and singular part of the measure $E(\cdot)$ is denoted by $E^{ac}(\cdot)$ and $E^s(\cdot)$, respectively. If $C \in [\mathcal{H}, \mathfrak{H}]$ is a Hilbert-Schmidt operator, then by [6, Lemma I.11]

$$\Sigma(\delta) := C^* E(\delta) C, \qquad \delta \in \mathcal{B}(\mathbb{R}),$$

is a trace class valued measure on $\mathcal{B}(\mathbb{R})$ of finite variation. This measure admits a unique decomposition

$$\Sigma(\cdot) = \Sigma^s(\cdot) + \Sigma^{ac}(\cdot)$$

into a singular measure $\Sigma^s(\cdot) = C^* E^s(\cdot) C$ and an absolutely continuous measure $\Sigma^{ac}(\cdot) = C^* E^{ac}(\cdot) C$. According to [6, Proposition I.13] the trace class valued function $\lambda \mapsto \Sigma(\lambda) := C^* E((-\infty, \lambda)) C$ admits a derivative $K(\lambda) := \frac{d}{d\lambda}\Sigma(\lambda) \geq 0$ in the trace class norm for a.e. $\lambda \in \mathbb{R}$ with respect to the Lebesgue measure $d\lambda$ and

$$\Sigma^{ac}(\delta) = \int_\delta K(\lambda) d\lambda, \quad \delta \in \mathcal{B}(\mathbb{R}),$$

holds. By $\mathcal{H}_\lambda := \overline{\mathrm{ran}\,(K(\lambda))} \subseteq \mathcal{H}$ we define a measurable family of subspaces in $\mathcal{H}$. Let $P(\lambda)$ be the orthogonal projection from $\mathcal{H}$ onto $\mathcal{H}_\lambda$ and define a measurable family of projections by

$$(Pf)(\lambda) := P(\lambda)f(\lambda), \qquad f \in L^2(\mathbb{R}, d\lambda, \mathcal{H}).$$

Then P is an orthogonal projection in $L^2(\mathbb{R}, d\lambda, \mathcal{H})$ and we denote the range of P by $L^2(\mathbb{R}, d\lambda, \mathcal{H}_\lambda)$. In the following we regard $L^2(\mathbb{R}, d\lambda, \mathcal{H}_\lambda)$ as the direct integral of the measurable family of subspaces $\{\mathcal{H}_\lambda\}_{\lambda\in\mathbb{R}}$.

Lemma A.1. *Let A, E, C and $K(\lambda)$ be as above and assume that the absolutely continuous subspace $\mathfrak{H}^{ac}(A)$ satisfies the condition*

$$\mathfrak{H}^{ac}(A) = \mathrm{clospan}\{E^{ac}(\delta)\mathrm{ran}\,(C) : \delta \in \mathcal{B}(\mathbb{R})\}.$$

Then the linear extension of the mapping

$$E^{ac}(\delta)Cf \mapsto \chi_\delta(\lambda)\sqrt{K(\lambda)}f \ \textit{ for a.e. } \lambda \in \mathbb{R}, \qquad f \in \mathcal{H}, \tag{A.1}$$

onto the dense subspace $\mathrm{span}\{E^{ac}(\delta)\mathrm{ran}\,(C) : \delta \in \mathcal{B}(\mathbb{R})\}$ *of* $\mathfrak{H}^{ac}(A)$ *admits a unique continuation to an isometric isomorphism from $\Phi : \mathfrak{H}^{ac}(E) \to L^2(\mathbb{R}, d\lambda, \mathcal{H}_\lambda)$ such that*

$$(\Phi E^{ac}(\delta)g)(\lambda) = \chi_\delta(\lambda)(\Phi g)(\lambda), \qquad g \in \mathfrak{H}^{ac}(A), \tag{A.2}$$

holds for any $\delta \in \mathcal{B}(\mathbb{R})$.

Proof. For $f \in \mathcal{H}$ and $\delta \in \mathcal{B}(\mathbb{R})$ we have

$$\|\chi_\delta(\cdot)\sqrt{K(\cdot)}f\|^2 = \int_\delta \|\sqrt{K(\lambda)}f\|^2_{\mathcal{H}} d\lambda = \|E^{ac}(\delta)Cf\|^2_{\mathfrak{H}}$$

and hence the extension of the mapping (A.1) onto the subspace

$$\mathrm{span}\{E^{ac}(\delta)\mathrm{ran}\,(C) : \delta \in \mathcal{B}(\mathbb{R})\}$$

is an isometry into $L^2(\mathbb{R}, d\lambda, \mathcal{H}_\lambda)$. Then the unique extension $\Phi : \mathfrak{H}^{ac}(A) \to L^2(\mathbb{R}, d\lambda, \mathcal{H}_\lambda)$ is isometric and it remains to show that Φ is onto. Suppose that there exists $h \in L^2(\mathbb{R}, d\lambda, \mathcal{H}_\lambda)$ such that

$$0 = (\Phi E^{ac}(\delta)Cf, h) = \int_\delta (\sqrt{K(\lambda)}f, h(\lambda))_{\mathcal{H}}\, d\lambda$$

holds for all $\delta \in \mathcal{B}(\mathbb{R})$ and $f \in \mathcal{H}$. This implies $(\sqrt{K(\lambda)}f, h(\lambda))_{\mathfrak{H}} = 0$ for a.e. $\lambda \in \mathbb{R}$ and hence $h(\lambda) \perp \mathcal{H}_\lambda$ for a.e. $\lambda \in \mathbb{R}$, thus $h(\lambda) = 0$ for a.e. $\lambda \in \mathbb{R}$. Hence Φ is surjective. The relation (A.2) for Φ follows from (A.1). □

From (A.2) we immediately get that the maximal multiplication operator $\mathcal{Q}$ in $L^2(\mathbb{R}, d\lambda, \mathcal{H}_\lambda)$,

$$\begin{aligned} (\mathcal{Q}f)(\lambda) &:= \lambda f(\lambda), \\ f \in \mathrm{dom}\,(\mathcal{Q}) &:= \{f \in L^2(\mathbb{R}, d\lambda, \mathcal{H}_\lambda) : \lambda f(\lambda) \in L^2(\mathbb{R}, d\lambda, \mathcal{H}_\lambda)\}. \end{aligned}$$

satisfies $\mathcal{Q}\Phi = \Phi A^{ac}$ and $\varphi(\mathcal{Q})\Phi = \Phi\varphi(A^{ac})$ for any bounded Borel measurable function $\varphi(\cdot) : \mathbb{R} \longrightarrow \mathbb{R}$. In other words, the direct integral $L^2(\mathbb{R}, d\lambda, \mathcal{H}_\lambda)$ forms a spectral representation of the absolutely continuous part A^{ac} of the selfadjoint operator A.

Suppose now that B is also a selfadjoint operator in $\mathfrak{H}$ and assume that the difference of the resolvents

$$(B-i)^{-1}-(A-i)^{-1}$$

is a trace class operator. Then the wave operators

$$W_\pm(B,A):=s-\lim_{t\to\pm\infty} e^{itB}e^{-itA}P^{ac}(A)$$

exist and are complete; cf., e.g., [6, Theorem I.1]. The scattering operator $S_{AB}:=W_+(B,A)^*W_-(B,A)$ regarded as an operator in $\mathfrak{H}^{ac}(A)$ is unitary and commutes with A. Therefore there is a measurable family $\{S_{AB}(\lambda)\}_{\lambda\in\mathbb{R}}$ of unitary operators $S_{AB}(\lambda)\in[\mathcal{H}_\lambda]$ such that S_{AB} is unitarily equivalent to the multiplication operator $\mathcal{S}_{AB}$ induced by $\{S_{AB}(\lambda)\}_{\lambda\in\mathbb{R}}$ in $L^2(\mathbb{R},d\lambda,\mathcal{H}_\lambda)$, that is, $\mathcal{S}_{AB}=\Phi S_{AB}\Phi^{-1}$. The measurable family $\{S_{AB}(\lambda)\}_{\lambda\in\mathbb{R}}$ is called the scattering matrix of the scattering system $\{A,B\}$.

The following theorem on the representation of the scattering matrix is an important ingredient in the proof of Theorem 3.1. A detailed proof of Theorem A.2 will appear in a forthcoming paper.

Theorem A.2. *Let A and B be selfadjoint operators in the separable Hilbert space $\mathfrak{H}$ and suppose that the resolvent difference admits the factorization*

$$(B-i)^{-1}-(A-i)^{-1}=\phi(A)CGC^*=QC^*,$$

where $C\in[\mathcal{H},\mathfrak{H}]$ is a Hilbert-Schmidt operator, $G\in[\mathcal{H}]$, $\phi(\cdot):\mathbb{R}\to\mathbb{R}$ is a bounded continuous function and $Q=\phi(A)CG$. Assume that the condition

$$\mathfrak{H}^{ac}(A)=\operatorname{clospan}\{E^{ac}(\delta)\operatorname{ran}(C):\delta\in\mathcal{B}(\mathbb{R})\} \tag{A.3}$$

is satisfied and let $K(\lambda)=\frac{d}{d\lambda}C^*E((-\infty,\lambda))C$ and $\mathcal{H}_\lambda=\overline{\operatorname{ran}(K(\lambda))}$ for a.e. $\lambda\in\mathbb{R}$.

Then $L^2(\mathbb{R},d\lambda,\mathcal{H}_\lambda)$ is a spectral representation of A^{ac} and the scattering matrix $\{S_{AB}(\lambda)\}_{\lambda\in\mathbb{R}}$ of the scattering system $\{A,B\}$ has the representation

$$S_{AB}(\lambda)=I_{\mathcal{H}_\lambda}+2\pi i(1+\lambda^2)^2\sqrt{K(\lambda)}Z(\lambda)\sqrt{K(\lambda)}\in[\mathcal{H}_\lambda] \tag{A.4}$$

for a.e. $\lambda\in\mathbb{R}$, where

$$Z(\lambda)=\frac{1}{\lambda+i}Q^*Q+\frac{\phi(\lambda)}{(\lambda+i)^2}G+Q_B(\lambda+i0) \tag{A.5}$$

and the limit $Q_B(\lambda+i0):=\lim_{\varepsilon\to+0}Q^*(B-\lambda-i\varepsilon)^{-1}Q$ is taken in the Hilbert-Schmidt norm.

Acknowledgment

The second author thanks the Weierstrass Institute of Applied Analysis and Stochastics in Berlin for financial support and hospitality.

References

[1] V.M. Adamjan and D.Z. Arov, *On a class of scattering operators and characteristic operator-functions of contractions*, Dokl. Akad. Nauk SSSR 160 (1965), 9–12.

[2] V.M. Adamjan and D.Z. Arov, *On scattering operators and contraction semigroups in Hilbert space*, Dokl. Akad. Nauk SSSR 165 (1965), 9–12.

[3] V.M. Adamjan and D.Z. Arov, *Unitary couplings of semi-unitary operators*, Akad. Nauk Armjan. SSR Dokl. 43 (1966) no. 5, 257–263.

[4] V.M. Adamjan and D.Z. Arov, *Unitary couplings of semi-unitary operators*, Mat. Issled. 1 (1966) vyp. 2, 3–64.

[5] V.M. Adamyan and B.S. Pavlov, *Null-range potentials and M.G. Krein's formula for generalized resolvents*, Zap. Nauchn. Semin. Leningr. Otd. Mat. Inst. Steklova 149 (1986) 7–23 (Russian); translation in J. Sov. Math. 42 no.2 (1988) 1537–1550.

[6] H. Baumgärtel and M. Wollenberg, *Mathematical Scattering Theory*, Akademie-Verlag, Berlin, 1983.

[7] J. Behrndt, M.M. Malamud, and H. Neidhardt, *Scattering theory for open quantum systems with finite rank coupling*, Math. Phys. Anal. Geom. 10 (2007), 313–358.

[8] J. Behrndt, M.M. Malamud, and H. Neidhardt, *Scattering matrices and Weyl functions*, Proc. London Math. Soc. 97 (2008), 568–598.

[9] J. Behrndt, M.M. Malamud, and H. Neidhardt, *Trace formulae for dissipative and coupled scattering systems*, Oper. Theory Adv. Appl. 188 (2008), 49–85.

[10] Ju.M. Berezans'kiĭ, *Expansions in Eigenfunctions of Selfadjoint Operators*, AMS Translations of Mathematical Monographs Vol. 17, Providence, R.I., 1968.

[11] J.F. Brasche, M.M. Malamud, and H. Neidhardt, *Weyl function and spectral properties of self-adjoint extensions*, Integral Equations Oper. Theory 43 (2002), 264–289.

[12] J. Brüning, V. Geyler, and K. Pankrashkin, *Spectra of self-adjoint extensions and applications to solvable Schrödinger operators*, Rev. Math. Phys. 20 (2008), 1–70.

[13] V.A. Derkach and M.M. Malamud, *On the Weyl function and Hermitian operators with gaps*, Sov. Math. Dokl. 35 (1987), 393–398.

[14] V.A. Derkach and M.M. Malamud, *Generalized resolvents and the boundary value problems for Hermitian operators with gaps*, J. Funct. Anal. 95 (1991), 1–95.

[15] V.A. Derkach and M.M. Malamud, *The extension theory of Hermitian operators and the moment problem*, J. Math. Sci. 73 (1995), 141–242.

[16] W.F. Donoghue, *Monotone Matrix Functions and Analytic Continuation*, Springer, Berlin-New York, 1974.

[17] J.B. Garnett, *Bounded Analytic Functions*, Academic Press, New York-London, 1981.

[18] V.I. Gorbachuk and M.L. Gorbachuk, *Boundary Value Problems for Operator Differential Equations*, Mathematics and its Applications (Soviet Series) 48, Kluwer Academic Publishers Group, Dordrecht, 1991.

[19] T. Kato, *Perturbation Theory for Linear Operators*, Grundlehren der Mathematischen Wissenschaften, Band 132, Springer, Berlin-New York, 1976.

[20] P. Koosis, *Introduction to H_p spaces.* LMS Lecture Note Series Vol. 40, Cambridge University Press, Cambridge, 1980.

[21] H. Langer and B. Textorius, *On generalized resolvents and Q-functions of symmetric linear relations (subspaces) in Hilbert space*, Pacific J. Math. 72 (1977), 135–165.

[22] P.D. Lax and R.S. Phillips, *Scattering Theory*, Academic Press, New York, 1967.

[23] M.M. Malamud, *On a formula for the generalized resolvents of a non-densely defined Hermitian operator*, Ukraïn. Mat. Zh. 44 (1992), 1658–1688.

[24] M.M. Malamud and S. M. Malamud, *Spectral theory of operator measures in a Hilbert space*, St. Petersburg Math. J. 15 (2004), 323–373.

[25] B.S. Pavlov, *Dilation theory and spectral analysis of nonselfadjoint differential operators*, In *Mathematical programming and related questions, Theory of operators in linear spaces (Russian)*, pages 3–69, Central. Èkonom. Mat. Inst. Akad. Nauk SSSR, Moscow, 1976.

[26] B.S. Pavlov, *Spectral analysis of a dissipative singular Schrödinger operator in terms of a functional model*, In *Partial differential equations, VIII*, volume 65 of Encyclopaedia Math. Sci., pages 87–153, Springer, Berlin, 1996.

[27] B. Sz.-Nagy and C. Foiaş, *Harmonic Analysis of Operators on Hilbert Space*, North-Holland Publishing Co., Amsterdam, 1970.

[28] J. Weidmann, *Lineare Operatoren in Hilberträumen. Teil II: Anwendungen*, B.G. Teubner, Stuttgart, 2003.

[29] D.R. Yafaev, *Mathematical Scattering Theory: General Theory*, AMS Translations of Mathematical Monographs Vol. 105, Providence, RI, 1992.

Jussi Behrndt
Technische Universität Berlin
Institut für Mathematik
MA 6–4, Straße des 17. Juni 136
D–10623 Berlin, Germany
e-mail: `behrndt@math.tu-berlin.de`

Mark M. Malamud
Donetsk National University
Department of Mathematics
Universitetskaya 24
83055 Donetsk, Ukraine
e-mail: `mmm@telenet.dn.ua`

Hagen Neidhardt
Weierstraß-Institut für
Angewandte Analysis und Stochastik
Mohrenstr. 39
D-10117 Berlin, Germany
e-mail: `neidhardt@wias-berlin.de`

Operator Theory:
Advances and Applications, Vol. 198, 87–95

Conditions for an Indefinite Sturm-Liouville Riesz Basis Property

Paul Binding and Andreas Fleige

To the memory of Peter Jonas

Abstract. We consider the Sturm-Liouville problem $-y'' = \lambda ry$ on $[-1,1]$ with Dirichlet boundary conditions and with an indefinite weight function r which changes sign at 0. We discuss several conditions known to be either necessary or sufficient for the eigenfunctions to form a Riesz basis of the Hilbert space $L_{2,|r|}(-1,1)$. Assuming that the odd part of r dominates the even part in a certain sense, we show that the above conditions (and also some new ones) are in fact all equivalent to this Riesz basis property.

Mathematics Subject Classification (2000). Primary 34B09, 34B24; Secondary 34L10.

Keywords. Indefinite Sturm-Liouville problem, Riesz basis.

1. Introduction

We shall compare, and in certain ways extend, some of the conditions in the literature on the Riesz Basis Property (RBP) for the Sturm-Liouville problem

$$-y'' = \lambda ry, \qquad y(-1) = y(1) = 0, \tag{1.1}$$

where $r \in L_\infty(-1,1)$. Then (1.1) has a countable set of simple eigenvalues, and the RBP means that the corresponding eigenfunctions can be normalised to form a Riesz basis [7] of the $|r|$-weighted Hilbert space $L_{2,|r|}(-1,1)$. We remark that we are adhering to the above setting (which could be generalised in various ways) since it is common ground for the literature under study.

For r of one sign, this problem goes back at least to the early 1800s, but we shall throughout assume that r is indefinite, specifically

$$xr(x) > 0 \quad \text{for} \quad \text{a.e. } x \in (-1,1). \tag{1.2}$$

Investigation of the corresponding RBP developed in the second half of the 1900s, motivated by problems in mathematical physics and probability leading to questions of full- and half- (or partial-) range completeness. We cite the work of Case (e.g., [3]) as seminal for some of the earlier work, that of Beals (e.g., [2]) as influencing many of the more recent contributions and three publications [4, 9, 10] which contain various explicit conditions equivalent to the RBP. By nature, such precise conditions are often difficult to check in specific cases, and attention has turned to more direct conditions which are either necessary or sufficient.

For example, Pyatkov [9, Theorem 2.2] gave a sufficient condition involving interpolation spaces and this was later reworked by Parfenov [8, eqn (7)] into an equivalent but more direct sufficient condition, which we shall label P, involving integrals of $|r|$ over subsets of $(-1,1)$ – see Section 2, Condition (b) et seq. A necessary condition, which we shall label V, was given by Volkmer [10, Equation (4.3)] as an inequality connecting three integrals depending on $|r|$ and functions h from a certain class D – see (3.1).

Volkmer employed condition V to show the existence of an odd weight function r without the RBP for (1.1), and V has been adapted by subsequent authors, e.g., in [1, 5], to give specific examples of such weights and also necessary conditions for the RBP depending more directly on r, again assumed to be odd. A further aspect of the problem was revealed by Parfenov [8] who showed (for odd weights r) that the RBP was actually equivalent to a one-sided version of P, to a necessary condition in [1], and to a new intermediate condition. This was complemented by [6] where a condition for 'non-odd' weights r was used to give equivalence of P and the RBP.

Here we shall continue the above investigations by extending all the results cited in the previous two paragraphs. For example, we weaken the non-oddness condition from [6] to (three forms of) 'odd-domination' (OD), where the odd part of r dominates the even part in some sense. We use these to prove equivalence of various conditions, some previously known only to be either necessary or sufficient for the RBP for odd r. In Section 2 we state precisely most of the conditions to be studied, and we give some simple relations between them. Section 3 contains some explicit estimates of the integrals used to define Condition V, for a particular sequence h_n from the class D above. Our principal results are in Section 4. In Theorem 4.1, some of the conditions in the literature are shown to be equivalent under two OD conditions (the weaker one being almost necessary for the single turning point condition (1.2) to hold). Under a stronger OD condition in Theorem 4.2, all of the conditions studied here, in particular conditions P and V, are shown to be equivalent to the RBP.

2. Formulation of conditions

For $f:(-1,1)\to\mathbb{R}$, we write

$$f^{\pm}(x)=\pm f(\pm x)\quad\text{for}\quad x\in(0,1)$$

so

$$f^e = \frac{1}{2}\left(f^+ - f^-\right) \quad \text{and} \quad f^o = \frac{1}{2}\left(f^+ + f^-\right) \tag{2.1}$$

are the even and odd parts of f on $(0,1)$.

We may motivate 'odd-domination' as follows. We note from (1.2) that $r^o \pm r^e = r^\pm$ are both positive. Indeed (1.2) is equivalent to

$$r^o > |r^e| \geq 0 \tag{2.2}$$

which shows that some form of OD is inherent in our standing assumption (1.2). Indeed, it follows from (2.2) that a function ρ exists satisfying $\rho(\varepsilon) \leq 1$ for all $\varepsilon \in (0,1]$ and

$$\int_0^x |r^e| \leq \rho(\varepsilon) \int_0^x r^o \quad \text{for all} \quad x \in (0,\varepsilon). \tag{2.3}$$

Developing this, we call r *weakly odd-dominated* **(Condition WOD)** if there is ρ satisfying (2.3) for all $\varepsilon \in (0,1]$ and such that

$$\rho(\varepsilon_0) < 1 \quad \text{for some} \quad \varepsilon_0 \in (0,1]. \tag{2.4}$$

From the above, this condition is 'almost' necessary for (1.2), and moreover it is easily seen that WOD implies the existence of a function ρ with (2.3) and $\rho(\varepsilon) < 1$ for *all* $\varepsilon \in (0,1]$. (Take, e.g., $\max\left(\rho(\varepsilon_0),\, (\int_0^\varepsilon r^o)^{-1}(\rho(\varepsilon_0)\int_0^{\varepsilon_0} r^o + \int_{\varepsilon_0}^\varepsilon r^o)\right)$ where ρ and ε_0 satisfy (2.3) and (2.4).)

We shall also consider stronger forms of this condition: r is *odd-dominated* **(Condition OD)** if ρ can be chosen to satisfy (2.3) and

$$\rho(\varepsilon) \to 0 \quad \text{as} \quad \varepsilon \to 0$$

and r is *strongly odd-dominated* **(Condition SOD)** if in addition

$$\rho(\varepsilon) = o\left(\varepsilon^{1/2}\right) \quad \text{as} \quad \varepsilon \to 0.$$

These conditions obviously include odd weight functions r, and may be compared with the 'non-odd' condition of [6, Equation (3.5)], viz.,

$$\|r^e\|_\varepsilon := \left(\int_0^\varepsilon |r^e|^2\right)^{1/2} \leq F(\varepsilon)\, I_\varepsilon \quad \text{for all} \quad \varepsilon \in (0,1], \tag{F}$$

for some continuous F satisfying $F(0) = 0$. Here and below we write

$$I_\varepsilon := \int_{-\varepsilon}^{\varepsilon} |r|. \tag{2.5}$$

The next result shows that all the above OD conditions are weaker than (F).

Proposition 2.1. (F) *implies SOD.*

Proof. Suppose (F) holds. Then by Cauchy-Schwartz and (2.1),

$$\int_0^\varepsilon |r^e| \le \|1\|_\varepsilon \|r^e\|_\varepsilon \le \varepsilon^{1/2} F(\varepsilon) \int_0^\varepsilon (r^- + r^+)$$
$$= \varepsilon^{1/2} F(\varepsilon) \int_0^\varepsilon 2r^o = o(\varepsilon^{1/2}) \int_0^\varepsilon r^o. \qquad \square$$

We turn now to some of the conditions that will be compared in Section 4. Parfenov [8, Theorem 6] lists four conditions, (a) being the RBP, that are equivalent for odd weight functions. For versions of the others, we use (2.5) and the analogous notation

$$I_\varepsilon^\pm := \int_0^\varepsilon r^\pm.$$

Condition (b): For some $\gamma, c > 0$, $0 < \eta \le \varepsilon \le 1$ implies

$$I_\eta \le c \left(\frac{\eta}{\varepsilon}\right)^\gamma I_\varepsilon.$$

We write (b)$^\pm$ for the corresponding conditions involving $I_\eta^\pm$ and $I_\varepsilon^\pm$. Parfenov's (b) is our (b)$^+$, and **Condition P** of Section 1 is our (b) but with I_η replaced by $\min\{I_\eta^-, I_\eta^+\}$. Parfenov showed [8, Theorem 3] that P was equivalent to an earlier condition which Pyatkov [9, Theorem 2.2] had shown to be sufficient for the RBP. In [6, Theorem 3.2], equivalence of P with the RBP was proved under condition (F) above.

Condition (c): For some $\mu \in (0, 1)$,

$$I_{\mu\varepsilon} \le I_\varepsilon / 2 \text{ for all } \varepsilon \in (0, 1).$$

We define (c)$^\pm$ analogously. Parfenov's (c) is our (c)$^+$.

Condition (d): There are no sequences a_n, b_n satisfying $0 < a_n < b_n \le 1$,

$$a_n / b_n \to 0 \quad \text{and} \quad I_{a_n} / I_{b_n} \to 1.$$

Again we define (d)$^\pm$ analogously, and Parfenov's (d) is our (d)$^+$. For odd weight functions r, this condition was shown in [1, Corollary 1] to be necessary for the RBP, and in [8, Theorem 6] this was improved to equivalence (also with (b)$^+$ and (c)$^+$).

The remaining RBP conditions that we consider will be defined in Section 3.

3. Volkmer's inequality

Volkmer's inequality [10, eqn (4.3)] takes the form

$$\left(\int_{-1}^1 |r|\,|g|^2\right)^2 \le K \int_{-1}^1 |h|^2 \int_{-1}^1 |g'|^2 \tag{3.1}$$

for some (fixed) $K > 0$ for all h from the class D of absolutely continuous functions on $[-1, 1]$ such that $g := h'/r$ is absolutely continuous on $[-1, 1]$ with

$$g(\pm 1) = 0 \quad \text{and} \quad g' \in L_2(-1, 1).$$

Volkmer [10, Section 4] showed that this condition, which we label **Condition V**, was necessary for the RBP for (1.1).

The aim of this section is to estimate the integrals in (3.1), for h from a particular sequence h_n. Given sequences a_n and b_n satisfying $0 < a_n < b_n < 1$, we take

$$g_n'(x) = \begin{cases} (a_n - b_n)^{-1} \operatorname{sgn} x & \text{if} \quad |x| \in (a_n, b_n) \\ 0 & \text{otherwise on } (-1,1) \end{cases}$$

with $g_n(-1) = 0$ (so $g_n(1) = 0$) and $h_n(x) = \int_{-1}^{x} r g_n$. Hence, g_n is a piecewise linear function vanishing outside $[-b_n, b_n]$ and it is constant 1 on $[-a_n, a_n]$. Related sequences have been employed in [1, 6, 8]. We note that $g = g_n$, $h = h_n$ satisfy the conditions required for V, so if r satisfies V then

$$J_n := \frac{\left(\int_{-1}^{1} |r|\, |g_n|^2\right)^2}{\int_{-1}^{1} |h_n|^2 \int_{-1}^{1} |g_n'|^2} \leq K. \tag{3.2}$$

We shall refer to (3.2) as **Condition $\mathbf{V_n}$**, K being independent of the sequences a_n and b_n as above.

Our first estimates are elementary.

Lemma 3.1.

$$\int_{-1}^{1} |g_n'|^2 = 2\,(b_n - a_n)^{-1} \quad \textit{and} \quad \int_{-1}^{1} |r|\, |g_n|^2 \geq A_n,$$

where

$$A_n := I_{a_n}^{-} = \int_0^{a_n} r^{-}.$$

Proof. The first statement is clear and for the second we use

$$0 \leq g_n \leq 1 \tag{3.3}$$

and $g_n = 1$ on $(-a_n, 0)$. □

The other integral is more involved, and we shall use the inequality

$$2 \int_0^x |r^e| \leq \delta_n \int_0^x r^- \quad \text{for all} \quad x \in (0, b_n) \tag{3.4}$$

as an aid. It will be seen in Section 4 that such numbers δ_n always exist under WOD.

Lemma 3.2. *Assume that* (3.4) *is satisfied for some* $\delta_n > 0$. *With* $B_n := I_{b_n}^{-}$ $(= \int_0^{b_n} r^-)$ *and* $D_n := B_n - A_n$ $(= \int_{a_n}^{b_n} r^-)$, *we have*

(i)	$\|h_n(x)\| = 0$	*if* $x \in (-1, -b_n)$
(ii)	$\|h_n(x)\| \leq D_n$	*if* $x \in (-b_n, -a_n)$
(iii)	$\|h_n(x)\| \leq B_n$	*if* $x \in (-a_n, 0)$
(iv)	$\|h_n(x)\| \leq \|h_n(-x)\| + \delta_n B_n$	*if* $x \in (0, 1)$.

Proof. (i) follows from $g = 0$ on $(-1, -b_n)$, and (ii), (iii) from (3.3). For (iv), we have

$$|h_n(x)| = \left| h_n(-x) + \int_0^x (r^+ - r^-)\, g_n \right| \le |h_n(-x)| + 2\int_0^x |r^e|\, g_n$$

since g_n is an even function. Thus it suffices to show that

$$2\int_0^x |r^e|\, g_n \le \delta_n B_n \text{ if } x \in (0,1). \tag{3.5}$$

We have for $x \in (0, b_n)$ by (3.3) and (3.4)

$$2\int_0^x |r^e|\, g_n \le 2\int_0^x |r^e| \le \delta_n \int_0^x r^- \le \delta_n \int_0^{b_n} r^- \le \delta_n B_n.$$

Since this is also true for $x \in (b_n, 1)$ where g_n vanishes, the proof of (3.5) is complete. □

We are now ready to estimate the third integral used to define J_n.

Corollary 3.3. *Under condition* (3.4) *we have*

$$\int_{-1}^1 |h_n|^2 \le 3\left(b_n D_n^2 + a_n B_n^2\right) + 2\delta_n^2 B_n^2.$$

Proof. From Lemma 3.2 (iv),

$$\int_{-1}^1 |h_n|^2 \le \int_{-1}^0 |h_n|^2 + \int_0^1 \left(2|h_n(-x)|^2 + 2\delta_n^2 B_n^2\right) dx = 3\int_{-1}^0 |h_n|^2 + 2\delta_n^2 B_n^2$$

and the result now follows from Lemma 3.2 (i)–(iii). □

4. Equivalences

Before we state our main results we first give some useful estimates. Recall that WOD implies that ρ can be chosen to satisfy (2.3) and $\rho(\varepsilon) < 1$ for all $\varepsilon \in (0,1]$.

Lemma 4.1. *Condition WOD implies*

$$\int_0^x r^+ \le \frac{1+\rho(\varepsilon)}{1-\rho(\varepsilon)} \int_0^x r^-, \tag{4.1}$$

$$\int_0^x r^- \le \frac{1+\rho(\varepsilon)}{1-\rho(\varepsilon)} \int_0^x r^+, \tag{4.2}$$

$$\int_0^x |r^e| \le \frac{\rho(\varepsilon)}{1-\rho(\varepsilon)} \int_0^x r^- \tag{4.3}$$

for all $x \in (0, \varepsilon)$ *and all* $\varepsilon \in (0,1]$.

Proof. (i) From (2.1) and (2.3) we obtain

$$\int_0^x (r^+ - r^-) \leq \int_0^x |r^+ - r^-| \leq \rho(\varepsilon) \int_0^x (r^+ + r^-) . \tag{4.4}$$

This implies (4.1) since $\rho(\varepsilon) < 1$ by WOD. Similarly, using $\int_0^x (r^- - r^+) \leq \int_0^x |r^+ - r^-|$ in (4.4) we obtain (4.2). From (2.1), (4.1) and (4.4) we have

$$2 \int_0^x |r^e| \leq \rho(\varepsilon) \int_0^x (r^+ + r^-) \leq \frac{2\rho(\varepsilon)}{1-\rho(\varepsilon)} \int_0^x r^-,$$

i.e., (4.3). □

Our results below will show how the degree of odd domination influences which RBP conditions are equivalent to each other. We start with the following conditions, defined in Section 2.

Theorem 4.2.

(i) *$P, (b), (b)^+$ and $(b)^-$ are all equivalent under WOD.*

(ii) *The conditions in* (i) *are also equivalent to* (c), (d) *and their* $^+$ *and* $^-$ *counterparts under OD.*

Proof. For two functions f_1, f_2 on $(0,1)$, write $f_1 \sim f_2$ if there exist numbers $a, \varepsilon > 0$ such that

$$af_1(x) \leq f_2(x) \leq a^{-1} f_1(x)$$

holds for all $x \in (0, \varepsilon)$, and write $f_1 \approx f_2$ if a can be taken arbitrarily close to 1 for sufficiently small ε.

Let $\int_0 r^\pm$ denote the integral as a function of the upper integration limit.

(i) Since (4.1) and (4.2) imply $\int_0 r^- \sim \int_0 r^+$, the result is clear.

(ii) Using OD we now obtain $\int_0 r^- \approx \int_0 r^+$, so (d) is equivalent to its $^+$ and $^-$ counterparts. For odd r, Parfenov [8, Theorem 6] notes that $(b)^+ \Rightarrow (c)^+ \Rightarrow (d)^+$ are obvious, and that remains true in our situation. Similarly, $(b)^- \Rightarrow (c)^- \Rightarrow (d)^-$ and (b)⇒(c)⇒(d). Moreover, [6, Lemma 3.1] proves (d) ⇒ P. Since OD implies WOD, part (i) completes the proof. □

The next result gives equivalence of all conditions considered, but at the price of a stronger form of odd domination.

Theorem 4.3. *Each condition in Theorem 4.2 is equivalent to V, V_n and the RBP for* (1.1) *under SOD.*

Proof. RBP ⇒ V is proved in [10, Section 4] and V ⇒ V_n is trivial.

$V_n \Rightarrow (d)^-$: Consider the functions g_n, h_n in Section 3, with A_n, B_n and D_n defined in Lemmas 3.1 and 3.2. If $(d)^-$ fails then we can take $a_n/b_n \to 0$ and $A_n/B_n \to 1$ so

$$a_n(b_n - a_n)^{-1} \to 0, \; b_n(b_n - a_n)^{-1} \to 1 \text{ and } D_n/A_n \to 0. \tag{4.5}$$

Since $b_n < 1$, $a_n/b_n \to 0$ implies $a_n \to 0$, and then $A_n/B_n \to 1$ implies

$$b_n \to 0. \tag{4.6}$$

By (4.3), estimate (3.4) holds with $\delta_n = 2\rho(b_n)/(1-\rho(b_n))$. Now we make use of SOD, together with (4.5) and (4.6), to see that

$$\frac{\delta_n^2}{b_n - a_n} \to 0. \tag{4.7}$$

Thus by Lemma 3.1, Corollary 3.3, (4.5) and (4.7), we obtain

$$J_n^{-1} = \frac{\int_{-1}^{1} |h_n|^2 \int_{-1}^{1} |g_n'|^2}{\left(\int_{-1}^{1} |r|\,|g_n|^2\right)^2} \le \frac{6\left(b_n D_n^2 + a_n B_n^2\right) + 4\delta_n^2 B_n^2}{(b_n - a_n)A_n^2} \to 0,$$

contradicting V_n.

(d)$^-$ $\Rightarrow$ P follows from Theorem 4.2 and P $\Rightarrow$ RBP follows from [8, Corollary 4]. □

Note that for the proof of Theorem 4.3 we used Condition V_n only for sequences a_n, b_n with $a_n/b_n \to 0$ and $A_n/B_n \to 1$. Thus we can combine the ideas of the previous two results to include the following weaker version of Condition V_n.

Condition (e): For each pair of sequences a_n, b_n with

$$0 < a_n < b_n \le 1, \quad a_n/b_n \to 0 \quad \text{and} \quad I_{a_n}/I_{b_n} \to 1,$$

the sequence J_n of (3.2) is bounded.

Note that the bound in (e) may depend on the sequences a_n and b_n. We define (e)$^\pm$ analogously to (d)$^\pm$.

Corollary 4.4. *Under SOD each condition in Theorems* 4.2 *and* 4.3 *is equivalent to* (e), *to* (e)$^+$ *and to* (e)$^-$.

Proof. $V_n \Rightarrow$ (e) is obvious, and equivalence of (e) with (e)$^+$ and (e)$^-$ follows as in Theorem 4.2. Finally the proof of Theorem 4.3 gives (e)$^- \Rightarrow$ (d)$^- \Rightarrow$ V$_n$. □

References

[1] N.L. Abasheeva, S.G. Pyatkov, *Counterexamples in indefinite Sturm-Liouville problems.* Siberian Adv. Math. 7, (1997), 1–8.

[2] R. Beals, *Indefinite Sturm-Liouville problems and half range completeness.* J. Differential Equations **56** (1985), 391–407.

[3] K.M. Case, *Elementary solutions of the transport equation and their applications.* Ann. Physics **9** (1960), 1–23.

[4] B. Ćurgus, B. Najman, *A Kreĭn space approach to elliptic eigenvalue problems with indefinite weights.* Differential Integral Equations **7** (1994), 1241–1252.

[5] A. Fleige, *A counterexample to completeness properties for indefinite Sturm-Liouville problems.* Math. Nachr. **190** (1998), 123–128.

[6] A. Fleige, *The Riesz Basis Property of an Indefinite Sturm-Liouville Problem with a Non Odd Weight Function.* Integral Equations Operator Theory **60** (2008), 237–246.

[7] I.C. Gohberg and M.G. Kreĭn, *Introduction to the theory of linear nonselfadjoint operators*, Translations of Mathematical Monographs, Vol. 18, American Mathematical Society, Providence, R.I., 1969.

[8] A.I. Parfenov, *On an Embedding Criterion for Interpolation Spaces and Application to Indefinite Spectral Problems.* Siberian Mathematical Journal, Vol. **44**, (2003), 638–644.

[9] S.G. Pyatkov, *Properties of eigenfunctions of linear sheaves.* Siberian Math. J. **30**, (1989), 587–597.

[10] H. Volkmer, *Sturm-Liouville problems with indefinite weights and Everitt's inequality.* Proc. Roy. Soc. Edinburgh Sect. A **126**, (1996), 1097–1112.

Paul Binding
Department of Mathematics and Statistics
University of Calgary
Calgary, Alberta, Canada T2N 1N4
e-mail: `binding@ucalgary.ca`

Andreas Fleige
Baroper Schulstraße 27 a
D-44225 Dortmund, Germany
e-mail: `andreas-fleige@versanet.de`

Operator Theory:
Advances and Applications, Vol. 198, 97–124

Convergence of Diagonal Padé Approximants for a Class of Definitizable Functions

Maxim Derevyagin and Vladimir Derkach

Dedicated to the memory of Peter Jonas

Abstract. Convergence of diagonal Padé approximants is studied for a class of functions which admit the integral representation $\mathfrak{F}(\lambda) = r_1(\lambda)\int_{-1}^{1}\frac{td\sigma(t)}{t-\lambda} + r_2(\lambda)$, where σ is a finite nonnegative measure on $[-1,1]$, r_1, r_2 are real rational functions bounded at ∞, and r_1 is nonnegative for real λ. Sufficient conditions for the convergence of a subsequence of diagonal Padé approximants of $\mathfrak{F}$ on $\mathbb{R}\setminus[-1,1]$ are found. Moreover, in the case when $r_1 \equiv 1$, $r_2 \equiv 0$ and σ has a gap (α,β) containing 0, it turns out that this subsequence converges in the gap. The proofs are based on the operator representation of diagonal Padé approximants of $\mathfrak{F}$ in terms of the so-called generalized Jacobi matrix associated with the asymptotic expansion of $\mathfrak{F}$ at infinity.

Mathematics Subject Classification (2000). Primary 30E05; Secondary 47A57.

Keywords. Generalized Nevanlinna function, definitizable function, Padé approximant, Pontryagin space, Jacobi matrix, orthogonal polynomials.

1. Introduction

Let $F(\lambda) = -\sum_{j=0}^{\infty}\frac{s_j}{\lambda^{j+1}}$ be a formal power series with $s_j \in \mathbb{R}$, and let L, M be positive integers. An $[L/M]$ Padé approximant for F is defined as a ratio

$$F^{[L/M]}(\lambda) = \frac{A^{[L/M]}(1/\lambda)}{B^{[L/M]}(1/\lambda)}$$

The research was supported by the Deutschen Akademischen Austauschdienst and the Technical University of Berlin.

of polynomials $A^{[L/M]}$, $B^{[L/M]}$ of formal degree L and M, respectively, such that $B^{[L/M]}(0) \neq 0$ and

$$\sum_{j=0}^{L+M-1} \frac{s_j}{\lambda^{j+1}} + F^{[L/M]}(\lambda) = O\left(\frac{1}{\lambda^{L+M+1}}\right), \quad \lambda \to \infty. \tag{1.1}$$

The classical Markov theorem [31] states that for every nonnegative measure σ on the interval $[-1,1]$ and the function

$$F(\lambda) = \int_{-1}^{1} \frac{d\sigma(t)}{t-\lambda} \tag{1.2}$$

with the Laurent expansion $F(\lambda) = -\sum_{j=0}^{\infty} \frac{s_j}{\lambda^{j+1}}$ at ∞ the diagonal Padé approximants $F^{[n/n]}$ exists for every $n \in \mathbb{N}$ and converge to F locally uniformly on $\mathbb{C} \setminus [-1,1]$. However, it should be noted that in the case when σ has a gap (α, β) in its support, the diagonal Padé approximants $F^{[n/n]}$ do not usually converge inside the gap (see [31]).

In [6] it was conjectured that for every function F holomorphic in a neighborhood of ∞ there is a subsequence of diagonal $[n/n]$ Padé approximants which converges to F locally uniformly in the neighborhood of ∞ (Padé hypothesis). In general, as was shown by D. Lubinsky [28] (see also [9]), this conjecture fails to hold, but for some classes of functions the Padé hypothesis is still true. For example, if F has the form

$$F(\lambda) = \int_{-1}^{1} \frac{d\sigma(t)}{t-\lambda} + r(\lambda),$$

where r is a rational function with poles outside of $[-1,1]$, the convergence of Padé approximants was proved by A. Gonchar [16] and E. Rakhmanov [33].

In [11, 12] we studied the Padé hypothesis in the class of generalized Nevanlinna functions introduced in [23] (see the definition at the beginning of Section 2), which contains, in particular, functions of the form

$$F(\lambda) = r_1(\lambda) \int_{-1}^{1} \frac{d\sigma(t)}{t-\lambda} + r_2(\lambda), \tag{1.3}$$

where:

(A1) σ is a finite nonnegative measure on $[-1,1]$;
(A2) $r_1 = q_1/w_1$ is a rational function, nonnegative for real λ ($\deg q_1 \leq \deg w_1$);
(A3) $r_2 = q_2/w_2$ is a real rational function such that $\deg q_2 < \deg w_2$.

Let F have the Laurent expansion $F(\lambda) = -\sum_{j=0}^{\infty} \frac{s_j}{\lambda^{j+1}}$ at ∞, and let $\mathcal{N}(\mathbf{s})$ be the set of all *normal indices* of the sequence $\mathbf{s} = \{s_i\}_{i=0}^{\infty}$, i.e., natural numbers $n_1 < n_2 < \cdots < n_j < \cdots$, for which

$$\det(s_{i+k})_{i,k=0}^{n_j-1} \neq 0, \quad j = 1, 2, \ldots. \tag{1.4}$$

As is known the sequence $\{n_j\}_{j=1}^{\infty}$ contains all natural n big enough. The Padé approximants for F were considered in [12] in connection with the theory of difference equations

$$\widetilde{b}_{j-1}u_{j-1} - p_j(\lambda)u_j + b_j u_{j+1} = 0, \quad j \in \mathbb{N}, \tag{1.5}$$

naturally related to the function F, where p_j are monic polynomials of degree $k_j = n_{j+1} - n_j$, $b_j > 0$, $\widetilde{b}_j$ are real numbers, such that $|\widetilde{b}_j| = b_j$, $j \in \mathbb{Z}_+ := \mathbb{N} \cup \{0\}$. It turned out that the diagonal Padé approximants for F exist for all $n = n_j$ and are calculated by the formula

$$F^{[n_j/n_j]}(\lambda) = -\frac{Q_j(\lambda)}{P_j(\lambda)},$$

where P_j, Q_j are polynomials of the first and the second type associated with the difference equation (1.5) (see [11]). In [12] it was shown that the sequence of diagonal Padé approximants for F converges to F locally uniformly in $\mathbb{C} \setminus ([-1,1] \cup \mathcal{P}(F))$, where $\mathcal{P}(F)$ is the set of poles of F. Subdiagonal Padé approximants $F^{[n_j/n_j-1]}$ for F exist if and only if

$$n_j \in \mathcal{N}_F := \{n_j \in \mathcal{N}(\mathbf{s}) : P_{j-1}(0) \neq 0\}.$$

The convergence of the sequence $\{F^{[n_j/n_j-1]}\}_{n_j \in \mathcal{N}_F}$ of subdiagonal Padé approximants for F on $\mathbb{C} \setminus \mathbb{R}$ was also proved.

In Theorem 3.5 of the present paper we improve the result of [12, Theorem 4.16] by pointing out sufficient condition for convergence of the sequence of subdiagonal Padé approximants $F^{[n/n-1]}$ to F in a neighborhood of ∞. In the previous notations this condition takes the form

(B) The sequence $\left\{b_{j-1}\frac{P_j(0)}{P_{j-1}(0)}\right\}_{n_j \in \mathcal{N}_F}$ is bounded.

The main part of the present paper is dedicated to the study of convergence of diagonal Padé approximants of definitizable functions with one "turning point". This class was introduced by P. Jonas in [18]. We postpone the exact definitions until Section 4 and we mention only that the typical representative of this class is the function

$$\mathfrak{F}(\lambda) = r_1(\lambda)\int_{-1}^{1}\frac{t\,d\sigma(t)}{t-\lambda} + r_2(\lambda), \tag{1.6}$$

where σ, r_1, r_2 satisfy the assumptions (A1), (A2) and

(A3′) $r_2 = q_2/w_2$ is a real rational function such that $\deg q_2 \leq \deg w_2$.

We prove that for every $\mathfrak{F}$ of the form (1.6) satisfying condition (B) the Padé hypothesis is still true. The idea is that the diagonal Padé approximants for $\mathfrak{F}$ are proportional to the subdiagonal Padé approximants for F. This fact was observed by A. Magnus in a formal setting [29] and its operator interpretation for the Nevanlinna class was given by B. Simon [34]. This observation and [12, Theorem 4.16] enable us to prove in Theorem 4.6 that the sequence of diagonal Padé approximants for $\mathfrak{F}$ converges to $\mathfrak{F}$ locally uniformly in $\mathbb{C} \setminus (\mathbb{R} \cup \mathcal{P}(\mathfrak{F}))$. Moreover, using the result of Theorem 3.5 we show that condition (B) is sufficient for the convergence of diagonal Padé approximants of $\mathfrak{F}$ in a neighborhood of ∞ (see Theorem 5.1).

In Theorem 5.5 we specify this result to the case when the function $\mathfrak{F}$ in (1.3) takes the form

$$\mathfrak{F}(\lambda) = \int_E \frac{td\sigma(t)}{t-\lambda}, \quad E = [-1, \alpha] \cup [\beta, 1], \tag{1.7}$$

where the measure σ has a gap (α, β) with $\alpha < 0 < \beta$. For this function one can observe a new effect, that the sequence of diagonal Padé approximants $\{\mathfrak{F}^{[n_j-1/n_j-1]}\}_{n_j \in \mathcal{N}_F}$ converges to $\mathfrak{F}$ in the gap (α, β). The proof of this result is based on the theory of generalized Jacobi matrices associated with generalized Nevanlinna functions and on the operator representation of the subdiagonal Padé approximants for the generalized Nevanlinna function $F(\lambda) := \mathfrak{F}(\lambda)/\lambda$. Moreover, in Theorem 5.5 we prove that for such a function condition (B) is also necessary and sufficient for the convergence of the sequence $\{\mathfrak{F}^{[n_j-1/n_j-1]}\}_{n_j \in \mathcal{N}_F}$ to $\mathfrak{F}$ in a neighborhood of ∞.

This theorem makes a bridge to the theory of classical orthogonal polynomials. In Proposition 5.7 we show that the condition (B) is in force, if 0 is not an accumulation point of zeros of polynomials P_n orthogonal with respect to σ. In the case when the measure σ in (1.7) satisfies the Szegö condition on each of the intervals $[-1, \alpha]$ and $[\beta, 1]$ we inspect the question: under what conditions 0 is not an accumulation point of zeros of polynomials P_n? In Proposition 5.9 we show that the results of E. Rakhmanov [33] can be applied to give a partial answer to this question and, hence, to find some sufficient conditions on α, β and σ for the existence of a subsequence of diagonal Padé approximants $\mathfrak{F}^{[n/n]}$ which converges to F locally uniformly in a neighborhood of ∞.

The paper is organized as follows. In Section 2 the basic facts concerning generalized Nevanlinna functions and their operator representations in terms of generalized Jacobi matrices are given. In Section 3 we state and improve some results from [12] on locally uniform convergence of subdiagonal Padé approximants for generalized Nevanlinna functions. In Section 4 we introduce the class $\mathbf{D}_{\kappa,-\infty}$ of definitizable functions with one "turning point", and find the formula connecting diagonal Padé approximants for $\mathfrak{F} \in \mathbf{D}_{\kappa,-\infty}$ with subdiagonal Padé approximants for generalized Nevanlinna function $F(\lambda) = \mathfrak{F}(\lambda)/\lambda$. In Section 5 we apply our results to subclasses of definitizable functions of the form (1.6) and (1.7).

This paper is dedicated to the memory of Peter Jonas. Discussions with him during several of our visits to Berlin have had a significant influence on the development of this paper.

2. Preliminaries

2.1. Moment problem in the class of generalized Nevanlinna functions

Let κ be a nonnegative integer. Recall that a function F, meromorphic in $\mathbb{C}_+ \cup \mathbb{C}_-$, is said to belong to the class $\mathbf{N}_\kappa$ if the domain of holomorphy $\rho(F)$ of the function

F is symmetric with respect to $\mathbb{R}$, $F(\bar{\lambda}) = \overline{F(\lambda)}$ for $\lambda \in \rho(F)$, and the kernel

$$\begin{cases} \mathsf{N}_F(\lambda,\omega) = \frac{F(\lambda)-\overline{F(\omega)}}{\lambda-\bar{\omega}}, & \lambda,\omega \in \rho(F); \\ \mathsf{N}_F(\lambda,\bar{\lambda}) = F'(\lambda), & \lambda \in \rho(F) \end{cases}$$

has κ negative squares on $\rho(F)$. The last statement means that for every $n \in \mathbb{N}$ and $\lambda_1, \lambda_2, \dots, \lambda_n \in \rho(F)$, $n \times n$ matrix $(\mathsf{N}_F(\lambda_i, \lambda_j))_{i,j=1}^n$ has at most κ negative eigenvalues (with account of multiplicities) and for some choice of $n, \lambda_1, \lambda_2, \dots, \lambda_n$ it has exactly κ negative eigenvalues (see [23]).

We will say (cf. [15]) that a generalized Nevanlinna function F belongs to the class $\mathbf{N}_{\kappa,-2n}$ if $F \in \mathbf{N}_\kappa$ and for some numbers $s_0, \dots, s_{2n} \in \mathbb{R}$ the following asymptotic expansion holds true

$$F(\lambda) = -\frac{s_0}{\lambda} - \frac{s_1}{\lambda^2} - \cdots - \frac{s_{2n}}{\lambda^{2n+1}} + o\left(\frac{1}{\lambda^{2n+1}}\right), \quad \lambda \widehat{\rightarrow} \infty, \tag{2.1}$$

where $\lambda \widehat{\rightarrow} \infty$ means that λ tends to ∞ nontangentially, that is inside the sector $\varepsilon < \arg \lambda < \pi - \varepsilon$ for some $\varepsilon > 0$. Let us set

$$\mathbf{N}_{\kappa,-\infty} := \bigcap_{n \geq 0} \mathbf{N}_{\kappa,-2n}.$$

In particular, every function of the form (1.3) where r_1, r_2, σ are subject to the assumptions (A1)–(A3), belongs to the class $\mathbf{N}_{\kappa,-\infty}$ for some $\kappa \in \mathbb{Z}_+$ (see [23]). Moreover, every generalized Nevanlinna function $F \in \mathbf{N}_{\kappa,-\infty}$ holomorphic at ∞ admits the representation (1.3) for some r_1, r_2, σ satisfying (A1)–(A3).

It will be sometimes convenient to use the following notation

$$F(\lambda) \sim -\sum_{j=0}^{\infty} \frac{s_j}{\lambda^{j+1}}, \quad \lambda \widehat{\rightarrow} \infty \tag{2.2}$$

to denote the validity of (2.1) for all $n \in \mathbb{N}$.

Problem $\mathbf{M}_\kappa(\mathbf{s})$. *Given are a nonnegative integer κ and a sequence $\mathbf{s} = \{s_j\}_{j=0}^\infty$ of real numbers, such that the matrices $S_n := (s_{i+j})_{i,j=0}^n$ are nondegenerate for all n big enough. Find a function $F \in \mathbf{N}_\kappa$, which has the asymptotic expansion* (2.2).

We say that the problem $\mathbf{M}_\kappa(\mathbf{s})$ is *determinate* if $\mathbf{M}_\kappa(\mathbf{s})$ has a unique solution. The problem $\mathbf{M}_\kappa(\mathbf{s})$ was considered in [24], where it was shown that the problem $\mathbf{M}_\kappa(\mathbf{s})$ is solvable if and only if the number of negative eigenvalues of the matrix S_n does not exceed κ for all $n \in \mathbb{N}$. The Schur algorithm for solving the problem $\mathbf{M}_\kappa(\mathbf{s})$ considered in [10] proceeds as follows.

Let $\mathcal{N}(\mathbf{s})$ be the set of all *normal indices* of the sequence $\mathbf{s}$, i.e., natural numbers n_j, for which

$$\det S_{n_j-1} \neq 0, \quad j = 1, 2, \dots. \tag{2.3}$$

If $F_0 := F$ is a generalized Nevanlinna function with the asymptotic expansion (2.2), then the function $-1/F_0$ can be represented as

$$-\frac{1}{F_0(\lambda)} = \varepsilon_0 p_0(\lambda) + b_0^2 F_1(\lambda), \tag{2.4}$$

where $\varepsilon_0 = \pm 1$, $b_0 > 0$, p_0 is a monic polynomial of degree $k_0 = n_1$ and F_1 is a generalized Nevanlinna function. Continuing this process one gets sequences $\varepsilon_j = \pm 1$, $b_j > 0$, $j \in \mathbb{Z}_+$ and a sequence of real monic polynomials p_j of degree $k_j = n_{j+1} - n_j$, such that F admits the following expansion into a P-fraction

$$-\frac{\varepsilon_0}{p_0(\lambda)} - \frac{\varepsilon_0\varepsilon_1 b_0^2}{p_1(\lambda)} - \cdots - \frac{\varepsilon_{N-1}\varepsilon_N b_{N-1}^2}{p_N(\lambda)} - \cdots . \tag{2.5}$$

A similar algorithm for a continued fraction expansion of a formal power series was proposed by A. Magnus in [29]. The objects ε_j, b_j, p_j are uniquely defined by the sequence $\mathbf{s} = \{s_j\}_{j=0}^{\infty}$ (see [10], [12]). The function F_1 in (2.4) is called the Schur transform of $F_0 \in \mathbf{N}_\kappa$ (cf. [5]).

2.2. Generalized Jacobi matrices

Let $p(\lambda) = p_k\lambda^k + \cdots + p_1\lambda + p_0$, $p_k = 1$ be a monic scalar real polynomial of degree k. Let us associate with the polynomial p its symmetrizator E_p and the companion matrix C_p, given by

$$E_p = \begin{pmatrix} p_1 & \dots & p_k \\ \vdots & \cdot^{\cdot^{\cdot}} & \\ p_k & & \end{pmatrix}, \quad C_p = \begin{pmatrix} 0 & \dots & 0 & -p_0 \\ 1 & & & -p_1 \\ & \ddots & & \vdots \\ & & 1 & -p_{k-1} \end{pmatrix}, \tag{2.6}$$

where all the non-specified elements are supposed to be equal to zero. As is known (see [17]), $\det(\lambda - C_p) = p(\lambda)$ and the spectrum $\sigma(C_p)$ of the companion matrix C_p is simple. The matrices E_p and C_p are related by

$$C_pE_p = E_pC_p^{\top}. \tag{2.7}$$

Definition 2.1. Let p_j be real monic polynomials of degree k_j

$$p_j(\lambda) = \lambda^{k_j} + p_{k_j-1}^{(j)}\lambda^{k_j-1} \cdots + p_1^{(j)}\lambda + p_0^{(j)},$$

and let $\varepsilon_j = \pm 1$, $b_j > 0$, $j \in \mathbb{Z}_+$. The tridiagonal block matrix

$$\mathcal{J} = \begin{pmatrix} A_0 & \widetilde{B}_0 & & \\ B_0 & A_1 & \widetilde{B}_1 & \\ & B_1 & A_2 & \ddots \\ & & \ddots & \ddots \end{pmatrix} \tag{2.8}$$

where $A_j = C_{p_j}$ and $k_{j+1} \times k_j$ matrices B_j and $k_j \times k_{j+1}$ matrices $\widetilde{B}_j$ are given by

$$B_j = \begin{pmatrix} 0 & \dots & b_j \\ \dots & \dots & \dots \\ 0 & \dots & 0 \end{pmatrix}, \ \widetilde{B}_j = \begin{pmatrix} 0 & \dots & \widetilde{b}_j \\ \dots & \dots & \dots \\ 0 & \dots & 0 \end{pmatrix}, \ \widetilde{b}_j = \varepsilon_j\varepsilon_{j+1}b_j, \ j \in \mathbb{Z}_+. \tag{2.9}$$

will be called a *generalized Jacobi matrix* associated with the sequences of polynomials $\{\varepsilon_j p_j\}_{j=0}^{\infty}$ and the numbers $\{b_j\}_{j=0}^{\infty}$.

Remark 2.2. Define an infinite matrix G by the equality

$$G = \operatorname{diag}(G_0, \dots, G_N, \dots), \quad G_j = \varepsilon_j E_{p_j}^{-1},\ j = 0, \dots, N \tag{2.10}$$

and let $\ell^2_{[0,\infty)}(G)$ be the space of ℓ^2-vectors with the inner product

$$[x, y] = (Gx, y)_{\ell^2_{[0,\infty)}}, \quad x, y \in \ell^2_{[0,\infty)}. \tag{2.11}$$

The inner product (2.11) is indefinite, if either $k_j > 1$ for some $j \in \mathbb{Z}_+$, or at least one ε_j is equal to -1. The space $\ell^2_{[0,\infty)}(G)$ is equivalent to a Kreĭn space (see [4]) if both G and G^{-1} are bounded in $\ell^2_{[0,\infty)}$. If one starts with the $\mathbf{N}_\kappa$-function F then $k_j = \varepsilon_j = 1$ for all j big enough. In this case, $\ell^2_{[0,\infty)}(G)$ is a Pontryagin space, and as follows from (2.7) the generalized Jacobi matrix $\mathcal{J}$ determines a symmetric operator S in the space $\ell^2_{[0,\infty)}(G)$ (see details in [11]). More general definition of generalized Jacobi matrix which is not connected with the Schur algorithm has been considered in [23].

Setting $\widetilde{b}_{-1} = \varepsilon_0$, define polynomials of the first kind (cf. [23]) $P_j(\lambda)$, $j \in \mathbb{Z}_+$, as solutions $u_j = P_j(\lambda)$ of the following system:

$$\widetilde{b}_{j-1} u_{j-1} - p_j(\lambda) u_j + b_j u_{j+1} = 0, \ \ j \in \mathbb{Z}_+, \tag{2.12}$$

with the initial conditions

$$u_{-1} = 0, \quad u_0 = 1. \tag{2.13}$$

Similarly, the polynomials of the second kind $Q_j(\lambda)$, $j \in \mathbb{Z}_+$, are defined as solutions $u_j = Q_j(\lambda)$ of the system (2.12) subject to the following initial conditions

$$u_{-1} = -1, \quad u_0 = 0. \tag{2.14}$$

It follows from (2.12) that P_j is a polynomial of degree $n_j = \sum\limits_{i=0}^{j-1} k_i$ with the leading coefficient $(b_0 \dots b_{j-1})^{-1}$ and Q_j is a polynomial of degree $n_j - k_0$ with the leading coefficient $\varepsilon_0 (b_0 \dots b_{j-1})^{-1}$. The equations (2.12) coincide with the three-term recurrence relations associated with P-fractions ([29], see also [19, Section 5.2]). The following statement is immediate from (2.12).

Proposition 2.3. ([12]). *Polynomials P_j and P_{j+1} (Q_j and Q_{j+1}) have no common zeros.*

The following connection between the polynomials of the first and second kinds P_j, Q_j and the shortened Jacobi matrices $\mathcal{J}_{[0,j]}$ can be found in [12, Proposition 3.3] (in the classical case see [8, Section 7.1.2]).

Proposition 2.4. *Polynomials P_j and Q_j can be found by the formulas*

$$P_j(\lambda) = (b_0 \dots b_{j-1})^{-1} \det(\lambda - \mathcal{J}_{[0,j-1]}), \tag{2.15}$$

$$Q_j(\lambda) = \varepsilon_0 (b_0 \dots b_{j-1})^{-1} \det(\lambda - \mathcal{J}_{[1,j-1]}). \tag{2.16}$$

Clearly, $\mathcal{J}_{[0,j-1]}$ is a symmetric operator in the subspace $\ell^2_{[0,j-1]}(G)$ of the indefinite inner product space $\ell^2_{[0,\infty)}(G)$, which consists of vectors

$$u = \{u_{ik}\}_{i=0,\dots,\infty}^{k=0,\dots,n_i-1}$$

such that $u_{ik} = 0$ for $i \geq j$. The m-function of the shortened matrix $\mathcal{J}_{[0,j-1]}$ is defined by

$$m_{[0,j-1]}(\lambda) = [(\mathcal{J}_{[0,j-1]} - \lambda)^{-1}e, e]. \tag{2.17}$$

Due to formulas (2.15), (2.16) it is calculated by

$$m_{[0,j-1]}(\lambda) = -\varepsilon_0 \frac{\det(\lambda - \mathcal{J}_{[1,j-1]})}{\det(\lambda - \mathcal{J}_{[0,j-1]})} = -\frac{Q_j(\lambda)}{P_j(\lambda)}. \tag{2.18}$$

Remark 2.5. Let us emphasize that the polynomials P_j and Q_j have no common zeros (see [11, Proposition 2.7]) and due to (2.18) the set of holomorphy of $m_{[0,j-1]}$ coincides with the resolvent set of $\mathcal{J}_{[0,j-1]}$.

Theorem 2.6 ([12]). *Let $F \in \mathbf{N}_{\kappa,-\infty}$ and the corresponding indefinite moment problem $M_\kappa(\mathbf{s})$ be determinate. Then:*

(i) *the generalized Jacobi matrix corresponding to F via* (2.5) *and* (2.8) *generates a selfadjoint operator $\mathcal{J}$ in $\ell^2_{[0,\infty)}(G)$ and*

$$F(\lambda) = [(\mathcal{J} - \lambda)^{-1}e, e]. \tag{2.19}$$

(ii) *the diagonal $[n_j/n_j]$ Padé approximants of $F(\lambda)$ coincide with $m_{[0,j-1]}(\lambda)$ and converge to $F(\lambda)$ locally uniformly on $\mathbb{C} \setminus \mathbb{R}$.*

The proof of this result is based on the fact that the compressed resolvents of $\mathcal{J}_{[0,j-1]}$ converge to the compressed resolvent of $\mathcal{J}$ (see [12, Theorem 4.8]). Theorem 2.6 contains as partial cases some results of A. Gonchar [16] and E. Rakhmanov [33], mentioned in Introduction, as well as the results of G.L. Lopes [27] concerning convergence of diagonal Padé approximants for rational perturbations of Stieltjes functions.

3. The convergence of subdiagonal Padé approximants

Let us consider the following finite generalized Jacobi matrix

$$\mathcal{J}_{[0,j]}(\tau) = \begin{pmatrix} A_0 & \widetilde{B}_0 & & \\ B_0 & \ddots & \ddots & \\ & \ddots & A_{j-1} & \widetilde{B}_{j-1} \\ & & B_{j-1} & A_j(\tau) \end{pmatrix}, \quad A_j(\tau) = \begin{pmatrix} 0 & \dots & 0 & -p_0^{(j)} + \tau \\ 1 & & & -p_1^{(j)} \\ & \ddots & & \vdots \\ & & 1 & -p_{k_j-1}^{(j)} \end{pmatrix}. \tag{3.1}$$

A vector $u = (u_{ik})_{i=0,\dots,j}^{k=0,\dots,n_i-1} \in \mathbb{C}^{n_j}$ is a left eigenvector of the matrix $\mathcal{J}_{[0,j]}(\tau)$, corresponding to the eigenvalue 0 if and only if $u_i = u_{i0}$, $i = 0,\dots,j$, satisfy the equations (2.12) for $i = 0,\dots,j-1$, and

$$\widetilde{b}_{j-1}u_{j-1} - (p_j(0) - \tau)u_j = 0, \tag{3.2}$$

$$u_{ik} = 0, \quad i = 0,\dots,j, \ k = 1,\dots,n_i - 1.$$

Therefore $u_i = P_i(0)$ $(i = 0,\dots,j)$. If $u_j = P_j(0) \neq 0$, then (3.2) yields

$$\tau = \tau_j := -\frac{\widetilde{b}_{j-1}P_{j-1}(0) - p_j(0)P_j(0)}{P_j(0)} = b_j\frac{P_{j+1}(0)}{P_j(0)}. \tag{3.3}$$

Proposition 3.1. *If $P_j(0) \neq 0$ then there exists a number $\tau_j \in \mathbb{R}$ such that*

$$0 \in \sigma_p(\mathcal{J}_{[0,j]}(\tau_j)). \tag{3.4}$$

Moreover, τ_j can be found by the formula (3.3).

Remark 3.2. We will use the notation $\mathcal{J}_{[0,j]}^{(K)}$ for the matrix $\mathcal{J}_{[0,j]}(\tau_j)$ with the property (3.4). In the case when the corresponding indefinite moment problem $M_\kappa(\mathbf{s})$ is determinate for all $\lambda \in \mathbb{C} \setminus (\mathbb{R} \cup \mathcal{P}(F))$ we have

$$m_{[0,j-1]}^{(K)}(\lambda) := [(\mathcal{J}_{[0,j-1]}^{(K)} - \lambda)^{-1}e, e] \to F(\lambda) = [(\mathcal{J} - \lambda)^{-1}e, e] \tag{3.5}$$

as $j \to \infty$ (see [12, Proposition 4.4]).

Theorem 3.3 ([12]). *Let a function $F \in \mathbf{N}_\kappa$ have the expansion* (2.1) *for every $n \in \mathbb{N}$, let the corresponding moment problem $\mathbf{M}_\kappa(\mathbf{s})$ be determinate, and let $\{n_j\}_{j=1}^\infty = \mathcal{N}(\mathbf{s})$ be the set of normal indices of the sequence $\mathbf{s} = \{s_i\}_{i=0}^\infty$. Then:*

(i) *The $[n_j/n_j-1]$ Padé approximant $F^{[n_j/n_j-1]}$ exists if and only if*

$$n_j \in \mathcal{N}_F := \{n_j \in \mathcal{N}(\mathbf{s}) : P_{j-1}(0) \neq 0\}; \tag{3.6}$$

(ii) *The sequence*

$$F^{[n_j/n_j-1]} = -\frac{Q_j(\lambda)P_{j-1}(0) - Q_{j-1}(\lambda)P_j(0)}{P_j(\lambda)P_{j-1}(0) - P_{j-1}(\lambda)P_j(0)}, \quad n_j \in \mathcal{N}_F,$$

converges to F locally uniformly in $\mathbb{C} \setminus (\mathbb{R} \cup \mathcal{P}(F))$.

Proof. We will sketch the proof of this theorem presented in [12].

Let $P_{j-1}(0) \neq 0$ and let $P_j^{(K)}$, $Q_j^{(K)}$ be the polynomials of the first and second kinds, respectively, associated with the matrix $\mathcal{J}_{[0,j-1]}^{(K)}$, that is

$$P_j^{(K)}(\lambda) = (b_0 \dots b_{j-1})^{-1}\det(\lambda - \mathcal{J}_{[0,j-1]}^{(K)}),$$
$$Q_j^{(K)}(\lambda) = \varepsilon_0(b_0 \dots b_{j-1})^{-1}\det(\lambda - \mathcal{J}_{[1,j-1]}^{(K)}).$$

Next, applying the determinant decomposition theorem one can obtain

$$P_j^{(K)}(\lambda) = P_j(\lambda) - \frac{\tau_{j-1}}{b_{j-1}} P_{j-1}(\lambda) = \frac{P_j(\lambda)P_{j-1}(0) - P_{j-1}(\lambda)P_j(0)}{P_{j-1}(0)},$$

$$Q_j^{(K)}(\lambda) = Q_j(\lambda) - \frac{\tau_{j-1}}{b_{j-1}} Q_{j-1}(\lambda) = \frac{Q_j(\lambda)P_{j-1}(0) - Q_{j-1}(\lambda)P_j(0)}{P_{j-1}(0)}.$$

It obviously follows from (3.5) that

$$m_{[0,j-1]}^{(K)}(\lambda) := \left[(\mathcal{J}_{[0,j-1]}^{(K)} - \lambda)^{-1} e, e\right] \sim -\sum_{i=0}^{\infty} \frac{s_i{}^{(K)}}{\lambda^{i+1}}, \quad \lambda \widehat{\to} \infty, \tag{3.7}$$

where $s_i{}^{(K)} = \left[(\mathcal{J}_{[0,j-1]}^{(K)})^i e, e\right]$. Using the form of the matrix $\mathcal{J}_{[0,j-1]}^{(K)}$ one gets

$$s_i^{(K)} = s_i \text{ if } i \leq 2n_j - 2,$$

$$s_{2n_j-1}^{(K)} = s_{2n_j-1} + (b_0 \dots b_{j-1})^2 \varepsilon_{j-1} \tau_{j-1}.$$

So, the function $m_{[0,j-1]}^{(K)}(\lambda)$ has the following asymptotic expansion

$$m_{[0,j-1]}^{(K)}(\lambda) = -\sum_{i=0}^{2n_j-2} \frac{s_i}{\lambda^{i+1}} + O\left(\frac{1}{\lambda^{2n_j}}\right), \quad \lambda \widehat{\to} \infty,$$

where the sequence $\{s_j\}_{j=0}^{\infty}$ corresponds to the generalized Jacobi matrix $\mathcal{J}$. On the other hand, due to (2.18)

$$m_{[0,j-1]}^{(K)}(\lambda) = -\frac{Q_j^{(K)}(\lambda)}{P_j^{(K)}(\lambda)}.$$

Further, setting

$$A^{[n_i/n_i-1]}\left(\frac{1}{\lambda}\right) = \left(\frac{1}{\lambda}\right)^{n_i} Q_i^{(K)}(\lambda), \quad B^{[n_i/n_i-1]}\left(\frac{1}{\lambda}\right) = \left(\frac{1}{\lambda}\right)^{n_i} P_i^{(K)}(\lambda),$$

for $i = 0, 1, \dots, j$ and taking into account the equality $P_j^{(K)}(0) = 0$, one obtains

$$m_{[0,j-1]}^{(K)}(\lambda) = \frac{A^{[n_j/n_j-1]}(1/\lambda)}{B^{[n_j/n_j-1]}(1/\lambda)}, \tag{3.8}$$

where

$$\deg A^{[n_j/n_j-1]} = n_j, \quad \deg B^{[n_j/n_j-1]} = n_j - 1, \quad B^{[n_j/n_j-1]}(0) = \frac{1}{b_0 \dots b_{j-1}} \neq 0.$$

Therefore, $m_{[0,j-1]}^{(K)}(\lambda)$ is the $[n_j/n_j-1]$ Padé approximant for the corresponding Hamburger series. This proves the first part of the theorem. The second statement rests on the fact mentioned in Remark 3.2. □

Remark 3.4. Condition $P_{j-1}(0) \neq 0$ is equivalent to

$$\begin{vmatrix} s_1 & \dots & s_{n_{j-1}} \\ \dots & \dots & \dots \\ s_{n_{j-1}} & \dots & s_{2n_{j-1}-1} \end{vmatrix} \neq 0. \tag{3.9}$$

It follows from Proposition 2.3 that the set $\mathcal{N}_F$ is infinite.

Example 1. Consider the following classical 2-periodic Jacobi matrix

$$\mathcal{J} = \begin{pmatrix} a_0 & b_0 & & \\ b_0 & a_1 & b_1 & \\ & b_1 & a_2 & \ddots \\ & & \ddots & \ddots \end{pmatrix}, \quad a_n = \frac{(-1)^n + 1}{2}, \quad b_n = 1, \quad n \in \mathbb{Z}_+.$$

The m-function corresponding to this Jacobi matrix can be found by using standard methods (see [32])

$$\varphi(\lambda) = ((\mathcal{J} - \lambda)^{-1} e, e)_{\ell^2_{[0,\infty)}} = \frac{\lambda - \lambda^2 + \sqrt{(\lambda^2 - \lambda - 2)^2 - 4}}{2(\lambda - 1)}$$

(here the branch of the square root is determined by the condition $\varphi(\lambda) = -1/\lambda + o(1/\lambda)$ as $\lambda \to \infty$). Therefore, the function $\varphi(\lambda)$ admits the integral representation

$$\varphi(\lambda) = \int_E \frac{d\sigma(t)}{t - \lambda},$$

where the support $E := \operatorname{supp} \sigma$ of the measure σ is contained in $[-2, 3]$. Since the $[n/n-1]$ Padé approximant is equal to $f^{[n/n-1]}(\lambda) = m^{(K)}_{[0,n-1]}(\lambda)$, its poles coincide with eigenvalues of the matrix $\mathcal{J}^{(K)}_{[0,n-1]}$. Let us show that the eigenvalue of the matrix $\mathcal{J}^{(K)}_{[0,2k]}$ with the largest absolute value tends to infinity as $k \to +\infty$. First we compute τ_n. Since $\tau_n = b_n P_{n+1}(0)/P_n(0) = -b_{n-1}\widetilde{b}_{n-1}/\tau_{n-1} + p_n(0)$, we have

$$\tau_n = -1/\tau_{n-1} - ((-1)^n + 1)/2.$$

Clearly, $\tau_0 = -1$. By induction, we have the following formulas

$$\tau_{2k} = -(k+1), \qquad \tau_{2k+1} = 1/(k+1).$$

Taking into account that $\mathcal{J}^{(K)}_{[0,2k]}$ is a self-adjoint matrix, one obtains

$$|\lambda_{\max}(\mathcal{J}^{(K)}_{[0,2k]})| = \|\mathcal{J}^{(K)}_{[0,2k]}\| \geq |(\mathcal{J}^{(K)}_{[0,2k]} e_{2k}, e_{2k})_{\ell^2}| = k,$$

where $\lambda_{\max}(\mathcal{J}^{(K)}_{[0,2k]})$ is the eigenvalue of the matrix $\mathcal{J}^{(K)}_{[0,2k]}$ with the largest absolute value. Therefore, $|\lambda_{\max}(\mathcal{J}^{(K)}_{[0,2k]})| \to +\infty$ as $k \to +\infty$. So, infinity is an accumulation point of the set of poles of the Padé approximants $f^{[n/n-1]}$ of the function φ holomorphic at infinity. (This example was given in [12] with several misprints.)

Under certain conditions, it is possible to say more about the convergence of the sequence $F^{[n/n-1]}$ on the real line.

Theorem 3.5. *Let F have the form*

$$F(\lambda) = r_1(\lambda)\int_{-1}^{1}\frac{d\sigma(t)}{t-\lambda} + r_2(\lambda) = -\sum_{j=0}^{\infty}\frac{s_j}{\lambda^{j+1}}, \quad |\lambda| > R,$$

where σ, r_1, r_2 satisfy the assumptions (A1)–(A3). *If the sequence $\{\tau_{j-1}\}_{n_j\in\mathcal{N}_F}$ is bounded, i.e.,*

$$\sup_{n_j\in\mathcal{N}_F}\left|b_{j-1}\frac{P_j(0)}{P_{j-1}(0)}\right| < \infty, \tag{3.10}$$

than there exists a constant $\varepsilon > 0$ such that the sequence $\{F^{[n_j/n_j-1]}\}_{n_j\in\mathcal{N}_F}$ converges to F locally uniformly in $\mathbb{C}\setminus([-1-\varepsilon, 1+\varepsilon]\cup\mathcal{P}(F))$.

Proof. It is obvious that F corresponds to the determinate moment problem $\mathbf{M}_\kappa(\mathbf{s})$. Moreover, the corresponding generalized Jacobi matrix $\mathcal{J}$ is a bounded linear operator. According to Theorem 3.3, the sequence $\{F^{[n_j/n_j-1]}\}_{n_j\in\mathcal{N}_F}$ converges to F locally uniformly in $\mathbb{C}\setminus(\mathbb{R}\cup\mathcal{P}(F))$. Due to (3.8) we have

$$F^{[n_j/n_j-1]}(\lambda) = m_{[0,j-1]}^{(K)}(\lambda) = \left[(\mathcal{J}_{[0,j-1]}^{(K)} - \lambda)^{-1}e, e\right]. \tag{3.11}$$

Since the sequence $\{\tau_{j-1}\}_{n_j\in\mathcal{N}_F}$ is bounded, one obtains

$$\|\mathcal{J}_{[0,j-1]}^{(K)}\| \le \|\mathcal{J}_{[0,j-1]}\| + |\tau_{j-1}| \le 1+\varepsilon, \quad n_j\in\mathcal{N}_F \tag{3.12}$$

for some $\varepsilon > 0$. It follows from the inequality

$$\|(\mathcal{J}_{[0,j-1]}^{(K)} - \lambda)^{-1}\| \le \frac{1}{|\lambda| - \|\mathcal{J}_{[0,j-1]}^{(K)}\|} \qquad (|\lambda| > \|\mathcal{J}_{[0,j-1]}^{(K)}\|)$$

and (3.12) that

$$\begin{aligned}\left|m_{[0,j-1]}^{(K)}(\lambda)\right| &= \left|\left((\mathcal{J}_{[0,j-1]}^{(K)} - \lambda)^{-1}e, Ge\right)_{\ell^2}\right| \\ &\le \|(\mathcal{J}_{[0,j-1]}^{(K)} - \lambda)^{-1}e\|_{\ell^2}\|Ge\|_{\ell^2} \le \frac{\|Ge\|_{\ell^2}}{|\lambda| - 1 - \varepsilon}.\end{aligned}$$

for $|\lambda| > 1+\varepsilon$. To complete the proof, it is sufficient to apply the Vitali theorem. □

Corollary 3.6. *If the sequence $\{\tau_{j-1}\}_{n_j\in\mathcal{N}_F}$ tends to 0 then the sequence $\left\{F^{[n_j/n_j-1]}\right\}_{n_j\in\mathcal{N}_F}$ converges to F locally uniformly in $\mathbb{C}\setminus([-1,1]\cup\mathcal{P}(F))$.*

Proof. The statement is implied by (2.17), (3.11), the relation

$$\|\mathcal{J}_{[0,j-1]} - \mathcal{J}_{[0,j-1]}^{(K)}\| \to 0, \quad j\to\infty,$$

and Theorem 2.6. □

Remark 3.7. The condition (3.10) can be reformulated in terms of the monic orthogonal polynomials

$$\widehat{P}_j(\lambda) = \frac{1}{\det S_{n_j-1}} \begin{vmatrix} s_0 & s_1 & \dots & s_{n_j} \\ \dots & \dots & \dots & \dots \\ s_{n_j-1} & s_{n_j} & \dots & s_{2n_j-1} \\ 1 & \lambda & \dots & \lambda^{n_j} \end{vmatrix},$$

which are connected with $P_j(\lambda)$ by the formulas $\widehat{P}_j(\lambda) = (b_0 \cdots b_{j-1})P_j(\lambda)$, $j \in \mathbb{N}$. Therefore, the condition (3.10) takes the form

$$\sup_{n_j \in \mathcal{N}_F} \left| \frac{\widehat{P}_j(0)}{\widehat{P}_{j-1}(0)} \right| < \infty. \tag{3.13}$$

Remark 3.8. It is clear from the proof of Theorem 3.5 that the existence of a converging subsequence of the $[n/n-1]$ Padé approximants follows from the existence of a bounded subsequence of $\{\tau_{j-1}\}_{n_j \in \mathcal{N}_F}$.

4. A class of definitizable functions and Padé approximants

4.1. Classes $\mathbf{D}_{\kappa,-\infty}$ and $\mathbf{D}^{\circ}_{\kappa,-\infty}$.

Definition 4.1. Let us say that a function $\mathfrak{F}$ meromorphic in $\mathbb{C}_+$ belongs to the class $\mathbf{D}_{\kappa,-\infty}$ if

$$F(\lambda) := \frac{\mathfrak{F}(\lambda)}{\lambda} \in \mathbf{N}_{\kappa,-\infty} \quad \text{and} \quad \mathfrak{F}(\lambda) = O(1), \quad \lambda \widehat{\rightarrow} \infty.$$

Clearly, every function $\mathfrak{F} \in \mathbf{D}_{\kappa,-\infty}$ is definitizable in the sense of [18]. Indeed, consider the factorization

$$\frac{\mathfrak{F}(\lambda)}{\lambda} = r^{-1}(\lambda)(r^{\sharp})^{-1}(\lambda)F_0(\lambda),$$

where r is a real rational function, $r^{\sharp}(\lambda) = \overline{r(\overline{\lambda})}$, and $F_0 \in \mathbf{N}_0$. Then

$$\frac{r(\lambda)r^{\sharp}(\lambda)}{\lambda}\mathfrak{F}(\lambda) = F_0(\lambda) \in \mathbf{N}_0$$

and, hence, $\mathfrak{F}$ is definitizable, $\dfrac{r(\lambda)r^{\sharp}(\lambda)}{\lambda}$ is definitizing multiplier.

It follows from (2.1) that every function $\mathfrak{F} \in \mathbf{D}_{\kappa,-\infty}$ admits the asymptotic expansion

$$\mathfrak{F}(\lambda) \sim -\mathfrak{s}_{-1} - \frac{\mathfrak{s}_0}{\lambda} - \frac{\mathfrak{s}_1}{\lambda^2} - \dots - \frac{\mathfrak{s}_{2n}}{\lambda^{2n+1}} - \dots, \quad \lambda \widehat{\rightarrow} \infty \tag{4.1}$$

which is connected with the asymptotic expansion (2.1) of $F(\lambda) = \frac{\mathfrak{F}(\lambda)}{\lambda}$ via the formulas

$$\mathfrak{s}_{j-1} = s_j, \ j \in \mathbb{Z}_+. \tag{4.2}$$

In what follows we use the Gothic script for all the notations associated with the $\mathbf{D}_{\kappa,-\infty}$ function and the Roman script for the $\mathbf{N}_{\kappa,-\infty}$ function to avoid confusion.

We also say that a function $\mathfrak{F}$ meromorphic in $\mathbb{C}_+$ belongs to the class $\mathbf{D}^\circ_{\kappa,-\infty}$ if

$$F(\lambda) := \frac{\mathfrak{F}(\lambda)}{\lambda} \in \mathbf{N}_{\kappa,-\infty}, \quad \mathfrak{F}(\lambda) = o(1), \quad \lambda \widehat{\to} \infty,$$

and the asymptotic expansion of the function $\mathfrak{F}$

$$\mathfrak{F}(\lambda) \sim -\frac{\mathfrak{s}_0}{\lambda} - \frac{\mathfrak{s}_1}{\lambda^2} - \cdots - \frac{\mathfrak{s}_{2n}}{\lambda^{2n+1}} - \cdots, \quad \lambda \widehat{\to} \infty \tag{4.3}$$

is *normalized* in a sense that the first nontrivial coefficient in (4.3) has modulus 1,

$$|\mathfrak{s}_{\mathfrak{n}_1 - 1}| = 1.$$

Let the set $\mathcal{N}(\mathfrak{s})$ of normal indices of the sequence $\mathfrak{s} = \{\mathfrak{s}_i\}_{i=0}^\infty$ corresponding to a function $\mathfrak{F} \in \mathbf{D}^\circ_{\kappa,-\infty}$ be defined by (2.3), that is

$$\mathcal{N}(\mathfrak{s}) = \{\mathfrak{n}_j : \det(\mathfrak{s}_{i+k})_{i,k=0}^{\mathfrak{n}_j - 1} \neq 0, \quad j = 1, 2, \ldots\}. \tag{4.4}$$

4.2. Normal indices of the $\mathbf{D}^\circ_{\kappa,-\infty}$ functions

Remind that the point ∞ is called a generalized pole of nonpositive type of $F \in \mathbf{N}_\kappa$ with multiplicity $\kappa_\infty(F)$, if

$$0 \leq \lim_{\lambda \widehat{\to} \infty} \frac{F(\lambda)}{\lambda^{2\kappa_\infty + 1}} < \infty, \quad -\infty \leq \lim_{\lambda \widehat{\to} \infty} \frac{F(\lambda)}{\lambda^{2\kappa_\infty - 1}} < 0. \tag{4.5}$$

Similarly, the point ∞ is called a generalized zero of nonpositive type of F with multiplicity $\pi_\infty(F)$, if

$$\infty \leq \lim_{\lambda \widehat{\to} \infty} \lambda^{2\pi_\infty + 1} F(\lambda) < 0, \quad 0 \leq \lim_{\lambda \widehat{\to} \infty} \lambda^{2\pi_\infty - 1} F(\lambda) < \infty. \tag{4.6}$$

It was shown in [24] that the multiplicity of ∞ as a generalized pole (zero) of nonpositive type of $F \in \mathbf{N}_\kappa$ does not exceed κ.

Lemma 4.2. *Let $\mathfrak{F} \in \mathbf{D}^\circ_{\kappa,-\infty}$, let the sequence $\mathfrak{s} = \{\mathfrak{s}_j\}_{j=0}^\infty$ be defined by the asymptotic expansion* (4.3), *and let $\mathcal{N}(\mathfrak{s}) = \{\mathfrak{n}_j\}_{j=1}^\infty$ be the set of normal indices of $\mathfrak{s}$. Then*

$$\mathfrak{n}_1 \leq 2\kappa. \tag{4.7}$$

Moreover, if $\mathfrak{n}_1 = 2\kappa$, then

$$\mathfrak{s}_{\mathfrak{n}_1 - 1} > 0. \tag{4.8}$$

Proof. Since $\pi_\infty(F) \leq \kappa$ it follows from (4.6), that

$$\infty \leq \lim_{\lambda \widehat{\to} \infty} \lambda^{2\pi_\infty(F) + 1} F(\lambda) < 0. \tag{4.9}$$

The normal index $\mathfrak{n}_1$ can be characterized by the relations $\mathfrak{s}_0 = \cdots = \mathfrak{s}_{\mathfrak{n}_1 - 2} = 0$, $\mathfrak{s}_{\mathfrak{n}_1 - 1} \neq 0$. Hence $F(\lambda) = \mathfrak{F}(\lambda)/\lambda$ has the asymptotic expansion

$$F(\lambda) \sim -\frac{\mathfrak{s}_{\mathfrak{n}_1 - 1}}{\lambda^{\mathfrak{n}_1 + 1}} - \cdots - \frac{\mathfrak{s}_{2n}}{\lambda^{2n+2}} - \cdots, \quad \lambda \widehat{\to} \infty \tag{4.10}$$

and (4.9) implies the inequality (4.7).

If equality prevails in (4.7) then $\pi_\infty(F) = \kappa$, the limit in (4.9) is finite and coincides with $-\mathfrak{s}_{\mathfrak{n}_1 - 1}$. This implies the inequality (4.8). □

Proposition 4.3. *Let* $\mathfrak{F} \in \mathbf{D}^{\circ}_{\kappa,-\infty}$ *and* $F(\lambda) = \mathfrak{F}(\lambda)/\lambda$ *have asymptotic expansions* (4.3), (2.2), *and let* $\mathcal{N}_F$, $\mathcal{N}(\mathbf{s})$, $\mathcal{N}(\mathfrak{s})$ *be defined by* (3.6), (2.3), (4.4). *Then*

$$\mathcal{N}_F = \mathcal{N}(\mathfrak{s}) \cap \mathcal{N}(\mathbf{s}).$$

Proof. Let $\mathcal{N}(\mathbf{s}) = \{n_j\}_{j=1}^{\infty}$. The statement is implied by (3.6) and the equality

$$\begin{vmatrix} \mathfrak{s}_0 & \dots & \mathfrak{s}_{n_j-1} \\ \dots & \dots & \dots \\ \mathfrak{s}_{n_j-1} & \dots & \mathfrak{s}_{2n_j-2} \end{vmatrix} = \begin{vmatrix} s_1 & \dots & s_{n_j} \\ \dots & \dots & \dots \\ s_{n_j} & \dots & s_{2n_j-1} \end{vmatrix} \neq 0,$$

which is immediate from (4.2) and (3.9). □

4.3. The Schur transform of the $\mathbf{D}^{\circ}_{\kappa,-\infty}$ functions

Let a function $\mathfrak{F} \in \mathbf{D}^{\circ}_{\kappa,-\infty}$ have the asymptotic expansion (4.3), let $\{\mathfrak{n}_j\}_{j=1}^{\infty}$ be the set of normal indices for $\mathfrak{s} = \{\mathfrak{s}_j\}_{j=0}^{\infty}$ and let $\mathfrak{S}_n = (\mathfrak{s}_{i+j})_{i,j=0}^{n}$. Let us set $\epsilon_0 = \operatorname{sign} \mathfrak{s}_{\mathfrak{n}_1-1}$,

$$\mathfrak{p}_0(\lambda) = \frac{1}{\det \mathfrak{S}_{\mathfrak{n}_1-1}} \begin{vmatrix} 0 & \dots & \mathfrak{s}_{\mathfrak{n}_1-1} & \mathfrak{s}_{\mathfrak{n}_1} \\ \vdots & \iddots & \iddots & \vdots \\ \mathfrak{s}_{\mathfrak{n}_1-1} & \mathfrak{s}_{\mathfrak{n}_1} & \dots & \mathfrak{s}_{\mathfrak{n}_1-1} \\ 1 & \lambda & \dots & \lambda^{\mathfrak{n}_1} \end{vmatrix}. \tag{4.11}$$

The Schur transform of the function $\mathfrak{F} \in \mathbf{D}^{\circ}_{\kappa,-\infty}$ is defined by the equality

$$-\frac{1}{\mathfrak{F}(\lambda)} = \epsilon_0 \mathfrak{p}_0(\lambda) + \mathfrak{b}_0^2 \widehat{\mathfrak{F}}(\lambda), \tag{4.12}$$

where $\mathfrak{b}_0$ is chosen in such a way that $\widehat{\mathfrak{F}}$ has a normalized expansion at ∞.

Theorem 4.4. *Let* $\mathfrak{F} \in \mathbf{D}^{\circ}_{\kappa,-\infty}$ *and let* $\widehat{\mathfrak{F}}$ *be the Schur transform of* $\mathfrak{F}$. *Then:*

(i) $\widehat{\mathfrak{F}} \in \mathbf{D}^{\circ}_{\kappa',-\infty}$ *for some* $\kappa' \le \kappa$;

(ii) *If* $\mathfrak{F} \in \mathbf{D}^{\circ}_{1,-\infty}$ *then* $\widehat{\mathfrak{F}} \in \mathbf{D}^{\circ}_{1,-\infty}$;

(iii) *The inverse Schur transform is given by*

$$\mathfrak{F}(\lambda) = -\frac{\epsilon_0}{\mathfrak{p}_0(\lambda) + \epsilon_0 \mathfrak{b}_0^2 \widehat{\mathfrak{F}}(\lambda)}. \tag{4.13}$$

Proof. (i) Direct calculations presented in [10, Lemmas 2.1, 2.4] show that $\widehat{\mathfrak{F}}$ admits the asymptotic expansion

$$\widehat{\mathfrak{F}}(\lambda) \sim -\frac{\mathfrak{s}_0^{(1)}}{\lambda} - \frac{\mathfrak{s}_1^{(1)}}{\lambda^2} - \dots - \frac{\mathfrak{s}_{2n}^{(1)}}{\lambda^{2n+1}} - \dots, \quad \lambda \widehat{\to} \infty \tag{4.14}$$

with some $\mathfrak{s}_j^{(1)} \in \mathbb{R}$, $j \in \mathbb{Z}_+$. Setting

$$G_1(\lambda) := \lambda \widehat{\mathfrak{F}}(\lambda) \tag{4.15}$$

one obtains from (4.12)

$$\mathfrak{b}_0^2 G_1(\lambda) + \epsilon_0 \lambda \mathfrak{p}_0(\lambda) = -\frac{1}{F(\lambda)} \in \mathbf{N}_{\kappa,-\infty}. \tag{4.16}$$

Since $\deg \epsilon_0\lambda\mathfrak{p}_0(\lambda) = \mathfrak{n}_1 + 1 \geq 2$ then ∞ is a generalized pole of nonpositive type of the polynomial $\epsilon_0\lambda\mathfrak{p}_0(\lambda)$ with multiplicity

$$\kappa_\infty(\epsilon_0\lambda\mathfrak{p}_0(\lambda)) \geq 1.$$

It follows from (4.14) and (4.15) that

$$\lim_{\lambda\widehat{\to}\infty} G_1(\lambda) = -\mathfrak{s}_0^{(1)} \tag{4.17}$$

and hence ∞ is not a generalized pole of nonpositive type of G_1. By [22, Satz 1.13] one obtains

$$\kappa(G_1) + \kappa(\epsilon_0\lambda\mathfrak{p}_0(\lambda)) = \kappa(-1/F) = \kappa,$$

and hence $G_1 \in \mathbf{N}_{\kappa'',-\infty}$ for some $\kappa'' \leq \kappa - 1$.

Consider the function

$$F_1(\lambda) := \frac{\widehat{\mathfrak{F}}(\lambda)}{\lambda} = \frac{G_1(\lambda)}{\lambda^2}.$$

It follows from (4.6) that the multiplicities of generalized zeros at ∞ of F_1 and G_1 are related as follows

$$\pi_\infty(F_1) = \pi_\infty(G_1) + 1. \tag{4.18}$$

So, by a theorem of M.G. Kreĭn and H. Langer [24, Theorem 3.5] $F_1 \in \mathbf{N}_{\kappa',-\infty}$, where $\kappa' = \kappa'' + 1 \leq \kappa$.

(ii) By Proposition 4.2 $\mathfrak{n}_1 \leq 2$ in the case $\kappa = 1$.

Assume first that $\mathfrak{n}_1 = 1$. Then $\deg \lambda\mathfrak{p}_0(\lambda) = 2$, $\kappa_\infty(\lambda\mathfrak{p}_0(\lambda)) = 1$, and hence $G_1 \in \mathbf{N}_{0,-\infty}$. Then it follows from (4.18) that $F_1 \in \mathbf{N}_{1,-\infty}$.

Let now $\mathfrak{n}_1 = 2$. Then $\deg \lambda\mathfrak{p}_0(\lambda) = 3$ and in view of (4.8) the leading coefficient of $\mathfrak{p}_0$ is positive. Therefore $\kappa_\infty(\lambda\mathfrak{p}_0(\lambda)) = 1$, and hence $G_1 \in \mathbf{N}_{0,-\infty}$ and $F_1 \in \mathbf{N}_{1,-\infty}$.

(iii) The last statement is checked by straightforward calculations. □

4.4. Diagonal Padé approximants of the function $\mathfrak{F} \in \mathbf{D}_{\kappa,-\infty}$

To prove the uniform convergence of diagonal Padé approximants for a function belonging to $\mathbf{D}_{\kappa,-\infty}$, we need the following lemma.

Lemma 4.5 (cf. [12], [29]). *Let $\mathfrak{F} \in \mathbf{D}_{\kappa,-\infty}$ and let $F(\lambda) := \mathfrak{F}(\lambda)/\lambda$. Then*

$$\mathfrak{F}^{[n-1/n-1]}(\lambda) = \lambda F^{[n/n-1]}(\lambda) \text{ for every } n \in \mathcal{N}_F. \tag{4.19}$$

Proof. Suppose that $n \in \mathcal{N}_F$. Then by Theorem 3.3 the Padé approximant $F^{[n/n-1]}$ exists and

$$F^{[n/n-1]}(\lambda) + \sum_{j=0}^{2n-2} \frac{s_j}{\lambda^{j+1}} = O(\lambda^{-2n}), \quad \lambda\widehat{\to}\infty. \tag{4.20}$$

Multiplying by λ one obtains

$$\lambda F^{[n/n-1]}(\lambda) + \sum_{j=0}^{2n-2} \frac{s_j}{\lambda^j} = O(\lambda^{-(2n-1)}), \quad \lambda\widehat{\to}\infty. \tag{4.21}$$

Now the first term in (4.21) can be represented as

$$\lambda F^{[n/n-1]}(\lambda) = \lambda \frac{A^{[n/n-1]}(1/\lambda)}{B^{[n/n-1]}(1/\lambda)},$$

where $\deg A^{[n/n-1]} \le n$, $\deg B^{[n/n-1]} \le n-1$, and $B^{[n/n-1]}(0) \neq 0$. Moreover, it follows from the asymptotic expansion (4.20) that $A^{[n/n-1]}(0) = 0$. Hence, $A_1\left(\frac{1}{\lambda}\right) = \lambda A^{[n/n-1]}\left(\frac{1}{\lambda}\right)$ is a polynomial in $\frac{1}{\lambda}$ of degree $\le n-1$. This proves that

$$\lambda F^{[n/n-1]}(\lambda) = \frac{A_1(1/\lambda)}{B^{[n/n-1]}(1/\lambda)},$$

where $\deg A_1 \le n-1$, $\deg B^{[n/n-1]} \le n-1$, and $B^{[n/n-1]}(0) \neq 0$. So, it follows from (4.21) that $\lambda F^{[n/n-1]}(\lambda)$ is the $[n-1/n-1]$ Padé approximant for $\mathfrak{F}$. □

Theorem 4.6. *Let $\mathfrak{F} \in \mathbf{D}_{\kappa,-\infty}$ and let*

$$F(\lambda) := \frac{1}{\lambda}\mathfrak{F}(\lambda) \sim -\sum_{j=0}^{\infty} \frac{s_j}{\lambda^{j+1}}, \quad \lambda \widehat{\rightarrow} \infty,$$

generate the determinate moment problem $\mathbf{M}_\kappa(\mathbf{s})$. Then the sequence of diagonal Padé approximants $\{\mathfrak{F}^{[n-1/n-1]}\}_{n\in\mathcal{N}_F}$ converges to $\mathfrak{F}$ locally uniformly on $\mathbb{C} \setminus (\mathbb{R} \cup \mathcal{P}(\mathfrak{F}))$.

Moreover, if condition (3.10) *is fulfilled for the function F of the form* (1.3) *then the sequence of diagonal Padé approximants converges to $\mathfrak{F}$ locally uniformly on $\mathbb{C} \setminus ([-1-\varepsilon, 1+\varepsilon] \cup \mathcal{P}(\varphi))$ for some $\varepsilon > 0$.*

Proof. It follows from Theorem 3.3 that the sequence $\{F^{[n/n-1]}\}_{n\in\mathcal{N}_F}$ converges to F locally uniformly on $\mathbb{C}\setminus(\mathbb{R}\cup\mathcal{P}(F))$. Since $\mathbb{R}\cup\mathcal{P}(\mathfrak{F}) = \mathbb{R}\cup\mathcal{P}(F)$ the statement on the convergence on $\mathbb{C} \setminus (\mathbb{R} \cup \mathcal{P}(\mathfrak{F}))$ is implied by Lemma 4.5.

Under the condition (3.10) the convergence on $\mathbb{C}\setminus([-1-\varepsilon, 1+\varepsilon]\cup\mathcal{P}(\varphi))$ for some $\varepsilon > 0$ is a consequence of Theorem 3.5. □

Remark 4.7. It should be noted that the above theorem and the appropriate variation of [7, Theorem 1.5.2] give us the possibility to make conclusions on the locally uniform convergence of Padé approximants for the function $\mathfrak{F}$ such that

$$\frac{\mathfrak{F}(\lambda)}{\lambda+\zeta} \in \mathbf{N}_{\kappa,-\infty}$$

for some $\zeta \in \mathbb{R}$ and $\mathfrak{F}(\lambda) = O(1)$ as $\lambda \widehat{\rightarrow} \infty$.

4.5. Generalized Jacobi matrix associated with the function $\mathfrak{F} \in \mathbf{D}^\circ_{\kappa,-\infty}$

Theorem 4.8. *Let $\mathfrak{F} \in \mathbf{D}^\circ_{\kappa,-\infty}$, let the sequence $\mathfrak{s} = \{\mathfrak{s}_j\}_{j=0}^\infty$ be defined by the asymptotic expansion* (4.3), *and let $\mathcal{N}(\mathfrak{s}) = \{\mathfrak{n}_i\}_{i=1}^\infty$ be the set of normal indices of $\mathfrak{s}$. Then:*

(i) *$\mathfrak{F}$ admits the expansion into the P-fraction*

$$-\frac{\epsilon_0}{\mathfrak{p}_0(\lambda)} - \frac{\epsilon_0\epsilon_1\mathfrak{b}_0^2}{\mathfrak{p}_1(\lambda)} - \cdots - \frac{\epsilon_{N-1}\epsilon_N\mathfrak{b}_{N-1}^2}{\mathfrak{p}_N(\lambda)} - \cdots, \tag{4.22}$$

where $\mathfrak{p}_i$ *are polynomials of degree* $\mathfrak{k}_i := \mathfrak{n}_{i+1} - \mathfrak{n}_i$ $(\leq 2\kappa)$, $\epsilon_i = \pm 1$, $\mathfrak{b}_i > 0$, $i \in \mathbb{Z}_+$;

(ii) *If* $\mathfrak{J}$ *is the generalized Jacobi matrix associated with the P-fraction* (4.22), *and* $\mathfrak{P}_i(\lambda)$, $\mathfrak{Q}_i(\lambda)$ *are given by*

$$\mathfrak{P}_i(\lambda) = (\mathfrak{b}_0 \dots \mathfrak{b}_{i-1})^{-1} \det(\lambda - \mathfrak{J}_{[0,i-1]}), \tag{4.23}$$

$$\mathfrak{Q}_i(\lambda) = \epsilon_0 (\mathfrak{b}_0 \dots \mathfrak{b}_{i-1})^{-1} \det(\lambda - \mathfrak{J}_{[1,i-1]}), \tag{4.24}$$

then the ith convergent to (4.22) *coincides with* $-\mathfrak{Q}_i(\lambda)/\mathfrak{P}_i(\lambda)$ *and is the* $[\mathfrak{n}_i/\mathfrak{n}_i]$ *Padé approximant for* $\mathfrak{F}(\lambda)$.

Proof. (i) It follows from Theorem 4.4 that any function $\mathfrak{F} \in \mathbf{D}^\circ_{\kappa,-\infty}$ can be represented as follows

$$\mathfrak{F}(\lambda) = -\frac{\epsilon_0}{\mathfrak{p}_0(\lambda) + \epsilon_0 \mathfrak{b}_0^2 \mathfrak{F}_1(\lambda)},$$

where $\mathfrak{p}_0$ is a monic polynomial of degree $\deg \mathfrak{p}_0 = \mathfrak{n}_1 \leq 2\kappa$ (see formula (4.7)), $\epsilon_0 = \pm 1$, $b_0 \in \mathbb{R}_+$, and $\mathfrak{F}_1 \in \mathbf{D}^\circ_{\kappa_1,-\infty}$ with $\kappa_1 \leq \kappa$. Further, one can apply Theorem 4.4 to $\mathfrak{F}_1$ and so on. Thus, the Schur algorithm leads to (4.22). To complete the proof, note that the relation $\deg \mathfrak{p}_i = \mathfrak{n}_{i+1} - \mathfrak{n}_i$ follows from [10, Corollary 3.6].
(ii) This part is proved in line with [11, Proposition 2.3] (see also [12]). □

Remark 4.9. Let $\widehat{\mathfrak{P}}_i$ be monic polynomials associated with $\mathfrak{P}_i$ by the equalities

$$\widehat{\mathfrak{P}}_i(\lambda) := (\mathfrak{b}_0 \dots \mathfrak{b}_{i-1}) \mathfrak{P}_i(\lambda), \quad \deg \mathfrak{P}_i = \mathfrak{n}_i, \quad i \in \mathbb{N}.$$

Then it follows from Theorem 3.5, Lemma 4.5, and Theorem 4.8 that $\widehat{\mathfrak{P}}_i$ are the Christoffel transformations of the polynomials $\widehat{P}_j$ corresponding to $F(\lambda) = \mathfrak{F}(\lambda)/\lambda$

$$\begin{aligned}\widehat{\mathfrak{P}}_i(\lambda) &= (b_0 \dots b_{j-1}) \left(P_j(\lambda) - \frac{P_j(0)}{P_{j-1}(0)} P_{j-1}(\lambda) \right) \frac{1}{\lambda} \\ &= \frac{\widehat{P}_j(\lambda)\widehat{P}_{j-1}(0) - \widehat{P}_{j-1}(\lambda)\widehat{P}_j(0)}{\widehat{P}_{j-1}(0)\lambda},\end{aligned}$$

such that $\mathfrak{n}_i = \deg \widehat{\mathfrak{P}}_i = \deg \widehat{P}_j = n_j - 1$, $n_j \in \mathcal{N}(\mathbf{s})$.

In the case when $\mathfrak{F} \in \mathbf{D}^\circ_{1,-\infty}$ one can simplify the form of the generalized Jacobi matrix $\mathfrak{J}$.

Proposition 4.10. *Let* $\mathfrak{F} \in \mathbf{D}^\circ_{1,-\infty}$ *satisfy the assumptions of Theorem* 4.8 *and let* $\mathfrak{J}$ *be the generalized Jacobi matrix associated with the P-fraction* (4.22). *Then* $\mathfrak{k}_i := \mathfrak{n}_{i+1} - \mathfrak{n}_i$ *is either* 1 *or* 2 *and the block matrix* $\mathfrak{A}_i$ *in* $\mathfrak{J}$ *takes the form*

$$\mathfrak{A}_i = \begin{cases} -\mathfrak{p}_0^{(i)}, & \text{if } \mathfrak{k}_i = 1; \\ \begin{pmatrix} 0 & -\mathfrak{p}_0^{(i)} \\ 1 & -\mathfrak{p}_1^{(i)} \end{pmatrix} & \text{if } \mathfrak{k}_i = 2, \end{cases}$$

where $\mathfrak{p}_0^{(i)}$, $\mathfrak{p}_1^{(i)}$ *are coefficients of the polynomials* $\mathfrak{p}_i$ *in* (4.22).

It may happen that the generalized Jacobi matrix $\mathfrak{J}$ is unbounded even in the case when $\mathfrak{F}$ is holomorphic at ∞ (see Example 2). It should be noted that bounded generalized Jacobi matrices associated to (4.22) were considered in [13].

5. Particular cases

5.1. The case when $\mathfrak{F}$ is holomorphic at ∞

Consider the function $\mathfrak{F}$ of the form

$$\mathfrak{F}(\lambda) = r_1(\lambda) \int_{-1}^{1} \frac{t d\sigma(t)}{t-\lambda} + r_2(\lambda), \tag{5.1}$$

where σ, r_1 and r_2 satisfy the assumptions (A1), (A2) and (A3′).

Theorem 5.1. *Let the function $\mathfrak{F}$ be of the form* (5.1) *and let $F(\lambda) = \mathfrak{F}(\lambda)/\lambda$. Then $\mathfrak{F} \in \mathbf{D}_{\kappa,-\infty}$ for some $\kappa \in \mathbb{Z}_+$ and the sequence of $[n-1/n-1]$ Padé approximants $\{\mathfrak{F}^{[n-1/n-1]}\}_{n\in\mathcal{N}_F}$ converges to $\mathfrak{F}$ locally uniformly in $\mathbb{C} \setminus (\mathbb{R} \cup \mathcal{P}(\varphi))$.*

Moreover, if the condition (3.10) *is fulfilled for the function F then the sequence $\{\mathfrak{F}^{[n-1/n-1]}\}_{n\in\mathcal{N}_F}$ converges to $\mathfrak{F}$ locally uniformly in $\mathbb{C} \setminus ([-1-\varepsilon, 1+\varepsilon] \cup \mathcal{P}(\varphi))$ for some $\varepsilon > 0$.*

Proof. The function $\mathfrak{F}(\lambda)/\lambda$ admits the representation

$$F(\lambda) = \frac{\mathfrak{F}(\lambda)}{\lambda} = r_1(\lambda) \int_a^b \frac{d\sigma(t)}{t-\lambda} + \widetilde{r}_2(\lambda),$$

where

$$\widetilde{r}_2(\lambda) = \frac{r_1(\lambda)}{\lambda} \int_a^b d\sigma(t) + \frac{r_2(\lambda)}{\lambda}.$$

Therefore $F \in \mathbf{N}_{\kappa,-\infty}$ (see [22]), and hence $\mathfrak{F} \in \mathbf{D}_{\kappa,-\infty}$.

The statements concerning convergence of the sequence of diagonal Padé approximants of $\mathfrak{F}$ are implied by Theorem 4.6. □

Remark 5.2. In fact, as is easily seen from [22], every function $\mathfrak{F} \in \mathbf{D}_{\kappa,-\infty}$ holomorphic at infinity admits the representation (5.1).

Example 2. Let $\theta \in \mathbb{R}$ be an irrational number and consider the function

$$\mathfrak{F}(\lambda) = \int_{-1+\cos\pi\theta}^{1+\cos\pi\theta} \frac{t d\sigma(t)}{t-\lambda}, \text{ where } d\sigma(t) = \frac{dt}{\sqrt{1-(t-\cos\pi\theta)^2}}.$$

Substitution $x = t - \cos\pi\theta$ leads to the equality

$$\mathfrak{F}(\lambda + \cos\pi\theta) = \int_{-1}^{1} \frac{(x+\cos\pi\theta)d\omega(x)}{x-\lambda}, \text{ where } d\omega(x) = \frac{dx}{\sqrt{1-x^2}}.$$

As was shown in [35] every point of $\mathbb{R}$ is an accumulation point of the set of poles of the diagonal Padé approximants for $\mathfrak{F}(\cdot + \cos\pi\theta)$. As a consequence, the diagonal Padé approximants for $\mathfrak{F}$ do not converge on $\mathbb{R}\setminus[-1+\cos\pi\theta, 1+\cos\pi\theta]$. Therefore, the corresponding generalized Jacobi matrix $\mathfrak{J}$ is unbounded.

However, there exists a subsequence of $\mathfrak{F}^{[n-1/n-1]}$ converging in a neighborhood of ∞. Indeed, applying Lemma 4.5 to the function $\mathfrak{F}-\gamma$ with $\gamma = \int_{1+\cos\pi\theta}^{-1+\cos\pi\theta} d\sigma(t)$, one obtains

$$\mathfrak{F}^{[n-1/n-1]}(\lambda) = \lambda F^{[n/n-1]}(\lambda) + \gamma, \quad \text{where } F(\lambda) = \int_{-1+\cos\pi\theta}^{1+\cos\pi\theta} \frac{d\sigma(t)}{t-\lambda}.$$

Clearly, the shifted Chebyshev polynomials $T_n(\cdot - \cos\pi\theta)$ are orthonormal with respect to σ. Consequently, we can calculate explicitly the coefficient

$$\tau_n = -\frac{T_{n+1}(\cos\pi\theta)}{2T_n(\cos\pi\theta)} = -\frac{\cos(n+1)\pi\{|\theta|\}}{2\cos n\pi\{|\theta|\}} = \frac{1}{2}(\cos\pi\{|\theta|\} - \sin\pi\{|\theta|\}\tan n\pi\{|\theta|\}),$$

where $n \in \mathbb{N}$ and $\{x\}$ denotes the fractional part of $x \in \mathbb{R}$. Since the set $\{\{n|\theta|\}\}_{n=0}^{\infty}$ is dense in $(0,1)$, there is a bounded subsequence of $\{\tau_n\}_{n=0}^{\infty}$ and thus, by Remark 3.8, there exists a subsequence of diagonal Padé approximants converging in a neighborhood of ∞.

Remark 5.3. Let us consider a function $\mathfrak{F}$ of the following form

$$\mathfrak{F}(\lambda) = \int_{-1}^{1} \frac{t d\sigma(t)}{t-\lambda}, \tag{5.2}$$

where σ is a nonnegative probability measure on $[-1,1]$. It is clear that

$$\mathfrak{F}(\lambda) = (\mathcal{J}(\mathcal{J}-\lambda)^{-1}e, e)_{\ell^2_{[0,\infty)}} = 1 + \lambda((\mathcal{J}-\lambda)^{-1}e, e)_{\ell^2_{[0,\infty)}},$$

where $\mathcal{J}$ is the classical Jacobi matrix constructed by the measure σ via the usual procedure [1]. Now, let us consider the following modified Padé approximant

$$\mathfrak{F}_*^{[n/n]}(\lambda) = (\mathcal{J}_{[0,n-1]}(\mathcal{J}_{[0,n-1]}-\lambda)^{-1}e, e) = \frac{P_n(\lambda) - \lambda Q_n(\lambda)}{P_n(\lambda)} \tag{5.3}$$

where P_n, Q_n are polynomials of the first and second kinds corresponding to the measure σ. It follows from the Markov theorem (as well as from the spectral decomposition theorem) that

$$\mathfrak{F}_*^{[n/n]} \to \mathfrak{F}$$

locally uniformly in $\mathbb{C}\setminus[-1,1]$. So, to avoid the phenomenon described in the above example one can use the modified Padé approximants (5.3) for the function $\mathfrak{F}$ of the form (5.2).

5.2. The case when supp σ has a gap

Assume now that $r_1(\lambda) \equiv r_2(\lambda) \equiv 1$ in (5.1) and the support E of the finite nonnegative measure σ is contained in the union of two intervals

$$E = [-1,\alpha] \cup [\beta, 1], \quad \alpha < 0 < \beta.$$

First, we will show that in this case the diagonal Padé approximants for $\mathfrak{F}$ have no poles inside the gap (α,β).

Proposition 5.4. *Let σ be a finite nonnegative measure on $E = [-1, \alpha] \cup [\beta, 1]$ and*

$$\mathfrak{F}(\lambda) = \int_E \frac{t d\sigma(t)}{t - \lambda}. \tag{5.4}$$

Then:

(i) $\mathfrak{F} \in \mathbf{D}^{\circ}_{1,-\infty}$;
(ii) *The polynomials $\mathfrak{P}_j$, $j \in \mathbb{N}$ have no zeros inside the gap (α, β);*
(iii) *The function*

$$\mathfrak{F}_0(\lambda) = \int_E \frac{t d\sigma(t)}{t - \lambda} - \gamma, \quad \gamma = \int_E d\sigma(t). \tag{5.5}$$

belongs to the class $\mathbf{D}_{0,-\infty}$.

Proof. (i) The first statement is implied by the equality

$$F(\lambda) = \frac{\mathfrak{F}(\lambda)}{\lambda} = \int_E \frac{d\sigma(t)}{t - \lambda} + \frac{1}{\lambda} \int_E d\sigma(t), \tag{5.6}$$

since $F \in \mathbf{N}_{1,-\infty}$.

(ii) Next, $\mathfrak{P}_j(0)$ coincides with the Hankel determinants

$$\mathfrak{P}_j(0) = \begin{vmatrix} \mathfrak{s}_1 & \dots & \mathfrak{s}_j \\ \dots & \dots & \dots \\ \mathfrak{s}_j & \dots & \mathfrak{s}_{2j-1} \end{vmatrix}$$

which are positive since $s_j = \mathfrak{s}_{j+1}$, $j = 0, 1, \dots$ are moments of the positive measure $t^2 d\sigma(t)$ with infinite support E.

Similarly, $\mathfrak{P}_j(\theta)$ coincides with the Hankel determinants

$$\mathfrak{P}_j(\theta) = \begin{vmatrix} \mathfrak{s}_1 - \theta\mathfrak{s}_0 & \dots & \mathfrak{s}_j - \theta\mathfrak{s}_{j-1} \\ \dots & \dots & \dots \\ \mathfrak{s}_j - \theta\mathfrak{s}_{j-1} & \dots & \mathfrak{s}_{2j-1} - \theta\mathfrak{s}_{2j-2} \end{vmatrix}$$

which are positive for $\theta \in (\alpha, \beta)$, since $s_j := \mathfrak{s}_{j+1} - \theta\mathfrak{s}_j$ are moments of the positive measure $t(t - \theta)d\sigma(t)$.

(iii) The third statement follows from the equality

$$F_0(\lambda) = \frac{\mathfrak{F}_0(\lambda)}{\lambda} = \int_E \frac{d\sigma(t)}{t - \lambda}. \tag{5.7}$$

□

Next, one can apply Theorem 3.5 to prove the convergence of diagonal Padé approximants for $\mathfrak{F}$ on the real line.

Theorem 5.5. *Let σ be a finite nonnegative measure on $E = [-1, \alpha] \cup [\beta, 1]$, let $\mathfrak{F}$ have the form (5.4) and let $\{\widehat{P}_j\}_{j=0}^{\infty}$ be the set of normalized polynomials orthogonal with respect to σ. Then:*

(i) *The sequence of diagonal Padé approximants $\{\mathfrak{F}^{[n-1/n-1]}\}_{n \in \mathcal{N}_F}$ converges to $\mathfrak{F}$ locally uniformly in $\mathbb{C} \setminus ((-\infty, \alpha] \cup [\beta, \infty))$;*

(ii) *The sequence* $\{\mathfrak{F}^{[n-1/n-1]}\}_{n\in\mathcal{N}_F}$ *converges to* $\mathfrak{F}$ *locally uniformly in*

$$\mathbb{C}\setminus([-1-\varepsilon,\alpha]\cup[\beta,1+\varepsilon]) \quad \textit{for some} \quad \varepsilon>0$$

if and only if the condition (3.13) *is fulfilled.*

Proof. (i) As follows from Lemma 4.5 and [12, Theorem 4.16] the Padé approximant ${\mathfrak{F}_0}^{[n-1/n-1]}$ takes the form

$$ {\mathfrak{F}_0}^{[n-1/n-1]}(\lambda) = \lambda F_0^{[n/n-1]}(\lambda) = \lambda((\mathcal{J}_{[0,n-1]}^{(K)}-\lambda)^{-1}e,e), \quad n=n_j\in\mathcal{N}_{F_0}, \quad (5.8)$$

where $\mathcal{J}$ is a classical Jacobi matrix corresponding to the measure σ, and $\mathcal{J}_{[0,n-1]}^{(K)}$ is defined in Remark 3.2. Let us emphasize that $\mathcal{J}_{[0,n-1]}^{(K)}$ is a classical Jacobi matrix since $F_0\in\mathbf{N}_0$. By Proposition 5.4 and Theorem 4.8 ${\mathfrak{F}_0}^{[n-1/n-1]}$ is holomorphic on $(\alpha,0)\cup(0,\beta)$ and hence by (5.8) and Remark 2.5 the set $(\alpha,0)\cup(0,\beta)$ is contained in $\rho(\mathcal{J}_{[0,n-1]})$. It follows from the spectral theorem that for arbitrary $\varepsilon>0$

$$\|(\mathcal{J}_{[0,n-1]}^{(K)}-\lambda)^{-1}\|\leq\frac{1}{\varepsilon} \quad \text{for all } \lambda\in(\alpha+\varepsilon,-\varepsilon)\cup(\varepsilon,\beta-\varepsilon).$$

Then by the Vitali theorem the sequence $\{{\mathfrak{F}_0}^{[n-1/n-1]}(\lambda)\}_{n\in\mathcal{N}_F}$ converges to

$$\mathfrak{F}_0(\lambda)=\lambda((\mathcal{J}-\lambda)^{-1}e,e)$$

locally uniformly on $\mathbb{C}\setminus((-\infty,\alpha]\cup\{0\}\cup[\beta,\infty))$. Moreover, for $\varepsilon>0$ small enough ${\mathfrak{F}_0}^{[n-1/n-1]}(\lambda)$ converges to $\mathfrak{F}_0(\lambda)$ uniformly on the circle $|\lambda|=\varepsilon$. Then by the mean value theorem

$${\mathfrak{F}_0}^{[n-1/n-1]}(0)=\frac{1}{2\pi}\int_0^{2\pi}{\mathfrak{F}_0}^{[n-1/n-1]}(\varepsilon e^{it})dt\to\frac{1}{2\pi}\int_0^{2\pi}\mathfrak{F}_0(\varepsilon e^{it})dt=\mathfrak{F}_0(0)$$

as $n\to\infty$. To complete the proof it remains to mention that the $[n-1/n-1]$ Padé approximants $\mathfrak{F}^{[n-1/n-1]}$, $n\in\mathcal{N}_F$ are connected with the $[n-1/n-1]$ Padé approximants $\mathfrak{F}_0^{[n-1/n-1]}$ by the equality

$$\mathfrak{F}^{[n-1/n-1]}=\mathfrak{F}_0^{[n-1/n-1]}+\gamma.$$

(ii) The sufficiency of the second statement is contained in Theorem 4.6. Let us prove the necessity by proving the inverse statement. So, suppose that

$$\sup_{n_j\in\mathcal{N}_F}\left|\frac{\widehat{P}_j(0)}{\widehat{P}_{j-1}(0)}\right|=\infty. \quad (5.9)$$

Due to (5.8) the poles of the Padé approximant ${\mathfrak{F}_0}^{[n-1/n-1]}$ coincide with eigenvalues of the matrix $\mathcal{J}_{[0,n-1]}^{(K)}$. Taking into account that $\mathcal{J}_{[0,n-1]}^{(K)}$ is a self-adjoint matrix, one obtains

$$|\lambda_{\max}(\mathcal{J}_{[0,n-1]}^{(K)})|=\|\mathcal{J}_{[0,n-1]}^{(K)}\|\geq|(\mathcal{J}_{[0,n-1]}^{(K)}e_{n-1},e_{n-1})|=\left|p_0^{(n-1)}-\frac{\widehat{P}_n(0)}{\widehat{P}_{n-1}(0)}\right|,$$

where $\lambda_{\max}(\mathcal{J}^{(K)}_{[0,n-1]})$ is the eigenvalue of the matrix $\mathcal{J}^{(K)}_{[0,n-1]}$ with the largest absolute value. Since the sequence $\{p_0^{(n-1)}\}_{n=1}^{\infty}$ is bounded, we have that infinity is an accumulation point of the set of all poles of Padé approximants ${\mathfrak{F}_0}^{[n-1/n-1]}$. $\square$

Remark 5.6. Let $\mathfrak{F}$ be a function having the following form

$$\mathfrak{F}(\lambda) = \int_{-1}^{1} \frac{\rho(t)dt}{t-\lambda},$$

where ρ is a nonvanishing on $[-1,1]$ complex-valued function. Under some assumptions on ρ, the locally uniform convergence of the diagonal Padé approximants for $\mathfrak{F}$ was proved by A. Magnus [30] (see also [35]). Using the technique of Riemann-Hilbert problems, the result was reproved by A.I. Aptekarev and W. van Assche [3].

Example 3. Let $\mathfrak{F}$ have the form (5.4) with an absolutely continuous measure $d\sigma(t) = \rho(t)dt$ such that $\rho(t)$ is an even function on $E = [-1,-\beta] \cup [\beta, 1]$ and $\rho(t) = 0$ for $t \in \mathbb{R} \setminus E$. Then the polynomials P_{2j+1} are odd (for instance, see [34, formula (5.90)]) and, therefore, $P_{2j+1}(0) = 0$ for all $j \in \mathbb{N}$. Hence the condition (3.10) is fulfilled and by Theorem 5.5 the Padé approximants $\mathfrak{F}^{[2j/2j]}$ converge to $\mathfrak{F}$ on $\mathbb{C}\setminus E$. This fact can be shown directly, since $\mathfrak{F}(\lambda)$ admits the representation

$$\mathfrak{F}(\lambda) = \varphi(\lambda^2), \quad \text{where } \varphi(\mu) = \int_{\beta^2}^{1} \frac{\sqrt{s}\rho(\sqrt{s})ds}{s-\mu} \in \mathbf{N}_0$$

and, hence, $\mathfrak{F}^{[2j/2j]}(\lambda) = \varphi^{[j/j]}(\lambda^2)$ converge to $\mathfrak{F}(\lambda) = \varphi(\lambda^2)$ for all $\lambda^2 \in \mathbb{C}\setminus[\beta^2, 1]$, or, equivalently, for all $\lambda \in \mathbb{C} \setminus E$.

Due to Remark 3.8 it is enough to find a bounded subsequence of $\{\tau_{j-1}\}_{n_j \in \mathcal{N}_F}$ to say that there exists a subsequence of diagonal Padé approximants of $\mathfrak{F}$ which converges locally uniformly in a neighborhood of ∞. In the following proposition we find a sufficient condition for the boundedness of a subsequence of $\{\tau_{j-1}\}_{n_j \in \mathcal{N}_F}$.

Proposition 5.7. *Let σ be a finite nonnegative measure on $E = [-1, \alpha] \cup [\beta, 1]$ and let $\{\widehat{P}_j\}_{j=0}^{\infty}$ be the set of monic polynomials orthogonal with respect to σ. Assume that 0 is not an accumulation point of zeros of a subsequence $\{\widehat{P}_{j_k}\}_{k=1}^{\infty}$. Then the sequence $\left\{\frac{\widehat{P}_{j_k+1}(0)}{\widehat{P}_{j_k}(0)}\right\}_{k=1}^{\infty}$ is bounded.*

Proof. The orthogonal polynomials $\widehat{P}_j$ satisfy the following recurrence relations

$$\lambda \widehat{P}_j(\lambda) = b_{j-1}^2 \widehat{P}_{j-1}(\lambda) + a_j \widehat{P}_j(\lambda) + \widehat{P}_{j+1}(\lambda), \quad b_j > 0, \quad a_j \in \mathbb{R},$$

which implies the following equality

$$\frac{\widehat{P}_{j+1}(\lambda)}{\widehat{P}_j(\lambda)} = \lambda - a_j + b_{j-1}^2 \left(- \frac{\widehat{P}_{j-1}(\lambda)}{\widehat{P}_j(\lambda)} \right). \tag{5.10}$$

It is well known (see [2]) that $-\frac{\widehat{P}_{j-1}(\lambda)}{\widehat{P}_j(\lambda)}$ belongs to $\mathbf{N}_0$ and hence there is a non-negative measure $\sigma^{(j)}$ such that

$$-\frac{\widehat{P}_{j-1}(\lambda)}{\widehat{P}_j(\lambda)} = \int_{-1}^{1} \frac{d\sigma^{(j)}(t)}{t-\lambda}.$$

Moreover $\sigma^{(j)}$ satisfies the condition

$$\int_{-1}^{1} d\sigma^{(j)}(t) = 1$$

because of the asymptotic relation $-\frac{\widehat{P}_{j-1}(\lambda)}{\widehat{P}_j(\lambda)} = -\frac{1}{\lambda} + o\left(\frac{1}{\lambda}\right)$ as $\lambda \to \infty$. Since the zeros of $\{\widehat{P}_{j_k}\}_{k=1}^{\infty}$ do not accumulate to 0, there exists $\delta > 0$ such that

$$(-\delta, \delta) \cap \operatorname{supp} \sigma^{(j_k)} = \emptyset, \quad k = 1, 2, \dots$$

So, we have the following estimate

$$\left| \frac{\widehat{P}_{j-1}(0)}{\widehat{P}_j(0)} \right| = \left| \int_{-1}^{1} \frac{d\sigma^{(j_k)}(t)}{t} \right| \leq \frac{1}{\delta}. \tag{5.11}$$

Finally, the boundedness of $\left\{ \frac{\widehat{P}_{j_k+1}(0)}{\widehat{P}_{j_k}(0)} \right\}_{k=1}^{\infty}$ follows from the boundedness of the sequences $\{a_j\}_{j=0}^{\infty}$, $\{b_j\}_{j=0}^{\infty}$, the estimate (5.11), and the equality (5.10). □

5.3. The case when σ satisfies the Szegö conditions

Now, a natural question arises: under what conditions 0 is not an accumulation point of zeros of a subsequence $\{P_{j_k}\}_{j=0}^{\infty}$? In this subsection the answer to this question is given for functions of the form (5.4) under an additional assumption that the measure $d\sigma(t) = \rho(t)dt$ satisfies the Szegö condition on each of the intervals $[-1, \alpha]$ and $[\beta, 1]$

$$\int_{-1}^{\alpha} \frac{\log \rho(t)}{\sqrt{(\alpha - t)(t+1)}} dt > -\infty, \quad \int_{\beta}^{1} \frac{\log \rho(t)}{\sqrt{(1-t)(t-\beta)}} dt > -\infty. \tag{5.12}$$

As is known the polynomials P_j have at most one zero in the interval (α, β). The information about accumulation points of these zeros can be formulated in terms of the harmonic measure $\omega(\lambda)$ of $[-1, \alpha]$ with respect to $\mathbb{C} \setminus E$, i.e., harmonic function on $\mathbb{C} \setminus E$ whose boundary values are equal 1 on $[-1, \alpha]$ and 0 on $[\beta, 1]$.

Remark 5.8. For more detailed and deep analysis of the behavior of zeros of orthogonal polynomials see [36] (see also [37]).

Assume first that $\omega(\infty)$ is an irrational number. Then by a theorem of E. Rakhmanov ([33, Theorem 0.2]) every point of (α, β) and, in particular, 0 is an accumulation point of zeros of a sequence $\{P_j\}_{j=0}^{\infty}$. However, since there is only one zero of P_j in the gap (α, β) it is possible to choose a subsequence of $\{P_j\}_{j=0}^{\infty}$ which zeros do not accumulate to 0. Further, as follows from Proposition 5.7 there is a subsequence of $[n/n]$ Padé approximants of $\mathfrak{F}$ which converges to $\mathfrak{F}$ locally uniformly on $\mathbb{C} \setminus ([-1-\varepsilon, \alpha] \cup [\beta, 1+\varepsilon])$ for some $\varepsilon > 0$.

Assume now that $\omega(\infty)$ is a rational number m/n, where $m, n \in \mathbb{N}$ and $gcd(m,n) = 1$. Then it follows from [33, formula (57)] that every accumulation point of zeros of polynomials $\{P_j\}_{j=1}^{\infty}$ in the interval (α, β) satisfies one of the equation

$$\omega_1(z) \equiv \frac{k}{n} (\mathrm{mod}\ 2), \quad k \in \mathbb{Z}, |k| \le n. \tag{5.13}$$

In the case when 0 is not a solution of the equation (5.13) it follows from Proposition 5.7 and Theorem 5.5 that the sequence $\{\mathfrak{F}^{[n_j-1/n_j-1]}\}_{n_j \in \mathcal{N}_f}$ of Padé approximants of $\mathfrak{F}$ converges to $\mathfrak{F}$ locally uniformly on $\mathbb{C} \setminus ([-1-\varepsilon, \alpha] \cup [\beta, 1+\varepsilon])$ for some $\varepsilon > 0$.

The harmonic measure of $[-1, \alpha]$ with respect to $\mathbb{C} \setminus E$ can be calculated explicitly (see [2]). Let real k be defined by

$$k^2 = \frac{2(\beta - \alpha)}{(1-\alpha)(1+\beta)}.$$

Consider the function $x = \mathrm{sn}\, w$ with the modulus k, and with the primitive periods $4K$, $2iK'$. As is known (see [2, p.190]) the mapping

$$z = \alpha + \frac{1-\alpha^2}{2\mathrm{sn}^2 \frac{K' \ln w}{\pi} + \alpha - 1} \tag{5.14}$$

maps conformally the ring

$$r := e^{-\frac{\pi K}{K'}} < |w| < 1 \tag{5.15}$$

onto the plane $\mathbb{C}$ with the cuts $[-1, \alpha] \cup [\beta, 1]$, moreover, the semicircle $|w| = 1$ $(\mathrm{Im}\ w \ge 0)$ is mapped onto the upper shore of the cut $[-1, \alpha]$.

As is well known (see [25]) the harmonic measure ω_R of the ring (5.15) has the form

$$\omega_R(w) = \frac{\ln|w| - \ln r}{\ln 1 - \ln r} = \frac{\ln|w| + \frac{\pi K}{K'}}{\frac{\pi K}{K'}}.$$

So, the harmonic measure of $[-1, \alpha]$ with respect to $\mathbb{C} \setminus E$ can be found by

$$\omega(z) = \frac{K'}{\pi K} \ln|w| + 1. \tag{5.16}$$

Let us choose $w_\infty \in (r, 1)$ and $w_0 \in (-1, r)$ such that

$$1 - 2\mathrm{sn}^2 \frac{K' \ln w_\infty}{\pi} = \alpha, \quad 1 - 2\mathrm{sn}^2 \frac{K' \ln w_0}{\pi} = \frac{1}{\alpha}. \tag{5.17}$$

Then the numbers w_∞, w_0 correspond to $z = \infty$ and $z = 0$ via (5.14). It follows from (5.14) that $\omega(\infty)$ is a rational number $\frac{m}{n}$ if and only if $\frac{K'}{\pi K} \ln w_\infty = \frac{m-n}{n}$, which in view of (5.16) is equivalent to

$$1 - 2\mathrm{sn}^2 K r = \alpha, \quad r \in Q \tag{5.18}$$

with $r = \frac{m-n}{n}$. Since $w_0 = -|w_0|$ one obtains from (5.17) and the reduction formula (see [2, Table XII])

$$1 - 2\mathrm{sn}^2 \frac{K' \ln w_0}{\pi} = 1 - \frac{2}{k^2 \mathrm{sn}^2(\frac{K' \ln |w_0|}{\pi} + iK')} = 1 - \frac{2}{k^2 \mathrm{sn}^2(\frac{K' \ln |w_0|}{\pi})}. \tag{5.19}$$

Hence one obtains that $\omega(0)$ is a rational number $\frac{m}{n}$ if and only if $\frac{K'}{\pi K} \ln |w_0|$ is a rational number, or, equivalently,

$$1 - \frac{2}{k^2 \mathrm{sn}^2(Kr')} = \frac{1}{\alpha} \tag{5.20}$$

for $r' = \frac{m-n}{n}$.

These calculations lead to the following

Proposition 5.9. *Let a finite nonnegative measure σ on $E = [-1, \alpha] \cup [\beta, 1]$ be absolutely continuous ($d\sigma(t) = \rho(t)dt$) and satisfies the Szegö conditions* (5.12). *Then:*

(i) *If α cannot be represented in the form* (5.18) *for some $r \in Q$ then there is a subsequence of $[n/n]$ Padé approximants of $\mathfrak{F}$ which converges to $\mathfrak{F}$ locally uniformly on $\mathbb{C} \setminus ([-1-\varepsilon, \alpha] \cup [\beta, 1+\varepsilon])$ for some $\varepsilon > 0$.*

(ii) *If α satisfies* (5.18) *for some $r = \frac{m}{n}$ with $m, n \in \mathbb{N}$ ($gcd(m,n) = 1$) and does not satisfies* (5.18) *for any $r' = \frac{k}{n}$ with $k \in \mathbb{N}$ $|k| \leq n$, then the sequence $\{\mathfrak{F}^{[n-1/n-1]}\}_{n \in \mathcal{N}_F}$ of Padé approximants of $\mathfrak{F}$ converges to $\mathfrak{F}$ locally uniformly on $\mathbb{C} \setminus ([-1-\varepsilon, \alpha] \cup [\beta, 1+\varepsilon])$ for some $\varepsilon > 0$.*

In this paper we considered the case of one "turning point". The following example shows that in the case of 2 "turning points" the behaviour of diagonal Padé approximants seems to be more complicated.

Example 4 ([35]). Let θ_1, θ_2, 1 ($0 < \theta_1 < \theta_2 < 1$) be rationally independent real, and let

$$F(\lambda) = \int_{-1}^{1} \frac{1}{t-\lambda} \frac{(t - \cos \pi\theta_1)(t - \cos \pi\theta_2)}{\sqrt{1-t^2}} dt.$$

Then all the diagonal Padé approximants $F^{[k/k]}$ exist, but do not converge locally uniformly on $\mathbb{C} \setminus \mathbb{R}$ since

$$\bigcap_{n=1}^{\infty} \overline{\bigcup_{k \geq n} \mathcal{P}(F^{[k/k]})} = \mathbb{C}.$$

References

[1] N.I. Achieser, *The classical moment problem*, Fizmatgiz, Moscow, 1961, 310 p.

[2] N.I. Achieser, *Elements of the theory of elliptic functions*, Nauka, Moscow, 1970.

[3] A.I. Aptekarev, W. van Assche, *Scalar and matrix Riemann-Hilbert approach to the strong asymptotics of Padé approximants and complex orthogonal polynomials with varying weight*, Journal of Approximation Theory, Vol. 129 (1994), 129–166.

[4] T.Ya. Azizov, I.S. Iokhvidov, *Foundations of the theory of linear operators in spaces with an indefinite metric*, Nauka, 1986; English translation: Linear operators in spaces with an indefinite metric, John Wiley & Sons, 1990.

[5] D. Alpay, A. Dijksma, H. Langer, The transformation of Issai Schur and related topics in an indefinite setting, Oper. Theory: Adv. Appl. Birkhäuser Verlag, Basel, Vol. 176 (2007), 1–98.

[6] Baker G., Gammel J.L., Wills J.G. *An investigation of the applicability of the Padé approximant method*, J. Math. Anal. Appl., Vol. 2, (1961), 405–418.

[7] G. Baker and P. Graves-Morris, *Padé approximants. Part I: Basic Theory*, Encyclopedia of Mathematics and Its Applications, vol. 13, Addison-Wesley, London, 1981.

[8] Ju.M. Berezanskii, *Expansions in Eigenfunctions of Self-Adjoint Operators*, Transl. Math. Monographs 17, Amer. Math. Soc., Providence, R.I., (1968).

[9] V.I. Buslaev, *On the Baker–Gammel–Wills conjecture in the theory of Padé approximants*, Sbornik: Mathematics V. 193:6 (2002), 811–823.

[10] M. Derevyagin, *On the Schur algorithm for indefinite moment problem*, Methods of Functional Analysis and Topology, Vol. 9 (2003), No. 2, 133–145.

[11] M. Derevyagin, V. Derkach, *Spectral problems for generalized Jacobi matrices*, Linear Algebra Appl., Vol. 382 (2004), 1–24.

[12] M. Derevyagin, V. Derkach, *On convergence of Padé approximants for generalized Nevanlinna functions.* Trans. Moscow Math. Soc., Vol. 68 (2007), 133–182.

[13] M. Derevyagin, *Generalized Jacobi operators in Krein spaces*, J. Math. Annal. Appl. 349 (2009) 568–582.

[14] V. Derkach, *On Kreĭn space symmetric linear relations with gaps*, Methods of Functional Analysis and Topology, Vol. 4 (1998), No. 2, 16–40.

[15] V.A.Derkach, S. Hassi, H.S.V. de Snoo, *Generalized Nevanlinna functions with polynomial asymptotic behaviour* Oper. Theory: Adv. Appl. Birkhäuser Verlag, Basel, Vol.122 (2001), 169–189.

[16] A.A. Gonchar, *On convergence of Padé approximants for some classes of meromorphic functions,* Mat. Sb. 97 (139) (1975), 607–629; English transl. in Math. USSR-Sb. 26 (1975).

[17] I. Gohberg, P. Lankaster and L. Rodman, *Matrix polynomials*, Academic Press, New York, 1982.

[18] P. Jonas, *Operator representations of definitizable functions*, Annales Acad. Sci. Fennicae, Series A. I. Mathematica, Vol. 25 (2000), 41–72.

[19] W.B. Jones and W.J. Thron, *Continued fractions*, Addison-Wesley Publishing Company, London-Amsterdam, 1980.

[20] T. Kato, *Perturbation theory for linear operators*, Springer-Verlag, 1966.

[21] M.G. Krein, *The theory of self-adjoint extensions of semibounded Hermitian operators and its applications. I.*, Mat. Sb., Vol. 20 (1947), No. 3, 431–495.

[22] M.G. Kreĭn and H. Langer, *Über einige Fortsetzungsprobleme, die eng mit der Theorie hermitescher Operatoren im Raume* Π_κ *zusammenhängen. I.*, Math. Nachr., Vol. 77 (1977), 187–236.

[23] M.G. Kreĭn and H. Langer, *On some extension problem which are closely connected with the theory of hermitian operators in a space Π_κ III. Indefinite analogues of the Hamburger and Stieltjes moment problems*, Beiträge zur Anal. Vol. 14 (1979), 25–40.

[24] M.G. Kreĭn and H. Langer, *Some propositions on analytic matrix functions related to the theory of operators in the space* Π_κ, Acta Sci. Math. (Szeged), Vol. 43 (1981), 181–205.

[25] N.F. Landkof, *Foundations of modern potential theory*, Springer-Verlag, Berlin, 1972, 424 p.

[26] H. Langer, *A characterization of generalized zeros of negative type of functions of the class* $\mathbf{N}_\kappa$, Oper. Theory Adv. Appl., Vol. 17 (1986), 201–212.

[27] G.L. Lopes, *On the convergence of the Padé approximants for meromorphic functions of Stieltjes type*, Mat. Sb. 111 (153) (1980), 308–316; English transl. in Math. USSR-Sb. 39 (1981).

[28] D.S. Lubinsky, *Rogers-Ramanujan and the Baker–Gammel–Wills (Padé) conjecture*, Ann. of Math., Vol. 157 (2003), 847–889.

[29] A. Magnus, *Certain continued fractions associated with the Padé table*, Math. Zeitschr., Vol. 78 (1962), 361–374.

[30] A.P. Magnus, *Toeplitz matrix techniques and convergence of complex weight Padé approximants*, J. Comp. and Applied Mathematics, Vol. 19 (1987), 23–38.

[31] A.A. Markov, *Two proofs of convergence of certain continued fractions*, Selected papers in the theory of continued fractions and functions least deviating from zero, Gostekhizdat, Moscow, 1948, 106–119.

[32] E.M. Nikishin and V.N. Sorokin, *Rational approximation and orthogonality*, Translations of Mathematical Monographs 92, Amer. Math. Soc., Providence, R.I., 1991.

[33] E.A. Rakhmanov, *Convergence of diagonal Padé approximants*, Mat. Sb. Vol. 104 (146) (1977), 271–291; English transl. in Math. USSR-Sb. 33 (1977).

[34] B. Simon, *The classical moment problem as a self-adjoint finite difference operator*, Advances in Mathematics Vol. 137 (1998), 82–203.

[35] H. Stahl, *On the divergence of certain Padé approximants and the behaviour of the associated orthogonal polynomials.* In: "Polynomes Ortogonaux et Applications", Lect. Notes Math. 1171. Springer Verlag. Heidelberg, 1983, 321–330.

[36] S.P. Suetin, *Approximation properties of the poles of diagonal Padé approximants for certain generalizations of Markov functions*, Mat. Sb., 2002, 193:12, 105–133.

[37] S.P. Suetin, *On polynomials orthogonal on several segments with indefinite weight*, Uspehi Matem. Nauk, 2005, 60:5(365), 181–182.

Maxim Derevyagin
Institute of Applied Mathematics and Mechanics
Donetsk, Ukraine
e-mail: `derevyagin.m@gmail.com`

Vladimir Derkach
Department of Mathematics
Donetsk University, Ukraine
e-mail: `derkach.v@gmail.com`

Operator Theory:
Advances and Applications, Vol. 198, 125–169

Approximation of $\mathcal{N}_\kappa^\infty$-functions II: Convergence of Models

Aad Dijksma, Annemarie Luger and Yuri Shondin

Dedicated to Peter Jonas, in memoriam

Abstract. This paper is a continuation of Part I, [9] in the list of references, where models for $\mathcal{N}_\kappa^\infty$-functions have been studied in detail. In the present paper we investigate the convergence of the corresponding models as a singular $\mathcal{N}_\kappa^\infty$-function is approximated by regular $\mathcal{N}_\kappa^\infty$-functions. This involves the theory about approximating an operator by operators acting in different spaces. In the last section an example related to the Bessel differential operator is worked out.

Mathematics Subject Classification (2000). Primary 47B25, 47B50, 47B32; Secondary 47A06.

Keywords. Generalized Nevanlinna function, realization, model, reproducing kernel space, Pontryagin space, self-adjoint operator, symmetric operator, linear relation, approximation, varying space, Bessel operator.

5. Introduction to Part II

5.1. By $\mathcal{N}_\kappa^{\nu\times\nu}$ ($\mathcal{N}_\kappa$ if $\nu = 1$) we denote the class of $\nu \times \nu$ matrix-valued generalized Nevanlinna functions with κ negative squares which means that the Nevanlinna kernel

$$K_N(z, w) = \frac{N(z) - N(w)^*}{z - w^*}$$

has κ negative squares. These functions can also be characterized via minimal models in a Pontryagin space with κ negative squares. Recall that a model (short for self-adjoint operator realization) of a function $N \in \mathcal{N}_\kappa^{\nu\times\nu}$ is the pair (A, Γ_z) or one of the triples $(\mathcal{P}, A, \Gamma_z)$ and (A, Γ_z, S), where A is a self-adjoint relation in a

The second author gratefully acknowledges support from the "Fond zur Förderung der wissenschaftlichen Forschung" (FWF, Austria), grant numbers P15540-N05 and J2540-N13.

Pontryagin space $\mathcal{P}$ with nonempty resolvent set $\rho(A)$, $\Gamma_z : \mathbb{C}^\nu \to \mathcal{P}$, $z \in \rho(A)$, is a Γ-field of mappings satisfying

$$\Gamma_z = (I_{\mathcal{P}} + (z-\zeta)(A-z)^{-1})\Gamma_\zeta, \quad \zeta,\, z \in \rho(A),$$

and

$$\frac{N(\zeta) - N(z)^*}{\zeta - z^*} = \Gamma_z^* \Gamma_\zeta, \quad \zeta,\, z \in \rho(A),\ z \neq \zeta^*,$$

and S is the symmetric relation defined by

$$S = \{\{f,g\} \in A \,|\, \Gamma_{z_0}^*(g - z_0^* f) = 0\},$$

the right-hand side being independent of $z_0 \in \rho(A)$. In terms of the model the function N can be written as

$$N(z) = N(z_0)^* + (z - z_0^*)\Gamma_{z_0}^*\big(I_{\mathcal{P}} + (z - z_0)(A-z)^{-1}\big)\Gamma_{z_0}, \quad z \in \rho(A).$$

A model exists for every $N \in \mathcal{N}_\kappa^{\nu\times\nu}$; it can always be chosen *minimal* which means that

$$\mathcal{P} = \overline{\operatorname{span}}\,\{\Gamma_z \mathbf{c} \mid z \in \rho(A), \mathbf{c} \in \mathbb{C}^\nu\}.$$

If the model is minimal then $\rho(A)$ coincides with hol$\,(N)$, the domain of holomorphy of N. Minimal models of a given function $N \in \mathcal{N}_\kappa^{\nu\times\nu}$ are unitarily equivalent. As an example of a minimal model, to be used in Subsection 7.4, we mention the *canonical* model, where (i) the state space is $\mathcal{L}(N)$, the reproducing kernel Pontryagin space with reproducing kernel K_N, (ii) the self-adjoint relation $A = A_N$ in $\mathcal{L}(N)$ is given by

$$A := \big\{\{f,g\} \in \mathcal{L}(N)^2 \,|\, \exists\, \mathbf{c} \in \mathbb{C}^\nu :\ g(\zeta) - \zeta f(\zeta) \equiv \mathbf{c}\big\} \tag{5.1}$$

and is determined by the fact that its resolvent operator is the difference-quotient operator:

$$((A-z)^{-1})f(\zeta) = \frac{f(\zeta) - f(z)}{\zeta - z}, \quad f \in \mathcal{L}(N),$$

and (iii) the Γ-field $\Gamma_z = \Gamma_{Nz}$ is given by

$$\big(\Gamma_z \mathbf{c}\big)(\zeta) = K_N(\zeta, z^*)\mathbf{c}, \quad \mathbf{c} \in \mathbb{C}^\nu,$$

see, for example, [5] and [9, Theorem 3.1]. Note that the adjoint of Γ_{z^*} is the evaluation operator at the point z, it will be denoted by E_z:

$$E_z f = \Gamma_{z^*}^* f = f(z), \quad f \in \mathcal{L}(N).$$

We recall from [9] that, by definition, a function N belongs to the class $\mathcal{N}_\kappa^\infty$ if $N \in \mathcal{N}_\kappa$ and in the minimal model of N the only eigenvalue of the representing relation A with a non-positive eigenvector is the point ∞. Equivalently, $N \in \mathcal{N}_\kappa^\infty$ if and only if it belongs to the class $\mathcal{N}_\kappa$ and admits the irreducible representation [9, (2.15)]:

$$N(z) = (z - z_0^*)^m n(z)(z - z_0)^m + p(z), \tag{5.2}$$

where $z_0 \in \mathrm{hol}\,(N)$, m is a nonnegative integer which is minimal, p is a real polynomial, and n belongs to $\mathcal{N}_0$ such that

$$\mathrm{hol}\,(n) = \mathrm{hol}\,(N), \qquad \lim_{y\to\infty} y^{-1}n(iy) = 0, \qquad \mathrm{Re}\,n(i) = 0.$$

It follows that $m \le \kappa$, and that (1) if $m = \kappa$, then $\deg p \le 2\kappa$, or $\deg p = 2\kappa + 1$ and the leading coefficient of p is > 0, and (2) if $m < \kappa$, then $\deg p = 2\kappa$, or $\deg p = 2\kappa \pm 1$ and the leading coefficient of p is $\gtrless 0$. In the sequel we write p as

$$p(z) = p_{2\kappa+1}z^{2\kappa+1} + p_{2\kappa}z^{2\kappa} + p_{2\kappa-1}z^{2\kappa-1} + \text{lower order terms}$$

and we assume throughout that κ is an integer ≥ 1. Recall that a function n belongs to $\mathcal{N}_0$ if and only if it has the integral representation:

$$n(z) = \alpha + \beta z + \int_{\mathbb{R}} \left(\frac{1}{t-z} - \frac{t}{t^2+1} \right) d\sigma(t), \tag{5.3}$$

where $\alpha, \beta \in \mathbb{R}$, $\beta \ge 0$, and the spectral function σ is nondecreasing and satisfies

$$\int_{\mathbb{R}} \frac{d\sigma(t)}{t^2+1} < \infty.$$

The Nevanlinna function n appearing in the irreducible representation (5.2) of $N \in \mathcal{N}_\kappa^\infty$ has the property that the linear term $\alpha + \beta z$ in its integral representation (5.3) is equal to 0, that is,

$$n(z) = \int_{\mathbb{R}} \left(\frac{1}{t-z} - \frac{t}{t^2+1} \right) d\sigma(t). \tag{5.4}$$

If $m > 0$ then the minimality of m implies that the measure $d\sigma$ is infinite:

$$\int_{\mathbb{R}} d\sigma(t) = \lim_{y\to\infty} y\mathrm{Im}\,n(iy) = \infty. \tag{5.5}$$

In the paper we shall mainly consider $\mathcal{N}_\kappa^\infty$-functions N for which in the irreducible representation (5.2) $m = \kappa$ or $m = 0$. In these cases N will be called *singular* or *regular* respectively; in the latter case ∞ is a regular critical point for the self-adjoint relation in the minimal model for N, see [9, formulas (2.19) and (2.20)].

The aim of this paper is to show that every singular function N from the class $\mathcal{N}_\kappa^\infty$ can be approximated by regular functions $N^{\{\ell\}}$, $\ell = 1, 2, \ldots$, from the same class and to investigate how the corresponding models constructed in [9] converge as $N^{\{\ell\}} \to N$, $\ell \to \infty$. In the study of this convergence we encounter the problem of approximating an operator by operators acting in different Pontryagin spaces. Such approximation problems in Banach and Hilbert space settings have been considered in [13, pp. 512, 513] and [22]; for approximations of this kind in an indefinite setting, see [18] and [17].

As a motivation to study the class $\mathcal{N}_\kappa^\infty$ serves the fact that functions from this class appear in connection with strongly (also called super) singular perturbations – even if realizations of such perturbations are constructed in a Hilbert space as in [14]. Moreover, when approximation questions come into play then functions from this class have good convergence properties, since the generalized poles not of positive type are, by assumption, fixed to one point, namely ∞. For related convergence problems in the class $\mathcal{N}_\kappa$ see [15]. The present paper was inspired by the recent articles [20] and [19], where approximations of strongly singular perturbations realized as operators in varying Pontryagin spaces were studied.

We summarize the contents of this paper. In Section 6 we study convergence of generalized Nevanlinna functions from $\mathcal{N}_\kappa^\infty$. We show in Theorem 6.5 that if the functions $N^{\{\ell\}} \to N$ in $\mathcal{N}_\kappa^\infty$, then the Nevanlinna functions and polynomials in the regularized representation of $N^{\{\ell\}}$ (for the definition see Subsection 5.2 below) also converge. We prove that every, in particular, every singular, $\mathcal{N}_\kappa^\infty$-function can be approximated locally uniformly by regular $\mathcal{N}_\kappa^\infty$-functions, see Theorem 6.8. In Section 7 we recall, following [17, 18], the definitions and results about strong approximation of operators in varying Pontryagin spaces. To apply this theory to our model approximation problem we make an assumption on the spectral measures, namely Hypothesis 7.5. This assumption is not very restrictive since every singular $\mathcal{N}_\kappa^\infty$-function can be approximated by regular functions satisfying this hypothesis, see Remark 7.6. The main results about approximation of models in this paper are given in Theorems 7.7, 7.9, and 7.10. In the first two theorems we consider the models recalled in the Subsection 5.2 below and in the third theorem the canonical models mentioned above. Finally, in Section 8 we consider an approximation problem associated with the Bessel differential expression announced in [8] and briefly described at the end of the Introduction in [9]. We construct an explicit family of regular boundary eigenvalue problems with eigenvalue depending boundary conditions which converge to the Bessel operator. The convergence is described in terms of Straus extensions, see Theorem 8.1, Theorem 8.6, and Theorem 8.7.

5.2. We recall the results from [9] used in the sequel. In [9] we decomposed a function $N \in \mathcal{N}_\kappa^\infty$ in various ways:

$$\begin{aligned} N(z) &= (z - z_0^*)^m n(z)(z - z_0)^m + p(z) \\ &= (z - z_0^*)^m (n(z) + q(z))(z - z_0)^m + r(z) && (5.6) \\ &= c_\Lambda^\#(z) n_\Lambda(z) c_\Lambda(z) + r_\Lambda(z) && (5.7) \\ &= (z - z_0^*)^\kappa n_r(z)(z - z_0)^\kappa + r_0(z). && (5.8) \end{aligned}$$

Here the first equality is the irreducible representation (5.2) of N and the second equality is obtained from the first by writing p as

$$p(z) = (z - z_0^*)^m q(z)(z - z_0)^m + r(z),$$

where q and r are real polynomials with $\deg r < 2m$. Evidently, if $m = 0$, then $q = p$, and if $m = \kappa$, then $q(z) = p_{2\kappa} + \kappa(z_0 + z_0^*)p_{2\kappa+1} + p_{2\kappa+1}z$. We call the expressions (5.7) and (5.8) *regularized representations* of N; the reason for this is explained in [9, Section 4]. In (5.7) $\Lambda = (\lambda_1, \ldots, \lambda_\kappa)$ is a κ-tuple of points from $\operatorname{hol}(N)$, r_Λ is a real polynomial with $\deg r_\Lambda < 2\kappa$,

$$c_\Lambda(z) = (z - \lambda_1) \cdots (z - \lambda_\kappa), \quad c^\#(z) = c(z^*)^*,$$

and, if R_λ stands for the difference-quotient operator

$$(R_\lambda f)(z) = \frac{f(z) - f(\lambda)}{z - \lambda}, \tag{5.9}$$

f being any scalar or vector function, then

$$n_\Lambda = R_{\lambda_1} R_{\lambda_2} \cdots R_{\lambda_\kappa} R_{\lambda_1^*} R_{\lambda_2^*} \cdots R_{\lambda_\kappa^*} N.$$

By [9, Lemma 4.1] n_Λ is a Nevanlinna function; its spectral function can be expressed explicitly in terms of the spectral function σ of n in (5.4). The equality (5.8) is obtained from (5.7) by specializing it to $\Lambda = (z_0, \ldots, z_0)$, then we write n_r and r_0 instead of n_Λ and r_Λ. If N is regular ($m = 0$) and n is given by (5.4), then according to [9, (4.5)]

$$n_r(z) = p_{2\kappa} + \kappa(z_0 + z_0^*)p_{2\kappa+1} + p_{2\kappa+1} z + \int_{\mathbb{R}} \frac{1}{t - z} \frac{d\sigma(t)}{|t - z_0|^{2\kappa}}. \tag{5.10}$$

If in (5.6) $m = \kappa$, then the decompositions (5.6) and (5.8) coincide, that is,

$$n_r(z) = n(z) + q(z), \quad r(z) = r_0(z).$$

In [9] we constructed models for $N \in \mathcal{N}_\kappa^\infty$ based on the representations (5.6) and (5.8). First we recall from [9, Theorem 3.2(ii)] the model associated with (5.6) for the case that N is singular, that is, $m = \kappa$, and in (5.2) the polynomial p has degree $\leq 2\kappa$, which implies that in (5.6) $q(z) = p_{2\kappa}$, the coefficient of $z^{2\kappa}$ in $p(z)$. Then we recall from [9, Theorem 4.3] the model associated with (5.8). These models will be used in Subsections 7.2 and 7.3.

(I) Let $N \in \mathcal{N}_\kappa^\infty$ be singular with irreducible representations (5.2) and (5.6):

$$N(z) = (z - z_0^*)^\kappa n(z)(z - z_0)^\kappa + p(z) = (z - z_0^*)^\kappa (n(z) + p_{2\kappa})(z - z_0)^\kappa + r(z),$$

where p and r are real polynomials with $\deg p \leq 2\kappa$ and $\deg r < 2\kappa$. As noted above the last representation coincides with the regularized representation. Consider the matrix-valued generalized Nevanlinna functions

$$\widetilde{N}(z) = \begin{pmatrix} n(z) & 0 \\ 0 & M(z) \end{pmatrix}, \quad M(z) = \begin{pmatrix} r(z) & (z - z_0^*)^\kappa \\ (z - z_0)^\kappa & 0 \end{pmatrix}.$$

Then in the reproducing kernel space $\widetilde{\mathcal{P}} = \mathcal{L}(\widetilde{N}) = \mathcal{L}(n) \oplus \mathcal{L}(M)$ with kernel $K_{\widetilde{N}}$ a minimal model for N is given by

$$(B, \Gamma_z, \widetilde{S}), \tag{5.11}$$

consisting of the self-adjoint relation

$$B = \big\{\{\widetilde{f}, \widetilde{g}\} \in \mathcal{L}(\widetilde{N}) \mid \exists \mathbf{h} \in \mathbb{C}^3 : \widetilde{g}(\zeta) - \zeta \widetilde{f}(\zeta) = (I_{\mathbb{C}^3} + \widetilde{N}(\zeta)\mathcal{B})\mathbf{h}\big\},$$

the defect function $\Gamma_z = K_{\widetilde{N}}(\,\cdot\,, z^*)v(z)$, and the symmetric relation

$$\widetilde{S} = \big\{\{\widetilde{f}, \widetilde{g}\} \in \mathcal{L}(\widetilde{N}) \mid \exists \mathbf{h} \in \mathbb{C}^3 \text{ with } h_2 = 0 : \widetilde{g}(\zeta) - \zeta \widetilde{f}(\zeta) = (\mathcal{A} + \widetilde{N}(\zeta)\mathcal{B})\mathbf{h}\big\},$$

where

$$v(z)=\begin{pmatrix}(z-z_0)^\kappa\\ 1\\ (z-z_0)^\kappa(n(z)+p_{2\kappa})\end{pmatrix},\quad \mathcal{A}=\begin{pmatrix}1&0&0\\0&0&0\\0&0&1\end{pmatrix},\quad \mathcal{B}=-\begin{pmatrix}0&0&1\\0&0&0\\1&0&0\end{pmatrix}.$$

The canonical self-adjoint extensions of $\widetilde{S}$ in $\widetilde{\mathcal{P}}$ can be put in one-to-one correspondence with the parameter $\alpha\in\mathbb{R}\cup\{\infty\}$ and denoted by $B^{\langle\alpha\rangle}$. They are described explicitly in [9, Theorem 3.3(ii)]; in particular, $B^{\langle\infty\rangle}=B$. In the sequel we only use the formula for their resolvents as finite-dimensional perturbations of the resolvent of the self-adjoint relation $A_{\widetilde{N}}$ (see (5.1)) in the canonical model of $\widetilde{N}$ (see [9, (4.24) and (4.25)]):

$$(B^{\langle\alpha\rangle}-z)^{-1}=(A_{\widetilde{N}}-z)^{-1}-\frac{1}{N(z)-\alpha}\Gamma_{\widetilde{N}z}K_\alpha(z)E_z,\quad \alpha\in\mathbb{R}, \tag{5.12}$$

and

$$(B-z)^{-1}=(A_{\widetilde{N}}-z)^{-1}+\Gamma_{\widetilde{N}z}K(z)E_z, \tag{5.13}$$

where

$$K_\alpha(z)=\begin{pmatrix}(z-z_0^*)^\kappa(z-z_0)^\kappa & (z-z_0)^\kappa & \alpha-r(z)\\ (z-z_0^*)^\kappa & 1 & (z-z_0^*)^\kappa(n(z)+p_{2\kappa})\\ \alpha-r(z) & (n(z)+p_{2\kappa})(z-z_0)^\kappa & (n(z)+p_{2\kappa})(\alpha-r(z))\end{pmatrix},$$

$$K(z)=\lim_{\alpha\to\infty}\frac{1}{\alpha}K_\alpha(z)=\begin{pmatrix}0&0&1\\0&0&0\\1&0&n(z)+p_{2\kappa}\end{pmatrix}.$$

(II) The model for $N\in\mathcal{N}_\kappa^\infty$ related to the regularized representation (5.8):

$$N(z)=(z-z_0^*)^\kappa n_r(z)(z-z_0)^\kappa+r_0(z)$$

with $\deg r_0<2\kappa$ is, because of the similarity with case (I), the same as in case I with the same matrices $\mathcal{A}$ and $\mathcal{B}$ but with $n(z)$ as well as $n(z)+p_{2\kappa}$ replaced by $n_r(z)$ and $r(z)$ replaced by $r_0(z)$. This model will only be applied when N is regular, that is, $m=0$. There is one important difference with the model in case (I), the model here, in case (II), need not be minimal, but as we shall see in Section 7 (see also Theorem 6.5), it has good approximation properties. In the sequel, related to this model we provide $\widetilde{N}$, M, $\widetilde{\mathcal{P}}$, B, Γ_{Nz}, v, $K_\alpha(z)$, $B^{\langle\alpha\rangle}$, etc. with an index r, that is, we denote them by $\widetilde{N}_r$, M_r, $\widetilde{\mathcal{P}}_r$, B_r, $\Gamma_{\widetilde{N}_r z}$, v_r, $K_{r,\alpha}(z)$, $B_r^{\langle\alpha\rangle}$, etc. In Section 7 we consider a sequence $N^{\{\ell\}}$, $\ell=1,2,\dots$, of regular $\mathcal{N}_\kappa^\infty$-functions, in this case without further explanation we write $\widetilde{N}_r^{\{\ell\}}$, $M_r^{\{\ell\}}$, $\widetilde{\mathcal{P}}_r^{\{\ell\}}$, $B_r^{\{\ell\}}$, $\Gamma_{\widetilde{N}_r,z}^{\{\ell\}}$, $v_r^{\{\ell\}}$, $K_{r,\alpha}^{\{\ell\}}(z)$, $B_r^{\{\ell\}\langle\alpha\rangle}$, etc. for the model of $N^{\{\ell\}}$.

We thank the referee for careful reading of the manuscript and useful critical comments.

6. Convergence in $\mathcal{N}_\kappa^\infty$

6.1. In this section we investigate the pointwise convergence of a sequence of functions from the class $\mathcal{N}_\kappa^\infty$. We start with sequences of functions from the class $\mathcal{N}_0$. The following result is well known; we give a complete proof as we could not find one in the literature. In part it goes back at least to [3], see also [23].

Lemma 6.1. *Let $n^{\{\ell\}}$ be a sequence of Nevanlinna functions which converges pointwise in $\mathcal{F}$ to a function n_0 on $\mathcal{F}$ as $\ell \to \infty$, where $\mathcal{F}$ is a subset of $\mathbb{C} \setminus \mathbb{R}$ with an accumulation point in $\mathbb{C} \setminus \mathbb{R}$, and set*

$$\mathcal{O} = \operatorname{int}\Big(\bigcap_{\ell \geq 1} \operatorname{hol}(n^{\{\ell\}}) \Big).$$

Then there is a unique Nevanlinna function n with the following properties:

(i) *n coincides with n_0 on $\mathcal{F}$ and $\mathcal{O} \subset \operatorname{hol}(n)$.*
(ii) *$n^{\{\ell\}}$ converges locally uniformly on $\mathcal{O}$ to n.*
(iii) *If $n^{\{\ell\}}$ and n have the integral representations* (5.3):

$$n^{\{\ell\}}(z) = \alpha^{\{\ell\}} + \beta^{\{\ell\}} z + \int_{\mathbb{R}} \left(\frac{1}{t-z} - \frac{t}{t^2+1} \right) d\sigma^{\{\ell\}}(t)$$

and

$$n(z) = \alpha + \beta z + \int_{\mathbb{R}} \left(\frac{1}{t-z} - \frac{t}{t^2+1} \right) d\sigma(t),$$

then as $\ell \to \infty$,

1. $\alpha^{\{\ell\}} \longrightarrow \alpha$*:*
2. $\beta^{\{\ell\}} + \displaystyle\int_{\mathbb{R}} \frac{d\sigma^{\{\ell\}}(t)}{t^2+1} \longrightarrow \beta + \int_{\mathbb{R}} \frac{d\sigma(t)}{t^2+1},$
3. $\sigma^{\{\ell\}}(\lambda) - \sigma^{\{\ell\}}(\mu) \longrightarrow \sigma(\lambda) - \sigma(\mu)$ *for all continuity points λ and μ of σ.*

Remark 6.2. (a) Lemma 6.1 implies that if for $\ell = 1, 2, \ldots$ the set $\operatorname{hol}(n^{\{\ell\}})$ contains an open subinterval (a, b) of the real axis, then also $(a, b) \subset \operatorname{hol}(n)$ and the sequence $n^{\{\ell\}}$ converges locally uniformly to n on $(a, b) \cup (\mathbb{C} \setminus \mathbb{R})$. This is a variant of [1, Problem 95].
(b) In general $\beta^{\{\ell\}} \not\longrightarrow \beta$. An example is given by the sequence constructed in the proof of Theorem 6.4 below. (Indeed in (6.7) below $\beta^{\{\ell\}} = 0$, whereas β need not be 0.)

Proof of Lemma 6.1. Since the functions $n^{\{\ell\}}$ are symmetric with respect to the real axis, without loss of generality we may and shall assume that $\mathcal{F} \subset \mathbb{C}^+$. Consider the Cayley transforms:

$$s^{\{\ell\}}(z) := \frac{n^{\{\ell\}}(z) - i}{n^{\{\ell\}}(z) + i} \quad \text{and} \quad s_0(z) := \frac{n_0(z) - i}{n_0(z) + i}.$$

Since the functions $s^{\{\ell\}}$ are holomorphic and bounded by 1 on $\mathbb{C}^+$ and on $\mathcal{F}$ pointwise convergent to s_0, by the Vitali-Porter theorem (see, for example, [2, Corollary 7.5]) s_0 can be extended by holomorphy to a function s bounded by 1 on $\mathbb{C}^+$ and the sequence $s^{\{\ell\}}$ converges locally uniformly on $\mathbb{C}^+$ to s. Applying the inverse Cayley transform:

$$n^{\{\ell\}}(z) = i\frac{1+s^{\{\ell\}}(z)}{1-s^{\{\ell\}}(z)} \quad \text{and} \quad n(z) := i\frac{1+s(z)}{1-s(z)}$$

and using that the functions $n^{\{\ell\}}$ are symmetric with respect to the real axis, we find that n is a Nevanlinna function which extends n_0 and that the sequence $n^{\{\ell\}}$ converges locally uniformly to n on $\mathbb{C}\setminus\mathbb{R}$. After the proof of item (iii) we will show that the inclusion $\mathcal{O} \subset \mathrm{hol}\,(n)$ holds and that the convergence is locally uniform on $\mathcal{O}$.

As to item (iii), because n is a Nevanlinna function it has an integral representation as in the theorem, and the first two limits are evident:

$$\alpha^{\{\ell\}} = \mathrm{Re}\, n^{\{\ell\}}(i) \longrightarrow \mathrm{Re}\, n(i) = \alpha, \tag{6.1}$$

$$\beta^{\{\ell\}} + \int_{\mathbb{R}} \frac{d\sigma^{\{\ell\}}(t)}{t^2+1} = \mathrm{Im}\, n^{\{\ell\}}(i) \longrightarrow \mathrm{Im}\, n(i) = \beta + \int_{\mathbb{R}} \frac{d\sigma(t)}{t^2+1}. \tag{6.2}$$

To prove item (iii) 3. note first that in the integral representations of $n^{\{\ell\}}$ and n without loss of generality we may assume $\sigma^{\{\ell\}}(0) = \sigma(0) = 0$. Otherwise replace $\sigma^{\{\ell\}}(t)$ by $\sigma^{\{\ell\}}(t) - \sigma^{\{\ell\}}(0)$ and $\sigma(t)$ by $\sigma(t) - \sigma(0)$ and observe that this replacement neither changes the integral representations for $n^{\{\ell\}}$ and n nor effects the differences in (iii) 3. Since the summands in the limit (6.2) are nonnegative, there exists a real number $K > 0$, such that

$$0 \le \int_{\mathbb{R}} \frac{d\sigma^{\{\ell\}}(t)}{t^2+1} \le K, \quad \ell = 1, 2, \ldots. \tag{6.3}$$

We claim that this implies that

$$|\sigma^{\{\ell\}}(x)| \le (x^2+1)K, \quad x \in \mathbb{R}, \ \ell = 1, 2, \ldots.$$

Indeed, this obviously holds for $x = 0$. For $\pm x > 0$ we have

$$\begin{aligned} |\sigma^{\{\ell\}}(x)| &= \pm\sigma^{\{\ell\}}(x) = \pm\int_0^x d\sigma^{\{\ell\}}(t) \\ &\le \pm\int_0^x \frac{x^2+1}{t^2+1} d\sigma^{\{\ell\}}(t) \le (x^2+1)\int_{\mathbb{R}} \frac{d\sigma^{\{\ell\}}(t)}{t^2+1} \le (x^2+1)K. \end{aligned}$$

Then by Helly's first theorem, formulated as in [4, Theorem 9.3.2], there exists a subsequence $\sigma^{\{\ell_j\}}$ and a nondecreasing function $\widetilde{\sigma}$ such that

$$|\widetilde{\sigma}(x)| \le (x^2+1)K, \quad x \in \mathbb{R},$$

and pointwise

$$\lim_{j\to\infty} \sigma^{\{\ell_j\}}(x) = \widetilde{\sigma}(x), \quad x \in \mathbb{R}.$$

For any $z \in \mathbb{C}\setminus\mathbb{R}$ and any $\varepsilon > 0$, there exists a number A, such that for all $a, b > A$

$$\left|\frac{1}{t-z}\right| < \varepsilon/K, \quad t \in (-\infty, -a) \cup (b, \infty),$$

and hence, on account of (6.3), for all $j = 1, 2, \ldots$

$$\int_{-\infty}^{-a} \left| \frac{1}{t-z} \right| \frac{d\sigma^{\{\ell_j\}}(t)}{t^2+1} < \varepsilon, \quad \int_{b}^{\infty} \left| \frac{1}{t-z} \right| \frac{d\sigma^{\{\ell_j\}}(t)}{t^2+1} < \varepsilon.$$

Because of these estimates, the generalization of Helly's second or integral theorem to an infinite interval, as formulated in [16, p. 500], yields

$$\lim_{j\to\infty} \int_{\mathbb{R}} \frac{1}{t-z} \frac{d\sigma^{\{\ell_j\}}(t)}{t^2+1} = \int_{\mathbb{R}} \frac{1}{t-z} \frac{d\widetilde{\sigma}(t)}{t^2+1}.$$

Together with the two limits (6.1) and (6.2) this implies that the function

$$n^{\{\ell_j\}}(z) = \alpha^{\{\ell_j\}} + z\left(\beta^{\{\ell_j\}} + \int_{\mathbb{R}} \frac{d\sigma^{\{\ell_j\}}(t)}{t^2+1}\right) + (z^2+1)\int_{\mathbb{R}} \frac{1}{t-z} \frac{d\sigma^{\{\ell_j\}}(t)}{t^2+1}$$

converges pointwise to the function

$$\alpha + z\left(\beta + \int_{\mathbb{R}} \frac{d\sigma(t)}{t^2+1}\right) + (z^2+1)\int_{\mathbb{R}} \frac{1}{t-z} \frac{d\widetilde{\sigma}(t)}{t^2+1}$$

as $j \to \infty$. On the other hand, by assumption, $\lim_{j\to\infty} n^{\{\ell_j\}}(z) = n(z)$ and thus

$$\int_{\mathbb{R}} \frac{1}{t-z} \frac{d\widetilde{\sigma}(t)}{t^2+1} = \int_{\mathbb{R}} \frac{1}{t-z} \frac{d\sigma(t)}{t^2+1}.$$

Applying the Stieltjes inversion formula we obtain that

$$\widetilde{\sigma}(\lambda) - \widetilde{\sigma}(\mu) = \sigma(\lambda) - \sigma(\mu)$$

for all continuity points λ and μ of σ. This completes the proof since every convergent subsequence of $\sigma^{\{\ell\}}(\lambda) - \sigma^{\{\ell\}}(\mu)$ tends to the same limit.
It remains to prove items (i) and (ii). We have

$$\mathcal{O} = (\mathbb{C} \setminus \mathbb{R}) \cup (\mathcal{O} \cap \mathbb{R}), \quad j = 1, 2, \ldots. \tag{6.4}$$

The last set on the right-hand side of (6.4) is a (countable) union of disjoint open intervals. Since on each such open interval every $\sigma^{\{\ell\}}$ is constant, it follows from (iii) 3. that σ is also constant there, hence $\mathcal{O} \cap \mathbb{R} \subset \operatorname{hol}(n)$. As also $\mathbb{C} \setminus \mathbb{R} \subset \operatorname{hol}(n)$ we have $\mathcal{O} \subset \operatorname{hol}(n)$ and the proof of item (i) is complete.
Now we prove (ii). We have shown above that it holds if $\mathcal{O} = \mathbb{C} \setminus \mathbb{R}$. We now assume that the inclusion $\mathbb{C} \setminus \mathbb{R} \subset \mathcal{O}$ is strict. This implies that $\mathcal{O}$ is a nonempty open connected subset of $\mathbb{C}$. Let $\mathcal{K}$ be a compact subset of $\mathcal{O}$. Write $n^{\{\ell\}}$ as

$$n^{\{\ell\}}(z) = \alpha^{\{\ell\}} + z\left(\beta^{\{\ell\}} + \int_{\mathbb{R}} \frac{d\sigma^{\{\ell\}}(t)}{t^2+1}\right) + (z^2+1)\int_{\mathbb{R}} \frac{1}{t-z} \frac{d\sigma^{\{\ell\}}(t)}{t^2+1}. \tag{6.5}$$

By item (iii) the first two summands on the right-hand side are bounded uniformly in $z \in \mathcal{K}$ and in $\ell = 1, 2, \ldots$. Since both $\beta^{\{\ell\}}$ and the integral in the second summand are nonnegative, the integral is uniformly bounded in ℓ as well: there exists a number $L > 0$ such that

$$\int_{\mathbb{R}} \frac{d\sigma^{\{\ell\}}(t)}{t^2+1} \leq L, \quad \ell = 1, 2, \ldots.$$

It follows that the last integral in (6.5) satisfies the inequality

$$\left|\int_{\mathbb{R}} \frac{1}{t-z}\frac{d\sigma^{\{\ell\}}(t)}{t^2+1}\right| = \left|\int_{\mathbb{R}\setminus\mathcal{O}} \frac{1}{t-z}\frac{d\sigma^{\{\ell\}}(t)}{t^2+1}\right| \leq L/\delta, \quad z\in\mathcal{K},\ \ell=1,2,\dots,$$

with $\delta := \operatorname{distance}(\mathcal{K},\mathbb{R}\setminus\mathcal{O}) > 0$. Thus we have shown that the functions $n^{\{\ell\}}$ are uniformly bounded on $\mathcal{K}$, and hence, since $\mathcal{K}$ is an arbitrary compact subset of $\mathcal{O}$, the functions $n^{\{\ell\}}$ are locally uniformly bounded on $\mathcal{O}$. As they converge pointwise on $\mathcal{F}\subset\mathcal{O}$, (ii) follows from the Vitali-Porter theorem. □

Definition 6.3. By $\mathcal{N}_0^f$ we denote the class of functions n of the form

$$n(z) = \gamma + \int_{\mathbb{R}} \frac{d\tau(t)}{t-z},$$

where $\gamma\in\mathbb{R}$ and τ is a bounded nondecreasing function with the property that all moments of $d\tau$ are finite, that is,

$$\int_{\mathbb{R}} t^j\,d\tau(t) < \infty, \quad j=0,1,\dots.$$

Evidently, $\mathcal{N}_0^f$ is a subset of $\mathcal{N}_0$.

Theorem 6.4. *For each $n\in\mathcal{N}_0$ there exists a sequence $f^{\{\ell\}}$ in $\mathcal{N}_0^f$ which converges to n locally uniformly on $\mathbb{C}\setminus\mathbb{R}$.*

Proof. Let $n\in\mathcal{N}_0$ be given with integral representation (5.3) :

$$n(z) = \alpha + \beta z + \int_{\mathbb{R}} \left(\frac{1}{t-z} - \frac{t}{t^2+1}\right) d\sigma(t),$$

and assume, without loss of generality, that the spectral function σ is left-continuous. For $\ell = 1,2,\dots$ define

$$\sigma^{\{\ell\}}(t) = \begin{cases} \sigma(-\ell) & \text{if } t\leq -\ell, \\ \sigma(t) & \text{if } -\ell<t<\ell, \\ \sigma(\ell) & \text{if } t\geq \ell \end{cases}$$

and the functions

$$f^{\{\ell\}}(z) = \alpha - \ell\beta + \int_{\mathbb{R}} \left(\frac{1}{t-z} - \frac{t}{t^2+1}\right) d\sigma^{\{\ell\}}(t) + \frac{\ell^2\beta}{\ell - z}. \tag{6.6}$$

Being the sum of Nevanlinna functions $f^{\{\ell\}}$ is itself a Nevanlinna function with integral representation

$$f^{\{\ell\}}(z) = \alpha^{\{\ell\}} + \beta^{\{\ell\}}z + \int_{\mathbb{R}} \left(\frac{1}{t-z} - \frac{t}{t^2+1}\right) d\tau^{\{\ell\}}(t), \tag{6.7}$$

where

$$\alpha^{\{\ell\}} = \alpha - \frac{\ell\beta}{\ell^2+1}, \quad \beta^{\{\ell\}} = 0,$$

and, if $1_{(\ell,\infty)}$ stands for the characteristic function of the open interval (ℓ,∞), the spectral function $\tau^{\{\ell\}}$ is given by

$$\tau^{\{\ell\}} = \sigma^{\{\ell\}} + \ell^2\beta 1_{(\ell,\infty)}.$$

Evidently, $f^{\{\ell\}} \in \mathcal{N}_0^f$:

$$f^{\{\ell\}}(z) = \gamma^{\{\ell\}} + \int_{\mathbb{R}} \frac{1}{t-z}\, d\tau^{\{\ell\}}(t) \quad \text{with} \quad \gamma^{\{\ell\}} = \alpha - \ell\beta - \int_{\mathbb{R}} \frac{t}{t^2+1} d\sigma^{\{\ell\}}(t).$$

We rewrite (6.6) as

$$f^{\{\ell\}}(z) = \alpha + \frac{\beta z}{1-z/\ell} + z\int_{\mathbb{R}} \frac{d\sigma^{\{\ell\}}(t)}{t^2+1} + (z^2+1)\int_{\mathbb{R}} \frac{1}{t-z}\frac{d\sigma^{\{\ell\}}(t)}{t^2+1}.$$

As $\ell \to \infty$, $\sigma^{\{\ell\}}(t) \to \sigma(t)$ for $t \in \mathbb{R}$ and therefore the summands on the right-hand side also converge:

$$\lim_{\ell\to\infty} \frac{\beta z}{1-z/\ell} = \beta z, \quad \lim_{\ell\to\infty} z\int_{\mathbb{R}} \frac{d\sigma^{\{\ell\}}(t)}{t^2+1} = z\int_{\mathbb{R}} \frac{d\sigma(t)}{t^2+1},$$

and, by Helly's second theorem, the last summand converges to

$$(z^2+1)\int_{\mathbb{R}} \frac{1}{t-z}\frac{d\sigma(t)}{t^2+1}.$$

Hence $f^{\{\ell\}}$ converges to n pointwise in $\mathbb{C}\setminus\mathbb{R}$ and, according to Lemma 6.1, this convergence is locally uniform. □

6.2. In this subsection we consider convergence in the class $\mathcal{N}_\kappa^\infty$. We first show that if $N^{\{\ell\}} \to N$ in $\mathcal{N}_\kappa^\infty$, then the Nevanlinna functions and polynomials in the regularized representation of $N^{\{\ell\}}$ converge to the Nevanlinna function and polynomial in the regularized representation of N.

Theorem 6.5. *Assume that $N^{\{\ell\}}$, $\ell = 1,2,\dots$, and N belong to $\mathcal{N}_\kappa^\infty$ and that $N^{\{\ell\}} \to N$ pointwise on $\mathbb{C}\setminus\mathbb{R}$ as $\ell\to\infty$, and set*

$$\mathcal{U} = \operatorname{int}\Big(\bigcap_{\ell\geq 1} \operatorname{hol}(N^{\{\ell\}})\Big).$$

Then the following statements hold:

(i) *$\mathcal{U} \subset \operatorname{hol}(N)$ and $N^{\{\ell\}}$ converges to N locally uniformly on $\mathcal{U}$.*

(ii) *If $N^{\{\ell\}}$ and N have the representations* (5.7):

$$N^{\{\ell\}}(z) = c_\Lambda^\#(z) n_\Lambda^{\{\ell\}}(z) c_\Lambda(z) + r_\Lambda^{\{\ell\}}(z), \;\; N(z) = c_\Lambda^\#(z) n_\Lambda(z) c_\Lambda(z) + r_\Lambda(z)$$

with $\Lambda \subset \mathcal{U}$, then $n_\Lambda^{\{\ell\}}$ converges locally uniformly to n_Λ on $\mathcal{U}$, and $r_\Lambda^{\{\ell\}}$ converges locally uniformly to r_Λ on $\mathbb{C}$.

Remark 6.6. The theorem implies that if for $\ell = 1,2,\dots$ the set $\operatorname{hol}(N^{\{\ell\}})$ contains an interval (a,b) of the real axis, then $(a,b) \subset \operatorname{hol}(N)$ and $N^{\{\ell\}}$ converges to N locally uniformly on $(a,b)\cup(\mathbb{C}\setminus\mathbb{R})$, compare with Remark 6.2.

Proof of Theorem 6.5. We prove (i) by first considering a special case of (ii); namely the case where $\Lambda = (\lambda_1, \ldots, \lambda_\kappa)$ consists of κ distinct points from $\mathbb{C} \setminus \mathbb{R}$. For this case the pointwise convergence of $N^{\{\ell\}}$ with representation (4.1) implies the pointwise convergence of the *real* polynomials $r_\Lambda^{\{\ell\}}$ of degree $\leq 2\kappa - 1$ to the real polynomial r_Λ degree $\leq 2\kappa - 1$ at the points λ_j, $j = 1, 2, \ldots, \kappa$. It follows that the convergence $r_\Lambda^{\{\ell\}} \to r_\Lambda$ is locally uniform on $\mathbb{C}$ and that the sequence $c_\Lambda^{\#} n_\Lambda^{\{\ell\}} c_\Lambda$ converges pointwise in $\mathbb{C} \setminus \mathbb{R}$ to $c_\Lambda^{\#} n_\Lambda c_\Lambda$. The latter implies that the sequence $n_\Lambda^{\{\ell\}}$ converges pointwise to n_Λ on the set $\mathcal{F} := \mathbb{C} \setminus \{\mathbb{R} \cup \{\lambda_1, \ldots, \lambda_\kappa, \lambda_1^*, \ldots, \lambda_\kappa^*\}\}$. By Lemma 6.1 with this set $\mathcal{F}$ and since in this case

$$\mathcal{U} = \operatorname{int}\Big(\bigcap_{\ell \geq 1} \operatorname{hol}(n_\Lambda^{\{\ell\}}) \Big),$$

we have that $\mathcal{U} \subset \operatorname{hol}(n_\Lambda)$ and the convergence $n_\Lambda^{\{\ell\}} \to n_\Lambda$ is locally uniform on $\mathcal{U}$. It now readily follows that $N^{\{\ell\}}$ converges to N locally uniformly on $\mathcal{U}$. This proves (i).

We now show (ii) for arbitrary $\Lambda \subset \mathcal{U}$. By (i) it follows that for $j = 1, 2, \ldots$ we have that

$$\frac{d^j}{dz^j} N^{\{\ell\}}(z) \longrightarrow \frac{d^j}{dz^j} N(z)$$

locally uniformly on $\mathcal{U}$. Hence if $\Lambda = (\lambda_1, \ldots, \lambda_\kappa)$ and for $j = 1, 2, \ldots, \kappa$ the number λ_j occurs k_j times in this sequence, then

$$\frac{d^i}{dz^i} r_\Lambda^{\{\ell\}}(\lambda_j) \longrightarrow \frac{d^i}{dz^i} r_\Lambda(\lambda_j), \quad i = 0, 1, \ldots, \widehat{k}_j,$$

where $\widehat{k}_j = 2k_j - 1$ or $k_j - 1$ depending on whether λ_j is in $\mathbb{R}$ or in $\mathbb{C} \setminus \mathbb{R}$. It follows that the sequence $r_\Lambda^{\{\ell\}}$ of real polynomials of degree $\leq 2\kappa - 1$ converges pointwise and hence locally uniformly on $\mathbb{C}$ to the real polynomial r_Λ. The arguments in the proof of item (i) can be repeated to prove that also $n_\Lambda^{\{\ell\}}$ converges locally uniformly on $\mathcal{U}$ to n_Λ. □

In the following corollary to Theorem 6.5 we spell out some limits. They will be used in the proof of Theorem 7.10.

Corollary 6.7. *Let $N \in \mathcal{N}_\kappa^\infty$ be singular with irreducible representation* (5.2)*:*

$$N(z) = (z - z_0^*)^\kappa n(z)(z - z_0)^\kappa + p(z),$$

where $\deg p \leq 2\kappa$ *and n has the integral representation* (5.4)*:*

$$n(z) = \int_{\mathbb{R}} \left(\frac{1}{t - z} - \frac{t}{t^2 + 1} \right) d\sigma(t).$$

For $\ell = 1, 2, \ldots$ let $N^{\{\ell\}} \in \mathcal{N}_\kappa^\infty$ be regular and given by

$$N^{\{\ell\}}(z) = n^{\{\ell\}}(z) + p^{\{\ell\}}(z),$$

where $\deg p^{\{\ell\}} \leq 2\kappa$ *and* $n^{\{\ell\}}$ *has the integral representation* (5.4)*:*

$$n^{\{\ell\}}(z) = \int_{\mathbb{R}} \left(\frac{1}{t-z} - \frac{t}{t^2+1} \right) d\sigma^{\{\ell\}}(t),$$

and set

$$\mathcal{U} = \operatorname{int}\Big(\bigcap_{\ell \geq 1} \operatorname{hol}(N^{\{\ell\}}) \Big).$$

Assume that $z_0 \in \mathcal{U}$ *and that* $\lim_{\ell\to\infty} N^{\{\ell\}} = N$ *locally uniformly on* $\mathcal{U}$. *Then*

$$\lim_{\ell\to\infty} \int_{\mathbb{R}} \frac{1}{t-z} \frac{d\sigma^{\{\ell\}}(t)}{|t-z_0|^{2\kappa}} + p_{2\kappa}^{\{\ell\}} = n(z) + p_{2\kappa} \text{ locally uniformly on } \mathcal{U},$$

$$\lim_{\ell\to\infty} p_{2\kappa-1}^{\{\ell\}} + \kappa(z_0+z_0^*)p_{2\kappa}^{\{\ell\}} + \int_{\mathbb{R}} \frac{d\sigma^{\{\ell\}}(t)}{|t-z_0|^{2\kappa}} = p_{2\kappa-1} + \kappa(z_0+z_0^*)p_{2\kappa},$$

in particular,

$$\lim_{\ell\to\infty} \int_{\mu}^{\lambda} \frac{d\sigma^{\{\ell\}}(t)}{|t-z_0|^{2\kappa}} = \sigma(\lambda) - \sigma(\mu) \text{ for all continuity points } \lambda, \mu \text{ of } \sigma,$$

$$\lim_{\ell\to\infty} \int_{\mathbb{R}} \frac{d\sigma^{\{\ell\}}(t)}{|t-z_0|^{2\kappa}} = +\infty, \quad \text{and} \quad \lim_{\ell\to\infty} p_{2\kappa-1}^{\{\ell\}} + \kappa(z_0+z_0^*)p_{2\kappa}^{\{\ell\}} = -\infty.$$

Proof. If we write N and $N^{\{\ell\}}$ in the regularized form with $\Lambda = (z_0, \dots, z_0)$:

$$N(z) = (z-z_0^*)^\kappa n_r(z)(z-z_0)^\kappa + r_0(z), \; N^{\{\ell\}}(z) = (z-z_0^*)^\kappa n_r^{\{\ell\}}(z)(z-z_0)^\kappa + r_0^{\{\ell\}}(z),$$

then

$$n_r(z) = n(z) + p_{2\kappa}, \quad n_r^{\{\ell\}}(z) = \int_{\mathbb{R}} \frac{1}{t-z} \frac{d\sigma^{\{\ell\}}(t)}{|t-z_0|^{2\kappa}} + p_{2\kappa}^{\{\ell\}}$$

(for the second equality see (5.10)). The first limit in the corollary, equivalent to

$$\lim_{\ell\to\infty} n_r^{\{\ell\}}(z) = n_r(z), \tag{6.8}$$

follows from Theorem 6.5. This theorem also implies the second limit, which is equivalent to

$$\lim_{\ell\to\infty} R_{z_0^*}^{\kappa-1} R_{z_0}^{\kappa} N^{\{\ell\}}(z)|_{z=z_0^*} = R_{z_0^*}^{\kappa-1} R_{z_0}^{\kappa} N(z)|_{z=z_0^*},$$

where R_λ stands for the difference-quotient operator (5.9). The third limit follows from (6.8) and Lemma 6.1 (iii). The fourth limit follows from the third and the fact that by (5.5) the measure $d\sigma$ is not finite. Finally, the last limit follows from the second and fourth limit. □

In the next theorem we show that every $\mathcal{N}_\kappa^\infty$-function can be approximated by regular $\mathcal{N}_\kappa^\infty$-functions with additional properties.

Theorem 6.8. *Each* $N \in \mathcal{N}_\kappa^\infty$ *with irreducible representation:*

$$N(z) = (z-z_0^*)^m n(z)(z-z_0)^m + p(z)$$

is the locally uniform limit in $\mathbb{C}\setminus\mathbb{R}$ *of a sequence of regular functions* $N^{\{\ell\}} \in \mathcal{N}_\kappa^\infty$ *with irreducible representation:*

$$N^{\{\ell\}} = n^{\{\ell\}} + p^{\{\ell\}},$$

where $p^{\{\ell\}}$ *is a real polynomial of degree* $\leq \max\{2m, \deg p\}$, $n^{\{\ell\}} \in \mathcal{N}_0^f$, *and the spectral measure* $d\sigma^{\{\ell\}}$ *of* $n^{\{\ell\}}$ *is uniformly absolutely continuous with respect to the spectral measure* $d\sigma$ *of* n*: there exists a number* $C > 0$ *such that for all integers* $\ell \geq 1$

$$|t - z_0|^{-2m} \frac{d\sigma^{\{\ell\}}(t)}{d\sigma(t)} \leq C \quad \textit{for almost all } t \in \mathbb{R} \textit{ with respect to } d\sigma.$$

Proof. By Theorem 6.4 there exists a sequence $f^{\{\ell\}} \in \mathcal{N}_0^f$ such that $f^{\{\ell\}} \to n$ locally uniformly in $\mathbb{C}\setminus\mathbb{R}$. Hence the sequence of generalized Nevanlinna functions

$$N^{\{\ell\}}(z) = (z - z_0^*)^m f^{\{\ell\}}(z)(z - z_0)^m + p(z)$$

converges to N locally uniformly on $\mathbb{C}\setminus\mathbb{R}$. If $f^{\{\ell\}}$ has the integral representation

$$f^{\{\ell\}}(z) = \gamma^{\{\ell\}} + \int_{\mathbb{R}} \frac{d\tau^{\{\ell\}}(t)}{t - z},$$

then also the function

$$n^{\{\ell\}}(z) = \gamma_0^{\{\ell\}} + \int_{\mathbb{R}} |t - z_0|^{2m} \frac{d\tau^{\{\ell\}}(t)}{t - z}$$

with

$$\gamma_0^{\{\ell\}} = -\int_{\mathbb{R}} \frac{t}{t^2 + 1} |t - z_0|^{2m} d\tau^{\{\ell\}}(t)$$

belongs to $\mathcal{N}_0^f$ and $\operatorname{Re} n^{\{\ell\}}(i) = 0$. It follows that

$$N^{\{\ell\}}(z) = (z - z_0^*)^m f^{\{\ell\}}(z)(z - z_0)^m + p(z) = n^{\{\ell\}}(z) + p^{\{\ell\}}(z),$$

where

$$p^{\{\ell\}}(z) = (z - z_0^*)^m \gamma^{\{\ell\}}(z - z_0)^m - \gamma_0^{\{\ell\}} - \int_{\mathbb{R}} \frac{|t - z_0|^{2m} - (z - z_0^*)^m (z - z_0)^m}{t - z} d\tau^{\{\ell\}}(t) + p(z)$$

is a real polynomial of degree at most $\max\{2m, \deg p\}$. The statement about the absolute continuity of the spectral measures follows from Theorem 6.4, its proof (with $\alpha = \beta = 0$), and the formulas given above. □

7. Approximation of models

7.1. We recall the definitions from [18, 17] concerning the approximation of an operator by operators acting in variable Pontryagin and Krein spaces. We begin with the Banach space case. Norms and inner products, positive definite as well as indefinite ones, will be denoted by $\|\cdot\|$ and $\langle\cdot,\cdot\rangle$, without, as here, or with index; it will be clear from the context on which spaces they are defined.

Definition 7.1. Let $\mathcal{B}$ and $\mathcal{B}^{\{\ell\}}$, $\ell = 1, 2, \dots$, be Banach spaces and let $\pi^{\{\ell\}}$ be linear mappings from $\mathcal{B}$ into $\mathcal{B}^{\{\ell\}}$. Then:

(i) The spaces $\mathcal{B}^{\{\ell\}}$ *approximate* $\mathcal{B}$ *with respect to* $\pi^{\{\ell\}}$, if for all $u \in \mathcal{B}$

$$\|\pi^{\{\ell\}}u\|_{\mathcal{B}^{\{\ell\}}} \to \|u\|_{\mathcal{B}} \quad \text{as } \ell \to \infty.$$

(ii) Elements $u^{\{\ell\}} \in \mathcal{B}^{\{\ell\}}$, $\ell = 1, 2, \dots$, *strongly approximate* an element $u \in \mathcal{B}$ *with respect to* $\pi^{\{\ell\}}$, in formula $u^{\{\ell\}} \xrightarrow{s} u$, if

$$\|u^{\{\ell\}} - \pi^{\{\ell\}}u\|_{\mathcal{B}^{\{\ell\}}} \to 0 \quad \text{as } \ell \to \infty.$$

(iii) Bounded linear operators $B^{\{\ell\}}$ on $\mathcal{B}^{\{\ell\}}$, $\ell = 1, 2, \dots$, *strongly approximate* a bounded linear operator B on $\mathcal{B}$ *with respect to* $\pi^{\{\ell\}}$, in formula $B^{\{\ell\}} \xrightarrow{s} B$, if for all $u \in \mathcal{B}$

$$B^{\{\ell\}}\pi^{\{\ell\}}u \xrightarrow{s} Bu \quad \text{as } \ell \to \infty.$$

If it is clear from the context we refrain from mentioning the operators $\pi^{\{\ell\}}$ with respect to which the approximation holds.

By the uniform boundedness principle the operators $\pi^{\{\ell\}}$ in Definition 7.1 (i) are uniformly bounded. If $\mathcal{B}$ and $\mathcal{B}^{\{\ell\}}$ are Hilbert spaces, denoted in this case by $\mathcal{H}$ and $\mathcal{H}^{\{\ell\}}$, respectively, then, by polarization, (i) implies that

$$\langle \pi^{\{\ell\}}u, \pi^{\{\ell\}}v\rangle_{\mathcal{H}^{\{\ell\}}} \to \langle u, v\rangle_{\mathcal{H}}, \quad u, v \in \mathcal{H}.$$

A converse also holds:

Theorem 7.2. *Let $\mathcal{H}^{\{\ell\}}$, $\ell = 1, 2, \dots$, and $\mathcal{H}$ be Hilbert spaces, let $\mathcal{D}$ be a dense linear manifold in $\mathcal{H}$, and for $\ell = 1, 2, \dots$ let $\pi^{\{\ell\}}$ be linear mappings from $\mathcal{D}$ into $\mathcal{H}^{\{\ell\}}$. If (a) the $\pi^{\{\ell\}}$'s are uniformly bounded on $\mathcal{D}$, that is, there is a real number $M > 0$ such that for all $u \in \mathcal{D}$ and all integers $\ell \geq 1$*

$$\|\pi^{\{\ell\}}u\|_{\mathcal{H}^{\{\ell\}}} \leq M\|u\|_{\mathcal{H}}$$

and (b) for all $u, v \in \mathcal{D}$

$$\langle \pi^{\{\ell\}}u, \pi^{\{\ell\}}v\rangle_{\mathcal{H}^{\{\ell\}}} \to \langle u, v\rangle_{\mathcal{H}},$$

then the spaces $\mathcal{H}^{\{\ell\}}$ approximate $\mathcal{H}$ with respect to the unique bounded continuation of each $\pi^{\{\ell\}}$ to all of $\mathcal{H}$.

In [17] this statement is given without the uniform boundedness condition for the $\pi^{\{\ell\}}$'s. The following example shows that this condition is needed. There the mappings $\pi^{\{\ell\}}$ are bounded but not uniformly bounded.

Example. Let $\mathcal{D}$ be the subset of $\mathcal{H} = L^2([0, \infty))$ consisting of all continuous functions with compact support, and let a_ℓ be a sequence of positive numbers such that

$$\lim_{\ell \to \infty} \frac{1}{\ell} \ln a_\ell = \infty.$$

Consider the spaces $\mathcal{H}^{\{\ell\}} = L^2_{\sigma^{\{\ell\}}}([0, a_\ell])$ with $d\sigma^{\{\ell\}}(t) = \left(1+\frac{t}{\ell}\right) dt$ and the mappings $\pi^{\{\ell\}} : \mathcal{H} \to \mathcal{H}^{\{\ell\}}$ given by

$$\left(\pi^{\{\ell\}} f\right)(t) = f(t) \quad \text{for } t \in [0, a_\ell].$$

Then for all $f, g \in \mathcal{D}$ it holds

$$\langle \pi^{\{\ell\}} f, \pi^{\{\ell\}} g\rangle_{\mathcal{H}^{\{\ell\}}} = \int_0^{a_\ell} g(t)^* f(t) \left(1 + \frac{t}{\ell}\right) dt \xrightarrow{\ell\to\infty} \int_0^{\infty} g(t)^* f(t) dt.$$

However, for $h(t) = \frac{1}{t+1} \in \mathcal{H}$

$$\langle \pi^{\{\ell\}} h, \pi^{\{\ell\}} h\rangle_{\mathcal{H}^{\{\ell\}}} = \int_0^{a_\ell} \left(1 + \frac{t}{\ell}\right) \frac{dt}{(t+1)^2} \xrightarrow{\ell\to\infty} \infty,$$

and hence the spaces $\mathcal{H}^{\{\ell\}}$ do not approximate $\mathcal{H}$ with respect to $\pi^{\{\ell\}}$.

Approximation in spaces with an indefinite metric is defined as follows.

Definition 7.3. Let $\mathcal{K}$ and $\mathcal{K}^{\{\ell\}}$, $\ell = 1, 2, \ldots$, be Krein spaces, let $\pi^{\{\ell\}} : \mathcal{K} \to \mathcal{K}^{\{\ell\}}$, $\ell = 1, 2, \ldots$, be linear mappings, and let J on $\mathcal{K}$ and $J^{\{\ell\}}$ on $\mathcal{K}^{\{\ell\}}$, $\ell = 1, 2, \ldots$, be fundamental symmetries. The spaces $\mathcal{K}^{\{\ell\}}$ *approximate* $\mathcal{K}$ *with respect to* $\pi^{\{\ell\}}$, $J^{\{\ell\}}$, and J, if for all $u \in \mathcal{K}$

$$\langle \pi^{\{\ell\}} u, \pi^{\{\ell\}} u\rangle_{\mathcal{K}^{\{\ell\}}} \to \langle u, u\rangle_{\mathcal{K}} \qquad \text{as } \ell \to \infty, \tag{7.1}$$

and

$$\langle J^{\{\ell\}} \pi^{\{\ell\}} u, \pi^{\{\ell\}} u\rangle_{\mathcal{K}^{\{\ell\}}} \to \langle Ju, u\rangle_{\mathcal{K}} \quad \text{as } \ell \to \infty. \tag{7.2}$$

We define, as usual, the norms in $\mathcal{K}$ and $\mathcal{K}^{\{\ell\}}$ as the norms of the associated Hilbert spaces, that is,

$$\|u\|_{\mathcal{K}} = \langle Ju, u\rangle_{\mathcal{K}}^{1/2},\ u \in \mathcal{K}, \quad \text{and} \quad \|v\|_{\mathcal{K}^{\{\ell\}}} = \langle J^{\{\ell\}} v, v\rangle_{\mathcal{K}^{\{\ell\}}}^{1/2},\ v \in \mathcal{K}^{\{\ell\}}.$$

With these norms $\mathcal{K}$ and $\mathcal{K}^{\{\ell\}}$ are Banach spaces and with the linear mappings $\pi^{\{\ell\}} : \mathcal{K} \to \mathcal{K}^{\{\ell\}}$, $\ell = 1, 2, \ldots$, having the properties (7.1) and (7.2), we define strong approximation of elements and operators in $\mathcal{K}$ as in Definition 7.1(ii) and (iii).
Evidently, if $\pi^{\{\ell\}}$ intertwines the fundamental symmetries, that is,

$$\pi^{\{\ell\}} J = J^{\{\ell\}} \pi^{\{\ell\}},$$

then the equalities (7.1) and (7.2) are equivalent.

In the case of Pontryagin spaces the following theorem holds, see [18, 17].

Theorem 7.4. *Let* $\mathcal{P}^{\{\ell\}}$, $\ell = 1, 2, \ldots$, *and* $\mathcal{P}$ *be Pontryagin spaces, let* $\mathcal{D}$ *be a dense linear manifold in* $\mathcal{P}$ *and let* $\pi^{\{\ell\}} : \mathcal{D} \to \mathcal{P}^{\{\ell\}}$, $\ell = 1, 2, \ldots$, *be linear mappings such that*

(i) $\operatorname{ind}_- \mathcal{P}^{\{\ell\}} = \operatorname{ind}_- \mathcal{P}$ *and*
(ii) $\lim_{\ell\to\infty} \langle \pi^{\{\ell\}} u, \pi^{\{\ell\}} v\rangle_{\mathcal{P}^{\{\ell\}}} = \langle u, v\rangle_{\mathcal{P}}$ *for all* $u, v \in \mathcal{D}$.

Let $\mathcal{D}_-$ be a maximal negative subspace of $\mathcal{D}$. Then $\mathcal{P} = (\mathcal{P} \ominus \mathcal{D}_-) \oplus \mathcal{D}_-$ and there exists an integer $\ell_0 \geq 1$ such that for all $\ell \geq \ell_0$

$$\mathcal{P}^{\{\ell\}} = \left(\mathcal{P}^{\{\ell\}} \ominus \pi^{\{\ell\}}(\mathcal{D}_-)\right) \oplus \pi^{\{\ell\}}(D_-)$$

are fundamental decompositions of the spaces $\mathcal{P}$ and $\mathcal{P}^{\{\ell\}}$. Denote by J and $J^{\{\ell\}}$ the corresponding fundamental symmetries and, in addition to (i) *and* (ii), *assume that*

(iii) *the mappings $\pi^{\{\ell\}}$, $\ell \geq \ell_0$, are uniformly bounded with respect to the Hilbert space inner products $\langle J^{\{\ell\}} \cdot, \cdot \rangle_{\mathcal{P}^{\{\ell\}}}$,*

or that

(iii)′ $\mathcal{D} = \mathcal{P}$.

Then the spaces $\mathcal{P}^{\{\ell\}}$, $\ell \geq \ell_0$, approximate $\mathcal{P}$ with respect to the bounded continuation of $\pi^{\{\ell\}}$ to all of $\mathcal{P}^{\{\ell\}}$ and the fundamental symmetries $J^{\{\ell\}}$ and J.

The theorem with conditions (i), (ii), and (iii)′ is due B. Najman, see [18, Theorem 2.1], the details of the proof of the theorem with conditions (i), (ii), and (iii) are left to the reader.

7.2. We now come to the model approximation problem. We use the same notation as in the Introduction, so we refrain from mentioning the details when referring to the irreducible and regularized representations of functions in $\mathcal{N}_\kappa^\infty$. Let N be a singular $\mathcal{N}_\kappa^\infty$-function with the irreducible and regularized representations:

$$N(z) = (z - z_0^*)^\kappa n(z)(z - z_0)^\kappa + p(z) = (z - z_0^*)^\kappa n_r(z)(z - z_0)^\kappa + r_0(z), \quad (7.3)$$

where n has the representation

$$n(z) = \int_{\mathbb{R}} \left(\frac{1}{t - z} - \frac{t}{t^2 + 1}\right) d\sigma(t), \qquad (7.4)$$

$\deg p \leq 2\kappa$, and hence

$$n_r(z) = n(z) + p_{2\kappa}, \quad r_0(z) = p(z) - p_{2\kappa}(z - z_0^*)^\kappa (z - z_0)^\kappa.$$

Assume that N is approximated on $\mathbb{C} \setminus \mathbb{R}$ by regular $\mathcal{N}_\kappa^\infty$-functions $N^{\{\ell\}}$, $\ell = 1, 2, \ldots$, with irreducible and regularized representations

$$N^{\{\ell\}}(z) = n^{\{\ell\}}(z) + p^{\{\ell\}}(z) = (z - z_0^*)^\kappa n_r^{\{\ell\}}(z)(z - z_0)^\kappa + r_0^{\{\ell\}}(z), \qquad (7.5)$$

where $\deg p^{\{\ell\}}$ is either equal to 2κ or to $2\kappa - 1$, in which case $p_{2\kappa-1}^{\{\ell\}}$ is < 0, $n^{\{\ell\}}$ has the integral representation:

$$n^{\{\ell\}}(z) = \int_{\mathbb{R}} \left(\frac{1}{t - z} - \frac{t}{t^2 + 1}\right) d\sigma^{\{\ell\}}(t), \qquad (7.6)$$

hence $n_r^{\{\ell\}}$ has the representation

$$n_r^{\{\ell\}}(z) = p_{2\kappa}^{\{\ell\}} + \int_{\mathbb{R}} \frac{1}{t - z} \frac{d\sigma^{\{\ell\}}(t)}{|t - z_0|^{2\kappa}}.$$

We make the following assumption.

Hypothesis 7.5. The spectral measures $d\sigma^{\{\ell\}}$ in (7.6) are uniformly absolutely continuous with respect to the spectral measure $d\sigma$ in (7.4) in the sense that there exists a number $C>0$ such that for integers $\ell \geq 1$

$$|t-z_0|^{-2\kappa}\frac{d\sigma^{\{\ell\}}(t)}{d\sigma(t)} \leq C \quad \text{almost everywhere with respect to } d\sigma.$$

Remark 7.6. By Theorem 6.8 there always exists an approximating sequence satisfying this hypothesis.

The model approximation problem is to describe the convergence $N^{\{\ell\}} \to N$ in terms of models. In Subsection 7.4 we consider the canonical models. In this subsection and Subsection 7.3 we choose the following models for N and its approximations $N^{\{\ell\}}$. For $N \in \mathcal{N}_\kappa^\infty$ we take the model $(B, K_{\widetilde{N}}(\,\cdot\,,z^*)v(z), \widetilde{S})$ from (5.11), which is a minimal model in the state space

$$\widetilde{\mathcal{P}} = \mathcal{L}(\widetilde{N}) = \mathcal{L}(n) \oplus \mathcal{L}(M). \tag{7.7}$$

The space $\mathcal{L}(n)$ is the Hilbert space spanned by the functions

$$\zeta \mapsto K_n(\zeta,w) = \frac{n(\zeta)-n(w)^*}{\zeta - w^*} = \int_{\mathbb{R}} \frac{1}{t-\zeta}\frac{1}{t-w^*}\,d\sigma(t), \quad w \in \mathbb{C}\setminus\mathbb{R},$$

and equipped with the inner product

$$\langle K_n(\zeta,w), K_n(\zeta,z)\rangle_{\mathcal{L}(n)} = K_n(z,w) = \int_{\mathbb{R}} \frac{1}{t-z}\frac{1}{t-w^*}\,d\sigma(t).$$

The space $\mathcal{L}(M)$ is a Pontryagin space with basis $u_1,\dots,u_\kappa,v_1,\dots,v_\kappa$ consisting of the 2-vector polynomials

$$u_j(\zeta) = R_{z_0}^j\begin{pmatrix} r_0(\zeta) \\ (\zeta-z_0)^\kappa\end{pmatrix}, \quad v_j(\zeta) = R_{z_0^*}^{\kappa-j+1}\begin{pmatrix}(\zeta-z_0^*)^\kappa \\ 0\end{pmatrix}, \quad j=1,\dots,\kappa,$$

and equipped with an inner product such that the Gram matrix $\widetilde{G}$ associated with this basis has the form

$$\widetilde{G} = \begin{pmatrix} G & I_{\mathbb{C}^\kappa} \\ I_{\mathbb{C}^\kappa} & 0\end{pmatrix},$$

where $G = (G_{ij})_{i,j=1}^\kappa$ with

$$\begin{aligned} G_{ij} &:= \langle u_j, u_i\rangle_{\mathcal{L}(\widetilde{N})} \\ &= \frac{1}{(j-1)!}\frac{1}{(i-1)!}\left(\frac{d}{dz}\right)^{j-1}\left(\frac{d}{dw^*}\right)^{i-1}\frac{r_0(z)-r_0(w^*)}{z-w^*}\bigg|_{z=w=z_0}, \end{aligned}$$

see [8, Lemma 5.1]. We define the subspaces $\widetilde{\mathcal{M}}_\pm$ of $\mathcal{L}(M)$ as the span of the 2-vector polynomials

$$u_{i\pm} := \pm u_i + \frac{1}{2}\sum_{j=1}^{\kappa}(I \mp G)_{ji}v_j, \quad i=1,\dots,\kappa. \tag{7.8}$$

Since these vector polynomials satisfy the orthogonality relations

$$\langle u_{j\pm}, u_{i\pm}\rangle_{\mathcal{L}(M)} = \pm\delta_{ij}, \quad \langle u_{j+}, u_{i-}\rangle_{\mathcal{L}(M)} = 0,$$

they form an orthonormal basis for $\widetilde{\mathcal{M}}_\pm$ and the decomposition

$$\mathcal{L}(M) = \widetilde{\mathcal{M}}_+ \oplus \widetilde{\mathcal{M}}_-$$

is a fundamental decomposition of $\mathcal{L}(M)$. If we denote by J_M the corresponding fundamental symmetry, then

$$\widetilde{J} = \begin{pmatrix} I & 0 \\ 0 & J_M \end{pmatrix} \tag{7.9}$$

is a fundamental symmetry for the space $\widetilde{P}$ from (7.7).

For the approximating function $N^{\{\ell\}}$, $\ell = 1, 2, \dots$, we consider the (not necessarily minimal) model

$$(B_r^{\{\ell\}}, K_{\widetilde{N}_r^{\{\ell\}}}(\,\cdot\,, z^*)v_r^{\{\ell\}}(z), \widetilde{S}_r^{\{\ell\}}) \tag{7.10}$$

in the state space

$$\widetilde{\mathcal{P}}_r^{\{\ell\}} = \mathcal{L}(\widetilde{N}_r^{\{\ell\}}) = \mathcal{L}(n_r^{\{\ell\}}) \oplus \mathcal{L}(M_r^{\{\ell\}}). \tag{7.11}$$

The space $\mathcal{L}(n_r^{\{\ell\}})$ and its inner product and the space $\mathcal{L}(M_r^{\{\ell\}})$, its basis, and the inner product with the corresponding Gram matrix and fundamental symmetry can be described as above with an evident change of notation: for example n, σ, r_0, G, and $u_{j\pm}$ are to be replaced by $n_r^{\{\ell\}}$, $\sigma^{\{\ell\}}$, $r_0^{\{\ell\}}$, $G^{\{\ell\}}$, and $u_{j\pm}^{\{\ell\}}$, etc.

Consider the set

$$\mathcal{D} = \left\{ \widetilde{f} = \begin{pmatrix} \sum_w' K_n(\,\cdot\,, w) f_w \\ \sum_{j=1}^\kappa (a_j u_j + b_j v_j) \end{pmatrix} \in \widetilde{\mathcal{P}} \,\middle|\, f_w, a_j, b_j \in \mathbb{C} \right\}, \tag{7.12}$$

where $\sum_w'$ stands for a finite sum over $w \in \mathbb{C} \setminus \mathbb{R}$, and define the mapping $\widetilde{\pi}^{\{\ell\}} : \mathcal{D} \to \widetilde{\mathcal{P}}_r^{\{\ell\}}$ by

$$\widetilde{\pi}^{\{\ell\}} \begin{pmatrix} \sum_w' K_n(\,\cdot\,, w) f_w \\ \sum_{j=1}^\kappa (a_j u_j + b_j v_j) \end{pmatrix} = \begin{pmatrix} \sum_w' K_{n_r^{\{\ell\}}}(\,\cdot\,, w) f_w \\ \sum_{j=1}^\kappa (a_j u_j^{\{\ell\}} + b_j v_j^{\{\ell\}}) \end{pmatrix}, \tag{7.13}$$

where $f_w, a_j, b_j \in \mathbb{C}$. If Hypothesis 7.5 holds, then $\widetilde{\pi}^{\{\ell\}}$ is well defined and bounded as we shall show in the proof of the following theorem; its bounded extension to all of $\widetilde{\mathcal{P}}$ will also be denoted by $\widetilde{\pi}^{\{\ell\}}$. Note that $\widetilde{\pi}^{\{\ell\}} u_j = u_j^{\{\ell\}}$ and $\widetilde{\pi}^{\{\ell\}} v_j = v_j^{\{\ell\}}$, but $\widetilde{\pi}^{\{\ell\}} u_{j\pm} \neq u_{j\pm}^{\{\ell\}}$ if $G^{\{\ell\}} \neq G$. This can be seen directly from (7.8). Applying the mapping $\widetilde{\pi}^{\{\ell\}}$ to (7.8) we obtain

$$\langle \widetilde{\pi}^{\{\ell\}} u_{j\pm}, \widetilde{\pi}^{\{\ell\}} u_{i\pm}\rangle_{\mathcal{L}(M_r^{\{\ell\}})} = \pm\delta_{ij} \pm (G^{\{\ell\}} - G)_{ij}, \quad \langle \widetilde{\pi}^{\{\ell\}} u_{j+}, \widetilde{\pi}^{\{\ell\}} u_{i-}\rangle_{\mathcal{L}(M_r^{\{\ell\}})} = 0.$$

Hence the subspaces $\widetilde{\pi}^{\{\ell\}}(\widetilde{\mathcal{M}}_\pm)$ are mutually orthogonal and, since, as we shall show, $G^{\{\ell\}} \to G$ as $\ell \to \infty$, there exists an integer $\ell_1 \geq 1$ such that for $\ell \geq \ell_1$ the subspaces $\widetilde{\pi}^{\{\ell\}}(\widetilde{\mathcal{M}}_+)$ and $\widetilde{\pi}^{\{\ell\}}(\widetilde{\mathcal{M}}_-)$ are positive and negative definite respectively, and

$$\mathcal{L}(M_r^{\{\ell\}}) = \widetilde{\pi}^{\{\ell\}}(\widetilde{\mathcal{M}}_+) \oplus \widetilde{\pi}^{\{\ell\}}(\widetilde{\mathcal{M}}_-). \tag{7.14}$$

If we denote by $J_r^{\{\ell\}}$ the corresponding fundamental symmetry, then for $\ell \geq \ell_1$

$$\widetilde{J}_r^{\{\ell\}} = \begin{pmatrix} I & 0 \\ 0 & J_r^{\{\ell\}} \end{pmatrix}$$

is a fundamental symmetry for the space $\widetilde{\mathcal{P}}^{\{\ell\}}$ in (7.11). It readily follows that

$$\widetilde{\pi}^{\{\ell\}}\widetilde{J} = \widetilde{J}_r^{\{\ell\}}\widetilde{\pi}^{\{\ell\}}. \tag{7.15}$$

With these preparations we can formulate the following theorem.

Theorem 7.7. *Assume that a singular $\mathcal{N}_\kappa^\infty$-function N of the form (7.3) with $\deg p \leq 2\kappa$ is approximated by regular $\mathcal{N}_\kappa^\infty$-functions $N^{\{\ell\}}$ of the form (7.5) with $\deg p^{\{\ell\}} \leq 2\kappa$ such that Hypothesis 7.5 holds. Then there is an integer $\ell_1 \geq 1$ such that the state spaces $\widetilde{\mathcal{P}}_r^{\{\ell\}}$ from (7.7) of the model (5.11) for $N^{\{\ell\}}$, $\ell \geq \ell_1$, approximate the state space $\widetilde{\mathcal{P}}$ from (7.11) of the model (7.10) for N with respect to the mappings $\widetilde{\pi}^{\{\ell\}}$ and the fundamental symmetries $\widetilde{J}_r^{\{\ell\}}$ and $\widetilde{J}$.*

Proof. Since the linear span of the kernel functions $K_n(\,\cdot\,,w)$ is dense in $\mathcal{L}(n)$ and the elements $u_1,\ldots,u_\kappa,v_1,\ldots,v_\kappa$ form a basis for $\mathcal{L}(M)$, the domain $\mathcal{D}$ in (7.12) is a dense subset of $\widetilde{\mathcal{P}}$. We prove that the mappings $\widetilde{\pi}^{\{\ell\}}$ are well defined on $\mathcal{D}$. Since they are diagonal with respect to the orthogonal decompositions (7.7) and (7.11), it suffices to prove it componentwise. We start with the compressions $\widetilde{\pi}^{\{\ell\}}\,|_{\mathcal{L}(n)\cap\mathcal{D}}$ to $\mathcal{L}(n)$. Let $f = \sum_w' K_n(\,\cdot\,,w)f_w$ with $f_w \in \mathbb{C}$ be an element of $\mathcal{L}(n)$. Then by Hypotheses 7.5 and with the notation $\widehat{f}(t) := \sum_w' f_w/(t-w)$ we find

$$\begin{aligned} \|\widetilde{\pi}^{\{\ell\}}\,|_{\mathcal{L}(n)\cap\mathcal{D}}\, f\|^2_{\mathcal{L}(n_r^{\{\ell\}})} &= \sum_{z,w}{}' K_{n_r^{\{\ell\}}}(z,w) f_z f_w^* \\ &= \int_{\mathbb{R}} |\widehat{f}(t)|^2 \frac{d\sigma^{\{\ell\}}(t)}{|t-z_0|^{2\kappa}} \leq C \int |\widehat{f}(t)|^2 d\sigma(t) = C\|f\|^2_{\mathcal{L}(n)}. \end{aligned}$$

Thus if $f = 0$, then also $\widetilde{\pi}^{\{\ell\}} f = 0$, hence the operators $\widetilde{\pi}^{\{\ell\}}\,|_{\mathcal{L}(n)\cap\mathcal{D}}$ are well defined. The inequality shows also that they are uniformly bounded, and so are their continuations to all of $\mathcal{L}(n)$. The compression $\widetilde{\pi}^{\{\ell\}}\,|_{\mathcal{L}(M)}$ maps the basis elements u_j and v_j of $\mathcal{L}(M)$ to the basis elements $u_j^{\{\ell\}}$ and $v_j^{\{\ell\}}$ of $\mathcal{L}(M_r^{\{\ell\}})$, respectively, $j = 1,\ldots,\kappa$, and hence the mapping is well defined.

According to Theorem 6.5 the pointwise convergence $N^{\{\ell\}} \to N$ implies the locally uniform convergence of the Nevanlinna functions: $n_r^{\{\ell\}} \to n + p_{2\kappa}$ and the polynomials: $r_0^{\{\ell\}} \to r_0$ and therefore the convergence of the kernels: $K_{n_r^{\{\ell\}}}(z,w) \to$

$K_n(z,w)$ and $K_{M_r^{\{\ell\}}}(z,w) \to K_M(z,w)$. Hence also

$$u_j^{\{\ell\}}(\zeta) = \frac{1}{(j-1)!}\frac{d^{j-1}}{dw^{*(j-1)}}K_{M_r^{\{\ell\}}}(\zeta,w)\big|_{w=z_0^*}\begin{pmatrix}1\\0\end{pmatrix} \to u_j(\zeta),$$
$$v_j^{\{\ell\}}(\zeta) = \frac{1}{(\kappa-j)!}\frac{d^{\kappa-j}}{dw^{*(\kappa-j)}}K_{M_r^{\{\ell\}}}(\zeta,w)\big|_{w=z_0}\begin{pmatrix}0\\1\end{pmatrix} \to v_j(\zeta),$$

and $G^{\{\ell\}} \to G$. Thus there exists an integer $\ell_1 \geq 1$ such that for $\ell \geq \ell_1$ (7.14) and (7.15) hold.

We prove the theorem by showing that (7.1) and (7.2) hold. By (7.15) the limits in these formulas are equivalent. Since the compressions of $\widetilde{\pi}^{\{\ell\}}$ to $\mathcal{L}(n)$ are uniformly bounded and $\mathcal{L}(M) \subset \mathcal{D}$, to prove (7.1) it suffices to show that it holds for the elements in $\mathcal{D}$, that is,

$$\lim_{\ell\to\infty} \langle \widetilde{\pi}^{\{\ell\}}\widetilde{f}, \widetilde{\pi}^{\{\ell\}}\widetilde{g}\rangle_{\widetilde{\mathcal{P}}_r^{\{\ell\}}} = \langle \widetilde{f}, \widetilde{g}\rangle_{\widetilde{\mathcal{P}}}, \quad \widetilde{f}, \widetilde{g} \in \mathcal{D}.$$

But this readily follows from the limits mentioned in the previous paragraph and the reproducing kernel property of the kernels $K_{n_r^{\{\ell\}}}(z,w)$ and $K_n(z,w)$. □

Corollary 7.8. *The sequences $u_i^{\{\ell\}}$ and $v_i^{\{\ell\}}$ in $\mathcal{L}(M_r^{\{\ell\}})$ are bounded: there exists a constant $C_1 > 0$ such that for all $\ell \geq \ell_1$ and $i = 1, 2, \ldots, \kappa$*

$$\|u_i^{\{\ell\}}\|_{\mathcal{L}(M_r^{\{\ell\}})} \leq C_1, \quad \|v_i^{\{\ell\}}\|_{\mathcal{L}(M_r^{\{\ell\}})} \leq C_1,$$

where $\|u\|^2_{\mathcal{L}(M_r^{\{\ell\}})} = \langle J_r^{\{\ell\}}u, u\rangle_{\mathcal{L}(M_r^{\{\ell\}})}$, $u \in \mathcal{L}(M_r^{\{\ell\}})$.

7.3. In the following theorem, where we use the same notation as in Theorem 7.7, we describe the approximation of the models in Subsection 7.2. Recall that $B^{\langle\alpha\rangle}$, $\alpha \in \mathbb{R}\cup\{\infty\}$, are the canonical self-adjoint extensions of the symmetric operator $\widetilde{S}$ in the space $\widetilde{\mathcal{P}}$ whose resolvents are given by (5.12) and (5.13), and that similar formulas hold for the self-adjoint extensions $B_r^{\langle\alpha\rangle}$ of $\widetilde{S}_r$ in $\widetilde{\mathcal{P}}_r$. By Theorem 7.7 the spaces $\widetilde{\mathcal{P}}_r^{\{\ell\}}$ approximate $\widetilde{\mathcal{P}}$ with respect to $\widetilde{\pi}^{\{\ell\}}$, $\widetilde{J}_r^{\{\ell\}}$, and $\widetilde{J}$.

Theorem 7.9. *Assume that a singular $\mathcal{N}_\kappa^\infty$-function N of the form (7.3) with $\deg p \leq 2\kappa$ is approximated by regular $\mathcal{N}_\kappa^\infty$-functions $N^{\{\ell\}}$ of the form (7.5) with $\deg p^{\{\ell\}} \leq 2\kappa$ such that Hypothesis 7.5 holds. Let α and α_ℓ, $\ell = \ell_1, \ell_1+1, \ldots$, be points in $\mathbb{R}\cup\{\infty\}$ and assume that $\alpha_\ell \to \alpha$ as $\ell \to \infty$. Then for all*

$$z \in \Big(\bigcap_{\ell\geq\ell_1} \rho(B_r^{\{\ell\}\langle\alpha_\ell\rangle})\Big) \cap \rho(B^{\langle\alpha\rangle}) \tag{7.16}$$

$\left(B_r^{\{\ell\}\langle\alpha_\ell\rangle} - z\right)^{-1} \xrightarrow{s} \left(B^{\langle\alpha\rangle} - z\right)^{-1}$ *with respect to $\widetilde{\pi}^{\{\ell\}}$, as $\ell \to \infty$.*

Proof. We have to prove that for z from (7.16)

$$\lim_{\ell\to\infty} \|\left(B_r^{\{\ell\}\langle\alpha_\ell\rangle} - z\right)^{-1}\widetilde{\pi}^{\{\ell\}}\widetilde{f} - \widetilde{\pi}^{\{\ell\}}\left(B^{\langle\alpha\rangle} - z\right)^{-1}\widetilde{f}\|_{\widetilde{\mathcal{P}}_r^{\{\ell\}}} = 0, \quad \widetilde{f} \in \widetilde{\mathcal{P}},$$

where $\|\cdot\|^2_{\widetilde{\mathcal{P}}_r^{\{\ell\}}} = \langle \widetilde{J}_r^{\{\ell\}}\cdot,\cdot\rangle_{\widetilde{\mathcal{P}}_r^{\{\ell\}}}$. For this we use the representations (5.12) and (5.13) for the resolvents and divide the proof into 3 steps. We assume throughout that the integer $\ell \geq \ell_1$. To avoid too many brackets, we write, for example, $R_z f(\zeta)$ instead of $(R_z f)(\zeta)$, where R_z is the difference-quotient operator (5.9); we hope that the reader will not be confused by this somewhat inconsistent notation.

1. First we prove that

$$\lim_{\ell\to\infty} \|(A_{\widetilde{N}_r^{\{\ell\}}} - z)^{-1}\widetilde{\pi}^{\{\ell\}}\widetilde{f} - \widetilde{\pi}^{\{\ell\}}(A_{\widetilde{N}} - z)^{-1}\widetilde{f}\|_{\widetilde{\mathcal{P}}_r^{\{\ell\}}} = 0, \quad \widetilde{f} \in \widetilde{\mathcal{P}}. \tag{7.17}$$

We claim that the mapping $\widetilde{\pi}^{\{\ell\}}|_{\mathcal{L}(n)}$ intertwines the difference-quotient operators in the spaces $\mathcal{L}(n_r^{\{\ell\}})$ and $\mathcal{L}(n)$:

$$(A_{n_r^{\{\ell\}}} - z)^{-1}\widetilde{\pi}^{\{\ell\}}|_{\mathcal{L}(n)} = \widetilde{\pi}^{\{\ell\}}|_{\mathcal{L}(n)}(A_n - z)^{-1}. \tag{7.18}$$

To see this, observe that

$$(A_n - z)^{-1}K_n(\zeta, w) = \frac{K_n(\zeta, w) - K_n(z, w)}{\zeta - z} = \frac{K_n(\zeta, z^*) - K_n(\zeta, w)}{z - w^*},$$

and that the right-hand side is the difference of two elements from $\mathcal{L}(n)$. Hence

$$\begin{aligned}\widetilde{\pi}^{\{\ell\}}|_{\mathcal{L}(n)}(A_n - z)^{-1}K_n(\zeta, w) &= \widetilde{\pi}^{\{\ell\}}|_{\mathcal{L}(n)}\frac{K_n(\zeta, z^*) - K_n(\zeta, w)}{z - w^*}\\ = \frac{K_{n_r^{\{\ell\}}}(\zeta, z^*) - K_{n_r^{\{\ell\}}}(\zeta, w)}{z - w^*} &= \frac{K_{n_r^{\{\ell\}}}(\zeta, w) - K_{n_r^{\{\ell\}}}(z, w)}{\zeta - z}.\end{aligned}$$

On the other hand

$$\begin{aligned}&(A_{n_r^{\{\ell\}}} - z)^{-1}\widetilde{\pi}^{\{\ell\}}|_{\mathcal{L}(n)}K_n(\zeta, w)\\ &\quad= (A_{n_r^{\{\ell\}}} - z)^{-1}K_{n_r^{\{\ell\}}}(\zeta, w) = \frac{K_{n_r^{\{\ell\}}}(\zeta, w) - K_{n_r^{\{\ell\}}}(z, w)}{\zeta - z},\end{aligned}$$

and thus (7.18) holds for all elements in the dense subset $\mathcal{D} \cap \mathcal{L}(n)$ and hence also on $\mathcal{L}(n)$. Next we calculate the action of the difference-quotient operator on the basis vectors u_j and v_j in $\mathcal{L}(M)$: We obtain for $j = 1, 2, \ldots, \kappa - 1$

$$\begin{aligned}R_z u_j(\zeta) &= R_z R_{z_0}^j \begin{pmatrix} \frac{r_0(\zeta)}{(\zeta - z_0)^\kappa} \end{pmatrix} = u_{j+1}(\zeta) + (z - z_0)u_{j+2}(\zeta) + \cdots\\ &\quad + (z - z_0)^{\kappa-j-1}u_\kappa(\zeta) + (z - z_0)^{\kappa-j}\begin{pmatrix} R_z R_{z_0}^\kappa r_0(\zeta) \\ 0 \end{pmatrix},\end{aligned}$$

and

$$R_z u_\kappa(\zeta) = \begin{pmatrix} R_z R_{z_0}^\kappa r_0(\zeta) \\ 0 \end{pmatrix}, \quad R_z v_1(\zeta) = 0,$$

and for $j = 2, 3, \ldots, \kappa$

$$\begin{aligned}R_z v_j(\zeta) &= R_z R_{z_0^*}^{\kappa-j+1}\begin{pmatrix} (\zeta - z_0^*)^\kappa \\ 0 \end{pmatrix}\\ &= v_{j-1}(\zeta) + (z - z_0^*)v_{j-2}(\zeta) + \cdots + (z - z_0^*)^{j-2}v_1(\zeta).\end{aligned}$$

Corresponding relations, but with the obvious change of r_0 to $r_0^{\{\ell\}}$, are obtained for the vectors $R_z u_j^{\{\ell\}}$ and $R_z v_j^{\{\ell\}}$. Since the functions v_j and $v_j^{\{\ell\}}$ coincide we obtain

$$(A_{M_r^{\{\ell\}}} - z)^{-1}\widetilde{\pi}^{\{\ell\}} v_j(\zeta) - \widetilde{\pi}^{\{\ell\}}(A_M - z)^{-1} v_j(\zeta) = 0$$

and

$$\begin{aligned}(A_{M_r^{\{\ell\}}} - z)^{-1}\widetilde{\pi}^{\{\ell\}} u_j(\zeta) &- \widetilde{\pi}^{\{\ell\}}(A_M - z)^{-1} u_j(\zeta)\\ &= (z - z_0)^{\kappa - j}\begin{pmatrix} R_z R_{z_0}^\kappa (r_0^{\{\ell\}}(\zeta) - r_0(\zeta)) \\ 0 \end{pmatrix}.\end{aligned}$$

The right-hand side is of the form $(z - z_0)^{\kappa-j}\sum_{j=1}^{\kappa-1} d_j^{\{\ell\}}(z) v_j^{\{\ell\}}$ with coefficients $d_j^{\{\ell\}}(z)$ tending to zero as $\ell \to \infty$, for each $j = 1, \dots, \kappa - 1$, since, by Theorem 6.5, $r_0^{\{\ell\}} \to r_0$ locally uniformly on $\mathbb{C}$. By Corollary 7.8 the sequences of elements $v_j^{\{\ell\}}$, $j = 1, \dots, \kappa$, are bounded and therefore

$$\lim_{\ell\to\infty} \left\| \left((A_{M_r^{\{\ell\}}} - z)^{-1}\widetilde{\pi}^{\{\ell\}} - \widetilde{\pi}^{\{\ell\}}(A_M - z)^{-1} \right) \begin{pmatrix} a \\ b \end{pmatrix} \right\|_{\mathcal{L}(M_r^{\{\ell\}})} = 0, \quad \begin{pmatrix} a \\ b \end{pmatrix} \in \mathcal{L}(M).$$

Together with relation (7.18) this proves the limit (7.17).

2. We now show that

$$\lim_{\ell\to\infty} \|\widetilde{\pi}^{\{\ell\}}\Gamma_{\widetilde{N}z}\mathbf{c} - \Gamma_{\widetilde{N}_r^{\{\ell\}}z}\mathbf{c}\|_{\widetilde{\mathcal{P}}_r^{\{\ell\}}} = 0, \quad \mathbf{c} \in \mathbb{C}^3, \tag{7.19}$$

and

$$\lim_{\ell\to\infty} \|(E_z - E_z^{\{\ell\}}\widetilde{\pi}^{\{\ell\}})\widetilde{f}\|_{\mathbb{C}^3}^2 = 0, \quad \widetilde{f} \in \widetilde{\mathcal{P}}. \tag{7.20}$$

To prove (7.19), we first note that

$$P_{\mathcal{L}(n_r^{\{\ell\}})}\left(\widetilde{\pi}^{\{\ell\}}\Gamma_{\widetilde{N}z} - \Gamma_{\widetilde{N}_r^{\{\ell\}}z}\right) = 0,$$

where $P_{\mathcal{L}(n_r^{\{\ell\}})}$ is the projection on $\mathcal{L}(n_r^{\{\ell\}})$. For the other components we have

$$\begin{aligned} &P_{\mathcal{L}(M_r^{\{\ell\}})}\left(\Gamma_{\widetilde{N}_r^{\{\ell\}}z} - \widetilde{\pi}^{\{\ell\}}\Gamma_{\widetilde{N}z}\right)\begin{pmatrix} c_2 \\ c_3 \end{pmatrix}(\zeta)\\ &\quad = c_2 R_z \begin{pmatrix} r_0^{\{\ell\}}(\zeta) \\ (\zeta - z_0)^\kappa \end{pmatrix} - c_2\widetilde{\pi}^{\{\ell\}} R_z \begin{pmatrix} r_0(\zeta) \\ (\zeta - z_0)^\kappa \end{pmatrix}\\ &\quad = c_2\left(u_1^{\{\ell\}}(\zeta) + \dots + (z - z_0)^{\kappa-1}u_\kappa^{\{\ell\}}(\zeta)\right) + c_2(z - z_0)^\kappa \begin{pmatrix} R_z R_{z_0}^\kappa r_0^{\{\ell\}}(\zeta) \\ 0 \end{pmatrix}\\ &\qquad - c_2\widetilde{\pi}^{\{\ell\}}\left(u_1^{\{\ell\}}(\zeta) + \dots + (z - z_0)^{\kappa-1}u_\kappa(\zeta)\right) - c_2(z - z_0)^\kappa\widetilde{\pi}^{\{\ell\}}\begin{pmatrix} R_z R_{z_0}^\kappa r_0(\zeta) \\ 0 \end{pmatrix}\\ &\quad = c_2(z - z_0)^\kappa \begin{pmatrix} R_z R_{z_0}^\kappa (r_0^{\{\ell\}}(\zeta) - r_0(\zeta)) \\ 0 \end{pmatrix}.\end{aligned}$$

The same argument as in 1. yields that this element goes to zero in the norm $\|\cdot\|_{\mathcal{L}(M_r^{\{\ell\}})}$, as $\ell \to \infty$. This shows (7.19). To prove (7.20), we write

$$\|(E_z - E_z^{\{\ell\}}\widetilde{\pi}^{\{\ell\}})\widetilde{f}\|_{\mathbb{C}^3}^2 = \sum_{s=1}^{3} |\langle (E_z - E_z^{\{\ell\}}\widetilde{\pi}^{\{\ell\}})\widetilde{f}, \mathbf{e}_s\rangle_{\mathbb{C}^3}|^2,$$

where $\mathbf{e}_1$, $\mathbf{e}_2$, and $\mathbf{e}_3$ are the unit column vectors in $\mathbb{C}^3$, which in this order, form the 3×3 identity matrix. For $s = 1, 2, 3$ we rewrite the inner products on the right-hand side in the form

$$\begin{aligned}\langle (E_z - E_z^{\{\ell\}}\widetilde{\pi}^{\{\ell\}})\widetilde{f}, \mathbf{e}_s\rangle_{\mathbb{C}^3} &= \langle \widetilde{f}, \Gamma_{\widetilde{N}z}\mathbf{e}_s\rangle_{\widetilde{\mathcal{P}}} - \langle \widetilde{\pi}^{\{\ell\}}\widetilde{f}, \Gamma_{\widetilde{N}_r^{\{\ell\}}z}\mathbf{e}_s\rangle_{\widetilde{\mathcal{P}}_r^{\{\ell\}}} \\ &= \langle \widetilde{f}, \Gamma_{\widetilde{N}z}\mathbf{e}_s\rangle_{\widetilde{\mathcal{P}}} - \langle \widetilde{\pi}^{\{\ell\}}\widetilde{f}, \widetilde{\pi}^{\{\ell\}}\Gamma_{\widetilde{N}z}\mathbf{e}_s\rangle_{\widetilde{\mathcal{P}}_r^{\{\ell\}}} - \langle \widetilde{\pi}^{\{\ell\}}\widetilde{f}, (\Gamma_{\widetilde{N}_r^{\{\ell\}}z} - \widetilde{\pi}^{\{\ell\}}\Gamma_{\widetilde{N}z})\mathbf{e}_s\rangle_{\widetilde{\mathcal{P}}_r^{\{\ell\}}}.\end{aligned}$$

Since the spaces $\widetilde{\mathcal{P}}^{\{\ell\}}$ approximate the space $\widetilde{\mathcal{P}}$ as $\ell \to \infty$ we have

$$\lim_{\ell\to\infty} \langle \widetilde{\pi}^{\{\ell\}}\widetilde{f}, \widetilde{\pi}^{\{\ell\}}\Gamma_{\widetilde{N}z}\mathbf{e}_s\rangle_{\widetilde{\mathcal{P}}_r^{\{\ell\}}} = \langle \widetilde{f}, \Gamma_{\widetilde{N}z}\mathbf{e}_s\rangle_{\widetilde{\mathcal{P}}}$$

and, by the uniform boundedness of $\widetilde{\pi}^{\{\ell\}}$ and (7.19), also

$$\lim_{\ell\to\infty} \langle \widetilde{\pi}^{\{\ell\}}\widetilde{f}, (\Gamma_{\widetilde{N}_r^{\{\ell\}}z} - \widetilde{\pi}^{\{\ell\}}\Gamma_{\widetilde{N}z})\mathbf{e}_s\rangle_{\widetilde{\mathcal{P}}_r^{\{\ell\}}} = 0.$$

From these limits (7.20) readily follows.

3. With the notations used in (5.12) we set for α, $\alpha_\ell \in \mathbb{R}$

$$H_\alpha(z) = \frac{1}{N(z)-\alpha} K_\alpha(z), \quad H_{\alpha_\ell}^{\{\ell\}}(z) = \frac{1}{N^{\{\ell\}}(z)-\alpha_\ell} K_{\alpha_\ell}^{\{\ell\}}(z),$$

for α, $\alpha_\ell = \infty$ these formulas have to be understood in the limit corresponding to the notation in (5.13). We also define

$$\widetilde{D}^{\{\ell\}}(z) = \big(B_r^{\{\ell\}\langle\alpha_\ell\rangle} - z\big)^{-1}\widetilde{\pi}^{\{\ell\}} - \widetilde{\pi}^{\{\ell\}}\big(B^{\langle\alpha\rangle} - z\big)^{-1}$$

for all z from (7.16). We have to show that for each such z

$$\lim_{\ell\to\infty} \|\widetilde{D}^{\{\ell\}}(z)\widetilde{f}\|_{\widetilde{\mathcal{P}}_r^{\{\ell\}}} = 0, \quad \widetilde{f} \in \widetilde{\mathcal{P}}. \tag{7.21}$$

Using (5.12) we find

$$\begin{aligned}\widetilde{D}^{\{\ell\}}(z) &= (A_{\widetilde{N}_r^{\{\ell\}}} - z)^{-1}\widetilde{\pi}^{\{\ell\}} - \widetilde{\pi}^{\{\ell\}}(A_{\widetilde{N}} - z)^{-1} \\ &\quad + \widetilde{\pi}^{\{\ell\}}\Gamma_{\widetilde{N}z}H_\alpha(z)E_z - \Gamma_{\widetilde{N}_r^{\{\ell\}}z}H_{\alpha_\ell}^{\{\ell\}}(z)E_z^{\{\ell\}}\widetilde{\pi}^{\{\ell\}} \\ &= (A_{\widetilde{N}_r^{\{\ell\}}} - z)^{-1}\widetilde{\pi}^{\{\ell\}} - \widetilde{\pi}^{\{\ell\}}(A_{\widetilde{N}} - z)^{-1} \\ &\quad + \big(\widetilde{\pi}^{\{\ell\}}\Gamma_{\widetilde{N}z} - \Gamma_{\widetilde{N}_r^{\{\ell\}}z}\big)H_\alpha(z)E_z \\ &\quad + \Gamma_{\widetilde{N}_r^{\{\ell\}}z}\Big(\big(H_\alpha(z) - H_{\alpha_\ell}^{\{\ell\}}(z)\big)E_z + H_{\alpha_\ell}^{\{\ell\}}(z)\big(E_z - E_z^{\{\ell\}}\widetilde{\pi}^{\{\ell\}}\big)\Big)\end{aligned}$$

and hence

$$\begin{aligned}\|\widetilde{D}^{\{\ell\}}(z)\widetilde{f}\|_{\widetilde{\mathcal{P}}_r^{\{\ell\}}} \le{} & \|(A_{\widetilde{N}_r^{\{\ell\}}} - z)^{-1}\widetilde{\pi}^{\{\ell\}}\widetilde{f} - \widetilde{\pi}^{\{\ell\}}(A_{\widetilde{N}} - z)^{-1}\widetilde{f}\|_{\widetilde{\mathcal{P}}_r^{\{\ell\}}} \\ & + \|(\widetilde{\pi}^{\{\ell\}}\Gamma_{\widetilde{N}z} - \Gamma_{\widetilde{N}_r^{\{\ell\}}z})H_\alpha(z)E_z\widetilde{f}\|_{\widetilde{\mathcal{P}}_r^{\{\ell\}}} \\ & + \|\Gamma_{\widetilde{N}_r^{\{\ell\}}z}(H_\alpha(z) - H_{\alpha_\ell}^{\{\ell\}}(z))E_z\widetilde{f}\|_{\widetilde{\mathcal{P}}_r^{\{\ell\}}} \\ & + \|\Gamma_{\widetilde{N}_r^{\{\ell\}}z}H_{\alpha_\ell}^{\{\ell\}}(z)(E_z - E_z^{\{\ell\}}\widetilde{\pi}^{\{\ell\}})\widetilde{f}\|_{\widetilde{\mathcal{P}}_r^{\{\ell\}}}.\end{aligned} \tag{7.22}$$

To prove (7.21) it suffices to show that the four summands tend to zero as $\ell \to \infty$:

(a) The first and second summands tend to zero by (7.17) and (7.19), respectively.

(b) To prove that the third summand tends to zero, we use the estimate

$$\begin{aligned}&\|\Gamma_{\widetilde{N}_r^{\{\ell\}}z}(H_\alpha(z) - H_{\alpha_\ell}^{\{\ell\}}(z))E_z\widetilde{f}\|^2_{\widetilde{\mathcal{P}}_r^{\{\ell\}}} \\ &\quad \le \sum_{s=1}^{3} \|\Gamma_{\widetilde{N}_r^{\{\ell\}}z}\mathbf{e}_s\|^2_{\widetilde{\mathcal{P}}_r^{\{\ell\}}} \sum_{s,t=1}^{3} |\langle (H_\alpha(z) - H_{\alpha_\ell}^{\{\ell\}}(z))\mathbf{e}_t, \mathbf{e}_s\rangle_{\mathbb{C}^3}|^2 \|\widetilde{f}(z)\|^2_{\mathbb{C}^3},\end{aligned} \tag{7.23}$$

where, as before, $\{\mathbf{e}_s\}_{s=1}^3$ denotes the standard basis in $\mathbb{C}^3$. We claim that the first factor on the right-hand side is uniformly bounded with respect to ℓ. To see this, note that it can be expressed as

$$\sum_{s=1}^{3} \|\Gamma_{\widetilde{N}_r^{\{\ell\}}z}\mathbf{e}_s\|^2_{\widetilde{\mathcal{P}}_r^{\{\ell\}}} = \frac{\operatorname{Im} n_r^{\{\ell\}}(z)}{\operatorname{Im} z} + \sum_{s=1}^{3} \|\Gamma_{\widetilde{N}_r^{\{\ell\}}z}\mathbf{e}_s\|^2_{\mathcal{L}(M_r^{\{\ell\}})}.$$

From the fundamental decomposition (7.14) where $\mathcal{M}_\pm$ is spanned by the vectors $u_{j\pm}^{\{\ell\}}$ from (7.8) it follows that

$$\begin{aligned}\|\Gamma_{\widetilde{N}_r^{\{\ell\}}z}\mathbf{e}_s\|^2_{\mathcal{L}(M_r^{\{\ell\}})} &= \sum_{j=1}^{\kappa} |\langle \Gamma_{\widetilde{N}_r^{\{\ell\}}z}\mathbf{e}_s, u_{j+}^{\{\ell\}}\rangle_{\mathcal{L}(M_r^{\{\ell\}})}|^2 + \sum_{j=1}^{\kappa} |\langle \Gamma_{\widetilde{N}_r^{\{\ell\}}z}\mathbf{e}_s, u_{j-}^{\{\ell\}}\rangle_{\mathcal{L}(M_r^{\{\ell\}})}|^2 \\ &= \sum_{j=1}^{\kappa} |\langle \mathbf{e}_s, u_{j+}^{\{\ell\}}(z^*)\rangle_{\mathbb{C}^3}|^2 + \sum_{j=1}^{\kappa} |\langle \mathbf{e}_s, u_{j-}^{\{\ell\}}(z^*)\rangle_{\mathbb{C}^3}|^2.\end{aligned}$$

For the last equality we used the reproducing kernel property. Hence

$$\sum_{s=1}^{3} \|\Gamma_{\widetilde{N}_r^{\{\ell\}}z}\mathbf{e}_s\|^2_{\widetilde{\mathcal{P}}_r^{\{\ell\}}} = \frac{\operatorname{Im} n_r^{\{\ell\}}(z)}{\operatorname{Im} z} + \sum_{j=1}^{\kappa} \left(\|u_{j+}^{\{\ell\}}(z^*)\|^2_{\mathbb{C}^2} + \|u_{j-}^{\{\ell\}}(z^*)\|^2_{\mathbb{C}^2} \right),$$

which implies the claim. As to the second factor on the right-hand side of (7.23), we have

$$\lim_{\ell\to\infty} \sum_{s,t=1}^{3} |\langle (H_\alpha(z) - H_{\alpha_\ell}^{\{\ell\}}(z))\mathbf{e}_t, \mathbf{e}_s\rangle_{\mathbb{C}^3}|^2 = 0,$$

since $\alpha_\ell \to \alpha$, $N^{\{\ell\}} \to N$, and the matrix elements of $K_{\alpha_\ell}^{\{\ell\}}(z)$ converge to the corresponding matrix elements of $K_\alpha(z)$. This implies that the third summand in (7.22) converges to zero.

(c) For the fourth summand in (7.22) we use the estimate

$$\|\Gamma_{\widetilde{N}_r^{\{\ell\}}z} H_{\alpha_\ell}^{\{\ell\}}(z)(E_z - E_z^{\{\ell\}}\widetilde{\pi}^{\{\ell\}})\widetilde{f}\|^2_{\widetilde{\mathcal{P}}_r^{\{\ell\}}}$$
$$\leq \sum_{s=1}^{3} \|\Gamma_{\widetilde{N}_r^{\{\ell\}}z}\mathbf{e}_s\|^2_{\widetilde{\mathcal{P}}_r^{\{\ell\}}} \sum_{s,t=1}^{3} |\langle H_{\alpha_\ell}^{\{\ell\}}(z)\mathbf{e}_t, \mathbf{e}_s\rangle_{\mathbb{C}^3}|^2 \|(E_z - E_z^{\{\ell\}}\widetilde{\pi}^{\{\ell\}})\widetilde{f}\|^2_{\mathbb{C}^3}.$$

As shown in (b), the first factor here is uniformly bounded in ℓ, and the same holds for the sequence $\sum_{s,t=1}^{3} |\langle H_{\alpha_\ell}^{\{\ell\}}(z)\mathbf{e}_t, \mathbf{e}_s\rangle_{\mathbb{C}^3}|^2$. Hence by (7.20) also the fourth summand in (7.22) goes to zero. □

7.4. In this subsection we prove the analogs of the model approximation Theorems 7.7 and 7.9 for the canonical models (see Subsection 5.1) for N in (7.3) with $\deg p \leq 2\kappa$ and $N^{\{\ell\}}$ in (7.5) with $\deg p^{\{\ell\}} \leq 2\kappa$. In accordance with (7.5) there are only two possibilities: either $\deg p^{\{\ell\}} = 2\kappa$ or $\deg p^{\{\ell\}} = 2\kappa - 1$ and then the leading coefficient $p_{2\kappa-1}^{\{\ell\}} < 0$. We set

$$\mathcal{P} = \mathcal{L}(N), \; P^{\{\ell\}} = \mathcal{L}(N^{\{\ell\}}), \quad \text{and} \quad \pi^{\{\ell\}} = \mathbf{V}_r^{\{\ell\}}\widetilde{\pi}^{\{\ell\}}\mathbf{V}^{-1} : \mathcal{P} \to \mathcal{P}^{\{\ell\}}, \tag{7.24}$$

where $\widetilde{\pi}^{\{\ell\}} : \widetilde{\mathcal{P}} \to \widetilde{\mathcal{P}}_r^{\{\ell\}}$ is the bounded extension of the mapping (7.13), $\mathbf{V} : \widetilde{\mathcal{P}} \to \mathcal{P}$ is the unitary transformation given by [9, (3.3)], and $\mathbf{V}_r^{\{\ell\}} : \widetilde{\mathcal{P}}_r^{\{\ell\}} \to \mathcal{P}^{\{\ell\}}$ is the surjective partial isometry described in [9, Lemma 4.2]. Here $\widetilde{\mathcal{P}}$ and $\widetilde{\mathcal{P}}_r^{\{\ell\}}$ are given by (7.7) and (7.11). The situation is illustrated in the following diagram:

$$\begin{array}{ccc}
\mathcal{P} = \mathcal{L}(N) & \xrightarrow{\;\pi^{\{\ell\}}\;} & \mathcal{P}^{\{\ell\}} = \mathcal{L}(N^{\{\ell\}}) \\
\Big\uparrow \mathbf{V} & & \Big\uparrow \mathbf{V}_r^{\{\ell\}} \\
\widetilde{\mathcal{P}} = \mathcal{L}(\widetilde{N}) & \xrightarrow{\;\widetilde{\pi}^{\{\ell\}}\;} & \widetilde{\mathcal{P}}_r^{\{\ell\}} = \mathcal{L}(\widetilde{N}_r^{\{\ell\}})
\end{array}$$

In order to obtain suitable fundamental decompositions in $\mathcal{P}$ and $\mathcal{P}^{\{\ell\}}$ we set

$$\mathcal{M}_- = \mathbf{V}\widetilde{\mathcal{M}}_-, \quad \mathcal{M}_-^{\{\ell\}} = \mathbf{V}_r^{\{\ell\}}\widetilde{\mathcal{M}}_-^{\{\ell\}} \text{ with } \widetilde{\mathcal{M}}_-^{\{\ell\}} = \widetilde{\pi}^{\{\ell\}}(\widetilde{\mathcal{M}}_-), \tag{7.25}$$

where $\widetilde{\mathcal{M}}_-$ is the negative definite subspace of $\widetilde{\mathcal{P}}$ spanned by the elements $u_{1-}, \ldots, u_{\kappa-}$ from (7.8). The subspace $\mathcal{M}_-$ is κ-dimensional and negative definite as $\dim \widetilde{M}_- = \kappa$ and $\mathbf{V}$ is unitary. If we denote by J the fundamental symmetry in $\mathcal{P}$ according to the fundamental decomposition

$$\mathcal{P} = (\mathcal{P} \ominus \mathcal{M}_-) \oplus \mathcal{M}_-,$$

then with $\widetilde{J}$ from (7.9) $J = \mathbf{V}\widetilde{J}\mathbf{V}^*$. The subspace $\mathcal{M}_-^{\{\ell\}}$ is κ-dimensional since $\dim \widetilde{\mathcal{M}}_-^{\{\ell\}} = \kappa$ and $\mathbf{V}_r^{\{\ell\}}$ is unitary if $\deg p^{\{\ell\}} = 2\kappa$ and a surjective partial isometry with $\ker \mathbf{V}_r^{\{\ell\}} \cap \widetilde{\mathcal{M}}_-^{\{\ell\}} = \{0\}$ if $\deg p^{\{\ell\}} = 2\kappa - 1$. Hence each $\mathcal{M}_-^{\{\ell\}}$ is negative definite if $\deg p^{\{\ell\}} = 2\kappa$. It will be shown in the proof of the following theorem that also in the case $\deg p^{\{\ell\}} = 2\kappa - 1$ the subspace $\mathcal{M}_-^{\{\ell\}}$ is negative definite for large enough ℓ. For such ℓ we denote by $J^{\{\ell\}}$ the fundamental symmetry in $\mathcal{P}^{\{\ell\}}$ with respect to the decomposition

$$\mathcal{P}^{\{\ell\}} = \left(\mathcal{P}^{\{\ell\}} \ominus \mathcal{M}_-^{\{\ell\}}\right) \oplus \mathcal{M}_-^{\{\ell\}}. \tag{7.26}$$

Denote by A and $A^{\{\ell\}}$ the representing self-adjoint relations in the canonical models of the functions N and $N^{\{\ell\}}$, respectively, as in (5.1). Furthermore, for $\alpha \in \mathbb{R} \cup \{\infty\}$ let $A^{\langle\alpha\rangle}$ and $A^{\{\ell\}\langle\alpha\rangle}$ be the self-adjoint relations

$$A^{\langle\alpha\rangle} = \begin{cases} \{\{f,g\} \in \mathcal{L}(N)^2 \,|\, \exists \mathbf{c} \in \mathbb{C} \colon\ g(\zeta) - \zeta f(\zeta) = N(\zeta)\mathbf{c}\}, & \alpha = 0, \\ \{\{f,g\} \in \mathcal{L}(N)^2 \,|\, \exists \mathbf{c} \in \mathbb{C} \colon\ g(\zeta) - \zeta f(\zeta) = (1 - \frac{1}{\alpha} N(\zeta))\mathbf{c}\}, & \text{otherwise}, \end{cases}$$

and

$$A^{\{\ell\}\langle\alpha\rangle} = \begin{cases} \{\{f,g\} \in \mathcal{L}(N^{\{\ell\}})^2 \,|\, \exists \mathbf{c} \in \mathbb{C} :\ g(\zeta) - \zeta f(\zeta) = N^{\{\ell\}}(\zeta)\mathbf{c}\}, & \alpha = 0, \\ \{\{f,g\} \in \mathcal{L}(N^{\{\ell\}})^2 \,|\, \exists \mathbf{c} \in \mathbb{C} :\ g(\zeta) - \zeta f(\zeta) = (1 - \frac{1}{\alpha} N^{\{\ell\}}(\zeta))\mathbf{c}\}, & \\ & \text{otherwise}, \end{cases}$$

where again formally $\frac{1}{\infty} = 0$. Here, for example, the $A^{\langle\alpha\rangle}$'s are a parametrization with parameter α of all canonical self-adjoint extensions of the symmetric operator

$$S = \{\{f,g\} \in \mathcal{L}(N)^2 \,|\, g(\zeta) - \zeta f(\zeta) = 0\}$$

in $\mathcal{L}(N)$. Recall from [8, Theorem 4.1 and its proof] and [9, Theorems 4.3 and 4.4] that $A^{\langle\alpha\rangle}$ and $A^{\{\ell\}\langle\alpha\rangle}$ are related to $B^{\langle\alpha\rangle}$ and $B^{\{\ell\}\langle\alpha\rangle}$ via the formulas:

$$A^{\langle\alpha\rangle} = \mathbf{V} B^{\langle\alpha\rangle} \mathbf{V}^{-1}, \quad A^{\{\ell\}\langle\alpha\rangle} = \mathbf{V}_r^{\{\ell\}} B_r^{\{\ell\}\langle\alpha\rangle} \mathbf{V}_r^{\{\ell\}*}. \tag{7.27}$$

Theorem 7.10. *Assume that a singular $\mathcal{N}_\kappa^\infty$-function N of the form (7.3) with $\deg p \le 2\kappa$ is approximated by regular $\mathcal{N}_\kappa^\infty$-functions $N^{\{\ell\}}$ of the form (7.5) with $\deg p^{\{\ell\}} \le 2\kappa$ such that Hypothesis 7.5 holds. Let α and α_ℓ be points in $\mathbb{R} \cup \{\infty\}$ such that $\lim_{\ell \to} \alpha_\ell = \alpha$. Then there is an integer $\ell_2 \ge 1$ such that the spaces $\mathcal{P}^{\{\ell\}}$, $\ell \ge \ell_2$, approximate the space $\mathcal{P}$ with respect to $\pi^{\{\ell\}}$, $J^{\{\ell\}}$, and J, and for each*

$$z \in \Big(\bigcap_{\ell \ge \ell_2} \rho\big(A^{\{\ell\}\langle\alpha_\ell\rangle}\big)\Big) \cap \rho\big(A^{\langle\alpha\rangle}\big) \tag{7.28}$$

$(A^{\{\ell\}\langle\alpha_\ell\rangle} - z)^{-1} \xrightarrow{s} (A^{\langle\alpha\rangle} - z)^{-1}$ *with respect to* $\pi^{\{\ell\}}$.

Proof. We use Theorem 7.4 to show the approximation of the spaces. Evidently, $\operatorname{ind}_- \mathcal{P}^{\{\ell\}} = \operatorname{ind}_- \mathcal{P} = \kappa$. Thus (i) holds. We show that (ii) and (iii)' hold, that is,

$$\lim_{\ell \to \infty} \langle \pi^{\{\ell\}} f, \pi^{\{\ell\}} g \rangle_{\mathcal{P}^{\{\ell\}}} = \langle f, g \rangle_{\mathcal{P}}, \quad f, g \in \mathcal{P}. \tag{7.29}$$

We split the proof into two parts.

Case 1: $\deg p^{\{\ell\}} = 2\kappa$. Then (7.29) holds, since $\mathbf{V}_r^{\{\ell\}}$ is unitary and hence

$$\begin{aligned} \lim_{\ell\to\infty} \langle \pi^{\{\ell\}} f, \pi^{\{\ell\}} g\rangle_{\mathcal{P}^{\{\ell\}}} &= \lim_{\ell\to\infty} \langle \mathbf{V}_r^{\{\ell\}} \widetilde{\pi}^{\{\ell\}} \mathbf{V}^{-1} f, \mathbf{V}_r^{\{\ell\}} \widetilde{\pi}^{\{\ell\}} \mathbf{V}^{-1} g\rangle_{\mathcal{P}^{\{\ell\}}} \\ &= \lim_{\ell\to\infty} \langle \widetilde{\pi}^{\{\ell\}} \mathbf{V}^{-1} f, \widetilde{\pi}^{\{\ell\}} \mathbf{V}^{-1} g\rangle_{\widetilde{\mathcal{P}}_r^{\{\ell\}}} = \lim_{\ell\to\infty} \langle \mathbf{V}^{-1} f, \mathbf{V}^{-1} g\rangle_{\widetilde{\mathcal{P}}} = \langle f, g\rangle_{\mathcal{P}}. \end{aligned}$$

Case 2: $\deg p^{\{\ell\}} = 2\kappa - 1$. Then $\mathbf{V}_r^{\{\ell\}}$ is a surjective partial isometry with

$$\ker \mathbf{V}_r^{\{\ell\}} = \operatorname{span}\{e_r^{\{\ell\}}\},$$

where $e_r^{\{\ell\}}$ denotes the "normalized" element $e_r^{\{\ell\}} = |p_{2\kappa-1}^{\{\ell\}}|^{-1/2}\eta_r^{\{\ell\}}$ with, as introduced in [9, Lemma 4.2],

$$\eta_r^{\{\ell\}} = \begin{pmatrix} -n_r^{\{\ell\}} & 0 & 1 \end{pmatrix}^\top.$$

There it is also shown that $n_r^{\{\ell\}} \in \mathcal{L}(n_r^{\{\ell\}})$,

$$\|n_r^{\{\ell\}}\|^2_{\mathcal{L}(n_r^{\{\ell\}})} = \int_{\mathbb{R}} \frac{d\sigma^{\{\ell\}}(t)}{|t-z_0|^{2\kappa}}, \quad \text{and} \quad \langle \eta_r^{\{\ell\}}, \eta_r^{\{\ell\}}\rangle_{\mathcal{L}(\widetilde{N}_r^{\{\ell\}})} = -p_{2\kappa-1}^{\{\ell\}} > 0.$$

We claim that

$$\lim_{\ell\to\infty} \langle e_r^{\{\ell\}}, \widetilde{\pi}^{\{\ell\}}\widetilde{f}\rangle_{\widetilde{\mathcal{P}}_r^{\{\ell\}}} = 0, \quad \widetilde{f} \in \widetilde{\mathcal{P}}. \tag{7.30}$$

For the elements $\widetilde{f} \in \widetilde{\mathcal{P}}$ of the form

$$\widetilde{f} = \widetilde{f}_{wabc} = \begin{pmatrix} K_n(\,\cdot\,,w)c \\ \sum_{j=1}^{\kappa}(a_j u_j + b_j v_j) \end{pmatrix}, \quad w \in \mathcal{U},\ a_i, b_i, c \in \mathbb{C},$$

(7.30) holds: According (7.13) we have that, as $\ell \to \infty$,

$$\langle e_r^{\{\ell\}}, \widetilde{\pi}^{\{\ell\}}\widetilde{f}\rangle_{\widetilde{\mathcal{P}}_r^{\{\ell\}}} = \frac{1}{|p_{2\kappa-1}^{\{\ell\}}|^{1/2}} \left\langle \begin{pmatrix} -n_r^{\{\ell\}} \\ 0 \\ 1 \end{pmatrix}, \begin{pmatrix} K_{n_r^{\{\ell\}}}(\,\cdot\,,w)c \\ \sum_{j=1}^{\kappa}(a_j u_j^{\{\ell\}} + b_j v_j^{\{\ell\}}) \end{pmatrix} \right\rangle_{\widetilde{\mathcal{P}}_r^{\{\ell\}}}$$

$$= \frac{1}{|p_{2\kappa-1}^{\{\ell\}}|^{1/2}} \left(-n_r^{\{\ell\}}(w)c^* + \left\langle \begin{pmatrix} 0 \\ 1 \end{pmatrix}, \sum_{j=1}^{\kappa}(a_j u_j^{\{\ell\}} + b_j v_j^{\{\ell\}}) \right\rangle_{\mathcal{L}(M_r^{\{\ell\}})} \right) \to 0,$$

because by Corollary 6.7 ($p_{2\kappa}^{\{\ell\}} = 0$): $p_{2\kappa-1}^{\{\ell\}} \to -\infty$, by assumption: $n_r^{\{\ell\}}(w) \to n(w)$ for $w \in \mathcal{U}$, and by Corollary 7.8: the second summand in the round brackets is bounded. Moreover,

$$\lim_{\ell\to\infty} \|e_r^{\{\ell\}}\|^2_{\widetilde{\mathcal{P}}_r^{\{\ell\}}} = \lim_{\ell\to\infty} \frac{1}{|p_{2\kappa-1}^{\{\ell\}}|} \left(\|n_r^{\{\ell\}}\|^2_{\mathcal{L}(n_r^{\{\ell\}})} + \left\| \begin{pmatrix} 0 \\ 1 \end{pmatrix} \right\|^2_{\mathcal{L}(M_r^{\{\ell\}})} \right) = 1.$$

Indeed, by Corollary 6.7 with $p_{2\kappa}^{\{\ell\}} = 0$,

$$\lim_{\ell\to\infty} \frac{1}{|p_{2\kappa-1}^{\{\ell\}}|} \|n_r^{\{\ell\}}\|^2_{\mathcal{L}(n_r^{\{\ell\}})} = \lim_{\ell\to\infty} \frac{1}{|p_{2\kappa-1}^{\{\ell\}}|} \int \frac{d\sigma^{\{\ell\}}(t)}{|t-z_0|^{2\kappa}} = 1$$

and, since the second summand in the round brackets is bounded in ℓ,

$$\lim_{\ell\to\infty} \frac{1}{|p_{2\kappa-1}^{\{\ell\}}|} \left\| \begin{pmatrix} 0 \\ 1 \end{pmatrix} \right\|^2_{\mathcal{L}(M_r^{\{\ell\}})} = 0. \tag{7.31}$$

Therefore, the sequence $\|e_r^{\{\ell\}}\|_{\widetilde{\mathcal{P}}_r^{\{\ell\}}}$ is bounded. This, the fact that the elements $\widetilde{f} = \widetilde{f}_{wabc}$ form a total set in $\widetilde{\mathcal{P}}$, and the fact that (7.30) holds for these elements, imply the claim (7.30).

Denote by $Q_r^{\{\ell\}}$ the orthogonal projection in $\widetilde{\mathcal{P}}_r^{\{\ell\}}$ on the one-dimensional subspace span$\{e_r^{\{\ell\}}\}$. Then for $\widetilde{f}, \widetilde{g} \in \widetilde{\mathcal{P}}$

$$\begin{aligned} &\lim_{\ell\to\infty} \langle \mathbf{V}_r^{\{\ell\}}\widetilde{\pi}^{\{\ell\}}\widetilde{f}, \mathbf{V}_r^{\{\ell\}}\widetilde{\pi}^{\{\ell\}}\widetilde{g}\rangle_{\mathcal{P}^{\{\ell\}}} \\ &\qquad = \lim_{\ell\to\infty} \langle (I - Q_r^{\{\ell\}})\widetilde{\pi}^{\{\ell\}}\widetilde{f}, \widetilde{\pi}^{\{\ell\}}\widetilde{g}\rangle_{\widetilde{\mathcal{P}}_r^{\{\ell\}}} = \langle \widetilde{f}, \widetilde{g}\rangle_{\widetilde{\mathcal{P}}} \end{aligned} \tag{7.32}$$

since, by (7.30),

$$\lim_{\ell\to\infty} \langle Q_r^{\{\ell\}}\widetilde{\pi}^{\{\ell\}}\widetilde{f}, \widetilde{\pi}^{\{\ell\}}\widetilde{g}\rangle_{\widetilde{\mathcal{P}}_r^{\{\ell\}}} = \lim_{\ell\to\infty} \langle \widetilde{\pi}^{\{\ell\}}\widetilde{f}, e_r^{\{\ell\}}\rangle_{\widetilde{\mathcal{P}}_r^{\{\ell\}}} \langle e_r^{\{\ell\}}, \widetilde{\pi}^{\{\ell\}}\widetilde{g}\rangle_{\widetilde{\mathcal{P}}_r^{\{\ell\}}} = 0.$$

Inserting $\widetilde{f} = \mathbf{V}^{-1}f, \widetilde{g} = \mathbf{V}^{-1}g$ and using that $\mathbf{V}$ is unitary we obtain (7.29).

Either because $\mathbf{V}_r^{\{\ell\}}$ is unitary or because of (7.32) we have

$$\lim_{\ell\to\infty} \langle \mathbf{V}_r^{\{\ell\}}\widetilde{\pi}^{\{\ell\}}u, \mathbf{V}_r^{\{\ell\}}\widetilde{\pi}^{\{\ell\}}v\rangle_{\mathcal{P}^{\{\ell\}}} = \langle u, v\rangle_{\widetilde{\mathcal{P}}}, \quad u, v \in \widetilde{\mathcal{M}}_-,$$

and this relation implies that there exists an integer $\ell_2 \geq \ell_1$ with ℓ_1 as in Theorems 7.7 and 7.9 such that for $\ell \geq \ell_2$ the space $\mathcal{M}_-^{\{\ell\}}$ in (7.25):

$$\mathcal{M}_-^{\{\ell\}} = \pi^{\{\ell\}}(\mathcal{M}_-)$$

is a κ-dimensional negative definite subspace of $\mathcal{P}^{\{\ell\}}$. Since for these ℓ the definition of $J^{\{\ell\}}$ as the fundamental symmetry in $\mathcal{P}^{\{\ell\}}$ with respect to the fundamental decomposition (7.26) is in accordance with the definition of the fundamental symmetries in Theorem 7.4, we have proved that the spaces $\mathcal{P}^{\{\ell\}}$, $\ell \geq \ell_2$, approximate the space $\mathcal{P}$ with respect to $\pi^{\{\ell\}}$, $J^{\{\ell\}}$, and J.

Next we show the approximation of the resolvents for z from (7.28), that is,

$$\lim_{\ell\to\infty} \|\big(A^{\{\ell\}\langle\alpha_\ell\rangle} - z\big)^{-1}\pi^{\{\ell\}}f - \pi^{\{\ell\}}\big(A^{\langle\alpha\rangle} - z\big)^{-1}f\|_{\mathcal{P}^{\{\ell\}}} = 0, \quad f \in \mathcal{P}. \tag{7.33}$$

Define the operator

$$D^{\{\ell\}}(z) = \big(A^{\{\ell\}\langle\alpha_\ell\rangle} - z\big)^{-1}\pi^{\{\ell\}} - \pi^{\{\ell\}}\big(A^{\langle\alpha\rangle} - z\big)^{-1}.$$

Then the definition (7.24) of $\pi^{\{\ell\}}$, and the relations (7.27) imply that for $f \in \mathcal{P}$

$$\begin{aligned} \|D^{\{\ell\}}(z)f\|_{\mathcal{P}^{\{\ell\}}} &= \|(A^{\{\ell\}\langle\alpha_\ell\rangle} - z)^{-1}\pi^{\{\ell\}}f - \pi^{\{\ell\}}(A^{\langle\alpha\rangle} - z)^{-1}f\|_{\mathcal{P}^{\{\ell\}}} \\ &= \|\mathbf{V}_r^{\{\ell\}}(B_r^{\{\ell\}\langle\alpha_\ell\rangle} - z)^{-1}\mathbf{V}_r^{\{\ell\}*}\pi^{\{\ell\}}f - \pi^{\{\ell\}}\mathbf{V}(B^{\langle\alpha\rangle} - z)^{-1}\mathbf{V}^{-1}f\|_{\mathcal{P}^{\{\ell\}}} \\ &= \|\mathbf{V}_r^{\{\ell\}}(B_r^{\{\ell\}\langle\alpha_\ell\rangle} - z)^{-1}\widetilde{\pi}^{\{\ell\}}\mathbf{V}^{-1}f - \mathbf{V}_r^{\{\ell\}}\widetilde{\pi}^{\{\ell\}}(B^{\langle\alpha\rangle} - z)^{-1}\mathbf{V}^{-1}f\|_{\mathcal{P}^{\{\ell\}}}. \end{aligned}$$

The last equality comes from the fact that $\mathbf{V}_r^{\{\ell\}*}\mathbf{V}_r^{\{\ell\}}$ is equal either to the identity operator I on $\widetilde{\mathcal{P}}_r^{\{\ell\}}$ or to the projection $I - Q_r^{\{\ell\}}$ in which case

$$(B_r^{\{\ell\}\langle\alpha_\ell\rangle} - z)^{-1}Q_r^{\{\ell\}} = 0.$$

We claim that there is a real number $C \geq 1$ such that for $\ell \geq \ell_2$

$$\|\mathbf{V}_r^{\{\ell\}} \widetilde{f}\|_{\mathcal{P}^{\{\ell\}}} \leq C \|\widetilde{f}\|_{\widetilde{\mathcal{P}}_r^{\{\ell\}}}, \quad \widetilde{f} \in \widetilde{\mathcal{P}}_r^{\{\ell\}}. \tag{7.34}$$

Assuming the claim to be true we find that

$$\begin{aligned}\|D^{\{\ell\}}(z)f\|_{\mathcal{P}^{\{\ell\}}} &\leq C\|(B_r^{\{\ell\}\langle\alpha_\ell\rangle} - z)^{-1}\widetilde{\pi}^{\{\ell\}}\mathbf{V}^{-1}f - \widetilde{\pi}^{\{\ell\}}(B^{\langle\alpha\rangle} - z)^{-1}\mathbf{V}^{-1}f\|_{\widetilde{\mathcal{P}}_r^{\{\ell\}}} \\ &= C\|\widetilde{D}^{\{\ell\}}(z)\mathbf{V}^{-1}f\|_{\widetilde{\mathcal{P}}_r^{\{\ell\}}}.\end{aligned}$$

By Theorem 7.9, we have $\lim_{\ell\to\infty} \|\widetilde{D}^{\{\ell\}}(z)\mathbf{V}^{-1}f\|_{\widetilde{\mathcal{P}}_r^{\{\ell\}}} = 0$ and therefore

$$\lim_{\ell\to\infty} \|D^{\{\ell\}}(z)f\|_{\mathcal{P}^{\{\ell\}}} = 0, \quad f \in \mathcal{P},$$

which implies (7.33).

It remains to prove the claim (7.34). We assume $\ell \geq \ell_2$ and set, see (7.25), $\mathcal{P}_-^{\{\ell\}} = \mathcal{M}_-^{\{\ell\}}$, $\mathcal{P}_+^{\{\ell\}} = \mathcal{P}^{\{\ell\}} \ominus \mathcal{M}_-^{\{\ell\}}$, $\widetilde{\mathcal{P}}_-^{\{\ell\}} = \widetilde{\mathcal{M}}_-^{\{\ell\}}$ and $\widetilde{\mathcal{P}}_+^{\{\ell\}} = \widetilde{\mathcal{P}}_r^{\{\ell\}} \ominus \widetilde{\mathcal{M}}_-^{\{\ell\}}$. Then

$$\mathcal{P}^{\{\ell\}} = \mathcal{P}_+^{\{\ell\}} \oplus \mathcal{P}_-^{\{\ell\}} \quad \text{and} \quad \widetilde{\mathcal{P}}_r^{\{\ell\}} = \widetilde{\mathcal{P}}_+^{\{\ell\}} \oplus \widetilde{\mathcal{P}}_-^{\{\ell\}} \tag{7.35}$$

are fundamental decompositions of $\mathcal{P}^{\{\ell\}}$ and $\widetilde{\mathcal{P}}_r^{\{\ell\}}$. As in the beginning of this proof we consider two cases.

Case 1: $\deg p^{\{\ell\}} = 2\kappa$. In this case the equality

$$\|\mathbf{V}_r^{\{\ell\}} f^{\{\ell\}}\|_{\mathcal{P}^{\{\ell\}}} = \|f^{\{\ell\}}\|_{\widetilde{\mathcal{P}}_r^{\{\ell\}}}, \quad f^{\{\ell\}} \in \widetilde{\mathcal{P}}_r^{\{\ell\}},$$

holds because $\mathbf{V}_r^{\{\ell\}}$ is unitary and has block-diagonal form with respect to the canonical decompositions (7.35). This proves (7.34) with $C = 1$.

Case 2: $\deg p^{\{\ell\}} = 2\kappa - 1$. Then, as mentioned before, $\mathbf{V}_r^{\{\ell\}}$ is a surjective partial isometry and we recall that $Q_r^{\{\ell\}} = I - \mathbf{V}_r^{\{\ell\}*}\mathbf{V}_r^{\{\ell\}}$ is the orthogonal projection in $\widetilde{\mathcal{P}}_r^{\{\ell\}}$ on the one-dimensional subspace $\ker \mathbf{V}_r^{\{\ell\}} = \operatorname{span}\{e_r^{\{\ell\}}\}$.

As $\mathcal{P}_-^{\{\ell\}} = \mathbf{V}_r^{\{\ell\}}\widetilde{\mathcal{P}}_-^{\{\ell\}}$, the mapping $\mathbf{V}_r^{\{\ell\}}$ admits the block-matrix representation

$$\mathbf{V}_r^{\{\ell\}} = \begin{pmatrix} V_{++}^{\{\ell\}} & 0 \\ V_{-+}^{\{\ell\}} & V_{--}^{\{\ell\}} \end{pmatrix} \tag{7.36}$$

with respect to the canonical decompositions (7.35) of $\mathcal{P}^{\{\ell\}}$ and $\widetilde{\mathcal{P}}_r^{\{\ell\}}$. We show that the norm of each of the three block-operators in (7.36) is uniformly bounded in $\ell \geq \ell_2$ and begin with $V_{--}^{\{\ell\}}$. For $\widetilde{f}_-^{\{\ell\}} \in \widetilde{\mathcal{P}}_-^{\{\ell\}}$ we have

$$\begin{aligned}\|V_{--}^{\{\ell\}}\widetilde{f}_-^{\{\ell\}}\|^2_{\mathcal{P}_-^{\{\ell\}}} &= -\langle (I - Q_r^{\{\ell\}})\widetilde{f}_-^{\{\ell\}}, \widetilde{f}_-^{\{\ell\}}\rangle_{\widetilde{\mathcal{P}}_r^{\{\ell\}}} \\ &= -\langle \widetilde{f}_-^{\{\ell\}}, \widetilde{f}_-^{\{\ell\}}\rangle_{\widetilde{\mathcal{P}}_-^{\{\ell\}}} + \langle Q_r^{\{\ell\}}\widetilde{f}_r^{\{\ell\}}, \widetilde{f}_-^{\{\ell\}}\rangle_{\widetilde{\mathcal{P}}_r^{\{\ell\}}} \\ &= \|\widetilde{f}_-^{\{\ell\}}\|^2_{\widetilde{\mathcal{P}}_-^{\{\ell\}}} - |\langle \widetilde{f}_-^{\{\ell\}}, e_r^{\{\ell\}}\rangle_{\widetilde{\mathcal{P}}_r^{\{\ell\}}}|^2 \end{aligned} \tag{7.37}$$

and hence

$$\|V_{--}^{\{\ell\}}\| \leq 1.$$

Now we estimate the norm of $V_{-+}^{\{\ell\}}$. For arbitrary $0 < \delta < 1$ there is an integer $\ell_3 \geq \ell_2$ such that for $\ell \geq \ell_3$

$$|\langle \widetilde{f}_-^{\{\ell\}}, e_r^{\{\ell\}} \rangle_{\widetilde{\mathcal{P}}_r^{\{\ell\}}}| \leq \delta \, \|\widetilde{f}_-^{\{\ell\}}\|_{\widetilde{\mathcal{P}}_-^{\{\ell\}}}, \quad \widetilde{f}_-^{\{\ell\}} \in \widetilde{\mathcal{P}}_-^{\{\ell\}}.$$

This can be proved by expanding $\widetilde{f}_-^{\{\ell\}}$ in the orthonormal basis $\{u_{j-}^{\{\ell\}}\}_{j=1}^\kappa$ of $\widetilde{\mathcal{P}}_-^{\{\ell\}}$ and then observing that the quantities

$$\langle u_{j-}^{\{\ell\}}, e_r^{\{\ell\}} \rangle_{\widetilde{\mathcal{P}}_r^{\{\ell\}}} = \frac{1}{|p_{2\kappa-1}^{\{\ell\}}|^{1/2}} \langle u_{j-}^{\{\ell\}}, \begin{pmatrix} 0 \\ 1 \end{pmatrix} \rangle_{\mathcal{L}(M_r^{\{\ell\}})}$$

can be made arbitrarily small for large enough ℓ (see (7.31)), say for $\ell \geq \ell_3 \geq \ell_2$. Using this and (7.37) we obtain that for $\ell \geq \ell_3$

$$\|V_{--}^{\{\ell\}} \widetilde{f}_-^{\{\ell\}}\|^2_{\mathcal{P}_-^{\{\ell\}}} \geq (1 - \delta^2) \|\widetilde{f}_-^{\{\ell\}}\|^2_{\widetilde{\mathcal{P}}_-^{\{\ell\}}}, \quad \widetilde{f}_-^{\{\ell\}} \in \widetilde{\mathcal{P}}_-^{\{\ell\}}.$$

We conclude that the norms of the inverse operators $(V_{--}^{\{\ell\}})^{-1}$, $\ell \geq \ell_3$, are bounded by $(1-\delta^2)^{-1/2}$. Combining this result with the fact that each κ-dimensional operator $V_{--}^{\{\ell\}}$, $\ell = \ell_2, \ell_2 + 1, \ldots, \ell_3 - 1$, is invertible (has trivial null space) we get that for some real number $C_1 > 0$ and all $\ell \geq \ell_2$

$$\|(V_{--}^{\{\ell\}})^{-1}\| \leq C_1. \tag{7.38}$$

Denote by $P_{\widetilde{\mathcal{P}}_\pm^{\{\ell\}}}$ the orthogonal projections in $\widetilde{\mathcal{P}}_r^{\{\ell\}}$ onto $\widetilde{\mathcal{P}}_\pm^{\{\ell\}}$. Then for elements $\widetilde{f}_+^{\{\ell\}} \in \widetilde{\mathcal{P}}_+^{\{\ell\}}$ and $\widetilde{g}_-^{\{\ell\}} \in \widetilde{\mathcal{P}}_-^{\{\ell\}}$ it holds

$$\begin{aligned} \langle P_{\widetilde{\mathcal{P}}_-^{\{\ell\}}} Q_r^{\{\ell\}} \widetilde{f}_+^{\{\ell\}}, \widetilde{g}_-^{\{\ell\}} \rangle_{\widetilde{\mathcal{P}}_-^{\{\ell\}}} &= -\langle (I - Q_r^{\{\ell\}}) \widetilde{f}_+^{\{\ell\}}, \widetilde{g}_-^{\{\ell\}} \rangle_{\widetilde{\mathcal{P}}_r^{\{\ell\}}} = \langle \mathbf{V}_r^{\{\ell\}} \widetilde{f}_+^{\{\ell\}}, \mathbf{V}_r^{\{\ell\}} \widetilde{g}_-^{\{\ell\}} \rangle_{\mathcal{P}^{\{\ell\}}} \\ &= -\langle V_{-+}^{\{\ell\}} \widetilde{f}_+^{\{\ell\}}, V_{--}^{\{\ell\}} \widetilde{g}_-^{\{\ell\}} \rangle_{\mathcal{P}_-^{\{\ell\}}} = -\langle V_{--}^{\{\ell\}*} V_{-+}^{\{\ell\}} \widetilde{f}_+^{\{\ell\}}, \widetilde{g}_-^{\{\ell\}} \rangle_{\widetilde{\mathcal{P}}_-^{\{\ell\}}} \end{aligned}$$

and hence

$$V_{-+}^{\{\ell\}} = -(V_{--}^{\{\ell\}*})^{-1} P_{\widetilde{\mathcal{P}}_-^{\{\ell\}}} Q_r^{\{\ell\}} P_{\widetilde{\mathcal{P}}_+^{\{\ell\}}}.$$

By (7.38) and since the norms of the projections are equal to 1 and the sequence $\|e_r^{\{\ell\}}\|_{\widetilde{\mathcal{P}}_r^{\{\ell\}}}$ is bounded, see (7.31), there is a real number $C_2 > 0$ such that for $\ell \geq \ell_2$

$$\|V_{-+}^{\{\ell\}}\| \leq \|(V_{--}^{\{\ell\}*})^{-1}\| \, \|Q_r^{\{\ell\}}\| = \|(V_{--}^{\{\ell\}})^{-1}\| \|e_r^{\{\ell\}}\|_{\widetilde{\mathcal{P}}_r^{\{\ell\}}} \leq C_2.$$

Finally, we consider the operator $V_{++}^{\{\ell\}}$. For $\widetilde{f}_+^{\{\ell\}} \in \widetilde{\mathcal{P}}_+^{\{\ell\}}$ we have

$$\langle V_{++}^{\{\ell\}} \widetilde{f}_+^{\{\ell\}}, V_{++}^{\{\ell\}} \widetilde{f}_+^{\{\ell\}} \rangle_{\mathcal{P}_+^{\{\ell\}}} = \langle V_{-+}^{\{\ell\}} \widetilde{f}_+^{\{\ell\}}, V_{-+}^{\{\ell\}} \widetilde{f}_+^{\{\ell\}} \rangle_{\mathcal{P}_-^{\{\ell\}}} + \langle (I - Q_r^{\{\ell\}}) \widetilde{f}_+^{\{\ell\}}, \widetilde{f}_+^{\{\ell\}} \rangle_{\widetilde{\mathcal{P}}_r^{\{\ell\}}}$$

and hence for some real number $C_3 > 0$ and all $\ell \geq \ell_2$

$$\|V_{++}^{\{\ell\}}\| \leq \|V_{-+}^{\{\ell\}}\| + \|(I - Q_r^{\{\ell\}})\|_{\widetilde{\mathcal{P}}_r^{\{\ell\}}} \leq C_3.$$

The obtained uniform bounds on the block matrix entries in the representation (7.36) of $\mathbf{V}_r^{\{\ell\}}$ imply (7.34) with $C = 1 + (C_2^2 + C_3^2)^{1/2}$. □

8. An example

8.1. In this section we discuss a concrete example in detail, namely an approximation problem connected with the Bessel operator related to the singular differential expression

$$b_B y(x) = -y''(x) + \frac{2}{x^2}\, y(x), \quad x \in \mathbb{R}^+ = (0, \infty), \tag{8.1}$$

which is of Bessel type with parameter $\nu = \frac{3}{2}$. The maximal and minimal realizations of b_B in the Hilbert space $\mathcal{H}_0 = L^2(\mathbb{R}^+)$ coincide and define a unique self-adjoint realization of b_B which we denote by L and call the Bessel operator. In [10], [11], and [14] with b_B there is associated the function

$$N(z) = \left(\sqrt{-z}\right)^3 = (-z)^{3/2}, \tag{8.2}$$

where

$$\sqrt{-z} = \sqrt{\rho}\, e^{\frac{1}{2}i(\theta - \pi)}, \quad z = \rho e^{i\theta},\ \rho > 0,\ 0 < \theta < 2\pi.$$

Writing $N(z)$ as $N(z) = z^2/\sqrt{-z}$, it follows, see, for example, [10], that N is an $\mathcal{N}_1$-function. Since its only generalized pole not of positive type is ∞, N belongs to $\mathcal{N}_1^\infty$and is singular. For $z_0 \in \mathbb{R}^- = (-\infty, 0)$ it can be written in the form

$$N(z) = (z - z_0)^2 n(z) + r(z) \tag{8.3}$$

with the Nevanlinna function

$$n(z) = \frac{\sqrt{-z} + \frac{1}{2}\sqrt{-z_0}}{(\sqrt{-z} + \sqrt{-z_0})^2} = \frac{1}{\pi} \int_0^\infty \frac{t^{3/2} dt}{(t - z)(t - z_0)^2} \tag{8.4}$$

and the polynomial

$$r(z) = -\frac{3}{2}(-z_0)^{1/2}(z - z_0) + (-z_0)^{3/2}. \tag{8.5}$$

It follows from (8.3) that N admits the decomposition (5.6):

$$N(z) = (z - z_0)^2\big((n(z) - \operatorname{Re} n(i)) + q(z)\big) + r(z)$$

with $m = \kappa = 1$, the normalized Nevanlinna function

$$n(z) - \operatorname{Re} n(i) = \frac{1}{\pi} \int_0^\infty \left(\frac{1}{t - z} - \frac{t}{t^2 + 1}\right) \frac{t^{3/2} dt}{(t - z_0)^2},$$

and the constant polynomial

$$q(z) \equiv p_2 = \operatorname{Re} n(i) = \frac{1}{\pi} \int_0^\infty \frac{t}{t^2 + 1} \frac{t^{3/2} dt}{(t - z_0)^2}.$$

Note that (8.3) is a decomposition of N of the form (5.8) with $n_r = n$ and $r_0 = r$.

In [10] the model for N was obtained by lifting the Bessel operator L to a self-adjoint relation in a Pontryagin space containing $\mathcal{H}_0$ as a subspace with finite co-dimension. In what follows we construct regular $\mathcal{N}_1^\infty$-functions $N^{\{\varepsilon\}}$ which converge to N as $\varepsilon \downarrow 0$ and satisfy Hypothesis 7.5 with $\ell = 1/\varepsilon$, see (8.25) below. Each $N^{\{\varepsilon\}}$ is a function related to the regular differential expression

$$b_B^{\{\varepsilon\}} y\,(x) = -y''(x) + \frac{2}{(x+\varepsilon)^2}\, y(x), \quad x \in \mathbb{R}^+, \quad \varepsilon > 0, \tag{8.6}$$

which formally converges to (8.1) as $\varepsilon \downarrow 0$. The results from Section 7 imply strong approximation of the model for N by the models for $N^{\{\varepsilon\}}$ in Pontryagin spaces. We prove that (1) the main self-adjoint operators/relations in these models are unitarily equivalent to linearizations of certain Straus operators associated with (8.1) and (8.6) in the Hilbert space $\mathcal{H}_0$ and eigenvalue dependent boundary conditions at $x = 0$ and (2) Straus extensions associated with $N^{\{\varepsilon\}}$strongly approximate, in the resolvent sense, the Straus extension associated with N. Recall that a (family of) Straus extension(s) $T(z)$ of a densely defined symmetric operator S in a Hilbert space $\mathcal{H}$ is the restriction of S^* to the range of a generalized resolvent $R(z) = P_{\mathcal{H}}(A - z)^{-1}|_{\mathcal{H}}$ in which A is a self-adjoint extension of S in a possibly larger Hilbert or Pontryagin space containing $\mathcal{H}$ as a subspace and $P_{\mathcal{H}}$ is the orthogonal projection onto $\mathcal{H}$. The operator/relation A is called a linearization of $T(z)$; it can always be chosen minimal, in which case A is unique up to unitary equivalence and $T(z)$ is defined for all $z \in \rho(A)$. The relation between $T(z)$ and $R(z)$ is given by

$$T(z) = R(z)^{-1} + z, \quad \text{or} \quad R(z) = (T(z) - z)^{-1}.$$

We will call the operator on the right-hand side of the last equality the resolvent of $T(z)$. For a characterization of Straus extensions and other details see [6]. We close this subsection with the characterization of a Straus extension related to the expression b_B. We choose a point $z_0 \in \mathbb{R}^-$ and fix it as a reference point in the sequel. For $z \in \mathbb{C} \setminus [0, \infty)$ we define the function $\varphi(z) = \varphi(\,\cdot\,, z)$ by

$$\varphi(x, z) = \left(\sqrt{-z} + \frac{1}{x}\right) \mathrm{e}^{-x\sqrt{-z}}, \quad x \in \mathbb{R}^+. \tag{8.7}$$

It satisfies the Bessel equation $(b_B - z)\varphi(x, z) = 0$, $x \in \mathbb{R}^+$, and is square integrable at ∞, but it does not belong to $\mathcal{H}_0$ due to the singularity at $x = 0$. However, for $z \in \mathbb{C} \setminus [0, \infty)$ the function $\varphi_0(z) = \varphi_0(\,\cdot\,, z)$ with

$$\varphi_0(x, z) = \begin{cases} \dfrac{\varphi(x, z) - \varphi(x, z_0)}{z - z_0}, & z \neq z_0, \\[2ex] \dfrac{\partial \varphi(x, z)}{\partial z}, & z = z_0, \end{cases} \qquad x \in \mathbb{R}^+, \tag{8.8}$$

belongs to $\mathcal{H}_0$ and it is used in the definition of the relations:

$$\begin{aligned} T^{\langle \alpha \rangle}(z) = \{\{h + c\varphi_0(w), Lh + wc\varphi_0(w)\} \mid h \in \operatorname{dom} L, c \in \mathbb{C}, \\ \langle (L - w)h, \varphi_0(w) \rangle_0 = -(n(w) + \tau_\alpha(z))c\}, \end{aligned} \tag{8.9}$$

where, as we shall show below, the set on the right-hand side is independent of the choice of $w \in \mathbb{R}^-$. Here n is defined by (8.4), α is a real number, and with r

from (8.5) τ_α is given by

$$\tau_\alpha(z) = \frac{r(z)-\alpha}{(z-z_0)^2}.$$

In the theorem we relate $T^{\langle\alpha\rangle}(z)$ to the operator $B^{\langle\alpha\rangle}$ in $\mathcal{L}(\widetilde{N})$ from (5.12) with $m=\kappa=1$,

$$\widetilde{N}(z) = \begin{pmatrix} n(z)-\operatorname{Re} n(i) & 0 & 0 \\ 0 & r(z) & z-z_0 \\ 0 & z-z_0 & 0 \end{pmatrix},$$

and $n(z)+p_{2\kappa}$ in the matrix formula for $K_\alpha(z)$ replaced by the right-hand side of (8.4).

Theorem 8.1.

(i) *The restriction*

$$S_0 = L \mid_{\{h\in \operatorname{dom} L \mid \langle (L-w)h,\varphi_0(w)\rangle_0 = 0\}}$$

with $w\in\mathbb{R}^-$ is a simple densely defined symmetric operator in $\mathcal{H}_0$ with defect indices $(1,1)$ and defect elements $\varphi_0(z)$, $z\in\mathbb{R}^-$.

(ii) *$T^{\langle\alpha\rangle}(z)$ is a Straus extension of S_0 in $\mathcal{H}_0$ and the minimal self-adjoint linearization of $T^{\langle\alpha\rangle}(z)$ is unitarily equivalent to the operator $B^{\langle\alpha\rangle}$ in the space $\mathcal{L}(\widetilde{N})$ from* (5.12) *as described above.*

Statement (ii) means that the resolvent $(T^{\langle\alpha\rangle}(z)-z)^{-1}$ is a generalized resolvent of S_0 and that there exists a unitary map $\mathbf{U}_0:\mathcal{H}_0\to\mathcal{L}(n)$ such that

$$\mathbf{U}_0(T^{\langle\alpha\rangle}(z)-z)^{-1} = P_{\mathcal{L}(n)}(B^{\langle\alpha\rangle}-z)^{-1}\mid_{\mathcal{L}(n)} \mathbf{U}_0. \tag{8.10}$$

Proof of Theorem 8.1. For $z\in\mathbb{R}^-$ the function $\varphi_0(x,z)$ is bounded at $x=0$ and hence square integrable in $\mathbb{R}^+$. Since $b_B\varphi_0(z)=\varphi(z)+z_0\varphi_0(z)$ does not belong to $\mathcal{H}_0$ it follows that

$$\varphi_0(z)\in\mathcal{H}_0\setminus\operatorname{dom}(L). \tag{8.11}$$

It is straightforward to check the equality

$$\varphi_0(z)-\varphi_0(w) = (z-w)(L-z)^{-1}\varphi_0(w). \tag{8.12}$$

The construction of S_0 is standard; in particular, from the last equality it follows that its domain is independent of the particular choice of the point $w\in\mathbb{R}^-$. The adjoint of S_0 is given by

$$S_0^* = L + \operatorname{span}\{\{\varphi_0(w), w\varphi_0(w)\}\}, \quad \text{direct sum},$$

which by (8.11) implies that S_0^* is an operator and S_0 is densely defined. For $z\in\mathbb{R}^-$ we have $\varphi_0(z)\perp\operatorname{ran}(S_0-z^*)$, hence $\varphi_0(z)\in\ker(S_0^*-z)$.

Since it is a restriction of S_0^*, the relation $T^{\langle\alpha\rangle}(z)$ from (8.9) is an operator. We calculate its resolvent to show that its definition is independent of the choice of $w \in \mathbb{R}^-$ and that it is a Straus extension of S_0. We set $(T^{\langle\alpha\rangle}(z) - z)^{-1} f = g$, which equivalently can be written as

$$\begin{aligned} g &= h + c\varphi_0(w) \\ f + zg &= Lh + wc\varphi_0(w) \end{aligned}$$

with

$$\big\langle (L-w)h, \varphi_0(w) \big\rangle_0 = -\big(n(w) + \tau_\alpha(z)\big)c. \tag{8.13}$$

The first two equations imply $h = (L-z)^{-1} f + c(z-w)(L-z)^{-1}\varphi_0(w)$ and hence

$$g = (L-z)^{-1} f + c\varphi_0(z).$$

Inserting h into (8.13) we obtain

$$c = -\frac{\langle f, \varphi_0(z^*)\rangle_0}{(z-w)\langle \varphi_0(z), \varphi_0(w)\rangle_0 + n(w) + \tau_\alpha(z)}.$$

We now use the relation

$$\langle \varphi_0(z), \varphi_0(w)\rangle_0 = \frac{n(z) - n(w)}{z - w} \tag{8.14}$$

(compare with [7, Section 4]), which for $w = z_0$ can be obtained directly from (8.7) and (8.8):

$$\begin{aligned}
&(z - z_0)\langle \varphi_0(z), \varphi_0(z_0)\rangle_0 \\
&= \frac{1}{2}\int_0^\infty \left(\sqrt{-z}\mathrm{e}^{-x(\sqrt{-z}+\sqrt{-z_0})} - \sqrt{-z_0}\mathrm{e}^{-2x\sqrt{-z_0}}\right) x\,dx \\
&\qquad + \frac{1}{2}\int_0^\infty \left(\mathrm{e}^{-x(\sqrt{-z}+\sqrt{-z_0})} - \mathrm{e}^{-2x\sqrt{-z_0}}\right) dx \\
&= \left(\frac{\sqrt{-z}}{(\sqrt{-z}+\sqrt{-z_0})^2} - \frac{\sqrt{-z_0}}{(2\sqrt{-z_0})^2}\right) + \left(\frac{1}{\sqrt{-z}+\sqrt{-z_0}} - \frac{1}{2\sqrt{-z_0}}\right) \\
&= \frac{\sqrt{-z} + 1/2\sqrt{-z_0}}{(\sqrt{-z}+\sqrt{-z_0})^2} - \frac{3/2\sqrt{-z_0}}{(2\sqrt{-z_0})^2} \\
&= n(z) - n(z_0)
\end{aligned}$$

and then for arbitrary $w \in \mathbb{R}^-$ from (8.12):

$$\begin{aligned}
n(z) - n(w) &= n(z) - n(z_0) + n(z_0) - n(w) \\
&= (z - z_0)\langle \varphi_0(z), \varphi_0(z_0)\rangle_0 + (z_0 - w)\langle \varphi_0(z_0), \varphi_0(w)\rangle_0 \\
&= (z - z_0)\big\langle \varphi_0(z), \varphi_0(w) + (z_0 - w)(L - z_0)^{-1}\varphi_0(w)\big\rangle_0 \\
&\qquad + (z_0 - w)\big\langle \varphi_0(z) + (z_0 - z)(L - z_0)^{-1}\varphi_0(z), \varphi_0(w)\big\rangle_0 \\
&= (z - w)\langle \varphi_0(z), \varphi_0(w)\rangle_0.
\end{aligned}$$

The resolvent formula

$$(T^{\langle\alpha\rangle}(z) - z)^{-1} = (L - z)^{-1} - \frac{\big\langle \,\cdot\,, \varphi_0(z^*)\big\rangle_0}{n(z) + \tau_\alpha(z)}\varphi_0(z) \tag{8.15}$$

now easily follows. It shows that the definition of $T^{\langle\alpha\rangle}(z)$ in (8.9) is independent of $w \in \mathbb{R}^-$ and that, since the right-hand side defines a generalized resolvent of S_0, $T^{\langle\alpha\rangle}(z)$ is a Straus extension of S_0.

To prove (8.10) we define the mapping $\mathbf{U}_0$

$$\mathbf{U}_0 : \mathcal{D}_0 := \operatorname{span}\{\varphi_0(w) \mid w \in \mathbb{C} \setminus [0,\infty)\} \to \mathcal{L}(n)$$

by

$$\mathbf{U}_0\varphi_0(w) = K_n(\,\cdot\,, w^*).$$

The domain and range of this mapping are dense in $\mathcal{H}_0$ and $\mathcal{L}(n)$, the former because S_0 is simple and the latter by definition of the reproducing kernel space. According to (8.14) and by the reproducing kernel property of the kernel K_n, $\mathbf{U}_0$ is isometric, hence well defined. If we denote its continuous continuation onto $\mathcal{H}_0$ also by $\mathbf{U}_0$, then $\mathbf{U}_0$ is unitary and using (8.14), (8.15), and [8, Theorem 4.6] we obtain (8.10). With the unitary mapping

$$\widetilde{\mathbf{U}}_0 : \mathcal{H}_0 \oplus \mathcal{L}\left(\begin{pmatrix} r(z) & z-z_0 \\ z-z_0 & 0 \end{pmatrix}\right) \to \mathcal{L}(n) \oplus \mathcal{L}\left(\begin{pmatrix} r(z) & z-z_0 \\ z-z_0 & 0 \end{pmatrix}\right) = \mathcal{L}(\widetilde{N})$$

defined as $\widetilde{\mathbf{U}}_0 = \begin{pmatrix} \mathbf{U}_0 & 0 \\ 0 & I \end{pmatrix}$ this further gives

$$(T^{\langle\alpha\rangle}(z) - z)^{-1} = P_{\mathcal{H}_0}\big(\widetilde{\mathbf{U}}_0^{-1} B^{\langle\alpha\rangle} \widetilde{\mathbf{U}}_0 - z\big)^{-1}\,|_{\mathcal{H}_0}\,.$$

Since the model $(B, K_{\widetilde{N}}(\,\cdot\,,z), \widetilde{S})$ associated with N is minimal this equality implies that every minimal self-adjoint linearization of the operator function $T^{\langle\alpha\rangle}(z)$ is unitarily equivalent to $B^{\langle\alpha\rangle}$. □

8.2. In this subsection we consider for small real numbers $\varepsilon > 0$ various operators and functions related to the regular differential expressions $b_B^{\{\varepsilon\}}$ defined by (8.6) and construct a family of regular $\mathcal{N}_1^\infty$-functions $N^{\{\varepsilon\}}$ which converges pointwise to the singular $\mathcal{N}_1^\infty$-function N from (8.2) and satisfies Hypothesis 7.5 with $\kappa = 1$ and $\ell = 1/\varepsilon$.

The minimal realization $S^{\{\varepsilon\}}$ of $b_B^{\{\varepsilon\}}$ in $\mathcal{H}_0$ is a symmetric densely defined operator and the maximal realization of $b_B^{\{\varepsilon\}}$ in $\mathcal{H}_0$ coincides with the adjoint $S^{\{\varepsilon\}*}$ of $S^{\{\varepsilon\}}$. The domain of $S^{\{\varepsilon\}*}$ is given by

$$\operatorname{dom} S^{\{\varepsilon\}*} = \big\{y \in \mathcal{H}_0 \mid y,\, y'\text{absolutely continuous},\, y'' \in \mathcal{H}_0\big\}.$$

Since $b_B^{\{\varepsilon\}}$ is regular at the left endpoint 0 and in limit point case at ∞, the defect indices of $S^{\{\varepsilon\}}$ are equal to $(1,1)$, and each self-adjoint extension of $S^{\{\varepsilon\}}$ can be obtained as the restriction of $S^{\{\varepsilon\}*}$ by a boundary condition of the form

$$y'(0) = \beta y(0), \quad \beta \in \mathbb{R} \cup \{\infty\}, \tag{8.16}$$

which if $\beta = \infty$ stands for the boundary condition $y(0) = 0$. We denote by $L_\beta^{\{\varepsilon\}}$ the self-adjoint extension corresponding to the parameter $\beta \in \mathbb{R} \cup \{\infty\}$. Note that $\mathbb{R}^- \subset \rho(L_\infty^{\{\varepsilon\}})$.

For $z \in \mathbb{C} \setminus [0, \infty)$ we consider the function $\varphi^{\{\varepsilon\}}(z) = \varphi^{\{\varepsilon\}}(\,\cdot\,, z)$ defined by

$$\varphi^{\{\varepsilon\}}(x, z) = \gamma(z, \varepsilon) \left(\sqrt{-z} + \frac{1}{x+\varepsilon} \right) \mathrm{e}^{-x\sqrt{-z}}, \quad x \in [0, \infty), \tag{8.17}$$

with

$$\gamma(z, \varepsilon) = \frac{1}{\varepsilon\sqrt{-z} + 1}.$$

The function $\varphi^{\{\varepsilon\}}(z)$ belongs to $\operatorname{dom} S^{\{\varepsilon\}*}$ and satisfies the differential equation $(S^{\{\varepsilon\}*} - z)\varphi^{\{\varepsilon\}}(z) = 0$. Moreover it has the initial value $\varphi^{\{\varepsilon\}}(0, z) = 1/\varepsilon$, and for $z, \zeta \in \mathbb{C}^+$ we have

$$\varphi^{\{\varepsilon\}}(z) - \varphi^{\{\varepsilon\}}(\zeta) = (z - \zeta)(L_\infty^{\{\varepsilon\}} - z)^{-1}\varphi^{\{\varepsilon\}}(\zeta). \tag{8.18}$$

As usual, the Titchmarsh-Weyl coefficient $m^{\{\varepsilon\}}$ for the pair $(L_\infty^{\{\varepsilon\}}, S^{\{\varepsilon\}})$ is defined as the value of the logarithmic derivative of $\varphi^{\{\varepsilon\}}(z)$ at $x = 0$:

$$m^{\{\varepsilon\}}(z) = \left(\varphi^{\{\varepsilon\}}(x, z)^{-1} \frac{d}{dx} \varphi^{\{\varepsilon\}}(x, z) \right)\Big|_{x=0}.$$

If we rescale $m^{\{\varepsilon\}}$ by setting

$$n^{\{\varepsilon\}} = \varepsilon^{-2} m^{\{\varepsilon\}},$$

then the equality

$$\langle \varphi^{\{\varepsilon\}}(z), \varphi^{\{\varepsilon\}}(\zeta) \rangle_0 = \frac{n^{\{\varepsilon\}}(z) - n^{\{\varepsilon\}}(\zeta^*)}{z - \zeta^*}, \quad z, \zeta \in \mathbb{C}^+, \tag{8.19}$$

holds, where $\langle\,\cdot\,,\,\cdot\,\rangle_0$ denotes the inner product in $\mathcal{H}_0 = L^2(\mathbb{R}^+)$. The equalities (8.18) and (8.19) imply that $(L_\infty^{\{\varepsilon\}}, \varphi^{\{\varepsilon\}}(z), S^{\{\varepsilon\}})$ is a model in $\mathcal{H}_0$ for the Nevanlinna function $n^{\{\varepsilon\}}$. The relation between the resolvents of $L_\beta^{\{\varepsilon\}}$ and $L_\infty^{\{\varepsilon\}}$ is given by the formula

$$(L_\beta^{\{\varepsilon\}} - z)^{-1} - (L_\infty^{\{\varepsilon\}} - z)^{-1} = -\frac{\langle\,\cdot\,, \varphi^{\{\varepsilon\}}(z^*)\rangle_0}{n^{\{\varepsilon\}}(z) - \varepsilon^{-2}\beta}\, \varphi^{\{\varepsilon\}}(z). \tag{8.20}$$

A direct calculation yields

$$n^{\{\varepsilon\}}(z) = -\varepsilon^{-2}\sqrt{-z} - \varepsilon^{-3}\gamma(z, \varepsilon) \tag{8.21}$$

and hence the Nevanlinna function $n^{\{\varepsilon\}}$ has the integral representation

$$n^{\{\varepsilon\}}(z) = \int_0^\infty \left(\frac{1}{t - z} - \frac{t}{t^2 + 1} \right) d\sigma^{\{\varepsilon\}}(t) + \operatorname{Re} n^{\{\varepsilon\}}(i),$$

where

$$d\sigma^{\{\varepsilon\}}(t) = \frac{1}{\pi} \operatorname{Im} n^{\{\varepsilon\}}(t + i0)\, dt = \frac{t^{3/2} dt}{\pi(\varepsilon^2 t + 1)}, \quad t \geq 0. \tag{8.22}$$

It follows from (8.21) that $n^{\{\varepsilon\}}$ has the asymptotic behavior

$$n^{\{\varepsilon\}}(z) = \varepsilon^{-1} z - \varepsilon^{-3} + (-z)^{3/2} + O(\varepsilon), \quad \varepsilon \downarrow 0. \tag{8.23}$$

To compensate the singular part we add the polynomial

$$q^{\{\varepsilon\}}(z) := -\varepsilon^{-1}z + \varepsilon^{-3} \tag{8.24}$$

to $n^{\{\varepsilon\}}$ and so obtain for $\varepsilon > 0$ the family of regular $\mathcal{N}_1^\infty$-functions

$$N^{\{\varepsilon\}}(z) = n^{\{\varepsilon\}}(z) + q^{\{\varepsilon\}}(z) = \gamma(z,\varepsilon)(-z)^{3/2}. \tag{8.25}$$

which converges to the singular $\mathcal{N}_1^\infty$-function $N(z)$ in (8.2) as $\varepsilon \downarrow 0$. Note that the irreducible representation of $N^{\{\varepsilon\}}$ is given by

$$N^{\{\varepsilon\}}(z) = \left(n^{\{\varepsilon\}}(z) - \operatorname{Re} n^{\{\varepsilon\}}(i)\right) + \left(q^{\{\varepsilon\}}(z) + \operatorname{Re} n^{\{\varepsilon\}}(i)\right).$$

Finally, if we denote by $d\sigma$ the spectral measure of the Nevanlinna function n in (8.4), then (8.22) implies

$$|t - z_0|^{-2}\frac{d\sigma^{\{\varepsilon\}}(t)}{d\sigma(t)} = \frac{1}{\varepsilon^2 t + 1} \leq 1$$

and hence Hypothesis 7.5 is satisfied.

Remark 8.2. The function N, which here is given as a limit of regularized Titchmarsh-Weyl coefficients, has also been associated with the differential expression (8.1) by other means. It has been obtained as an analytic continuation of a Q-function for small parameter $\nu < 1$ to the value $\nu = \frac{3}{2}$ (see [10]) or as a generalized Titchmarsh-Weyl coefficient (see [11] or [12]). Moreover, it appears in connection with a certain super singular perturbation of the Bessel operator, see [10] or also [14] for a corresponding “model” (in a slightly different sense than used here) involving a differential operator in a Hilbert space, which includes also functions which are not square integrable at 0.

8.3. In this subsection we digress for a moment. We present two theorems and make some observations to show that the irreducible representations of the functions $N^{\{\varepsilon\}}$ in (8.25) are not good enough to lead to an approximation theorem for Straus extensions. In the next subsections we consider the regularized representations of $N^{\{\varepsilon\}}$ and prove an approximation theorem, see Theorem 8.7 below.

As to the first theorem recall that L is the unique self-adjoint realization of the Bessel expression b_B (see (8.1)) in $\mathcal{H}_0$ and $L_\beta^{\{\varepsilon\}}$ is the self-adjoint restriction of the maximal realization of $b_B^{\{\varepsilon\}}$ (see (8.6)) in $\mathcal{H}_0$ determined by the boundary condition (8.16).

Theorem 8.3. *For every $\beta \in \mathbb{R} \cup \{\infty\}$ the self-adjoint operators $L_\beta^{\{\varepsilon\}}$ converge to the self-adjoint operator L in the strong resolvent sense as $\varepsilon \downarrow 0$.*

Proof. Denote by $\mathcal{D}$ the subspace of $\mathcal{H}_0$ consisting of all absolutely continuous functions g, such that $g' \in \mathcal{H}_0$ and $g(0) = 0$. Note that the domains of the quadratic forms corresponding to $L_\infty^{\{\varepsilon\}}$ as well as to L coincide with $\mathcal{D}$. Let $\{\varepsilon_j\}_{j=1}^\infty$ be a monotonously decreasing sequence of positive numbers which converges to zero. One sees directly that

$$\langle b_B^{\{\varepsilon_{j+1}\}} f, f\rangle_0 \le \langle b_B^{\{\varepsilon_j\}} f, f\rangle_0 \le \langle b_B f, f\rangle_0, \quad f \in \mathcal{D}.$$

In this situation [21, Theorem 3.1] shows that the corresponding operators $L_\infty^{\{\varepsilon_j\}}$ converge in the strong resolvent sense to a self-adjoint operator $\widehat{L}$ in $\mathcal{H}_0$ as $j \to \infty$. As the self-adjoint realization of b_B in $\mathcal{H}_0$ is unique, it follows that $\widehat{L} = L$. This shows the strong convergence $(L_\infty^{\{\varepsilon\}} - z)^{-1} \to (L - z)^{-1}$ for every $z \in \mathbb{C} \setminus \mathbb{R}$. Now consider $\beta \in \mathbb{R}$. The norm of the rank one operator on the right-hand side of (8.20) is by (8.19) equal to

$$\|\varphi^{\{\varepsilon\}}(z)\|^2 \frac{1}{|n^{\{\varepsilon\}}(z) - \varepsilon^{-2}\beta|} = \frac{|\operatorname{Im} n^{\{\varepsilon\}}(z)|}{|\operatorname{Im} z||n^{\{\varepsilon\}}(z) - \varepsilon^{-2}\beta|}, \quad z \in \mathbb{C} \setminus \mathbb{R}.$$

By (8.23) it tends to 0 as $\varepsilon \downarrow 0$. For $\beta \in \mathbb{R}$ the theorem now follows from (8.20). □

It can be shown that the theorem still holds if one allows the real parameter β to depend on ε. The theorem illustrates that for a singular differential expression which is in the "limit point case" at both endpoints of the interval one cannot influence the realization in the limit by varying only the parameter in the boundary conditions of the approximations. This phenomenon is connected with the fact that the limit of Nevanlinna functions is again a Nevanlinna function. The additive decomposition of $N^{\{\varepsilon\}}$ in (8.25) might suggest to consider boundary conditions which also depend on the spectral parameter z. This we do in the following paragraph.

For $\alpha \in \mathbb{R}$ consider the restriction $T^{\{\varepsilon\}\langle\alpha\rangle}(z)$ of the maximal operator $S^{\{\varepsilon\}*}$ associated with the expression $b_B^{\{\varepsilon\}}$ which is determined by the boundary condition

$$y'(0) = \beta(\varepsilon, z)y(0) \quad \text{with} \quad \beta(\varepsilon, z) = \varepsilon^2\big(\alpha - q^{\{\varepsilon\}}(z)\big),$$

where $q^{\{\varepsilon\}}(z)$ is given by (8.24). A standard calculation shows that

$$(T^{\{\varepsilon\}\langle\alpha\rangle}(z) - z)^{-1} = (L_\infty^{\{\varepsilon\}} - z)^{-1} - \frac{\langle\,\cdot\,, \varphi^{\{\varepsilon\}}(z^*)\rangle_0}{N^{\{\varepsilon\}}(z) - \alpha}\varphi^{\{\varepsilon\}}(z), \tag{8.26}$$

where $\varphi^{\{\varepsilon\}}(z)$ and $N^{\{\varepsilon\}}$ are given by (8.17) and (8.25). According to M.G. Krein's formula this relation also shows that $(T^{\{\varepsilon\}\langle\alpha\rangle}(z) - z)^{-1}$ is a compressed resolvent of some self-adjoint extension (in a larger Pontryagin space) of the minimal operator $S^{\{\varepsilon\}}$, that is, $T^{\{\varepsilon\}\langle\alpha\rangle}(z)$ is a Straus extension of $S^{\{\varepsilon\}}$. The following theorem is an analog of Theorem 8.1. Its proof is similar to that of Theorem 8.1 and therefore omitted. The operator $B^{\{\varepsilon\}\langle\alpha\rangle}$ in the theorem is described in [9, Theorem 3.3(iii)] and [8, Theorem 4.7(ii)] and

$$\widetilde{N}^{\{\varepsilon\}} = \begin{pmatrix} n^{\{\varepsilon\}} - \operatorname{Re} n^{\{\varepsilon\}}(i) & 0 \\ 0 & q^{\{\varepsilon\}} + \operatorname{Re} n^{\{\varepsilon\}}(i) \end{pmatrix}$$

so that $\mathcal{L}(\widetilde{N}^{\{\varepsilon\}}) = \mathcal{L}(n^{\{\varepsilon\}}) \oplus \mathcal{L}(q^{\{\varepsilon\}})$.

Theorem 8.4. *The minimal self-adjoint linearization of* $T^{\{\varepsilon\}\langle\alpha\rangle}(z)$ *is unitarily equivalent to the operator* $B^{\{\varepsilon\}\langle\alpha\rangle}$ *in* $\mathcal{L}(\widetilde{N}^{\{\varepsilon\}})$.

Remark 8.5. Relation (8.26) implies that every minimal linearization of $T^{\{\varepsilon\}\langle\alpha\rangle}(z)$ is a representing relation for the generalized Nevanlinna function

$$-\frac{1}{N^{\{\varepsilon\}}(z)-\alpha}.$$

Theorem 8.4 implies that the corresponding model is actually minimal, which otherwise does not follow directly.

Consider a real function $\alpha(\varepsilon)$ and assume that $\alpha(\varepsilon)\to\alpha\in\mathbb{R}\cup\{\infty\}$ as $\varepsilon\downarrow 0$. Then $N^{\{\varepsilon\}}(z)\to N(z)$ and Theorem 7.10 implies the convergence of the resolvents of $B^{\{\varepsilon\}\langle\alpha(\varepsilon)\rangle}$ in the sense of strong approximation. Therefore, by Theorem 8.4, also the resolvents of the corresponding minimal linearizations of $T^{\{\varepsilon\}\langle\alpha(\varepsilon)\rangle}(z)$ converge in some Pontryagin spaces. However, the resolvents $(T^{\{\varepsilon\}\langle\alpha(\varepsilon)\rangle}(z)-z)^{-1}$ in $\mathcal{H}_0$ do not converge as can be seen from (8.26): the functions $\varphi^{\{\varepsilon\}}(z)$ do not remain square integrable at the origin when $\varepsilon\downarrow 0$, which reflects the fact that the summands in the decomposition $N^{\{\varepsilon\}}=n^{\{\varepsilon\}}+q^{\{\varepsilon\}}$ do not converge separately.

8.4. The objects of interest in this section are summarized in the following table.

	approximating objects		target objects
differential expressions:	$b_B^{\{\varepsilon\}}$		b_B
	$\updownarrow$		$\updownarrow$
convergent modifications of Titchmarsh-Weyl functions:	$N^{\{\varepsilon\}}$	$\xrightarrow{\varepsilon\to 0}$	N
minimal realizations in Π_1-spaces:	$B^{\{\varepsilon\}\langle\alpha\rangle}$, $\mathcal{L}(\widetilde{N}^{\{\varepsilon\}})$		$B^{\langle\alpha\rangle}$, $\mathcal{L}(\widetilde{N})$
regularized, not necessarily minimal, realizations in Π_1-spaces:	$B_r^{\{\varepsilon\}\langle\alpha\rangle}$, $\mathcal{L}(\widetilde{N}_r^{\{\varepsilon\}})$	$\xrightarrow{\varepsilon\to 0}$	$B^{\langle\alpha\rangle}$, $\mathcal{L}(\widetilde{N})$

The operators $B^{\langle\alpha\rangle}$ and $B^{\{\varepsilon\}\langle\alpha\rangle}$ appeared in Subsections 8.1 and 8.3 as self-adjoint linearizations of Straus operators related to the functions N and $N^{\{\varepsilon\}}$, see Theorem 8.1 and Theorem 8.4. In this subsection the operator $B_r^{\{\varepsilon\}\langle\alpha\rangle}$ will play a role. Recall from item (II) at the end of Section 5 that this operator was constructed using the regularized representation of $N^{\{\varepsilon\}}$. As before $z_0\in\mathbb{R}^-$ is a fixed reference point. The regularized representation (5.8) of $N^{\{\varepsilon\}}$ in (8.25) reads as

$$N^{\{\varepsilon\}}(z)=(z-z_0)^2 n_r^{\{\varepsilon\}}(z)+r_0^{\{\varepsilon\}}(z),$$

where

$$n_r^{\{\varepsilon\}}(z) = \gamma^2(z_0,\varepsilon)\gamma(z,\varepsilon)\frac{\sqrt{-z}+\frac{1}{2}\sqrt{-z_0}+\frac{1}{2}\varepsilon\sqrt{-z_0}\sqrt{-z}}{(\sqrt{-z}+\sqrt{-z_0})^2}$$

and

$$r_0^{\{\varepsilon\}}(z) = \gamma^2(z_0,\varepsilon)\left(\varepsilon z_0 - \frac{3}{2}\sqrt{-z_0}\right)(z-z_0) + \gamma(z_0,\varepsilon)(-z_0)^{3/2}.$$

This representation will be used to obtain a converging family of operators in $\mathcal{H}_0$. We shall use the notation as in Subsection 8.2, in particular the notation around (8.16) and (8.17). For $\alpha \in \mathbb{R}$ we denote by $T_r^{\{\varepsilon\}\langle\alpha\rangle}(z)$ the z dependent rank one perturbation of $L_\infty^{\{\varepsilon\}}$ in $\mathcal{H}_0$ defined by

$$T_r^{\{\varepsilon\}\langle\alpha\rangle}(z) = L_\infty^{\{\varepsilon\}} + \frac{\langle\,\cdot\,,\varphi^{\{\varepsilon\}}(z_0)\rangle_0}{\tau_\alpha^{\{\varepsilon\}}(z)}\,\varphi^{\{\varepsilon\}}(z_0)$$

with

$$\tau_\alpha^{\{\varepsilon\}}(z) = \frac{r_0^{\{\varepsilon\}}(z)-\alpha}{(z-z_0)^2}.$$

The following theorem is the analog of Theorem 8.1. The function $\varphi_0^{\{\varepsilon\}}(z) = \varphi_0^{\{\varepsilon\}}(\,\cdot\,,z)$, $z \in \mathbb{C}\setminus[0,\infty)$, with

$$\varphi_0^{\{\varepsilon\}}(x,z) = \begin{cases} \dfrac{\varphi^{\{\varepsilon\}}(x,z)-\varphi^{\{\varepsilon\}}(x,z_0)}{z-z_0}, & z \neq z_0, \\[2ex] \dfrac{\partial\varphi^{\{\varepsilon\}}(x,z)}{\partial z}, & z = z_0, \end{cases} \qquad x \in \mathbb{R}^+,$$

is the analog of the function in (8.8), it belongs to $\operatorname{dom} L_\infty^{\{\varepsilon\}}$ and satisfies

$$\varphi_0^{\{\varepsilon\}}(z) = (L_\infty^{\{\varepsilon\}} - z)^{-1}\varphi^{\{\varepsilon\}}(z_0). \tag{8.27}$$

Theorem 8.6.

(i) *The restriction*

$$\begin{aligned} S_0^{\{\varepsilon\}} &= L_\infty^{\{\varepsilon\}}\,|_{\{h\in\operatorname{dom} L_\infty^{\{\varepsilon\}}\,|\,\langle(L_\infty^{\{\varepsilon\}}-w)h,\varphi_0^{\{\varepsilon\}}(w)\rangle_0=0\}} \\ &= L_\infty^{\{\varepsilon\}}\,|_{\{h\in\operatorname{dom} L_\infty^{\{\varepsilon\}}\,|\,\langle h,\varphi^{\{\varepsilon\}}(z_0)\rangle_0=0\}} \end{aligned}$$

with $w \in \mathbb{R}^-$ is a symmetric operator in $\mathcal{H}_0$ with defect indices $(1,1)$.

(ii) *$T_r^{\{\varepsilon\}\langle\alpha\rangle}(z)$ is a Straus extension of $S_0^{\{\varepsilon\}}$ and the operator $B_r^{\{\varepsilon\}\langle\alpha\rangle}$ in $\mathcal{L}(\widetilde{N}^{\{\varepsilon\}})$ is unitarily equivalent to a self-adjoint linearization of $T_r^{\{\varepsilon\}\langle\alpha\rangle}(z)$.*

Note, the extension $B_r^{\{\varepsilon\}\langle\alpha\rangle}$ is non-minimal, [9, Theorem 4.3 and Theorem 4.4]. The second part of (ii) means that there is a unitary operator $\mathbf{U}_0^{\{\varepsilon\}} : \mathcal{H}_0 \to \mathcal{L}(n_r^{\{\varepsilon\}})$ such that

$$\mathbf{U}_0^{\{\varepsilon\}}(T_r^{\{\varepsilon\}\langle\alpha\rangle}(z)-z)^{-1} = P_{\mathcal{L}(n_r^{\{\varepsilon\}})}(B_r^{\{\varepsilon\}\langle\alpha\rangle}-z)^{-1}\,|_{\mathcal{L}(n_r^{\{\varepsilon\}})}\,\mathbf{U}_0^{\{\varepsilon\}}. \tag{8.28}$$

Proof of Theorem 8.6. The proof of (i) is straightforward and omitted. We only prove part (ii). A standard calculation and (8.27) yield that

$$(T_r^{\{\varepsilon\}\langle\alpha\rangle}(z)-z)^{-1}=(L_\infty^{\{\varepsilon\}}-z)^{-1}-\frac{\langle\,\cdot\,,\varphi_0^{\{\varepsilon\}}(z^*)\rangle_0}{\langle\varphi_0^{\{\varepsilon\}}(z),\varphi^{\{\varepsilon\}}(z_0)\rangle_0+\tau_\alpha^{\{\varepsilon\}}(z)}\varphi_0^{\{\varepsilon\}}(z). \quad (8.29)$$

To rewrite the denominator in (8.29) we note that the relation (8.19)

$$\langle\varphi^{\{\varepsilon\}}(z),\varphi^{\{\varepsilon\}}(\zeta)\rangle_0=\frac{n^{\{\varepsilon\}}(z)-n^{\{\varepsilon\}}(\zeta^*)}{z-\zeta^*}$$

also holds for negative real points z and ζ. If $z=\zeta$ then the difference-quotient on the right-hand side becomes a derivative, hence

$$\langle\varphi_0^{\{\varepsilon\}}(z),\varphi^{\{\varepsilon\}}(z_0)\rangle_0=\frac{1}{(z-z_0)^2}(n^{\{\varepsilon\}}(z)-n^{\{\varepsilon\}}(z_0))-\frac{1}{z-z_0}n^{\{\varepsilon\}\prime}(z_0). \quad (8.30)$$

Applying twice the difference-quotient operator to the identity

$$n^{\{\varepsilon\}}(z)+q^{\{\varepsilon\}}(z)=(z-z_0)^2n_r^{\{\varepsilon\}}(z)+r^{\{\varepsilon\}}(z)$$

we find that (8.30) equals $n_r^{\{\varepsilon\}}(z)$. Hence, after inserting the formula for $\tau_\alpha^{\{\varepsilon\}}$, the resolvent formula (8.29) becomes

$$(T_r^{\{\varepsilon\}\langle\alpha\rangle}(z)-z)^{-1}=(L_\infty^{\{\varepsilon\}}-z)^{-1}-(z-z_0)^2\frac{\langle\,\cdot\,,\varphi_0^{\{\varepsilon\}}(z^*)\rangle_0}{N^{\{\varepsilon\}}(z)-\alpha}\varphi_0^{\{\varepsilon\}}(z).$$

This formula shows that $T_r^{\{\varepsilon\}\langle\alpha\rangle}(z)$ is a Straus extension of $S_0^{\{\varepsilon\}}$. It also implies (8.28). To see this, define the mapping

$$\mathbf{U}_r^{\{\varepsilon\}}:\mathcal{D}_0^{\{\varepsilon\}}:=\operatorname{span}\{\varphi_0^{\{\varepsilon\}}(w)\mid w\in\mathbb{C}\setminus\mathbb{R}^+\}\to\mathcal{L}(n_r^{\{\varepsilon\}})$$

by

$$\mathbf{U}_r^{\{\varepsilon\}}\varphi_0^{\{\varepsilon\}}(z)=K_{n_r^{\{\varepsilon\}}}(\,\cdot\,,z^*).$$

The domain and range of this operator are dense. To see the former, let $x\perp\mathcal{D}_0^{\{\varepsilon\}}$. Then for all $w\in\mathbb{C}\setminus R^+$

$$0=\langle x,\varphi_0^{\{\varepsilon\}}(w)\rangle=\langle(L_\infty^{\{\varepsilon\}}-z_0)^{-1}x,\varphi^{\{\varepsilon\}}(w)\rangle$$

implies $(L_\infty^{\{\varepsilon\}}-z_0)^{-1}x=0$, hence $x=0$. The relation

$$\langle\varphi_0^{\{\varepsilon\}}(w),\varphi_0^{\{\varepsilon\}}(z^*)=K_{n_r^{\{\varepsilon\}}}(z,w^*),$$

which follows similarly as above, implies that $\mathbf{U}_r^{\{\varepsilon\}}$ preserves the inner product and hence $\mathbf{U}_r^{\{\varepsilon\}}$ is well defined. It follows that the bounded continuation of $\mathbf{U}_r^{\{\varepsilon\}}$ to all of $\mathcal{H}_0$ is unitary; it will also be denoted by $\mathbf{U}_r^{\{\varepsilon\}}$. Now (8.28) follows by applying [9, Remark 4.5(c)] with

$$N=N^{\{\varepsilon\}},\quad n_r=n_r^{\{\varepsilon\}},\text{ and }\widetilde{N}_r=\begin{pmatrix}n_r^{\{\varepsilon\}}&0\\0&M_r\end{pmatrix}\text{ with }M_r=\begin{pmatrix}r^{\{\varepsilon\}}&z-z_0^*\\z-z_0&0\end{pmatrix}.$$

□

8.5. In this subsection we prove the strong convergence of the resolvents of the Straus extensions $T_r^{\{\varepsilon\}\langle\alpha(\varepsilon)\rangle}(z)$ in Theorem 8.6 related to $N^{\{\varepsilon\}}$ from (8.25) to the resolvent of the Straus extension $T^{\langle\alpha\rangle}(z)$ in Theorem 8.1 related to N from (8.2) as $\varepsilon \downarrow 0$. Here $\alpha(\varepsilon)$ is a real function of ε which converges to $\alpha \in \mathbb{R} \cup \{\infty\}$ as $\varepsilon \downarrow 0$. Although both extensions act in the same space $\mathcal{H}_0$ we have to introduce the unitary mapping

$$U_\varepsilon = (\mathbf{U}_r^{\{\varepsilon\}})^{-1}\widetilde{\pi}_1^{\{\varepsilon\}}\mathbf{U}_0,$$

where the unitary mappings $\mathbf{U}_0$ and $\mathbf{U}_r^{\{\varepsilon\}}$ are from (8.10) and (8.28) and

$$\widetilde{\pi}_1^{\{\varepsilon\}} : \mathcal{L}(n) \to \mathcal{L}(n_r^{\{\epsilon\}})$$

is the restriction of the mapping $\widetilde{\pi}^{\{\varepsilon\}}$ in (7.13) with (ℓ replaced by ε) to the first component. Since

$$\langle U_\varepsilon \varphi_0(z), U_\varepsilon \varphi_0(w)\rangle = \frac{n_r^{\{\varepsilon\}}(z) - n_r^{\{\varepsilon\}}(w^*)}{z - w^*}$$

converges to

$$\frac{n(z) - n(w^*)}{z - w^*} = \langle \varphi_0(z), \varphi_0(w)\rangle,$$

Lemma 7.2 implies that $\mathcal{H}_0$ approximates $\mathcal{H}_0$ with respect to U_ε in the sense of Definition 7.1 as $\varepsilon \downarrow 0$. The following approximation theorem now follows from Theorem 8.6, Theorem 8.1, and Theorem 7.9.

Theorem 8.7. *Let $\alpha(\varepsilon)$ be a real function which converges to $\alpha \in \mathbb{R}\cup\{\infty\}$ as $\varepsilon \downarrow 0$. Then for $z \in \mathbb{C}\setminus\mathbb{R}$, the resolvents $(T_r^{\{\varepsilon\}\langle\alpha(\varepsilon)\rangle}(z) - z)^{-1}$ approximate the resolvent $(T^{\langle\alpha\rangle}(z) - z)^{-1}$ strongly with respect to U_ε as $\varepsilon \downarrow 0$, that is,*

$$\lim_{\varepsilon\downarrow 0} \left\|(T_r^{\{\varepsilon\}\langle\alpha(\varepsilon)\rangle}(z) - z)^{-1}U_\varepsilon f - U_\varepsilon(T^{\langle\alpha\rangle}(z) - z)^{-1} f\right\|_{\mathcal{H}_0} = 0, \quad f \in \mathcal{H}_0.$$

Remark 8.8. The resolvent $(T^{\langle\alpha\rangle}(z) - z)^{-1}$ also allows another interpretation. Namely, in [10] the functions $\varphi(z)$ and N were interpreted as defect function and Q–function of a symmetric operator S and a canonical extension H^∞ of S in a Pontryagin space $\Pi \supset \mathcal{H}_0$ with $\mathrm{ind}_-\Pi = 1$. It holds $\sigma(H^\infty) = \sigma(L) \cup \{\infty\}$ and the compression of the resolvent of H^∞ to $\mathcal{H}_0$ coincides with the resolvent of L. Note that the models $(B, K_{\widetilde{N}}(\,\cdot\,, z^*)v(z), \widetilde{S})$ and $(H^\infty, \varphi(\cdot, z), S)$ for N are unitarily equivalent since they are both minimal. In connection with the second model there is a one parameter family of self-adjoint extensions $H^{\langle\alpha\rangle}$, $\alpha \in \mathbb{R}$, of S such that $H^{\langle\alpha\rangle} \sim B^{\langle\alpha\rangle}$ for each α. The operator $H^{\langle\alpha\rangle}$ admits the following interpretation $(T^{\langle\alpha\rangle}(z) - z)^{-1} = P_{\mathcal{H}_0}(H^{\langle\alpha\rangle} - z)^{-1}\,|_{\mathcal{H}_0}$.

Especially simple models of N and $N^{\{\varepsilon\}}$ are obtained by taking the limit of the reference point z_0 along the negative half-axis, $z_0 \uparrow 0$. The formulas of n and r, $n_r^{\{\varepsilon\}}$ and $r_0^{\{\varepsilon\}}$ show that this limit is admissible. Then for the ingredients of the model of N in [9, Theorem 3.2 (ii)], we obtain

$$n(z) = \frac{1}{\sqrt{-z}}, \quad r(z) = 0, \quad v(z) = \begin{pmatrix} z & 1 & -\sqrt{-z} \end{pmatrix}^\top,$$

$$\widetilde{N}_r(z) = \begin{pmatrix} \frac{1}{\sqrt{-z}} & 0 \\ 0 & M_r(z) \end{pmatrix}, \quad M_r(z) = \begin{pmatrix} 0 & z \\ z & 0 \end{pmatrix}.$$

and the ingredients of the model of $N^{\{\varepsilon\}}$ become

$$n_r^{\{\varepsilon\}}(z) = \frac{\gamma(z,\varepsilon)}{\sqrt{-z}}, \quad r_0^{\{\varepsilon\}}(z) = 0, \quad v_r^{\{\varepsilon\}}(z) = \begin{pmatrix} z & 1 & -\sqrt{-z}\gamma(z,\varepsilon) \end{pmatrix}^\top,$$

$$\widetilde{N}_r^{\{\varepsilon\}}(z) = \begin{pmatrix} \gamma(z,\varepsilon)/\sqrt{-z} & 0 \\ 0 & M_r^{\{\varepsilon\}}(z) \end{pmatrix}, \quad M_r^{\{\varepsilon\}}(z) = \begin{pmatrix} 0 & z \\ z & 0 \end{pmatrix}.$$

References

[1] L. de Branges, *Hilbert spaces of entire functions*, Prentice-Hall, Englewood Cliffs, N.J., 1968.

[2] R.B. Burckel, *An introduction to classical complex analysis*, Vol. 1, Mathematische Reihe **64**, Birkhäuser, Basel, 1979.

[3] E.A. Coddington and R.G. Gilbert, *Generalized resolvents of ordinary differential operators*, Trans. Amer. Math. Soc. **93**(2) (1959), 216–241.

[4] E.A. Coddington and N. Levinson, *Theory of ordinary differential equations*, McGraw-Hill, New York, 1955.

[5] A. Dijksma, H. Langer, A. Luger, and Yu. Shondin, *Minimal realizations of scalar generalized Nevanlinna functions related to their basic factorization*, Oper. Theory Adv. Appl. **154**, Birkhäuser, Basel, 2004, 69–90.

[6] A. Dijksma, H. Langer, and H.S.V. de Snoo, *Selfadjoint Π_κ-extensions of symmetric subspaces: an abstract approach to boundary problems with spectral parameter in the boundary conditions*, Integral Equations Operator Theory **7**(4) (1984), 459–515; Addendum: Integral Equations Operator Theory **7**(6) (1984), 905.

[7] A. Dijksma, H. Langer, Yu. Shondin, and C. Zeinstra, *Self-adjoint operators with inner singularities and Pontryagin spaces*, Oper. Theory Adv. Appl. **118**, Birkhäuser, Basel, 2000, 105–175.

[8] A. Dijksma, A. Luger, and Yu. Shondin, *Minimal models for $\mathcal{N}_\kappa^\infty$-functions*, Oper. Theory Adv. Appl. **163**, Birkhäuser, Basel, 2005, 97–134.

[9] A. Dijksma, A. Luger, and Yu. Shondin, *Approximation of $\mathcal{N}_\kappa^\infty$-functions I: models and regularization*, Oper. Theory Adv. Appl. **188**, Birkhäuser, Basel, 2008, 87–112.

[10] A. Dijksma and Yu. Shondin, *Singular point-like perturbations of the Bessel operator in a Pontryagin space*, J. Differential Equations **164** (2000), 49–91.

[11] C. Fulton, *Titchmarsh-Weyl m-functions for Second-order Sturm-Liouville Problems with two singular endpoints*, Math. Nachr. **281** (2008), 1418–1475.

[12] F. Gesztesy and M. Zinchenko, *On spectral theory for Schrödinger operators with strongly singular potentials*, Math. Nachr. **279**(9-10) (2006), 1041–1082.

[13] T. Kato, *Perturbation theory for linear operators*, Die Grundlehren der mathematischen Wissenschaften **132**, Springer, Heidelberg, 1966.

[14] P. Kurasov and A. Luger, *An operator theoretic interpretation of the generalized Titchmarsh-Weyl coefficient for a singular Sturm-Liouville problem*, submitted.

[15] H. Langer and A. Luger, *Convergence of generalized Nevanlinna functions*, in preparation.

[16] B.M. Levitan and I.S. Sargsjan, *Introduction to Spectral theory*, Transl. Math. Monographs **39**, Amer. Math. Soc., Providence, R.I., 1975.

[17] H. Langer and B. Najman, *Perturbation theory for definizable operators in Krein spaces*, J. Operator Theory **9** (1983), 247–317.

[18] B. Najman, *Perturbation theory for selfadjoint operators in Pontrjagin spaces*, Glasnik Mat. **15** (1980) 351–370.

[19] Yu. Shondin, *On approximation of high order singular perturbations*, J. Phys. A: Math. Gen. **38** (2005), 5023–5039.

[20] O.Yu. Shvedov, *Approximations for strongly singular evolution equations*, J. Funct. Anal. **210**(2) (2004), 259–294.

[21] B. Simon, *A canonical decomposition for quadratic forms with applications to monotone convergence theorems*, J. Funct. Anal. **28**(3) (1978), 377–385.

[22] F. Stummel, *Diskrete Konvergenz linearer Operatoren* I, Math. Annal. **190** (1970), 45–92; II, Math. Z. **141** (1975), 231–264.

[23] V.A. Yavrian, *On M.G. Krein's spectral shift function for canonical systems of differential equations*, Oper. Theory Adv. Appl. **117**, Birkhäuser, Basel, 2000, 393–417.

Aad Dijksma
Department of Mathematics
University of Groningen
P.O. Box 407
9700 AK Groningen, The Netherlands
e-mail: `dijksma@rug.nl`

Annemarie Luger
Department of Mathematics
Lund Institute of Technology
Box 118
SE-221 00 Lund, Sweden
e-mail: `luger@maths.lth.se`

Yuri Shondin
Department of theoretical Physics
State Pedagogical University
Str. Ulyanova 1
Nizhny Novgorod GSP 37, 603950 Russia
e-mail: `shondin@sinn.ru`

Operator Theory:
Advances and Applications, Vol. 198, 171–183

Nonnegative Extensions via Embeddings

K.-H. Förster and M.M. Nafalska

Dedicated to the memory of our friend Peter Jonas

Abstract. For a given densely defined nonnegative operator in a Hilbert space $\mathcal{H}$ we give a representation of all nonnegative selfadjoint extensions with the help of the embedding operator from the form domain of the respective extension into $\mathcal{H}$. In particular, we discuss the class of extremal extensions.

Mathematics Subject Classification (2000). Primary 47B25; Secondary 47B65.

Keywords. Nonnegative selfadjoint extension, Friedrichs extension, Kreĭn-von Neumann extension, extremal extension, energy space.

1. Introduction

Let A be a densely defined nonnegative operator in the Hilbert space $\{\mathcal{H}, (\cdot,\cdot)\}$. In this note we give a complete answer to following questions, see Theorem 3.1:

1. If $\tilde{A} \geq 0$ is a selfadjoint extension of A does there exist a Hilbert space $\{\mathfrak{L}, (\cdot,\cdot)_{\mathfrak{L}}\}$ that is embedded in $\{\mathcal{H}, (\cdot,\cdot)\}$ such that $\tilde{A} = (i_{\mathfrak{L}}^{-1})^* i_{\mathfrak{L}}^{-1} - I$?
2. If $\{\mathfrak{L}, (\cdot,\cdot)_{\mathfrak{L}}\}$ is embedded in $\{\mathcal{H}, (\cdot,\cdot)\}$ when is $(i_{\mathfrak{L}}^{-1})^* i_{\mathfrak{L}}^{-1} - I$ a nonnegative selfadjoint extension of A?

In [15] similar questions were considered for semibounded relations with finite defect indices. In addition, we discuss these questions for the class of extremal extensions of A.

This paper is organized as follows. In Section 2 we introduce the notations and collect some results used throughout this paper. In Section 3 we give answers to the questions above via contractive embeddings of the energy space of the Friedrichs extension into $\{\mathfrak{L}, (\cdot,\cdot)_{\mathfrak{L}}\}$ and contractive embeddings of $\{\mathfrak{L}, (\cdot,\cdot)_{\mathfrak{L}}\}$ into the energy space of the Kreĭn-von Neumann extension of A. In Section 4 we consider these questions for extremal extensions of A; the answer is given by isometric embeddings. In Section 5 we illustrate the results by regular Sturm-Liouville operators without potential.

M.M. Nafalska gratefully acknowledges the support from the German Academic Exchange Service (DAAD), project D/08/08852.

2. Nonnegative extensions of nonnegative operators

We say that the Hilbert space $\mathcal{X}$ is embedded in the Hilbert space $\mathcal{Y}$, and write $\mathcal{X} \hookrightarrow \mathcal{Y}$, if $\mathcal{X}$ is a vector subspace of $\mathcal{Y}$ and the identity operator $i_{\mathcal{X},\mathcal{Y}} : \mathcal{X} \to \mathcal{Y}$, $x \mapsto x$, is continuous. Later on we consider several Hilbert spaces $\mathcal{X}$ embedded into a fixed Hilbert space $\mathcal{H}$. In this case we agree to write $i_{\mathcal{X}}$ instead of $i_{\mathcal{X},\mathcal{H}}$.

The operator $(i_{\mathcal{L}}^{-1})^* i_{\mathcal{L}}^{-1}$ appearing in the representation of A in Theorem 3.1 was motivated by [18, page 12], where to a continuous embedding $i_{\mathcal{X},\mathcal{Y}}$ from $\{\mathcal{X}, (\cdot,\cdot)_{\mathcal{X}}\}$ into $\{\mathcal{Y}, (\cdot,\cdot)_{\mathcal{Y}}\}$ a positive definite operator S in $\mathcal{Y}$ with domain

$$\operatorname{dom} S = \{ i_{\mathcal{X},\mathcal{Y}} u \in \mathcal{X} \hookrightarrow \mathcal{Y} \mid i_{\mathcal{X},\mathcal{Y}} v \mapsto (u,v)_{\mathcal{X}} \text{ is continuous for all } v \in \mathcal{X} \}$$

and

$$(u,v)_{\mathcal{X}} = (S i_{\mathcal{X},\mathcal{Y}} u, i_{\mathcal{X},\mathcal{Y}} v)_{\mathcal{Y}}, \quad u,v \in \mathcal{X}, \tag{2.1}$$

was defined. From (2.1) it follows immediately that $S = (i_{\mathcal{X},\mathcal{Y}}^{-1})^* i_{\mathcal{X},\mathcal{Y}}^{-1}$.

In the following, $(\cdot,\cdot)_A$ denotes the inner product induced by the graph norm of an operator A between Hilbert spaces.

Let $A \geq 0$ be a densely defined operator in the Hilbert space $\{\mathcal{H}, (\cdot,\cdot)\}$. In his famous work [17], M.G. Kreĭn presented a characterization of all nonnegative selfadjoint extensions of the operator A. He showed the existence of two nonnegative selfadjoint operators A_N and A_F such that a selfadjoint operator $\tilde{A} \geq 0$ is an extension of A if and only if it satisfies the inequalities

$$A_N \leq \tilde{A} \leq A_F, \tag{2.2}$$

where (2.2) can be defined via the associated forms from Kato's representation theorems, see also [1]. More precisely, for selfadjoint operators $B, C \geq 0$ the inequality $B \leq C$ is satisfied if $\operatorname{dom} C^{1/2} \subset \operatorname{dom} B^{1/2}$ and for $f \in \operatorname{dom} C^{1/2}$ we have $\|B^{1/2} f\| \leq \|C^{1/2} f\|$. In the terminology from above this means that the embedding of $\{\operatorname{dom} C^{1/2}, (\cdot,\cdot)_{C^{1/2}}\}$ into $\{\operatorname{dom} B^{1/2}, (\cdot,\cdot)_{B^{1/2}}\}$ is contractive. Since $\operatorname{dom} C$ is a core of $C^{1/2}$ an equivalent criterion is

$$\operatorname{dom} C \subset \operatorname{dom} B^{1/2}, \qquad \|B^{1/2} f\|^2 \leq (Cf, f), \quad \text{for all } f \in \operatorname{dom} C. \tag{2.3}$$

Also the inequality

$$(C+I)^{-1} \leq (B+I)^{-1}$$

may serve as an equivalent definition for $B \leq C$, cf. [16]. In the case of nonnegative relations a generalized version of (2.2) holds, see [7].

The operator A_N is called the Kreĭn-von Neumann extension of A and was firstly constructed by J. von Neumann in the case that A is positive definite, cf. [20]. A_F is called the Friedrichs extension of A. Let t be the nonnegative form defined by

$$t[f,g] = (Af, g), \quad f,g \in \operatorname{dom} A = \operatorname{dom} t. \tag{2.4}$$

Corresponding to a construction of K. Friedrichs, cf. [8], A_F is the operator associated to the closure $\bar{t}$ of t in the sense of [16]. Moreover, from the Representation

Theorems in [16] it follows that $\operatorname{dom} A$ is a core for $A_F^{1/2}$ and

$$\bar{t}[f,g] = (A_F^{1/2} f, A_F^{1/2}), \qquad f,g \in \operatorname{dom} A_F^{1/2} = \operatorname{dom} \bar{t},$$

so that $\operatorname{dom} A_F$ is the set of all $f \in \operatorname{dom} A^*$ for which there exists a sequence $(f_n) \subset \operatorname{dom} A$ with

$$f_n \to f, \ \big(A(f_n - f_m), f_n - f_m\big) \to 0, \quad n,m \to \infty.$$

An analogous characterization of the domain of the Kreĭn-von Neumann extension A_N of A was proven in [2]: $\operatorname{dom} A_N$ is the set of all $f \in \operatorname{dom} A^*$ for which there exists a sequence $(f_n) \subset \operatorname{dom} A$ with

$$Af_n \to A^* f, \ \big(A(f_n - f_m), f_n - f_m\big) \to 0, \quad n,m \to \infty.$$

Replacing A by $\tilde{A}$ in (2.4), where $\tilde{A} \geq 0$ is a selfadjoint extension of A, again due to the Representation Theorems, it follows that $\tilde{A}[f,g] := \bar{t}[f,g] = (\tilde{A}^{1/2} f, \tilde{A}^{1/2} f)$, for $f,g \in \operatorname{dom} \tilde{A}^{1/2} = \operatorname{dom} \bar{t}$ and $\operatorname{dom} \tilde{A}$ is a core for $\tilde{A}^{1/2}$.

If S is a closable densely defined operator from the Hilbert space $\mathcal{H}$ into the Hilbert space $\mathcal{K}$ we can also consider the closable form

$$s[f,g] = (Sf, Sg), \quad f,g \in \operatorname{dom} S = \operatorname{dom} s. \tag{2.5}$$

Clearly, the form s is closed if and only the same is true for the operator S. Moreover, we have

$$\bar{s}[f,g] = (\overline{S} f, \overline{S} g), \quad f,g \in \operatorname{dom} \overline{S} = \operatorname{dom} \bar{s} \tag{2.6}$$

and $S^* \overline{S} = S^* S^{**}$ is the nonnegative selfadjoint operator associated to the form $\bar{s}$, so that $S^* S^{**}[f,g] = \bar{s}[f,g] = (S^* S^{**} f, g)$, $f,g \in \operatorname{dom} S^* S^{**}$, and we have

$$\operatorname{dom} S^{**} = \operatorname{dom} \bar{s} = \operatorname{dom} (S^* S^{**})^{1/2}. \tag{2.7}$$

In [5], [22], [23], [24], [25] it was shown that the Friedrichs extension A_F and the Kreĭn-von Neumann extension A_N of A admit a factorization of the form $S^* S^{**}$, cf. [7], [14] for the case of nonnegative relations and [12], [13] for the case of the sum of two nonnegative operators or relations. For that purpose an auxiliary Hilbert space $\mathcal{H}_A$ was defined as the completion of the range of A with respect to the norm induced by the inner product

$$\langle Af, Ag \rangle = (Af, g), \quad f,g \in \operatorname{dom} A.$$

The operators $Q : \mathcal{H} \supset \operatorname{dom} A \to \mathcal{H}_A$ with $f \mapsto Af$ and $J : \mathcal{H}_A \supset \operatorname{ran} A \to \mathcal{H}$ with $Af \mapsto Af$ are closable and satisfy $Q^{**} \subset J^*$ and $A = JQ$. Furthermore, the Friedrichs and the Kreĭn-von Neumann extension are given by

$$A_F = Q^* Q^{**} \qquad \text{and} \qquad A_N = J^{**} J^*. \tag{2.8}$$

Together with (2.7) it follows that $\operatorname{dom} A_F^{1/2} = \operatorname{dom} Q^{**}$, $\operatorname{dom} A_N^{1/2} = \operatorname{dom} J^*$ and the graph norms of $A_N^{1/2}$ ($A_F^{1/2}$) and J^* (Q^{**}, respectively) coincide. Moreover, (2.2) implies that every nonnegative selfadjoint extensions $\tilde{A}$ of A satisfies $\operatorname{dom} Q^{**} \subset \operatorname{dom} \tilde{A}^{1/2} \subset \operatorname{dom} J^*$.

In [5], [6] particularly the class of extremal extensions of the operator A was discussed. These extensions are images under the Cayley transform of extreme points of an operator interval, cf. [6], [11, Proposition 4.19]. By definition, a selfadjoint extension $\tilde{A} \geq 0$ of A is called extremal if

$$\inf \left\{ \left(\tilde{A}(h-f), h-f \right) \mid f \in \operatorname{dom} A \right\} = 0, \qquad h \in \operatorname{dom} \tilde{A}, \tag{2.9}$$

cf. the last remark in [3]. The following theorem which can be found in [5, Theorem 4.4] gives a factorization of all extremal extensions of the operator A. Moreover, it provides a useful criterion for extremality.

Theorem 2.1. *Let $A \geq 0$ be a densely defined operator in $\mathcal{H}$. Then the following statements are equivalent:*

(i) *$\tilde{A}$ is an extremal extension of A;*
(ii) *$\tilde{A} = \tilde{A}_{\mathfrak{L}} := J^*|^*_{\mathfrak{L}} J^*|^{**}_{\mathfrak{L}}$, where $\mathfrak{L}$ is a subspace of $\mathcal{H}$ satisfying* $\operatorname{dom} A \subset \mathfrak{L} \subset \operatorname{dom} J^*$;
(iii) $A_N[f,g] = \tilde{A}[f,g]$, $f, g \in \operatorname{dom} \tilde{A}^{1/2}$.

An extension of these and further results to nonnegative relations can be found in [10], [11], [14]. The case of sectorial extremal extensions of sectorial operators and relations was discussed in [3], [4].

3. Nonnegative extensions via contractive embeddings

In this section we give a characterization of all nonnegative selfadjoint extensions of a densely defined nonnegative operator by means of some Hilbert space $\mathfrak{L}$ that lies between $\operatorname{dom} A_F^{1/2}$ and $\operatorname{dom} A_N^{1/2}$.

In the case of semibounded relations a representation of the Friedrichs extension analogous to (E2) in Theorem 3.1 was given in [15, Theorem 3.5]. Under the additional assumption that the original relation is closed and has finite defect indices a statement similar to (E1) was shown, cf. [15, Theorem 3.9, Corollary 3.15]. Under these conditions the problem 2 of the introduction was solved, cf. [15, Proposition 3.14].

Theorem 3.1. *Let $A \geq 0$ be a densely defined operator in $\{\mathcal{H}, (\cdot,\cdot)\}$. Then $\tilde{A} \geq 0$ is a selfadjoint extension of A if and only if there exists a Hilbert space $\{\mathfrak{L}, (\cdot,\cdot)_{\mathfrak{L}}\}$ that is embedded in $\{\mathcal{H}, (\cdot,\cdot)\}$ satisfying the following conditions:*

(E1) $\left\{\operatorname{dom} A_F^{1/2}, (\cdot,\cdot)_{A_F^{1/2}}\right\} \hookrightarrow \left\{\mathfrak{L}, (\cdot,\cdot)_{\mathfrak{L}}\right\} \hookrightarrow \left\{\operatorname{dom} A_N^{1/2}, (\cdot,\cdot)_{A_N^{1/2}}\right\}$ *and both embeddings are contractive;*
(E2) *$\tilde{A}$ has the representation $\tilde{A} = (i_{\mathfrak{L}}^{-1})^* i_{\mathfrak{L}}^{-1} - I$, where $i_{\mathfrak{L}}$ denotes the embedding operator from the Hilbert space $\mathfrak{L}$ into the Hilbert space $\mathcal{H}$.*

If both conditions are satisfied then the following assertions hold:

(a) *$i_{\mathfrak{L}}$ is contractive;*
(b) $\{\mathfrak{L}, (\cdot,\cdot)_{\mathfrak{L}}\} = \{\operatorname{dom} \tilde{A}^{1/2}, (\cdot,\cdot)_{\tilde{A}^{1/2}}\}$;
(c) *The first embedding in* (E1) *is isometric.*

Proof. In the following $\operatorname{ran} i_{\mathfrak{L}}$ and $\mathfrak{L}$ are identified. Let $\tilde{A} \geq 0$ be a selfadjoint extension of A. We define the Hilbert space

$$\{\mathfrak{L}, (\cdot,\cdot)_{\mathfrak{L}}\} := \{\operatorname{dom} \tilde{A}^{1/2}, (\cdot,\cdot)_{\tilde{A}^{1/2}}\}.$$

According to $A_N \leq \tilde{A} \leq A_F$, we conclude that $\mathfrak{L}$ satisfies the first condition. Since

$$\|f\|^2_{\tilde{A}^{1/2}} = \|f\|^2 + \|\tilde{A}^{1/2} f\|^2 \geq \|f\|^2, \quad f \in \operatorname{dom} \tilde{A}^{1/2},$$

the embedding operator

$$i_{\mathfrak{L}} : \mathfrak{L} \to \mathcal{H}, \quad f \mapsto f,$$

is contractive. Furthermore, $i_{\mathfrak{L}}$ is closed and has dense range which coincides with $\operatorname{dom} \tilde{A}^{1/2}$. This implies that its inverse $i_{\mathfrak{L}}^{-1}$ is a closed densely defined operator. Due to (2.5)–(2.7) it follows that the operator

$$S := (i_{\mathfrak{L}}^{-1})^* i_{\mathfrak{L}}^{-1}$$

is densely defined nonnegative and selfadjoint with

$$\operatorname{dom} S^{1/2} = \operatorname{dom} (i_{\mathfrak{L}}^{-1}) = \operatorname{ran} i_{\mathfrak{L}} = \mathfrak{L} = \operatorname{dom} \tilde{A}^{1/2}.$$

Its domain is given by

$$\begin{aligned} \operatorname{dom} S &= \left\{ f \in \operatorname{dom} (i_{\mathfrak{L}}^{-1}) \;\middle|\; i_{\mathfrak{L}}^{-1} f \in \operatorname{dom} \left((i_{\mathfrak{L}}^{-1})^* \right) \right\} \\ &= \left\{ f \in \operatorname{dom} \tilde{A}^{1/2} \;\middle|\; f \in \operatorname{dom} \left((i_{\mathfrak{L}}^{-1})^* \right) \right\}. \end{aligned}$$

We show that $\operatorname{dom} \tilde{A} \subset \operatorname{dom} S$. Recall that for $g \in \mathfrak{L}$ we have $i_{\mathfrak{L}} g = g \in \mathcal{H}$. Hence, for all $g \in \operatorname{dom} \tilde{A}^{1/2} = \mathfrak{L}$, $f \in \operatorname{dom} \tilde{A}$, it follows that

$$\begin{aligned} |(i_{\mathfrak{L}}^{-1} g, f)_{\mathfrak{L}}| &= |(g, f)_{\mathfrak{L}}| = |(g, f) + (\tilde{A}^{1/2} g, \tilde{A}^{1/2} f)| = |(g, f + \tilde{A} f)| \\ &\leq \|g\| \, \|f + \tilde{A} f\|. \end{aligned}$$

This implies $f \in \operatorname{dom} \left((i_{\mathfrak{L}}^{-1})^* \right)$ and, therefore, $f \in \operatorname{dom} S$. Now let $f, g \in \operatorname{dom} \tilde{A}$. Then we have

$$\begin{aligned} (Sf, g) &= \left((i_{\mathfrak{L}}^{-1})^* i_{\mathfrak{L}}^{-1} f, g \right) = (i_{\mathfrak{L}}^{-1} f, i_{\mathfrak{L}}^{-1} g)_{\mathfrak{L}} = (f, g)_{\mathfrak{L}} \\ &= (f, g) + (\tilde{A}^{1/2} f, \tilde{A}^{1/2} g) = (f, g) + (\tilde{A} f, g) \\ &= \left((\tilde{A} + I) f, g \right). \end{aligned}$$

This implies $Sf = (\tilde{A} + I) f$ for all $f \in \operatorname{dom} \tilde{A}$. It follows that

$$\tilde{A} + I \subset S = S^* \subset (\tilde{A} + I)^* = \tilde{A} + I.$$

We conclude that $\tilde{A} + I = S$. Thus, $\{\mathfrak{L}, (\cdot,\cdot)_{\mathfrak{L}}\} = \{\operatorname{dom} \tilde{A}^{1/2}, (\cdot,\cdot)_{\tilde{A}^{1/2}}\}$ and $\tilde{A}$ satisfy both conditions.

Now the converse implication is shown. Let (E1) and (E2) be fulfilled. Observe that the embedding

$$i_N : \{\operatorname{dom} A_N^{1/2}, (\cdot,\cdot)_{A_N^{1/2}}\} \to \mathcal{H}, \quad f \mapsto f,$$

is contractive with dense range. Hence the embedding

$$i_{\mathfrak{L}} : \{\mathfrak{L}, (\cdot,\cdot)_{\mathfrak{L}}\} \to \mathcal{H}, \quad i_{\mathfrak{L}} = i_N \circ i_{\mathfrak{L},N},$$

has the same properties; here $i_{\mathfrak{L},N}$ denotes the embedding from $\{\mathfrak{L}, (\cdot,\cdot)_{\mathfrak{L}}\}$ into $\{\operatorname{dom} A_N^{1/2}, (\cdot,\cdot)_{A_N^{1/2}}\}$. This implies (a) and, moreover, $S := (i_{\mathfrak{L}}^{-1})^* i_{\mathfrak{L}}^{-1}$ is a nonnegative selfadjoint operator that has the form domain

$$\operatorname{dom} S^{1/2} = \operatorname{dom}(i_{\mathfrak{L}}^{-1}) = \mathfrak{L} \subset \operatorname{dom} A_N^{1/2}.$$

Since $i_{\mathfrak{L}}$ is contractive, it follows that for $f \in \operatorname{dom} S \subset \operatorname{dom} S^{1/2}$, we have

$$(Sf, f) = ((i_{\mathfrak{L}}^{-1})^* i_{\mathfrak{L}}^{-1} f, f) = (i_{\mathfrak{L}}^{-1} f, i_{\mathfrak{L}}^{-1} f)_{\mathfrak{L}} = (f, f)_{\mathfrak{L}} \geq (f, f). \tag{3.1}$$

Hence, $S - I$ is nonnegative. Next it is shown that

$$A_N \leq S - I \leq A_F, \tag{3.2}$$

which implies that the operator $S - I$ is an extension of A, cf. the remark above (2.2). Let $f \in \operatorname{dom} A_F^{1/2} \subset \mathfrak{L} \subset \operatorname{dom} A_N^{1/2}$. According to (3.1), for $f \in \operatorname{dom} S^{1/2} = \mathfrak{L}$, the equality

$$\|S^{1/2} f\| = \|f\|_{\mathfrak{L}},$$

holds true. Thus, for $f \in \operatorname{dom} A_F$, we have

$$\begin{aligned}(A_F + I)[f] = ((A_F + I)f, f) &= \|f\|^2 + \|A_F^{1/2} f\|^2 = \|f\|^2_{A_F^{1/2}} \\ &\geq \|f\|^2_{\mathfrak{L}} = \|S^{1/2} f\|^2 = S[f].\end{aligned}$$

Since $\operatorname{dom} A_F$ is a core of the form $(A_F + I)[\,\cdot\,]$ and $\operatorname{dom} A_F^{1/2} = \operatorname{dom}\big((A_F + I)^{1/2}\big)$, we conclude the following inequality:

$$S[f] \leq (A_F + I)[f], \quad f \in \operatorname{dom} A_F^{1/2}.$$

Together with the fact that $\operatorname{dom} A_F^{1/2} \subset \operatorname{dom} S^{1/2}$ this implies $S \leq A_F + I$. Since $S - I$ and A_F are nonnegative selfadjoint operators, the inequality $S \leq A_F + I$ is equivalent to

$$S - I \leq A_F.$$

Now we will show the left inequality in (3.2). Let $f \in \operatorname{dom} S \subset \operatorname{dom} A_N^{1/2}$. According to (3.1), we have

$$(Sf, f) = \|f\|^2_{\mathfrak{L}} \geq \|f\|^2_{A_N^{1/2}} = \|A_N^{1/2} f\|^2 + (f, f). \tag{3.3}$$

Thus, $(S - I)[f] \geq A_N[f]$, $f \in \operatorname{dom} S$. Since $\operatorname{dom} S$ is a core of the form $(S - I)[\,\cdot\,]$, this implies

$$A_N \leq S - I.$$

From (2.2) it follows that $\tilde{A} := S - I$ is a nonnegative selfadjoint extension of A. Therefore,

$$\operatorname{dom} \tilde{A}^{1/2} = \operatorname{dom} i_{\mathfrak{L}}^{-1} = \operatorname{ran} i_{\mathfrak{L}} = \mathfrak{L}.$$

This implies (b). Since $\operatorname{dom} A$ is a core of $A_F^{1/2}$, together with the fact that we have $(A_F^{1/2} f, A_F^{1/2} g) = (\tilde{A}^{1/2} f, \tilde{A}^{1/2} g)$, $f, g \in \operatorname{dom} A$, for every nonnegative selfadjoint extension $\tilde{A}$ of A it follows that

$$\tilde{A}[f,g] = A_F[f,g], \quad f,g \in \operatorname{dom} A_F^{1/2}. \tag{3.4}$$

Hence, the left embedding in condition (E1) from Theorem 3.1 is isometric if $\{\mathfrak{L}, (\cdot,\cdot)_{\mathfrak{L}}\} = \{\operatorname{dom} \tilde{A}^{1/2}, (\cdot,\cdot)_{\tilde{A}^{1/2}}\}$ which implies (c). This completes the proof. □

In the following we denote the nonnegative selfadjoint extension of A constructed in Theorem 3.1 by

$$\tilde{A}(\mathfrak{L}) = (i_{\mathfrak{L}}^{-1})^* i_{\mathfrak{L}}^{-1} - I, \tag{3.5}$$

where $\{\mathfrak{L}, (\cdot,\cdot)_{\mathfrak{L}}\}$ is a Hilbert space that is embedded in $\mathcal{H}$ and that satisfies condition (E1) from Theorem 3.1.

From Theorem 3.1, (2.5)–(2.7) we obtain the following statements concerning the Friedrichs and the Kreĭn-von Neumann extension of A. Partially, they correspond to [15, Theorem 3.5, Theorem 3.9, Proposition 3.14, Corollary 3.15].

Corollary 3.2. *Let $A \geq 0$ be a densely defined operator in $\{\mathcal{H}, (\cdot,\cdot)\}$. Then:*

(i) $\{\mathfrak{L}, (\cdot,\cdot)_{\mathfrak{L}}\} = \{\operatorname{dom} A_F^{1/2}, (\cdot,\cdot)_{A_F^{1/2}}\}$ *if and only if* $\tilde{A}(\mathfrak{L}) = A_F$;

(ii) $\{\mathfrak{L}, (\cdot,\cdot)_{\mathfrak{L}}\} = \{\operatorname{dom} A_N^{1/2}, (\cdot,\cdot)_{A_N^{1/2}}\}$ *if and only if* $\tilde{A}(\mathfrak{L}) = A_N$.

The next statement gives a property of the resolvent of the operator $\tilde{A}(\mathfrak{L})$.

Corollary 3.3. *Let $A \geq 0$ be a densely defined operator in $\{\mathcal{H}, (\cdot,\cdot)\}$ and let $\{\mathfrak{L}, (\cdot,\cdot)_{\mathfrak{L}}\}$ be a Hilbert space which is embedded in $\{\mathcal{H}, (\cdot,\cdot)\}$ satisfying condition* (E1) *from Theorem* 3.1. *Then the identity $\left(\tilde{A}(\mathfrak{L}) + I\right)^{-1} = i_{\mathfrak{L}} i_{\mathfrak{L}}^*$ is valid, where $i_{\mathfrak{L}}$ denotes the embedding from $\mathfrak{L}$ to $\mathcal{H}$.*

Proof. Observe that

$$\tilde{A}(\mathfrak{L}) + I = (i_{\mathfrak{L}}^{-1})^* i_{\mathfrak{L}}^{-1} = (i_{\mathfrak{L}}^*)^{-1} i_{\mathfrak{L}}^{-1} = (i_{\mathfrak{L}} i_{\mathfrak{L}}^*)^{-1}.$$

Since $\operatorname{ran} i_{\mathfrak{L}} = \mathfrak{L}$ is dense in $\mathcal{H}$ the operator $i_{\mathfrak{L}}^*$ is injective and $(i_{\mathfrak{L}}^*)^{-1} = (i_{\mathfrak{L}}^{-1})^*$. The third equality is clear since the injectivity of the operators $i_{\mathfrak{L}}$ and $i_{\mathfrak{L}}^*$ implies the same for the product $i_{\mathfrak{L}} i_{\mathfrak{L}}^*$. Hence, $\left(\tilde{A}(\mathfrak{L}) + I\right)^{-1} = i_{\mathfrak{L}} i_{\mathfrak{L}}^*$, as required. □

Remark 3.4. Corollary 3.3 implies that $\left(\tilde{A}(\mathfrak{L}) + I\right)^{-1}$ is compact if and only if the embedding operator $i_{\mathfrak{L}}$ is compact which yields an abstract variation of Rellich's Criterion. For more results concerning the relation of compact resolvents and selfadjoint extensions of nonnegative operators (or relations) see [11].

Remark 3.5. In particular, if $A = A^* \geq 0$ then we obtain for the embedding i of $\{\operatorname{dom} A^{1/2}, (\cdot,\cdot)_{A^{1/2}}\}$ into $\{\mathcal{H}, (\cdot,\cdot)\}$ the representation (3.5) and $(A+I)^{-1} = (i^{-1})^* i^{-1}$.

Corollary 3.3 gives the following characterization of all nonnegative selfadjoint extensions of a densely defined nonnegative operator.

Corollary 3.6. *Let $A \geq 0$ be a densely defined operator and let $\tilde{A} \geq 0$ be selfadjoint in $\{\mathcal{H}, (\cdot,\cdot)\}$. Then $\tilde{A}$ is an extension of A if and only if*

$$\|i_F{}^* f\|_{A_F^{1/2}} \leq \|i_{\mathfrak{L}}{}^* f\|_{\mathfrak{L}} \leq \|i_N{}^* f\|_{A_N^{1/2}}, \quad f \in \mathcal{H},$$

where $i_F, i_{\mathfrak{L}}$ and i_N denote the embeddings from

$$\{\operatorname{dom} A_F^{1/2}, (\cdot,\cdot)_{A_F^{1/2}}\},\ \{\mathfrak{L}, (\cdot,\cdot)_{\mathfrak{L}}\} = \{\operatorname{dom} \tilde{A}^{1/2}, (\cdot,\cdot)_{\tilde{A}^{1/2}}\} \text{ and } \{\operatorname{dom} A_N^{1/2}, (\cdot,\cdot)_{A_N^{1/2}}\},$$

respectively, into the Hilbert space $\{\mathcal{H}, (\cdot,\cdot)\}$.

Proof. Since for every nonnegative selfadjoint operator $\tilde{A}$ the inequalities $A_N \leq \tilde{A} \leq A_F$ are equivalent to

$$(I + A_F)^{-1} \leq (I + \tilde{A})^{-1} \leq (I + A_N)^{-1},$$

it follows from Lemma 3.3 that $i_F i_F^* \leq i_{\mathfrak{L}} i_{\mathfrak{L}}^* \leq i_N i_N^*$. Now we conclude the statement from (2.2). □

From the proof of Theorem 3.1 we obtain another characterization of the ordering of nonnegative selfadjoint extensions of A which was defined in Section 2.

Proposition 3.7. *Let $A \geq 0$ be a densely defined operator in $\{\mathcal{H}, (\cdot,\cdot)\}$ and let $\{\mathfrak{L}_1, (\cdot,\cdot)_{\mathfrak{L}_1}\}$ and $\{\mathfrak{L}_2, (\cdot,\cdot)_{\mathfrak{L}_2}\}$ be Hilbert spaces which are embedded in $\{\mathcal{H}, (\cdot,\cdot)\}$ satisfying condition* (E1) *from Theorem* 3.1. *Then:*

(i) *$\tilde{A}(\mathfrak{L}_1) \leq \tilde{A}(\mathfrak{L}_2)$ if and only if $\{\mathfrak{L}_2, (\cdot,\cdot)_{\mathfrak{L}_2}\} \hookrightarrow \{\mathfrak{L}_1, (\cdot,\cdot)_{\mathfrak{L}_1}\}$ such that the embedding is contractive;*

(ii) *The operators $\tilde{A}(\mathfrak{L}_1)$ and $\tilde{A}(\mathfrak{L}_2)$ coincide if and only if $\mathfrak{L}_1 = \mathfrak{L}_2$ and $(\cdot,\cdot)_{\mathfrak{L}_1} = (\cdot,\cdot)_{\mathfrak{L}_2}$.*

4. Extremal extensions via contractive embeddings

In this section we give a sufficient and necessary condition for the Hilbert spaces $\{\mathfrak{L}, (\cdot,\cdot)_{\mathfrak{L}}\}$ such that the nonnegative selfadjoint extensions $\tilde{A}(\mathfrak{L})$ of A which we constructed in Section 3 are extremal. Further, we give the relation between $\tilde{A}(\mathfrak{L})$ and the extremal extension $\tilde{A}_{\mathfrak{L}}$ from Theorem 2.1.

Let $A \geq 0$ be a densely defined operator in $\{\mathcal{H}, (\cdot,\cdot)\}$. Recall that every nonnegative selfadjoint extension of A has the representation

$$\tilde{A}(\mathfrak{L}) = (i_{\mathfrak{L}}^{-1})^* i_{\mathfrak{L}}^{-1} - I, \tag{4.1}$$

where $\{\mathfrak{L}, (\cdot,\cdot)_{\mathfrak{L}}\}$ is a Hilbert space embedded in $\mathcal{H}$ such that both of the following embeddings are contractive:

$$\{\operatorname{dom} A_F^{1/2}, (\cdot,\cdot)_{A_F^{1/2}}\} \hookrightarrow \{\mathfrak{L}, (\cdot,\cdot)_{\mathfrak{L}}\} \hookrightarrow \{\operatorname{dom} A_N^{1/2}, (\cdot,\cdot)_{A_N^{1/2}}\}, \tag{4.2}$$

and $i_{\mathfrak{L}}$ denotes the embedding operator from $\mathfrak{L}$ into $\mathcal{H}$. Actually, $\{\mathfrak{L}, (\cdot,\cdot)_{\mathfrak{L}}\} = \{\operatorname{dom} \tilde{A}(\mathfrak{L})^{1/2}, (\cdot,\cdot)_{\tilde{A}(\mathfrak{L})^{1/2}}\}$. Further, call to mind that for every subspace $\mathfrak{L}$ of $\mathcal{H}$

with $\operatorname{dom} A \subset \mathfrak{L} \subset \operatorname{dom} A_N^{1/2}$,

$$\tilde{A}_{\mathfrak{L}} = J^*|_{\mathfrak{L}}^* J^*|_{\mathfrak{L}}^{**}$$

is an extremal extension of A, cf. Theorem 2.1.

Proposition 4.1. *Let $A \geq 0$ be a densely defined operator in $\{\mathcal{H}, (\cdot,\cdot)\}$ and let $\{\mathfrak{L}, (\cdot,\cdot)_{\mathfrak{L}}\}$ be a Hilbert space that is embedded in $\{\mathcal{H}, (\cdot,\cdot)\}$ satisfying condition* (E1) *from Theorem* 3.1. *Then:*

(i) *The inequality $\tilde{A}_{\mathfrak{L}} \leq \tilde{A}(\mathfrak{L})$ holds true;*

(ii) *Let $\hat{A} \geq 0$ be a selfadjoint extension of A satisfying $\tilde{A}_{\mathfrak{L}} \leq \hat{A} \leq \tilde{A}(\mathfrak{L})$. Then $\hat{A} = \tilde{A}_{\mathfrak{L}}$ if and only if $\hat{A}$ is extremal;*

(iii) *If $\mathfrak{L}$ is closed with respect to the graph norm of $A_N^{1/2}$ then we have*

$$\operatorname{dom} \tilde{A}(\mathfrak{L})^{1/2} = \operatorname{dom} \tilde{A}_{\mathfrak{L}}^{1/2}.$$

Proof. (i) Since $\tilde{A}_{\mathfrak{L}} = J^*|_{\mathfrak{L}}^* J^*|_{\mathfrak{L}}^{**}$, it follows that

$$\operatorname{dom} \tilde{A}_{\mathfrak{L}}^{1/2} = \overline{\mathfrak{L}}^{\|\cdot\|_{J^*}} \supseteq \mathfrak{L} = \operatorname{dom} \tilde{A}(\mathfrak{L})^{1/2}. \tag{4.3}$$

In addition, we have

$$\tilde{A}(\mathfrak{L})[f] \geq A_N[f] = \tilde{A}_{\mathfrak{L}}[f], \quad f \in \operatorname{dom} \tilde{A}(\mathfrak{L})^{1/2},$$

cf. (2.2) and Theorem 2.1. This implies $\tilde{A}_{\mathfrak{L}} \leq \tilde{A}(\mathfrak{L})$.

(ii) Clearly, if $\hat{A} = \tilde{A}_{\mathfrak{L}}$ then from Theorem 2.1 it follows that $\hat{A}$ is an extremal extension of A. Conversely, let $\hat{A}$ be an extremal extension of A such that $\tilde{A}_{\mathfrak{L}} \leq \hat{A} \leq \tilde{A}(\mathfrak{L})$. Then

$$\mathfrak{L} = \operatorname{dom} \tilde{A}(\mathfrak{L})^{1/2} \subset \operatorname{dom} \hat{A}^{1/2} \subset \operatorname{dom} \tilde{A}_{\mathfrak{L}}^{1/2} = \overline{\mathfrak{L}}^{\|\cdot\|_{J^*}}. \tag{4.4}$$

From this we conclude that $\operatorname{dom} \hat{A}^{1/2}$ is closed with respect to the graph norm of J^*. Hence, (4.4) implies $\operatorname{dom} \hat{A}^{1/2} = \operatorname{dom} \tilde{A}_{\mathfrak{L}}^{1/2}$. Together with the fact that $\|\tilde{A}_{\mathfrak{L}}^{1/2} f\| = \|\hat{A}^{1/2} f\|$, $f \in \operatorname{dom} \tilde{A}_{\mathfrak{L}}^{1/2}$, which is valid since $\tilde{A}_{\mathfrak{L}}$ and $\hat{A}$ are extremal extensions of A, we obtain $\tilde{A}_{\mathfrak{L}} = \hat{A}$.

(iii) Let $\mathfrak{L}$ be closed with respect to the graph norm of $A_N^{1/2}$. Then (2.8) and (4.3) yield $\operatorname{dom} \tilde{A}(\mathfrak{L})^{1/2} = \operatorname{dom} \tilde{A}_{\mathfrak{L}}^{1/2}$. □

The next theorem characterizes all extremal extensions of A via (4.1) and (4.2) for a special Hilbert space $\{\mathfrak{L}, (\cdot,\cdot)_{\mathfrak{L}}\}$. It is essentially a consequence of Theorem 2.1.

Theorem 4.2. *Let $A \geq 0$ be a densely defined operator in $\{\mathcal{H}, (\cdot,\cdot)\}$. Then $\tilde{A}$ is an extremal extension of A if and only if there exists a Hilbert space $\{\mathfrak{L}, (\cdot,\cdot)_{\mathfrak{L}}\}$ embedded in $\{\mathcal{H}, (\cdot,\cdot)\}$ which satisfies conditions* (E1) *and* (E2) *from Theorem* 3.1, *such that the right embedding in* (E1) *is isometric. Then we have $\tilde{A} = \tilde{A}(\mathfrak{L}) = \tilde{A}_{\mathfrak{L}}$.*

Proof. Let $\tilde{A}$ be an extremal extension of A. As in the proof of Theorem 3.1 we define

$$\{\mathfrak{L}, (\cdot,\cdot)_{\mathfrak{L}}\} := \{\operatorname{dom} \tilde{A}^{1/2}, (\cdot,\cdot)_{\tilde{A}^{1/2}}\}.$$

According to Theorem 2.1 we have $\|f\|_{A_N^{1/2}} = \|f\|_{\tilde{A}^{1/2}}$, $f \in \operatorname{dom} \tilde{A}^{1/2}$. This implies that the embedding operator

$$i_{\mathfrak{L},N} : \{\operatorname{dom} \tilde{A}^{1/2}, (\cdot,\cdot)_{\tilde{A}^{1/2}}\} \to \{\operatorname{dom} A_N^{1/2}, (\cdot,\cdot)_{A_N^{1/2}}\}$$

is isometric. Conversely, let $\{\mathfrak{L}, (\cdot,\cdot)_{\mathfrak{L}}\}$ be a Hilbert space that is embedded in $\mathcal{H}$ satisfying conditions (E1) and (E2) from Theorem 3.1. In addition, let

$$\|f\|_{\mathfrak{L}} = \|f\|_{A_N^{1/2}}, \quad f \in \mathfrak{L}.$$

Hence, we have equality in (3.3), which leads to $\tilde{A}[f] = A_N[f]$, $f \in \operatorname{dom} \tilde{A}^{1/2}$. Thus, $\tilde{A}$ is extremal. Since $\mathfrak{L}$ is closed with respect to the graph norm of $A_N^{1/2}$ it follows from Lemma 4.1 that $\operatorname{dom} \tilde{A}^{1/2} = \operatorname{dom} \tilde{A}_{\mathfrak{L}}^{1/2}$. Together with the fact that both extensions, $\tilde{A}$ and $\tilde{A}_{\mathfrak{L}}$, are extremal, we conclude that they coincide. □

According to (3.4), actually both embeddings in condition (E1) from Theorem 3.1 are isometric if $\tilde{A}$ is an extremal extension of A.

5. An example

Let $\mathcal{H} = L^2(I)$, where $I = (a, b)$ is a finite interval. Further, let p be a real-valued function with $p > 0$ almost everywhere and $p \in L^1_{\text{loc}}(I)$. Moreover, assume that the function $p^{-1} := \frac{1}{p}$ belongs to $L^1(I)$. Then the regular Sturm-Liouville operator without potential

$$Af = -(pf')',$$

defined on the domain

$$\operatorname{dom} A = \{f \in L^2(I) \mid f, pf' \in AC(I),\ (pf')' \in L^2(I),\\ f(a) = f(b) = (pf')(a) = (pf')(b) = 0\}$$

is closed, densely defined, and nonnegative, cf. [26]. It is well known that its Friedrichs extension A_F is determined by Dirichlet boundary conditions which is also true in the general case where A is a semibounded Sturm-Liouville operator with potential, cf. [21]. In a forthcoming paper we will show that

$$\operatorname{dom} A_F^{1/2} = \left\{ f \in L^2(I) \;\middle|\; f \in AC(I),\ p^{1/2} f' \in L^2(I),\ f(a) = f(b) = 0 \right\},$$

$$\|A_F^{1/2} f\|^2 = \int_a^b p(x)|f'(x)|^2 dx, \quad f \in \operatorname{dom} A_F^{1/2},$$

and

$$\operatorname{dom} A_N^{1/2} = \left\{ f \in L^2(I) \;\middle|\; f \in AC(I),\ p^{1/2} f' \in L^2(I) \right\},$$
$$\|A_N^{1/2} f\|^2 = \int_a^b p(x)\,|f'(x)|^2 dx - \frac{|f(b)-f(a)|^2}{F_{p^{-1}}(b) - F_{p^{-1}}(a)}, \quad f \in \operatorname{dom} A_N^{1/2},$$

cf. [19]; here $F_{p^{-1}}$ is a primitive of the function p^{-1}. Moreover, we have

$$\operatorname{dom} A_N^{1/2} = \operatorname{dom} A_F^{1/2} \dotplus \operatorname{span}\{\mathbf{1}, F_{p^{-1}}\},$$

where $\mathbf{1} : I \to I$, $x \mapsto 1$. These considerations together with (2.3) and (3.4) imply that $\mathfrak{L}$ is the form domain of a nonnegative selfadjoint extension $\tilde{A}(\mathfrak{L})_{\alpha,\beta}$ of A, where $\alpha, \beta \in \mathbb{C}$, if and only if

$$\mathfrak{L} = \operatorname{dom} A_F^{1/2} \dotplus \operatorname{span}\{\alpha + \beta F_{p^{-1}}\}$$

and

$$\int_a^b p(x)\,|f'(x)|^2 dx - \frac{|f(b)-f(a)|^2}{F_{p^{-1}}(b) - F_{p^{-1}}(a)} = \|A_N^{1/2} f\|^2 \le \|\tilde{A}(\mathfrak{L})^{1/2} f\|^2 = (A^* f, f),$$

for $f \in \mathfrak{L} \cap \operatorname{dom} A^*$, and for $f \in \operatorname{dom} A$, we have

$$(Af, f) = \|\tilde{A}(\mathfrak{L})^{1/2} f\|^2, \tag{5.1}$$

cf. (2.2), (2.3), (3.4). To obtain (5.1) we used (2.4) and the fact that $\operatorname{dom} A$ is a core of $A_F^{1/2}$. Therefore, $\mathfrak{L}$ is the form domain of a nonnegative selfadjoint extension $\tilde{A}(\mathfrak{L})_{\alpha,\beta}$ of A if and only if (5.1) is valid and for $f \in \mathfrak{L} \cap \operatorname{dom} A^*$, we have

$$\frac{|f(b)-f(a)|^2}{F_{p^{-1}}(b) - F_{p^{-1}}(a)} \ge (pf')(b)\overline{f(b)} - (pf')(a)\overline{f(a)}. \tag{5.2}$$

$\mathfrak{L}$ is the form domain of an extremal extension $\tilde{A}(\mathfrak{L})_{\alpha,\beta}$ of A if and only if we have equality in (5.2).

This implies that the nonnegative selfadjoint extensions according to the following boundary conditions are extremal:

- $f(a) = f(b) = 0$ (Dirichlet boundary conditions);
- $(pf)'(a) = (pf)'(b) = \frac{f(b)-f(a)}{F_{p^{-1}}(b) - F_{p^{-1}}(a)}$ (Kreĭn-von Neumann boundary conditions);
- $f(a) = f(b)$, $(pf)'(a) = (pf)'(b)$ (periodic boundary conditions).

Whereas the nonnegative selfadjoint extensions according to the following boundary conditions are not extremal:

- $(pf)'(a) = (pf)'(a) = 0$ (Neumann boundary conditions);
- $f(a) = -f(b)$, $(pf)'(a) = -(pf)'(b)$ (semi-periodic boundary conditions).

A boundary triplet for A^* is given by $\{\mathbb{C}^2, \Gamma_0, \Gamma_1\}$, where the mappings $\Gamma_0, \Gamma_1 :$ $\operatorname{dom} A^* \to \mathbb{C}^2$ are defined by

$$\Gamma_0 f = \begin{pmatrix} f(a) \\ f(b) \end{pmatrix}, \quad f \in \operatorname{dom} A^*,$$

$$\Gamma_1 f = \begin{pmatrix} (pf')(a) - \frac{f(b)-f(a)}{F_{p-1}(b)-F_{p-1}(a)} \\ -(pf')(b) + \frac{f(b)-f(a)}{F_{p-1}(b)-F_{p-1}(a)} \end{pmatrix}, \quad f \in \operatorname{dom} A^*,$$

cf. [19]. The definition of a boundary triplet (also called boundary value space) can be found for example in [3], [5], [9]. Since $\ker \Gamma_0 = \operatorname{dom} A_F$ and $\ker \Gamma_1 = \operatorname{dom} A_N$ this is a basic (or fundamental) boundary triplet, cf. [3, page 8]. Therefore, all extremal extensions $\tilde{A}_{\alpha,\beta}$ of A can be parametrized by means of this triplet, cf. the last remark in [3] or [5, Theorem 7.1]. More precisely, apart from A_F and A_N, they are restrictions of A^* corresponding to the boundary conditions

$$\beta f(a) = \alpha f(b),$$
$$\alpha \left((pf')(a) - \frac{f(b) - f(a)}{F_{p^{-1}}(b) - F_{p^{-1}}(a)} \right) = \overline{\beta} \left((pf')(b) - \frac{f(b) - f(a)}{F_{p^{-1}}(b) - F_{p^{-1}}(a)} \right),$$

where $\alpha \in \mathbb{R}, \beta \in \mathbb{C}$ and $\alpha^2 + |\beta|^2 = 1$. The corresponding form domains are given by

$$\operatorname{dom} \tilde{A}_{\alpha,\beta}^{1/2} = \operatorname{dom} A_F^{1/2} \dotplus \operatorname{span} \left\{ (\beta - \alpha) F_{p^{-1}}(\cdot) + \alpha F_{p^{-1}}(b) - \beta F_{p^{-1}}(a) \right\},$$

cf. [19]. In the case that $p = 1$ the extremal extensions of A and their associated sesquilinear forms were given in [5].

References

[1] N.I. Achieser, I.M. Glasmann, *Theorie der linearen Operatoren im Hilbert-Raum.* Verlag Harri Deutsch, Thun, 1981.

[2] T. Ando, K. Nishio, *Positive selfadjoint extensions of positive symmetric operators.* Tôhoku Math. J. **22** (1970), 65–75.

[3] Yu.M. Arlinskiĭ, *Positive spaces of boundary values and sectorial extensions of a nonnegative symmetric operator.* Ukr. Math. J. **40** No. 1 (1988), 8–14.

[4] Yu.M. Arlinskiĭ, *Extremal extensions of sectorial linear relations*, Matematychniĭ Studiĭ **7**, No. 1, (1997), 81–96.

[5] Yu.M. Arlinskiĭ, S. Hassi, Z. Sebestyén, H.S.V. de Snoo, *On the class of extremal extensions of a nonnegative operator.* Oper. Theory: Adv. Appl. (B. Sz.-Nagy memorial volume) **127** (2001), 41–81.

[6] Yu.M. Arlinskiĭ, E. Tsekanovskiĭ, *Quasi selfadjoint contractive extensions of hermitian contractions*, Teor. Funkts., Funkts. Anal. Prilozhen **50**, (1988), 9–16.

[7] E.A. Coddington, H.S.V. de Snoo, *Positive selfadjoint extensions of positive symmetric subspaces.*, Math. Z. **159** (1978), 203–214.

[8] K. Friedrichs, *Spektraltheorie halbbeschränkter Operatoren und Anwendungen auf die Spektralzerlegung von Differentialoperatoren.* Math. Ann. **109** (1934), 465–487.

[9] V.I. Gorbachuk, M.L. Gorbachuk, *Boundary value problems for operator differential equations*, Kluwer Academic Publishers, 1991.

[10] S. Hassi, *On the Friedrichs and the Kreĭn-von Neumann extension of nonnegative relations.* Acta Wasaensia No. **122** (2004), 37–54.

[11] S. Hassi, M. Malamud, H.S.V. de Snoo, *On Kreĭn's extension theory of nonnegative operators.* Math. Nachr. **274–275** (2004), 40–73.

[12] S. Hassi, A. Sandovici, H.S.V. de Snoo, H. Winkler, *Form sums of nonnegative selfadjoint operators*, Acta Math. Hung. **111**, (2006), 81–105.

[13] S. Hassi, A. Sandovici, H.S.V. de Snoo, H. Winkler, *Extremal extensions for the sum of nonnegative selfadjoint relations*, Proc. Am. Math. Soc. **135**, No. 10, (2007), 3193–3204.

[14] S. Hassi, A. Sandovici, H.S.V. de Snoo, H. Winkler, *A general factorization approach to the extension theory of nonnegative operators and relations.* J. Operator Theory **58** (2007), 351–386.

[15] M. Kaltenbäck, H. Winkler, H. Woracek, *Symmetric relations of finite negativity.* Operator Theory Adv. Appl. **162** (2006), 191–210.

[16] T. Kato, *Perturbation theory for linear operators.* Springer-Verlag, 1995.

[17] M.G. Kreĭn, *Theory of selfadjoint extensions of semibounded operators and its applications I, II.*, Mat. Sb. **20, 21** (1947), 431–495, 365–404.

[18] J.L. Lions, E. Magènes, *Problèmes aux limites non homogènes et applications, Volume* 1. Dunond Paris, 1968.

[19] M.M. Nafalska, *Extremal Extensions of Nonnegative Operators with Applications.* Doctoral Thesis, Technische Universität Berlin, 2008.

[20] J. von Neumann, *Allgemeine Eigenwerttheorie hermitescher Funktionaloperatoren.* Math. Ann. **102** (1929), 49–131.

[21] H.-D. Niessen, A. Zettl, *The Friedrichs extension of regular ordinary differential operators.* Proc. R. Soc. Edinb., Sect. A **114** (1990), 229–236.

[22] V. Prokaj, Z. Sebestyén, *On extremal positive operator extensions.* Acta Sci. Math. (Szeged) **62**, No. 3-4 (1996), 485–491.

[23] V. Prokaj, Z. Sebestyén, *On Friedrichs extensions of operators.* Acta Sci. Math. (Szeged) **62** (1996), 243–246.

[24] Z. Sebestyén, E. Sikolya, *On Kreĭn-von Neumann and Friedrichs extension.* Acta Sci. Math. (Szeged) **69** (2003), 323–336.

[25] Z. Sebestyén, J. Stochel, *Restrictions of positive self-adjoint operators.* Acta Sci. Math. (Szeged) **55** (1991), 149–154.

[26] J. Weidmann, *Lineare Operatoren in Hilberträumen, Teil II: Anwendungen.* B.G. Teubner Verlag, 2003.

K.-H. Förster and M.M. Nafalska
Technische Universität Berlin
Straße des 17. Juni 136,
D-10623 Berlin, Germany
e-mail: `foerster@math.tu-berlin.de`
`nafalska@math.tu-berlin.de`

Operator Theory:
Advances and Applications, Vol. 198, 185–210

Fredholm Properties of Unbounded Operators on Interpolation Spaces

Kerstin Günther

Dedicated to the memory of Peter Jonas

Abstract. We study Fredholm properties and properties on the continuous invertibility of unbounded operators S_0, S_1, S_Δ and S_Σ, which are connected to a linear operator between compatible couples. If S_0 and S_1 are everywhere defined and bounded, then we obtain the operators usually considered in the classical interpolation theory. As an example, we study differential operators on different L^p-spaces induced by the same differential expression.

Mathematics Subject Classification (2000). 46B70, 46M35, 47A53, 47A55, 47A10.

Keywords. Abstract Sobolev spaces, interpolation theory, induced operators, unbounded operators, Fredholm properties, continuous invertibility.

1. Introduction

In this paper, we examine unbounded operators on interpolation spaces. A well-known class of such operators are differential operators on different L^p-spaces corresponding to the same differential expression (see [Gol66], [BL76]). This shows that there exist results on this topic, but Y. Chen seems to be the first author, who studied these operators systematically from an abstract point of view (see [Che01]).

For two compatible couples (E_0, E_1) and (F_0, F_1) and a not necessarily bounded operator $S : E_\Sigma \supseteq \mathrm{D}(S) \longrightarrow F_\Sigma$, we consider the operators S_0, S_1, S_Δ and S_Σ connected to S in the following way: S_0, S_1 and S_Δ are the induced operators (of S to the corresponding spaces, see [BL76]) and S_Σ is the 'sum' of S_0 and S_1. If S_0 and S_1 are everywhere defined and bounded, then the operators mentioned above coincide with the operators usually considered in the classical interpolation theory and the operators S_Σ and S are equal. In our case, these operators are not bounded.

Of particular interest will be the Fredholm properties and properties on the continuous invertibility of the operators S_0, S_1, S_Δ and S_Σ. Some of the results are known, when the operators are bounded. A well-known procedure for obtaining results on an unbounded (closable) operator from the theory of bounded operators is, to consider its abstract Sobolev space. The abstract Sobolev space corresponding to a closable operator is the domain of the closure of this operator provided with its graph norm.

In Section 2, we recall the concept of the abstract Sobolev space (cf. [EN00]) and the induced operators and state some fundamental results, which will be used throughout this paper.

In Section 3, we examine the operators S_0, S_1, S_Δ and S_Σ. Of particular interest will be the relation between the operators S_Σ and S.

In Section 4, we construct interpolation morphisms on the compatible couple of the abstract Sobolev spaces corresponding to S_0 and S_1. This is the main step to obtain results on unbounded operators from the classical interpolation theory for bounded operators.

In Section 5 and Section 6, we extend known results on the Fredholm properties and properties on the continuous invertibility of the operators S_0, S_1, S_Δ and S_Σ (see [Kra96] for bounded operators or [Che01]).

In the last section, we consider differential operators as an example. We introduce differential operators on the intersection and the sum of two L^p-spaces and prove that their induced operators to certain L^p-spaces are the well-known differential operators. It is shown that many known results for the classical differential operators (see [Gol66]) can be extended to differential operators on the intersection or the sum of two L^p-spaces. Moreover, we obtain results on the Fredholm properties of these differential operators. Finally, we prove that the operators S_Σ and S coincide in this example.

2. Abstract Sobolev spaces and induced operators

Firstly, we introduce the abstract Sobolev space of a closable operator, which is the domain of the closure of this operator provided with the corresponding graph norm.

Definition 2.1. *Let E, F be Banach spaces and $S : E \supseteq \mathrm{D}(S) \longrightarrow F$ be linear and closable. We define the abstract Sobolev space G_S by*

$$G_S := \{\mathrm{D}(\overline{S}),\ \|\cdot\|_{\overline{S}}\},$$

where $\|\cdot\|_{\overline{S}}$ denotes the graph norm of $\overline{S}$, i.e., $\|x\|_{\overline{S}} = \|x\|_E + \|\overline{S}x\|_F$ for all $x \in \mathrm{D}(\overline{S})$. The corresponding linear operator $i_S : G_S \longrightarrow E$ is defined by

$$i_S u := u, \quad u \in G_S.$$

We have the following situation.

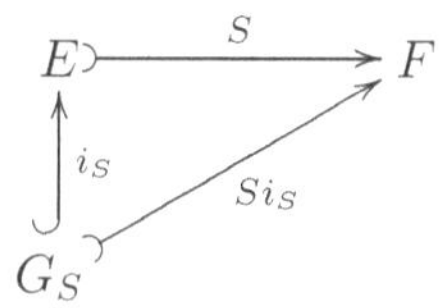

Obviously, the abstract Sobolev space is a Banach space. Let E, F be Banach spaces and $S : E \supseteq \mathrm{D}(S) \longrightarrow F$ be linear and closable. The operator i_S is injective and bounded. From the definition of G_S and i_S, we obtain that Si_S is bounded on its domain and closable. It holds

$$\overline{Si_S} = \overline{S}i_S = \overline{S}i_{\overline{S}}, \quad \mathrm{N}(S) = i_S\{\mathrm{N}(Si_S)\}, \quad \mathrm{R}(S) = \mathrm{R}(Si_S). \tag{1}$$

In the next proposition, we describe the connection between the semi-Fredholm properties of S and Si_S (see [Kat66, p. 230] for the definition of semi-Fredholm, Fredholm, nullity, deficiency and the index). This proposition is an immediate consequence of (1).

Proposition 2.2. *Let E, F be Banach spaces and $S : E \supseteq \mathrm{D}(S) \longrightarrow F$ be linear and closed. The operator S is semi-Fredholm (Fredholm) if and only if Si_S is semi-Fredholm (Fredholm); in this case, the nullities* $\mathrm{nul}(S)$, $\mathrm{nul}(Si_S)$ *or the deficiencies* $\mathrm{def}(S)$, $\mathrm{def}(Si_S)$ *(the indices* $\mathrm{ind}(S)$, $\mathrm{ind}(Si_S)$*) are finite and equal.*

In particular, S is continuously invertible if and only if Si_S is continuously invertible. In this case, the operator Si_S is an isomorphism.

Now, we introduce the concept of induced operators. A continuous embedding is an everywhere defined, linear, injective and bounded operator between two Banach spaces.

Definition 2.3. *Suppose E, F, $\check{E}$ and $\check{F}$ are Banach spaces and $i_E : \check{E} \longrightarrow E$, $i_F : \check{F} \longrightarrow F$ are continuous embeddings. Let $S : E \supseteq \mathrm{D}(S) \longrightarrow F$ be linear. We define the linear operator $\check{S}_{\check{E},\check{F}} : \check{E} \supseteq \mathrm{D}(\check{S}_{\check{E},\check{F}}) \longrightarrow \check{F}$ by*

$$\mathrm{D}(\check{S}_{\check{E},\check{F}}) := \{\check{x} \in \check{E} : i_E\check{x} \in \mathrm{D}(S) \text{ and } Si_E\check{x} \in \mathrm{R}(i_F)\},$$
$$\check{S}_{\check{E},\check{F}}\check{x} := \check{y} \quad \text{if and only if} \quad Si_E\check{x} = i_F\check{y}$$

for all $\check{x} \in \mathrm{D}(\check{S}_{\check{E},\check{F}})$ and call it the induced operator of S corresponding to $\check{E}$ and $\check{F}$ (and i_E, i_F).

The situation is shown in the following commutative diagram.

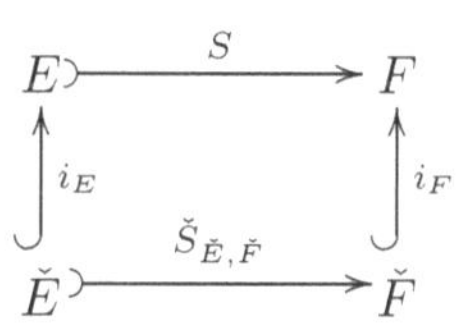

Let E, F, $\check{E}$, $\check{F}$ be Banach spaces, $i_E : \check{E} \longrightarrow E$, $i_F : \check{F} \longrightarrow F$ be continuous embeddings and $S : E \supseteq \mathrm{D}(S) \longrightarrow F$ be linear. Since i_F is injective, the induced operator is well defined. It holds

$$i_E\{\mathrm{N}(\check{S}_{\check{E},\check{F}})\} = \mathrm{N}(S) \cap \mathrm{R}(i_E), \quad i_F\{\mathrm{R}(\check{S}_{\check{E},\check{F}})\} \subseteq \mathrm{R}(S) \cap \mathrm{R}(i_F)$$

and the induced operator is maximal, i.e., suppose $R : \check{E} \supseteq \mathrm{D}(R) \longrightarrow \check{F}$ is linear. Then $i_F R \subseteq S i_E$ if and only if $R \subseteq \check{S}_{\check{E},\check{F}}$.

From the definition of the induced operator, we obtain the following proposition.

Proposition 2.4. *Suppose E, F, $\check{E}$ and $\check{F}$ are Banach spaces and $i_E : \check{E} \longrightarrow E$, $i_F : \check{F} \longrightarrow F$ are continuous embeddings. Let $S : E \supseteq \mathrm{D}(S) \longrightarrow F$ be linear.*

(i) *If $T : E \supseteq \mathrm{D}(T) \longrightarrow F$ is linear such that $S \subseteq T$, i.e., S is a restriction of T, then $\check{S}_{\check{E},\check{F}} \subseteq \check{T}_{\check{E},\check{F}}$.*
(ii) *If S is injective, then the operator $\check{S}_{\check{E},\check{F}}$ is injective and $(\check{S^{-1}})_{\check{F},\check{E}} = (\check{S}_{\check{E},\check{F}})^{-1}$ on $\mathrm{D}((\check{S^{-1}})_{\check{F},\check{E}}) = \mathrm{R}(\check{S}_{\check{E},\check{F}})$.*
(iii) *If S is closed, then $\check{S}_{\check{E},\check{F}}$ is closed. If S is closable, then $\check{S}_{\check{E},\check{F}}$ is closable and $\overline{\check{S}_{\check{E},\check{F}}} \subseteq \check{\overline{S}}_{\check{E},\check{F}}$.*

If the induced operator of an operator S is closed, then the operator S is not necessarily closed, as the following example shows. For an interval $I \subseteq \mathbb{R}$, the Lebesgue measure λ and $1 \leq p \leq \infty$, let $(L^p(I), \ \|\cdot\|_{L^p})$ be the corresponding L^p-space (see [DS67, p. 241] for the definition). For $k \in \mathbb{N}$, we denote by $C^k(\overline{I})$ the set of all scalar-valued functions f on I such that f is k-times continuously differentiable on I and $f^{(j)}$ has a continuous extension to $\overline{I}$ for all $j \in \{0, 1, \ldots, k\}$. Analogously, the space $C(\overline{I})$ is the set of all functions f on I such that f has a continuous extension to $\overline{I}$.

Example 2.5. Assume $I = (-1, 1)$, $E := F := L^2(I)$, $\check{E} := \check{F} := C(\overline{I})$ and i_E, i_F are the corresponding continuous embeddings. Let $S : L^2(I) \supseteq \mathrm{D}(S) \longrightarrow L^2(I)$ be defined by

$$\begin{aligned} \mathrm{D}(S) &:= C^1(\overline{I}), \\ Sf &:= f', \quad f \in \mathrm{D}(S). \end{aligned}$$

Then $\mathrm{D}(\check{S}_{\check{E},\check{F}}) = C^1(\overline{I})$ and $\check{S}_{\check{E},\check{F}} f = f'$ for all $f \in \mathrm{D}(\check{S}_{\check{E},\check{F}})$. It is well known that $\check{S}_{\check{E},\check{F}}$ is closed and S is not closed.

Now, we describe the induced operator of $S i_S$ to the spaces $G_{\check{S}_{\check{E},\check{F}}}$ and $\check{F}$. Suppose E, F, $\check{E}$ and $\check{F}$ are Banach spaces and $i_E : \check{E} \longrightarrow E$, $i_F : \check{F} \longrightarrow F$ are continuous embeddings. Let $S : E \supseteq \mathrm{D}(S) \longrightarrow F$ be linear and closable.

We consider the following commutating diagram,

$$\begin{array}{ccccc} G_S & \xhookrightarrow{\ i_S\ } & E & \xrightarrow{\ S\ } & F \\ \uparrow{\scriptstyle i_{G_S}} & & \uparrow{\scriptstyle i_E} & & \uparrow{\scriptstyle i_F} \\ G_{\check{S}_{\check{E},\check{F}}} & \xhookrightarrow{\ i_{\check{S}_{\check{E},\check{F}}}\ } & \check{E} & \xrightarrow{\ \check{S}_{\check{E},\check{F}}\ } & \check{F} \end{array}$$

where the linear operator $i_{G_S} : G_{\check{S}_{\check{E},\check{F}}} \longrightarrow G_S$ is defined by

$$i_{G_S}\check{u} := u \quad \text{if and only if} \quad u \in G_S \text{ and } i_E i_{\check{S}_{\check{E},\check{F}}}\check{u} = i_S u, \quad \check{u} \in G_{\check{S}_{\check{E},\check{F}}}.$$

The operator i_{G_S} is everywhere defined, injective and, since it holds

$$\|i_{G_S}\check{u}\|_{\overline{S}} = \|i_E i_{\check{S}_{\check{E},\check{F}}}\check{u}\|_E + \|i_F \overline{\check{S}_{\check{E},\check{F}}} i_{\check{S}_{\check{E},\check{F}}}\check{u}\|_F \leq \max\{\|i_E\|, \|i_F\|\} \cdot \|\check{u}\|_{\overline{\check{S}_{\check{E},\check{F}}}}$$

for all $\check{u} \in G_{\check{S}_{\check{E},\check{F}}}$, the operator i_{G_S} is bounded. From the definition of the operators, we obtain that i_{G_S} is the induced operator of i_E (with respect to $G_{\check{S}_{\check{E},\check{F}}}$ and G_S) and

$$\check{(Si_S)}_{G_{\check{S}_{\check{E},\check{F}}},\check{F}} = \check{S}_{\check{E},\check{F}} i_{\check{S}_{\check{E},\check{F}}}. \tag{2}$$

Now, we consider conjugates of induced operators.

Let S be a linear and closed operator between Banach spaces and D be a subspace of the domain of S such that $\overline{S_{|D}} = S$. Then D is said to be a core of S.

Theorem 2.6. *Suppose E, F, $\check{E}$ and $\check{F}$ are Banach spaces and $i_E : \check{E} \longrightarrow E$, $i_F : \check{F} \longrightarrow F$ are continuous embeddings with dense range. Assume $S : E \supseteq \mathrm{D}(S) \longrightarrow F$ is linear and closed such that S and $\check{S}_{\check{E},\check{F}}$ are densely defined. Let $\check{((\check{S}_{\check{E},\check{F}})')}_{F',E'}$ be the induced operator of $(\check{S}_{\check{E},\check{F}})'$ with respect to $(i_F)'$ and $(i_E)'$.*

(i) *It holds $S' \subseteq \check{((\check{S}_{\check{E},\check{F}})')}_{F',E'}$.*

(ii) *Assume S is closed. If $S' = \check{((\check{S}_{\check{E},\check{F}})')}_{F',E'}$, then $i_E\{\mathrm{D}(\check{S}_{\check{E},\check{F}})\}$ is a core of S.*

We have the following situation.

$$\begin{array}{ccc} E' & \xleftarrow{\ S'\ } & F' \\ \downarrow{\scriptstyle (i_E)'} & & \downarrow{\scriptstyle (i_F)'} \\ (\check{E})' & \xleftarrow{\ (\check{S}_{\check{E},\check{F}})'\ } & (\check{F})' \end{array}$$

Note that, since the operators i_E and i_F have dense range, their conjugates are injective.

Proof. (i) Let $y' \in \mathrm{D}(S')$ and $x' \in E'$ such that $S'y' = x'$. Since

$$\begin{aligned} \langle \check{S}_{\check{E},\check{F}}\check{x}, (i_F)'y'\rangle &= \langle i_F\check{S}_{\check{E},\check{F}}\check{x}, y'\rangle = \langle Si_E\check{x}, y'\rangle \\ &= \langle i_E\check{x}, S'y'\rangle = \langle \check{x}, (i_E)'S'y'\rangle \end{aligned}$$

for $\check{x} \in \mathrm{D}(\check{S}_{\check{E},\check{F}})$, we conclude that $(i_F)'y' \in \mathrm{D}((\check{S}_{\check{E},\check{F}})')$ and $(\check{S}_{\check{E},\check{F}})'(i_F)'y' = (i_E)'S'y'$. Thus $y' \in \mathrm{D}(\check{((\check{S}_{\check{E},\check{F}})')}_{F',E'})$ and $\check{((\check{S}_{\check{E},\check{F}})')}_{F',E'}y' = S'y'$.

(ii) It is well known that the mapping $H : (E' \times F')/(\mathrm{G}(S))^{\perp} \longrightarrow (\mathrm{G}(S))'$ defined by

$$\langle (x, Sx), H([(x', y')]_{(\mathrm{G}(S))^{\perp}}) \rangle := \langle x, x' \rangle + \langle Sx, y' \rangle$$

for all $x' \in E', y' \in F', (x, Sx) \in \mathrm{G}(S)$ is an isomorphism. Now, assume $x' \in E'$ and $y' \in F'$ such that

$$0 = \langle (i_E\check{x}, i_F\check{S}\check{x}), H([(x', y')]_{(\mathrm{G}(S))^{\perp}}) \rangle = \langle i_E\check{x}, x' \rangle + \langle S i_E \check{x}, y' \rangle$$

for all $(\check{x}, \check{S}_{\check{E},\check{F}}\check{x}) \in \mathrm{G}(\check{S}_{\check{E},\check{F}})$. Firstly, we show that $y' \in \mathrm{D}(S')$. The mapping

$$\check{x} \mapsto \langle i_E\check{x}, x' \rangle = -\langle S i_E \check{x}, y' \rangle = -\langle \check{S}_{\check{E},\check{F}}\check{x}, (i_F)'y' \rangle$$

is bounded on $\mathrm{D}(\check{S}_{\check{E},\check{F}})$. Thus $(i_F)'y' \in \mathrm{D}((\check{S}_{\check{E},\check{F}})')$ and $(\check{S}_{\check{E},\check{F}})'(i_F)'y' = -(i_E)'x'$. Therefore $y' \in \mathrm{D}(\check{((\check{S}_{\check{E},\check{F}})')}_{F',E'}) = \mathrm{D}(S')$. Hence

$$0 = \langle i_E\check{x}, x' \rangle + \langle S i_E\check{x}, y' \rangle = \langle i_E\check{x}, x' + S'y' \rangle$$

for all $\check{x} \in \mathrm{D}(\check{S}_{\check{E},\check{F}})$. Since $x \mapsto \langle x, x' + S'y' \rangle$ is bounded on E and $i_E\{\mathrm{D}(\check{S}_{\check{E},\check{F}})\}$ is dense in E, we get that $\langle x, x' + S'y' \rangle = 0$ for all $x \in E$. Hence $x' = -S'y'$ and we obtain that $(x', y') \in (\mathrm{G}(S))^{\perp}$. It follows that $i_E\{\mathrm{D}(\check{S}_{\check{E},\check{F}})\}$ is a core of S. □

3. The unbounded operators $(S_0, S_1)_\Sigma$, S_0, S_1, S_Δ and S_Σ

In this section, we introduce the operators $(S_0, S_1)_\Sigma$, S_0, S_1, S_Δ, S_Σ and examine their properties. We are especially interested in the relation between S_Σ and S.

In further considerations in this paper, the induced operators will always be constructed with continuous embeddings, which correspond to continuous inclusions: Assume $\check{E}$, $\check{F}$, E and F are Banach spaces such that $\check{E} \subseteq E$ and $\check{F} \subseteq F$ with continuous inclusions. Let i_E and i_F be the corresponding embeddings, i.e., $i_E : \check{E} \longrightarrow E$ and $i_F : \check{F} \longrightarrow F$ with

$$i_E\check{x} := \check{x}, \quad \check{x} \in \check{E},$$
$$i_F\check{y} := \check{y}, \quad \check{y} \in \check{F}.$$

Assume $S : E \supseteq \mathrm{D}(S) \longrightarrow F$ is linear. Then the induced operator $\check{S}_{\check{E},\check{F}}$ will always correspond to i_E and i_F.

As usual in the classical interpolation theory, we will identify elements of the domain with corresponding elements of the range of these continuous embeddings to simplify the notation, i.e., we identify $i_E\check{x}$ with $\check{x}$ for $\check{x} \in \check{E}$.

For a compatible couple (E_0, E_1) (i.e., E_0, E_1 are Banach spaces, which can be continuously embedded into a Hausdorff topological vector space), we define in the well-known way the Banach space $E_\Delta := E_0 \cap E_1$ with the norm

$$\|x_\Delta\|_{E_\Delta} := \max\{\|x_\Delta\|_{E_0}, \|x_\Delta\|_{E_1}\}, \quad x_\Delta \in E_\Delta$$

and the Banach space $E_\Sigma := E_0 + E_1$ with the norm

$$\|x_\Sigma\|_{E_\Sigma} := \inf_{\substack{x_\Sigma = x_0 + x_1 \\ x_0 \in E_0,\ x_1 \in E_1}} \{\|x_0\|_{E_0} + \|x_1\|_{E_1}\}, \quad x_\Sigma \in E_\Sigma.$$

Then $E_\Delta \subseteq E_j \subseteq E_\Sigma$ with continuous inclusions, where $j \in \{0, 1\}$.

Now, we introduce $(S_0, S_1)_\Sigma$, the induced operators S_0, S_1 and S_Δ and the 'sum' S_Σ of S_0 and S_1.

Definition 3.1. *Let (E_0, E_1), (F_0, F_1) be compatible couples. Assume $S_0 : E_0 \supseteq \mathrm{D}(S_0) \longrightarrow F_0$ and $S_1 : E_1 \supseteq \mathrm{D}(S_1) \longrightarrow F_1$ are linear such that S_0 and S_1 agree on $\mathrm{D}(S_0) \cap \mathrm{D}(S_1)$ (the values of these operators are considered in F_Σ). The linear operators $(S_0, S_1)_\Delta : E_\Delta \supseteq \mathrm{D}((S_0, S_1)_\Delta) \longrightarrow F_\Delta$ and $(S_0, S_1)_\Sigma : E_\Sigma \supseteq \mathrm{D}((S_0, S_1)_\Sigma) \longrightarrow F_\Sigma$ are defined by*

$$\begin{aligned}
&\mathrm{D}((S_0, S_1)_\Delta) := \mathrm{D}(S_0) \cap \mathrm{D}(S_1),\\
&(S_0, S_1)_\Delta x_\Delta := S_0 x_\Delta = S_1 x_\Delta, \quad x_\Delta \in \mathrm{D}(S_0) \cap \mathrm{D}(S_1),\\
&\mathrm{D}((S_0, S_1)_\Sigma) := \mathrm{D}(S_0) + \mathrm{D}(S_1),\\
&(S_0, S_1)_\Sigma (x_0 + x_1) := S_0 x_0 + S_1 x_1, \quad x_0 \in \mathrm{D}(S_0),\ x_1 \in \mathrm{D}(S_1).
\end{aligned}$$

From $S_0 x_\Delta = S_1 x_\Delta$ for all $x_\Delta \in \mathrm{D}(S_0) \cap \mathrm{D}(S_1)$, we see that $(S_0, S_1)_\Sigma$ is well defined.

For simplifying the notation, we introduce the following operators S_0, S_1, S_Δ and S_Σ. Unless otherwise stated, these operators will be related to S as in the next definition throughout this paper.

Definition 3.2. *Let (E_0, E_1), (F_0, F_1) be compatible couples and $S : E_\Sigma \supseteq \mathrm{D}(S) \longrightarrow F_\Sigma$ be linear. We define the linear operators*

$$\begin{aligned}
S_j &: E_j \supseteq \mathrm{D}(S_j) \longrightarrow F_j \quad \text{with} \quad S_j := \check{S}_{E_j, F_j}, \quad j \in \{0, 1, \Delta\},\\
S_\Sigma &: E_\Sigma \supseteq \mathrm{D}(S_\Sigma) \longrightarrow F_\Sigma \quad \text{with} \quad S_\Sigma := (S_0, S_1)_\Sigma,
\end{aligned}$$

where the induced operators are constructed with the continuous embeddings, which correspond to the continuous inclusions.

We have the following situation,

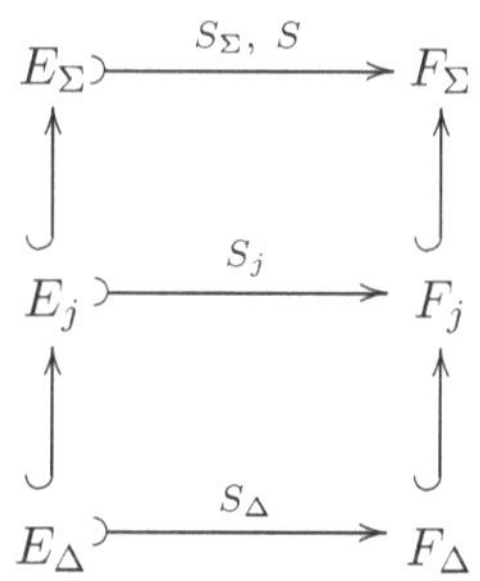

where $j \in \{0, 1\}$ and the injective operators in the diagram correspond to the continuous inclusions. The assumptions in Definition 3.2 imply that S_0 and S_1 coincide on $\mathrm{D}(S_0) \cap \mathrm{D}(S_1)$. Thus S_Σ is well defined.

In the next two propositions, we collect some properties of the operators introduced above. The results are immediate consequences of the definition of the operators and will be used several times in this paper.

Proposition 3.3. *Let* (E_0, E_1), (F_0, F_1) *be compatible couples and the operator* $S : E_\Sigma \supseteq \mathrm{D}(S) \longrightarrow F_\Sigma$ *be linear. It holds*

(i) $S_\Sigma \subseteq S$,
(ii) $\breve{(S_\Sigma)}_{E_k,F_k} = S_k$ *for* $k \in \{0, 1, \Delta, \Sigma\}$,
(iii) $\breve{(S_j)}_{E_\Delta,F_\Delta} = S_\Delta = (S_0, S_1)_\Delta$ *for* $j \in \{0, 1\}$.

Proposition 3.4. *Let* (E_0, E_1), (F_0, F_1) *be compatible couples and the operator* $S : E_\Sigma \supseteq \mathrm{D}(S) \longrightarrow F_\Sigma$ *be linear. For* $j \in \{0, 1\}$, *we have*

$$\begin{aligned} &\mathrm{D}(S_0) \cap \mathrm{D}(S_1) = \mathrm{D}(S_\Delta), \\ &\mathrm{N}(S_\Delta) = \mathrm{N}(S_j) \cap E_\Delta = \mathrm{N}(S_\Sigma) \cap E_\Delta = \mathrm{N}(S) \cap E_\Delta, \\ &\mathrm{N}(S_j) = \mathrm{N}(S_\Sigma) \cap E_j = \mathrm{N}(S) \cap E_j, \\ &\mathrm{N}(S_0) \cap \mathrm{N}(S_1) = \mathrm{N}(S_\Delta), \quad \mathrm{N}(S_0) + \mathrm{N}(S_1) \subseteq \mathrm{N}(S_\Sigma), \\ &\mathrm{R}(S_0) \cap \mathrm{R}(S_1) \supseteq \mathrm{R}(S_\Delta), \quad \mathrm{R}(S_0) + \mathrm{R}(S_1) = \mathrm{R}(S_\Sigma). \end{aligned}$$

If S is injective (closable), then Proposition 2.4 (ii), (iii) and Proposition 3.3 (i) yield that S_j is injective (closable) for $j \in \{0, 1, \Delta, \Sigma\}$. If S is closed, then S_0, S_1 and S_Δ are closed, see Proposition 2.4 (iii), and S_Σ is closable by Proposition 3.3 (i). But if S is closed, then S_Σ is not closed in general, see Example 3.6.

Now, we investigate the operators S_Σ and S. In the next example, these operators coincide.

Example 3.5. Let $E_0 := F_0 := L^{p_0}(\mathbb{R})$ and $E_1 := F_1 := L^{p_1}(\mathbb{R})$, where $1 \le p_0, p_1 \le \infty$. Assume $q : \mathbb{R} \longrightarrow \mathbb{C}$ is measurable, invertible with $q(x) \neq 0$ for all $x \in \mathbb{R}$ and $q^{-1} \in L^\infty(\mathbb{R})$. Furthermore, let $S : L^{p_0}(\mathbb{R}) + L^{p_1}(\mathbb{R}) \supseteq \mathrm{D}(S) \longrightarrow L^{p_0}(\mathbb{R}) + L^{p_1}(\mathbb{R})$ be defined by

$$\begin{aligned} &\mathrm{D}(S) := \{f \in L^{p_0}(\mathbb{R}) + L^{p_1}(\mathbb{R}) : qf \in L^{p_0}(\mathbb{R}) + L^{p_1}(\mathbb{R})\}, \\ &Sf := qf, \quad f \in \mathrm{D}(S). \end{aligned}$$

Let $j \in \{0, 1\}$. We have

$$\mathrm{D}(S_j) = \{f \in L^{p_j}(\mathbb{R}) : qf \in L^{p_j}(\mathbb{R})\}.$$

Let $f \in \mathrm{D}(S)$. Choose $g_j \in L^{p_j}(\mathbb{R})$ such that $qf = g_0 + g_1$. Since $q^{-1}g_j \in L^{p_j}(\mathbb{R})$, we have $q^{-1}g_j \in \mathrm{D}(S_j)$. Therefore $q^{-1}g_0 + q^{-1}g_1 \in \mathrm{D}(S_\Sigma)$. Since $q^{-1}g_0 + q^{-1}g_1 = f$, it follows that $f \in \mathrm{D}(S_\Sigma)$. Hence $S_\Sigma = S$ by Proposition 3.3 (i).

In general, the operators S_Σ and S are not equal as the next example demonstrates.

Example 3.6. Assume (E_0, E_1) is a compatible couple such that $E_1 \subset E_0$ with $\|x\|_{E_1} \ge \|x\|_{E_0}$ for all $x \in E_1$ (e.g., $E_j := L^{p_j}((0, 1))$, $j \in \{0, 1\}$, and $1 \le p_0 <$

$p_1 \le \infty$) and $(F_0, F_1) := (E_1, E_0)$. If we choose $S := \mathrm{id}_{E_0}$, then

$$S_0 : E_0 \supseteq \mathrm{D}(S_0) \longrightarrow E_1 \text{ with } \mathrm{D}(S_0) = E_1,\ S_0 x_0 = x_0 \text{ for } x_0 \in \mathrm{D}(S_0),$$
$$S_1 : E_1 \longrightarrow E_0 \text{ with } \mathrm{D}(S_1) = E_1,\ S_1 x_1 = x_1 \text{ for } x_1 \in \mathrm{D}(S_1).$$

Therefore

$$S_\Sigma : E_0 \supseteq \mathrm{D}(S_\Sigma) \longrightarrow E_0 \text{ with } \mathrm{D}(S_\Sigma) = E_1,\ S_\Sigma x_\Sigma = x_\Sigma \text{ for } x_\Sigma \in \mathrm{D}(S_\Sigma)$$

and S do not coincide.

In the previous example, the operators S, S_1 and $S_\Delta = \mathrm{id}_{E_1}$ are bounded. But S_0 is only closed (cf. Proposition 2.4 (iii)). Since $S_\Sigma \subseteq S$, see Proposition 3.3, we know that S_Σ is closable.

From the following remark, we get a criterion such that S_Σ is equal to S.

Remark 3.7. *Let U, V be linear operators from a Banach space E into a Banach space F such that $U \subseteq V$. If $\mathrm{N}(U) = \mathrm{N}(V)$ and $\mathrm{R}(U) = \mathrm{R}(V)$, then $U = V$ on $\mathrm{D}(U) = \mathrm{D}(V)$. Indeed, let $x \in \mathrm{D}(V)$. Then there exists $\tilde{x} \in \mathrm{D}(U)$ such that $U\tilde{x} = Vx$. Thus $\tilde{x} - x \in \mathrm{N}(V) = \mathrm{N}(U)$. Hence $x \in \mathrm{D}(U)$. From above, it follows that $\mathrm{N}(U) = \mathrm{N}(V)$ and $\mathrm{R}(U) = \mathrm{R}(V)$ if and only if $U = V$ on $\mathrm{D}(U) = \mathrm{D}(V)$.*

Now, assume (E_0, E_1), (F_0, F_1) are compatible couples and the operator $S : E_\Sigma \supseteq \mathrm{D}(S) \longrightarrow F_\Sigma$ is linear. Then $\mathrm{N}(S_\Sigma) = \mathrm{N}(S)$ and $\mathrm{R}(S_\Sigma) = \mathrm{R}(S)$ if and only if $S_\Sigma = S$ by Proposition 3.3 (i) and the remark above.

In the next theorem, we have a conditions such that S_Σ and S are equal.

Theorem 3.8. *Let (E_0, E_1), (F_0, F_1) be compatible couples and $S : E_\Sigma \supseteq \mathrm{D}(S) \longrightarrow F_\Sigma$ be linear.*

(i) *If S_0 and S_1 are everywhere defined and bounded, then S_Δ and $S_\Sigma = S$ are everywhere defined and bounded and it holds*

$$\|S_\Delta\| \le \max\{\|S_0\|, \|S_1\|\},$$
$$\|S_\Sigma\| = \|S\| \le \max\{\|S_0\|, \|S_1\|\}.$$

(ii) *If S_Δ and S_Σ are everywhere defined and bounded, then S_0 and S_1 are everywhere defined and bounded.*

Proof. (i) Assume S_0 and S_1 are everywhere defined and bounded. From Proposition 3.4, we obtain $\mathrm{D}(S_\Delta) = E_\Delta$ and $\mathrm{D}(S_\Sigma) = E_\Sigma$. Since $S_\Sigma \subseteq S$ by Proposition 3.3 (i), it follows that $\mathrm{D}(S) = E_\Sigma$ and $S_\Sigma = S$. With the definition of the operators, the inequalities for the norms are obtained.

(ii) Assume S_Δ and S_Σ are everywhere defined and bounded. Let $j \in \{0, 1\}$. Since S_Σ is closed, it follows that S_j is closed by Proposition 2.4 (iii) and Proposition 3.3 (ii). Assume $x_0 \in E_0$. Since $\mathrm{D}(S_0) + \mathrm{D}(S_1) = \mathrm{D}(S_\Sigma) = E_\Sigma$, there exists $\tilde{x}_j \in \mathrm{D}(S_j)$ such that $x_0 = \tilde{x}_0 + \tilde{x}_1$. We get $\tilde{x}_1 \in E_\Delta = \mathrm{D}(S_\Delta) \subseteq \mathrm{D}(S_0)$. Thus $\mathrm{D}(S_0) = E_0$. Similarly, it is possible to show that $\mathrm{D}(S_1) = E_1$. Hence we conclude that S_0 and S_1 are bounded. □

Let (E_0, E_1), (F_0, F_1) be compatible couples and $S : E_\Sigma \longrightarrow F_\Sigma$ be everywhere defined, linear and bounded such that S_0 and S_1 are everywhere defined and bounded. Then the pair (S_0, S_1) is said to be an interpolation morphism (corresponding to S with respect to (E_0, E_1) and (F_0, F_1)). Its norm is defined by

$$\|(S_0, S_1)\|_{\mathrm{Mor}} := \max\{\|S_0\|, \|S_1\|\}.$$

The following proposition will be useful in the next three sections.

Proposition 3.9. *Let (E_0, E_1), (F_0, F_1) be compatible couples. Assume $S_0 : E_0 \longrightarrow F_0$ and $S_1 : E_1 \longrightarrow F_1$ are everywhere defined, linear and bounded such that the restrictions $S_{0|E_\Delta}$ and $S_{1|E_\Delta}$ are equal (the values of these operators are considered in F_Σ).*

(i) *The operator $(S_0, S_1)_\Sigma$ is linear, everywhere defined and bounded and, for $j \in \{0, 1\}$, it holds $\check{((S_0, S_1)_\Sigma)}_{E_j,F_j} = S_j$.*
(ii) *We have $\check{(S_j)}_{E_\Delta,F_\Delta} = \check{((S_0, S_1)_\Sigma)}_{E_\Delta,F_\Delta}$ for $j \in \{0, 1\}$.*
(iii) *The pair*

$$(S_0, S_1) = \left(\check{((S_0, S_1)_\Sigma)}_{E_0,F_0}, \check{((S_0, S_1)_\Sigma)}_{E_1,F_1}\right)$$

is an interpolation morphism (corresponding to $(S_0, S_1)_\Sigma$).

Proof. (i) Obviously, $(S_0, S_1)_\Sigma$ is linear. From the definition of the operators, we obtain that $\check{((S_0, S_1)_\Sigma)}_{E_j,F_j} = S_j$ for $j \in \{0, 1\}$. Thus $(S_0, S_1)_\Sigma$ is everywhere defined and bounded, see Theorem 3.8 (i).

(ii) For $j \in \{0, 1\}$, it holds

$$\check{(S_j)}_{E_\Delta,F_\Delta} = \check{\left(\check{((S_0, S_1)_\Sigma)}_{E_j,F_j}\right)}_{E_\Delta,F_\Delta} = \check{((S_0, S_1)_\Sigma)}_{E_\Delta,F_\Delta}$$

by (i) and Proposition 3.3 (iii).

(iii) This follows from (i). □

4. Compatible couples of abstract Sobolev spaces and related interpolation morphisms

In this section, we construct interpolation morphisms on the compatible couple of the abstract Sobolev spaces corresponding to S_0 and S_1. With these interpolation morphisms, it is possible to generalize results of the classical interpolation theory to unbounded operators.

Let (E_0, E_1) and (F_0, F_1) be compatible couples and $S : E_\Sigma \supseteq \mathrm{D}(S) \longrightarrow F_\Sigma$ be linear such that S_Σ is closed. Then the induced operators S_0, S_1 and S_Δ are closed, see Proposition 2.4 (iii) and Proposition 3.3 (ii). The pair of Banach spaces (G_{S_0}, G_{S_1}) is a compatible couple since it can be continuously embedded in the space E_Σ. Thus the intersection and the sum of G_{S_0} and G_{S_1} in the next theorem are well defined.

Theorem 4.1. *Let (E_0, E_1) and (F_0, F_1) be compatible couples and $S : E_\Sigma \supseteq \mathrm{D}(S) \longrightarrow F_\Sigma$ be linear such that S_Σ is closed. Then it holds*

(i) $G_{S_\Delta} = G_{S_0} \cap G_{S_1}$ *with equivalent norms,*
(ii) $G_{S_0} + G_{S_1} = G_{S_\Sigma}$ *with equivalent norms.*

Proof. (i) From Proposition 3.4, we know that $\mathrm{D}(S_\Delta) = \mathrm{D}(S_0) \cap \mathrm{D}(S_1)$. Since

$$\|u\|_{S_\Delta} \geq \|u\|_{G_{S_0} \cap G_{S_1}}$$

for all $u \in G_{S_\Delta}$, we obtain (i).

(ii) It holds $\mathrm{D}(S_0) + \mathrm{D}(S_1) = \mathrm{D}(S_\Sigma)$ by definition and

$$\|u_0\|_{S_0} + \|u_1\|_{S_1} \geq \|u_0 + u_1\|_{S_\Sigma}$$

for all $u_0 \in G_{S_0}$ and $u_1 \in G_{S_1}$. Hence (ii) is proved. □

The norms in Theorem 4.1 (i) and (ii) are not equal in general. Indeed, for $E_0 := F_1 := (\mathbb{C}, |\cdot|)$, $E_1 := F_0 := (\mathbb{C}, 2|\cdot|)$ and $S := \mathrm{id}_{E_\Sigma}$, it holds $G_{S_0} = G_{S_1} = (\mathbb{C}, 3|\cdot|)$, $G_{S_\Delta} = (\mathbb{C}, 4|\cdot|)$ and $G_{S_\Sigma} = (\mathbb{C}, 2|\cdot|)$.

Now, we consider interpolation morphisms on compatible couples of abstract Sobolev spaces. In the following theorem, the values of the operators are considered in E_Σ and F_Σ, respectively.

Theorem 4.2. *Let* (E_0, E_1) *and* (F_0, F_1) *be compatible couples and* $S : E_\Sigma \supseteq \mathrm{D}(S) \longrightarrow F_\Sigma$ *be linear such that* S_Σ *is closed. Then*

(i) $i_{S_0|G_{S_0} \cap G_{S_1}} = i_{S_1|G_{S_0} \cap G_{S_1}}$ *and* (i_{S_0}, i_{S_1}) *is an interpolation morphism,*
(ii) $S_0 i_{S_0|G_{S_0} \cap G_{S_1}} = S_1 i_{S_1|G_{S_0} \cap G_{S_1}}$ *and* $(S_0 i_{S_0}, S_1 i_{S_1})$ *is an interpolation morphism.*

Proof. (i) Obviously, $i_{S_0|G_{S_0} \cap G_{S_1}} = i_{S_1|G_{S_0} \cap G_{S_1}}$. Since i_{S_0} and i_{S_1} are everywhere defined and bounded, (i) follows from Proposition 3.9 (iii).

(ii) It holds

$$S_0 i_{S_0|G_{S_0} \cap G_{S_1}} = S_\Sigma i_{S_\Sigma|G_{S_0} \cap G_{S_1}} = S_1 i_{S_1|G_{S_0} \cap G_{S_1}}.$$

Since the operators $S_0 i_{S_0}$ and $S_1 i_{S_1}$ are everywhere defined and bounded, we obtain (ii) from Proposition 3.9 (iii). □

From the previous theorem, it follows that the operators studied in Lemma 4.3 are well defined.

Lemma 4.3. *Let* (E_0, E_1), (F_0, F_1) *be compatible couples and* $S : E_\Sigma \supseteq \mathrm{D}(S) \longrightarrow F_\Sigma$ *be linear such that* S_Σ *is closed. It holds*

(i) $(i_{S_0}, i_{S_1})_\Sigma = i_{S_\Sigma}$ *and* $(S_0 i_{S_0}, S_1 i_{S_1})_\Sigma = S_\Sigma i_{S_\Sigma}$,
(ii) $\check{(}(S_0 i_{S_0}, S_1 i_{S_1})_\Sigma)_{G_{S_j}, F_j} = S_j i_{S_j}$ *for* $j \in \{0, 1, \Delta\}$.

Proof. (i) This follows from Theorem 4.1 (ii) and the definition of the operators.

(ii) It holds $\check{(}S_j)_{E_\Delta, F_\Delta} = S_\Delta$ for $j \in \{0, 1\}$ by Proposition 3.3 (iii). Thus (2) and Proposition 3.9 (ii) yield

$$\check{(}(S_0 i_{S_0}, S_1 i_{S_1})_\Sigma)_{G_{S_\Delta}, F_\Delta} = S_\Delta i_{S_\Delta}.$$

Using Proposition 3.9 (i), we obtain the other equalities. □

5. Fredholm properties of the operators S_0, S_1, S_Δ and S_Σ

In this section, we investigate the relation between the Fredholm properties of S_0, S_1, S_Δ and S_Σ. Some of the results are known, when the operators are bounded, see [Kra96, p. 51, Corollary 3.4.5].

We need the following two lemmata to obtain results on the Fredholm properties of the bounded operators S_0, S_1, S_Δ and S_Σ. Lemma 5.1 was proved in [Sch87, p. 13, Lemma 2.1] for an everywhere defined and bounded linear operator S such that (S_0, S_1) is an interpolation morphism. It is possible to prove this lemma similarly for arbitrary linear operators.

Lemma 5.1. *Let (E_0, E_1), (F_0, F_1) be compatible couples and $S : E_\Sigma \supseteq \mathrm{D}(S) \longrightarrow F_\Sigma$ be linear. It holds $\mathrm{N}(S_0)+\mathrm{N}(S_1) = \mathrm{N}(S_\Sigma)$ if and only if $\mathrm{R}(S_\Delta) = \mathrm{R}(S_0)\cap\mathrm{R}(S_1)$.*

In [Che01, p. 260, Theorem 3.1], it was proved that S_Σ is compact under the assumptions of the next lemma. Similarly, the proof of the compactness of S_Δ is straightforward.

Lemma 5.2. *Let (E_0, E_1), (F_0, F_1) be compatible couples and $S : E_\Sigma \longrightarrow F_\Sigma$ be everywhere defined, linear and bounded. If S_0 and S_1 are compact, then S_Δ and S_Σ are compact.*

Let E and F be Banach spaces. We denote by $\mathrm{B}(E, F)$ the set of all everywhere defined and bounded linear operators from E into F. If $E = F$, we write $\mathrm{B}(E)$ for short. Assume $S : E \supseteq \mathrm{D}(S) \longrightarrow F$ is a Fredholm operator. Then there exists an everywhere defined and bounded linear operator $T : F \longrightarrow E$ with $\mathrm{R}(T) \subseteq \mathrm{D}(S)$ such that there exist compact operators $K_E \in \mathrm{B}(E)$ and $K_F \in \mathrm{B}(F)$ with

$$TS = \mathrm{id}_E + K_E \text{ on } \mathrm{D}(S), \qquad ST = \mathrm{id}_F + K_F.$$

The operator T is called Fredholm-inverse of S. Furthermore, we denote by $\mathrm{FR}(E, F)$ the set of all finite-rank operators U, i.e., $U \in \mathrm{B}(E, F)$ with $\dim \mathrm{R}(U) < \infty$. If $E = F$, we write $\mathrm{FR}(E)$ for short. It is well known that the compact operators $K_E \in \mathrm{B}(E)$ and $K_F \in \mathrm{B}(F)$ above can be replaced by finite-rank operators $F_E \in \mathrm{FR}(E)$ and $F_F \in \mathrm{FR}(F)$, respectively.

Definition 5.3. *Suppose (E_0, E_1), (F_0, F_1) are compatible couples and $S : E_\Sigma \longrightarrow F_\Sigma$ is linear such that (S_0, S_1) is an interpolation morphism. If there exist*

- *a linear operator $T : F_\Sigma \longrightarrow E_\Sigma$ such that (T_0, T_1) is an interpolation morphism and*
- *operators $U \in \mathrm{FR}(E_\Sigma)$, $V \in \mathrm{FR}(F_\Sigma)$ such that*

$$TS + U = \mathrm{id}_{E_\Sigma},$$
$$ST + V = \mathrm{id}_{F_\Sigma},$$

then (S_0, S_1) is said to be Fredholm with respect to (E_0, E_1) and (F_0, F_1).

The next theorem extends a result of M. Krause, see [Kra96, p. 51, Corollary 3.4.5].

Theorem 5.4. *Let (E_0, E_1), (F_0, F_1) be compatible couples such that E_Δ is dense in both E_0, E_1 and F_Δ is dense in both F_0, F_1. Assume $S : E_\Sigma \longrightarrow F_\Sigma$ is linear such that (S_0, S_1) is an interpolation morphism. Then the following statements are equivalent.*

(a) *S_Δ, S_Σ are Fredholm operators and*

$$\operatorname{ind}(S_\Delta) = \operatorname{ind}(S_\Sigma),$$

(b) *S_0, S_1 and S_Δ are Fredholm operators and*

$$\operatorname{ind}(S_0) = \operatorname{ind}(S_1) = \operatorname{ind}(S_\Delta), \quad \operatorname{nul}(S_0) = \operatorname{nul}(S_1) = \operatorname{nul}(S_\Delta),$$

(c) *S_0, S_1 and S_Σ are Fredholm operators and*

$$\operatorname{ind}(S_0) = \operatorname{ind}(S_1) = \operatorname{ind}(S_\Sigma), \quad \operatorname{nul}(S_0) = \operatorname{nul}(S_1) = \operatorname{nul}(S_\Sigma),$$

(d) *S_0, S_1 are Fredholm operators and*

$$\operatorname{ind}(S_0) = \operatorname{ind}(S_1), \quad \operatorname{nul}(S_0) = \operatorname{nul}(S_1),$$
$$\mathrm{R}(S_\Sigma) \cap F_\Delta = \mathrm{R}(S_\Delta),$$

(e) *S_0, S_1 are Fredholm operators and there exist Fredholm-inverses S_0^F and S_1^F of S_0 and S_1, respectively, such that $S^F_{0\,|F_\Delta} = S^F_{1\,|F_\Delta}$,*

(f) *S_0, S_1 are Fredholm operators and there exist Fredholm-inverses S_0^F and S_1^F of S_0 and S_1, respectively, such that (S_0^F, S_1^F) is an interpolation morphism,*

(g) *(S_0, S_1) is Fredholm with respect to (E_0, E_1) and (F_0, F_1).*

(h) *Let D be a dense subspace of F_Δ. Then*

$$\mathrm{N}(S_0) = \mathrm{N}(S_1) = \mathrm{N}(S_\Delta) = \mathrm{N}(S_\Sigma),$$

where $\operatorname{nul}(S_k) < \infty$, and there exists a finite-dimensional subspace $H \subseteq D$ with $\mathrm{R}(S_k) \oplus H = F_k$ for $k \in \{0, 1, \Delta, \Sigma\}$.

If S_0^F and S_1^F are Fredholm-inverses of S_0 and S_1, respectively, such that $S^F_{0\,|F_\Delta} = S^F_{1\,|F_\Delta}$, then $\check{(}(S_0^F, S_1^F)_\Sigma)_{F_\Delta, E_\Delta}$ and $(S_0^F, S_1^F)_\Sigma$ are Fredholm-inverses of S_Δ and S_Σ, respectively, and

$$\|\check{(}(S_0^F, S_1^F)_\Sigma)_{F_\Delta, E_\Delta}\| \leq \|(S_0^F, S_1^F)\|_{\mathrm{Mor}}, \quad \|(S_0^F, S_1^F)_\Sigma\| \leq \|(S_0^F, S_1^F)\|_{\mathrm{Mor}}.$$

Proof. The equivalence of the statements (a), (d), (g) and (h) follows immediately from [Kra96, p. 51, Corollary 3.4.5].

(h)$\Longrightarrow$(e) For $k \in \{0, 1, \Sigma\}$, the operators S_k are Fredholm operators. Assume $M_k \subseteq E_k$ such that

$$\mathrm{N}(S_k) \oplus M_k = E_k, \quad k \in \{0, 1, \Sigma\}.$$

Let $j \in \{0, 1\}$. Since $\mathrm{N}(S_j) = \mathrm{N}(S_\Sigma)$, we have $M_j \subseteq M_\Sigma$. The restriction of S_j to M_j with the range space $\mathrm{R}(S_j)$ is an isomorphism. Hence $S_j^F : F_j \longrightarrow E_j$ defined by

$$S_j^F(y_j + h_j) := x_j \quad \text{if and only if} \quad x_j \in M_j \text{ and } S_j x_j = y_j,$$

where $y_j \in \mathrm{R}(S_j)$ and $h_j \in H$, is well defined and a Fredholm-inverse of S_j.

Now, choose $y \in F_\Delta$. Then there exist $x_j \in M_j$ and $h_j \in H$ such that $S_j x_j + h_j = y$. The fact that $\mathrm{R}(S_\Sigma) \oplus H = F_\Sigma$ implies $S_0 x_0 = S_1 x_1$. Since $x_j \in M_j \subseteq M_\Sigma$ and S_Σ is injective on M_Σ, we conclude that $x_0 = x_1$. Thus $S_0^F y = x_j = S_1^F y$.

(e)$\Longrightarrow$(f) From Theorem 3.8 (iii), we get this implication.

(f)$\Longrightarrow$(b) Let S_0^F and S_1^F be Fredholm-inverses of S_0 and S_1, respectively, such that (S_0^F, S_1^F) is an interpolation morphism. Then $S_0^F{}_{|F_\Delta} = S_1^F{}_{|F_\Delta}$. Proposition 3.9 (ii) and (iii) yield that

$$\check{(}(S_0^F, S_1^F)_\Sigma)_{F_\Delta, E_\Delta} = \check{(}S_0^F)_{F_\Delta, E_\Delta} = \check{(}S_1^F)_{F_\Delta, E_\Delta}$$

is everywhere defined and bounded. Let $K_{E_j} \in \mathrm{B}(E_j)$ and $K_{F_j} \in \mathrm{B}(F_j)$ be compact such that $S_j^F S_j = \mathrm{id}_{E_j} + K_{E_j}$ and $S_j S_j^F = \mathrm{id}_{F_j} + K_{F_j}$, $j \in \{0,1\}$. Since

$$x_\Delta + K_{E_0} x_\Delta = \check{(}S_0^F)_{F_\Delta, E_\Delta} S_\Delta x_\Delta = \check{(}S_1^F)_{F_\Delta, E_\Delta} S_\Delta x_\Delta = x_\Delta + K_{E_1} x_\Delta, \quad x_\Delta \in E_\Delta,$$

we obtain that $K_{E_0 | E_\Delta} = K_{E_1 | E_\Delta}$. Thus (K_{E_0}, K_{E_1}) is an interpolation morphism, see Proposition 3.9 (iii). Proposition 3.9 (ii) and Lemma 5.2 imply that

$$\check{(}(K_{E_0}, K_{E_1})_\Sigma)_{E_\Delta, E_\Delta} = \check{(}K_{E_0})_{E_\Delta, E_\Delta} = \check{(}K_{E_1})_{E_\Delta, E_\Delta}$$

is compact. It is possible to show similarly that $\check{(}K_{F_0})_{F_\Delta, F_\Delta} = \check{(}K_{F_1})_{F_\Delta, F_\Delta}$ is compact and

$$S_\Delta \check{(}S_j^F)_{F_\Delta, E_\Delta} = \mathrm{id}_{F_\Delta} + \check{(}K_{F_j})_{F_\Delta, F_\Delta}, \quad j \in \{0,1\}.$$

Thus S_Δ is a Fredholm operator with Fredholm-inverse $\check{(}S_0^F)_{F_\Delta, E_\Delta} = \check{(}S_1^F)_{F_\Delta, E_\Delta}$. It holds

$$-\mathrm{ind}(S_\Delta) \geq -\mathrm{ind}(S_j) = \mathrm{ind}(S_j^F) \geq \mathrm{ind}(\check{(}S_j^F)_{F_\Delta, E_\Delta}) = -\mathrm{ind}(S_\Delta)$$

and $\mathrm{nul}(S_\Delta) = \mathrm{nul}(S_j)$ by [Kra96, p. 44, Lemma 3.4.2], where $j \in \{0,1\}$. Hence (b) is proved.

(b)$\Longrightarrow$(c) From [Kra96, p. 44, Lemma 3.4.2], we know that there exists a subspace $H \subseteq F_\Delta$ such that

$$\mathrm{R}(S_k) \oplus H = F_k, \quad k \in \{0, 1, \Delta\}.$$

For $0 \neq y \in \mathrm{R}(S_0) \cap \mathrm{R}(S_1)$, it follows that $y \notin H$ and therefore $y \in \mathrm{R}(S_\Delta)$. Thus $\mathrm{R}(S_\Delta) = \mathrm{R}(S_0) \cap \mathrm{R}(S_1)$, see Proposition 3.4. Since $\mathrm{N}(S_0) = \mathrm{N}(S_\Delta) = \mathrm{N}(S_1)$ by Proposition 3.4, we conclude that $\mathrm{N}(S_0) = \mathrm{N}(S_1) = \mathrm{N}(S_\Sigma)$ from Lemma 5.1. Moreover, it holds $\mathrm{R}(S_\Sigma) + H = F_\Sigma$, see again Proposition 3.4. Thus S_Σ is Fredholm. Then there exists a subspace $H_\Sigma \subseteq H$ such that

$$\mathrm{R}(S_\Sigma) \oplus H_\Sigma = F_\Sigma,$$

see [Kra96, p. 44, Lemma 3.4.2]. To prove that $H = H_\Sigma$, assume $y \in H \backslash H_\Sigma$. Then $y \in \mathrm{R}(S_\Sigma)$. Since $\mathrm{R}(S_\Sigma) = \mathrm{R}(S_0) + \mathrm{R}(S_1)$ by Proposition 3.4, there exists $y_j \in \mathrm{R}(S_j)$, $j \in \{0,1\}$, with $y_0 + y_1 = y$. Assume that $y_1 \neq 0$. (For $y_1 = 0$, we obtain $y \in \mathrm{R}(S_0)$ and therefore $y \notin H$; a contradiction.) Then $y \in F_\Delta$ implies

$y_1 = y - y_0 \in F_0$. Since $y_1 \notin H$, we have $y_1 \in \mathrm{R}(S_0)$. Thus $y \in \mathrm{R}(S_0)$ and therefore $y \notin H$. This is a contradiction. Hence $H = H_\Sigma$ and (c) is proved.

(c)$\Longrightarrow$(d) Assume $M_k \subseteq E_k$ such that

$$\mathrm{N}(S_k) \oplus M_k = E_k, \quad k \in \{0, 1, \Sigma\}.$$

Let $j \in \{0,1\}$. It holds $\mathrm{N}(S_j) = \mathrm{N}(S_\Sigma)$ by Proposition 3.4. Thus $M_j \subseteq M_\Sigma$. From [Kra96, p. 44, Lemma 3.4.2], we know that there exists a subspace $H \subseteq F_\Delta$ such that

$$\mathrm{R}(S_k) \oplus H = F_k, \quad k \in \{0, 1, \Sigma\}.$$

It follows that $\mathrm{R}(S_\Delta) + H \subseteq F_\Delta$. Now, let $y \in F_\Delta$. Then there exist $x_j \in M_j$ and $h_j \in H$ such that $S_j x_j + h_j = y$. The equality $\mathrm{R}(S_\Sigma) \oplus H = F_\Sigma$ yields $S_0 x_0 = S_1 x_1$ and $h_0 = h_1$. Since $M_j \subseteq M_\Sigma$ and S_Σ is injective on M_Σ, we obtain $x_0 = x_1 \in E_\Delta$. It follows that $S_\Delta x_j + h_j = y$. Therefore $\mathrm{R}(S_\Delta) + H = F_\Delta$. Since $\mathrm{R}(S_\Delta) \cap H \subseteq \mathrm{R}(S_j) \cap H = \{0\}$, we conclude that $\mathrm{R}(S_\Delta) \oplus H = F_\Delta$. To obtain (d), it is possible to conclude similarly as in the proof of the implication (1)$\Longrightarrow$(2) of [Kra96, p. 46, Theorem 3.4.4].

Now, let $j \in \{0,1\}$ and assume S_j^F is a Fredholm-inverse of S_j such that $S_{0\,|F_\Delta}^F = S_{1\,|F_\Delta}^F$. Then $\check{(S_j^F)}_{F_\Delta, E_\Delta}$ is a Fredholm-inverse of S_Δ, see the proof of the implication (f)$\Longrightarrow$(b). Proposition 3.9 (ii) yields that

$$\check{(S_j^F)}_{F_\Delta, E_\Delta} = \check{((S_0^F, S_1^F)_\Sigma)}_{F_\Delta, E_\Delta}$$

is a Fredholm-inverse of S_Δ. Assume K_{E_j} and K_{F_j} are defined as in the proof of the implication (f)$\Longrightarrow$(b). Then (K_{E_0}, K_{E_1}) is an interpolation morphism and $(K_{E_0}, K_{E_1})_\Sigma$ is compact by Lemma 5.2. We have

$$\begin{aligned}(S_0^F, S_1^F)_\Sigma S_\Sigma (x_0 + x_1) &= x_0 + K_{E_0} x_0 + x_1 + K_{E_1} x_1 \\ &= (x_0 + x_1) + (K_{E_0}, K_{E_1})_\Sigma (x_0 + x_1)\end{aligned}$$

for $x_0 \in E_0$ and $x_1 \in E_1$. Similarly, we obtain $S_\Sigma (S_0^F, S_1^F)_\Sigma = \mathrm{id}_{F_\Sigma} + (K_{F_0}, K_{F_1})_\Sigma$ and $(K_{F_0}, K_{F_1})_\Sigma$ is compact. Thus $(S_0^F, S_1^F)_\Sigma$ is a Fredholm-inverse of S_Σ. The inequalities of the norms follow from Theorem 3.8 (i). □

Now, we discuss Fredholm properties of the unbounded operators S_0, S_1, S_Δ and S_Σ.

Theorem 5.5. *Let (E_0, E_1), (F_0, F_1) be compatible couples such that F_Δ is dense in both F_0, F_1. Assume $S : E_\Sigma \supseteq \mathrm{D}(S) \longrightarrow F_\Sigma$ is linear such that S_Σ is closed and $\mathrm{D}(S_\Delta)$ is a core of S_0 and S_1. Then the following statements are equivalent.*

(a′) *S_Δ, S_Σ are Fredholm operators and*

$$\mathrm{ind}(S_\Delta) = \mathrm{ind}(S_\Sigma),$$

(b′) *S_0, S_1 and S_Δ are Fredholm operators and*

$$\mathrm{ind}(S_0) = \mathrm{ind}(S_1) = \mathrm{ind}(S_\Delta), \quad \mathrm{nul}(S_0) = \mathrm{nul}(S_1) = \mathrm{nul}(S_\Delta),$$

(c′) *S_0, S_1 and S_Σ are Fredholm operators and*

$$\operatorname{ind}(S_0) = \operatorname{ind}(S_1) = \operatorname{ind}(S_\Sigma), \quad \operatorname{nul}(S_0) = \operatorname{nul}(S_1) = \operatorname{nul}(S_\Sigma),$$

(d′) *S_0, S_1 are Fredholm operators and*

$$\operatorname{ind}(S_0) = \operatorname{ind}(S_1), \quad \operatorname{nul}(S_0) = \operatorname{nul}(S_1),$$
$$\mathrm{R}(S_\Sigma) \cap F_\Delta = \mathrm{R}(S_\Delta),$$

(e′) *S_0, S_1 are Fredholm operators and there exist Fredholm-inverses S_0^F and S_1^F of S_0 and S_1, respectively, such that $S^F_{0\,|F_\Delta} = S^F_{1\,|F_\Delta}$,*

(f′) *S_0, S_1 are Fredholm operators and there exist Fredholm-inverses S_0^F and S_1^F of S_0 and S_1, respectively, such that (S_0^F, S_1^F) is an interpolation morphism,*

(g′) *$(S_0 i_{S_0}, S_1 i_{S_1})$ is Fredholm with respect to (G_{S_0}, G_{S_1}) and (F_0, F_1).*

(h′) *Let D be a dense subspace of F_Δ. Then*

$$\mathrm{N}(S_0) = \mathrm{N}(S_1) = \mathrm{N}(S_\Delta) = \mathrm{N}(S_\Sigma),$$

where $\operatorname{nul}(S_k) < \infty$, and there exists a finite-dimensional subspace $H \subseteq D$ with $\mathrm{R}(S_k) \oplus H = F_k$ for $k \in \{0, 1, \Delta, \Sigma\}$.

Proof. For $k \in \{0, 1, \Delta\}$, it holds

$$(S_0 i_{S_0}, S_1 i_{S_1})_\Sigma = S_\Sigma i_{S_\Sigma},$$
$$\check{(}(S_0 i_{S_0}, S_1 i_{S_1})_\Sigma)_{G_{S_k}, F_k} = S_k i_{S_k}$$

by Lemma 4.3. From Proposition 2.2, we know that $S_l i_{S_l}$ is Fredholm if and only if S_l is Fredholm; in this case, the nullities, the deficiencies and the indices of $S_l i_{S_l}$ and S_l coincide, $l \in \{0, 1, \Delta, \Sigma\}$.

Let $j \in \{0, 1\}$ and assume S_j is a Fredholm operator with Fredholm-inverse S_j^F such that $S^F_{0\,|F_\Delta} = S^F_{1\,|F_\Delta}$. Note that this is equivalent to the fact that (S_0^F, S_1^F) is an interpolation morphism by Proposition 3.9 (iii). Then

$$(i_{S_0})^{-1} S^F_{0\,|F_\Delta} = (i_{S_1})^{-1} S^F_{1\,|F_\Delta},$$

which is equivalent to the property that $((i_{S_0})^{-1} S_0^F, (i_{S_1})^{-1} S_1^F)$ is an interpolation morphism, see Proposition 3.9 (iii). Let $F_{E_j} \in \mathrm{FR}(E_j)$ and $F_{F_j} \in \mathrm{FR}(F_j)$ be the corresponding finite-rank operators such that

$$S_j^F S_j = \mathrm{id}_{E_j} + F_{E_j} \text{ on } \mathrm{D}(S_j), \quad S_j S_j^F = \mathrm{id}_{F_j} + F_{F_j}.$$

It holds $\mathrm{R}(S_j^F) \subseteq \mathrm{D}(S_j)$. Thus $F_{E_j}\{\mathrm{D}(S_j)\} \subseteq \mathrm{D}(S_j)$. Then $(i_{S_j})^{-1} F_{E_j} i_{S_j}$ is everywhere defined, closed and therefore bounded. Moreover, $(i_{S_j})^{-1} F_{E_j} i_{S_j}$ is a finite-rank operator. Since

$$(i_{S_j})^{-1} S_j^F S_j i_{S_j} = \mathrm{id}_{G_{S_j}} + (i_{S_j})^{-1} F_{E_j} i_{S_j}, \quad S_j i_{S_j} (i_{S_j})^{-1} S_j^F = \mathrm{id}_{F_j} + F_{F_j},$$

the operator $(i_{S_j})^{-1} S_j^F$ is a Fredholm-inverse of $S_j i_{S_j}$.

Conversely, assume $S_j i_{S_j}$ is a Fredholm operator with Fredholm-inverse $(S_j i_{S_j})^F$ such that $(S_0 i_{S_0})^F{}_{|F_\Delta} = (S_1 i_{S_1})^F{}_{|F_\Delta}$. This is equivalent to the property that $((S_0 i_{S_0})^F, (S_1 i_{S_1})^F)$ is an interpolation morphism, see Proposition 3.9 (iii).

Let $m \in \{0, 1, \Sigma\}$. From Theorem 5.4, we obtain that $S_\Sigma i_{S_\Sigma}$ is Fredholm and there exists a subspace $H \subseteq F_\Delta$ such that $\mathrm{R}(S_m i_{S_m}) \oplus H = F_m$. Thus $\mathrm{R}(S_m) \oplus H = F_m$, see (1). Moreover, we have

$$\mathrm{N}(S_0 i_{S_0}) = \mathrm{N}(S_1 i_{S_1}) = \mathrm{N}(S_\Sigma i_{S_\Sigma})$$

with $\mathrm{nul}(S_m i_{S_m}) < \infty$ by Theorem 5.4. This implies $\mathrm{N}(S_0) = \mathrm{N}(S_1) = \mathrm{N}(S_\Sigma)$ with $\mathrm{nul}(S_m) < \infty$, see again (1). Hence S_m is Fredholm and there exists $M_m \subseteq E_m$ such that

$$\mathrm{N}(S_m) \oplus M_m = E_m.$$

The restriction of S_m to $M_m \cap \mathrm{D}(S_m)$ with range space $\mathrm{R}(S_m)$ is an isomorphism. Thus

$$S_j^F(y_j + h) := x_j \quad \text{if and only if} \quad x_j \in M_j \cap \mathrm{D}(S_j) \text{ and } S_j x_j = y_j,$$

where $y_j \in \mathrm{R}(S_j)$, $h \in H$, is well defined and a Fredholm-inverse of S_j. Now, it is possible to conclude similarly as in the proof of the implication (h)$\Longrightarrow$(e) of Theorem 5.4 that $S_0^F{}_{|F_\Delta} = S_1^F{}_{|F_\Delta}$, which is equivalent to the fact that (S_0^F, S_1^F) is an interpolation morphism by Proposition 3.9 (iii).

Since $\mathrm{D}(S_\Delta)$ is a core of S_0 and S_1, the abstract Sobolev space G_{S_Δ} is dense in both G_{S_0} and G_{S_1}. Then the theorem follows from (1) and Theorem 5.4 applied to the interpolation morphism $(S_0 i_{S_0}, S_1 i_{S_1})$ corresponding to the compatible couples (G_{S_0}, G_{S_1}) and (F_0, F_1). □

Note that it is possible to obtain further results from the previous theorem with Proposition 2.2 and (1).

6. Continuous invertibility of the operators S_0, S_1, S_Δ and S_Σ

In this section, we obtain results on the relation between the continuous invertibility of S_0, S_1, S_Δ and S_Σ. There are similarities to the results on the relation between the Fredholm properties of these operators, see Theorem 5.5. But note that we do not need any additional assumptions unlike in Theorem 5.5.

The following lemma is needed to prove Proposition 6.2. M. Krause has proved this lemma for an everywhere defined and bounded linear operator S such that (S_0, S_1) is an interpolation morphism, see [Kra96, p. 40, Lemma 3.3.1]. The proof as in [Kra96, p. 40, Lemma 3.3.1] works similarly for the following lemma.

Lemma 6.1. *Let (E_0, E_1), (F_0, F_1) be compatible couples and $S : E_\Sigma \supseteq \mathrm{D}(S) \longrightarrow F_\Sigma$ be linear. If $\mathrm{R}(S_\Sigma) \cap F_\Delta = \mathrm{R}(S_\Delta)$, then $\mathrm{N}(S_\Sigma) = \mathrm{N}(S_0) + \mathrm{N}(S_1)$.*

The first part of the main theorem of this section, Theorem 6.4, is an immediate consequence of the next two propositions.

Proposition 6.2. *Let (E_0, E_1), (F_0, F_1) be compatible couples and the operator $S : E_\Sigma \supseteq \mathrm{D}(S) \longrightarrow F_\Sigma$ be linear. The following statements are equivalent.*

(i) *S_Δ and S_Σ are injective and surjective,*
(ii) *S_0, S_1 and S_Δ are injective and surjective,*

(iii) S_0, S_1 *and* S_Σ *are injective and surjective,*
(iv) S_0, S_1 *are injective and surjective and it holds* $\mathrm{R}(S_\Sigma) \cap F_\Delta = \mathrm{R}(S_\Delta)$,
(v) S_0, S_1 *are injective and surjective and it holds* ${S_0}^{-1}{}_{|F_\Delta} = {S_1}^{-1}{}_{|F_\Delta}$.

Proof. (i)$\Longrightarrow$(ii) Since S_Σ is injective, we know that S_0 and S_1 are injective, see Proposition 2.4 (ii) and Proposition 3.3 (ii).

To show that S_0 is surjective, let $y_0 \in F_0 \subseteq F_\Sigma$. Since S_Σ is surjective, there exists $x_\Sigma \in E_\Sigma$ such that $S_\Sigma x_\Sigma = y_0$. Let $x_0 \in \mathrm{D}(S_0)$ and $x_1 \in \mathrm{D}(S_1)$ such that $x_0 + x_1 = x_\Sigma$. From $y_0 = S_\Sigma(x_0 + x_1) = S_0 x_0 + S_1 x_1$, it follows that $S_1 x_1 \in F_\Delta$. Since S_Δ is surjective, there exists $x_\Delta \in \mathrm{D}(S_\Delta)$ such that $S_1 x_1 = S_\Delta x_\Delta = S_1 x_\Delta$. The operator S_1 is injective. Therefore $x_1 = x_\Delta$. Then we have $y_0 = S_0 x_0 + S_1 x_1 = S_0 x_0 + S_0 x_\Delta$. Thus S_0 is surjective.

To show that S_1 is surjective, we proceed as above.

Similarly, the proof of the implications (ii)$\Longrightarrow$(iii) (with Proposition 3.4), (iii)$\Longrightarrow$(v), (v)$\Longrightarrow$(iv) and (iv)$\Longrightarrow$(i) (with Proposition 3.4 and Lemma 6.1) is straightforward. □

Proposition 6.3. *Let* (E_0, E_1), (F_0, F_1) *be compatible couples and the operator* $S : E_\Sigma \supseteq \mathrm{D}(S) \longrightarrow F_\Sigma$ *be linear. Assume* S_j *is injective and surjective for* $j \in \{0, 1, \Delta, \Sigma\}$. *Then the following statements are equivalent.*

(i) $(S_0)^{-1} \in \mathrm{B}(F_0, E_0)$ *and* $(S_1)^{-1} \in \mathrm{B}(F_1, E_1)$,
(ii) $(S_\Delta)^{-1} \in \mathrm{B}(F_\Delta, E_\Delta)$ *and* $(S_\Sigma)^{-1} \in \mathrm{B}(F_\Sigma, E_\Sigma)$.

Proof. Assume (i) holds. From the definition of the norms on E_Δ, F_Δ, E_Σ and F_Σ, we obtain

$$\|(S_\Delta)^{-1}\| \le \max\{\|(S_0)^{-1}\|, \|(S_1)^{-1}\|\},$$
$$\|(S_\Sigma)^{-1}\| \le \max\{\|(S_0)^{-1}\|, \|(S_1)^{-1}\|\}.$$

Conversely, assume (ii) holds. Then S_Σ is closed. Proposition 2.4 (iii) and Proposition 3.3 (ii) yield that S_0 and S_1 are closed. Since S_0 and S_1 are injective and surjective, (i) follows. □

The following theorem generalizes [Kra96, p. 40, Proposition 3.3.2]. Note that we do not need – unlike in Theorem 5.5 – any assumptions on the density of F_Δ in F_0 and F_1 nor that $\mathrm{D}(S_\Delta)$ is a core of S_0 and S_1.

Theorem 6.4. *Let* (E_0, E_1), (F_0, F_1) *be compatible couples and* $S : E_\Sigma \supseteq \mathrm{D}(S) \longrightarrow F_\Sigma$ *be linear. The following statements are equivalent.*

(a″) S_Δ *and* S_Σ *are continuously invertible,*
(b″) S_0, S_1 *and* S_Δ *are continuously invertible,*
(c″) S_0, S_1 *and* S_Σ *are continuously invertible,*
(d″) S_0, S_1 *are continuously invertible and* $\mathrm{R}(S_\Sigma) \cap F_\Delta = \mathrm{R}(S_\Delta)$,
(e″) S_0, S_1 *are continuously invertible and* $(S_0)^{-1}{}_{|F_\Delta} = (S_1)^{-1}{}_{|F_\Delta}$,

(f″) *S_0, S_1 are continuously invertible and $((S_0)^{-1}, (S_1)^{-1})$ is an interpolation morphism,*

(g″) *S_0, S_1 are closed, $S_0 i_{S_0}$, $S_1 i_{S_1}$ are continuously invertible and the pair $((S_0 i_{S_0})^{-1}, (S_1 i_{S_1})^{-1})$ is an interpolation morphism with respect to the compatible couples (F_0, F_1) and (G_{S_0}, G_{S_1}).*

If one of the statements (a″)–(g″) *holds, then S_k is closed, the operator $S_k i_{S_k}$ is an isomorphism for $k \in \{0, 1, \Delta, \Sigma\}$ and it holds*

$$\check{(}((S_0)^{-1}, (S_1)^{-1})_\Sigma)_{F_\Delta, E_\Delta} = (S_\Delta)^{-1} \quad \textit{with} \quad \|(S_\Delta)^{-1}\| \leq \|((S_0)^{-1}, (S_1)^{-1})\|_{\mathrm{Mor}},$$

$$((S_0)^{-1}, (S_1)^{-1})_\Sigma = (S_\Sigma)^{-1} \quad \textit{with} \quad \|(S_\Sigma)^{-1}\| \leq \|((S_0)^{-1}, (S_1)^{-1})\|_{\mathrm{Mor}}.$$

Proof. From Proposition 6.2 and Proposition 6.3, we obtain the equivalence of the statements (a″)–(e″). The statements (e″) and (f″) are equivalent, see Proposition 3.9 (iii). From Proposition 2.2, we know that S_j is continuously invertible if and only if S_j is closed and $S_j i_{S_j}$ is continuously invertible, where $j \in \{0, 1\}$; in this case, we have

$$(S_0)^{-1}{}_{|F_\Delta} = (S_1)^{-1}{}_{|F_\Delta} \iff (S_0 i_{S_0})^{-1}{}_{|F_\Delta} = (S_1 i_{S_1})^{-1}{}_{|F_\Delta}.$$

From Proposition 3.9 (iii), we know that $(S_0 i_{S_0})^{-1}{}_{|F_\Delta} = (S_1 i_{S_1})^{-1}{}_{|F_\Delta}$ is equivalent to the fact that $((S_0 i_{S_0})^{-1}, (S_1 i_{S_1})^{-1})$ is an interpolation morphism.

Now, assume one of the statements (a″)–(g″) holds. Then S_k is closed and, by Proposition 2.2, $S_k i_{S_k}$ is an isomorphism for $k \in \{0, 1, \Delta, \Sigma\}$. We have

$$\check{(}((S_0)^{-1}, (S_1)^{-1})_\Sigma)_{F_\Delta, E_\Delta} = \check{(}(S_j)^{-1})_{F_\Delta, E_\Delta} = (\check{(}S_j)_{F_\Delta, E_\Delta})^{-1} = S_\Delta^{-1}, \quad j \in \{0, 1\},$$

see Proposition 2.4 (ii), Proposition 3.3 (iii) and Proposition 3.9 (ii). Since $\mathrm{R}(S_0) + \mathrm{R}(S_1) = \mathrm{R}(S_\Sigma)$ by Proposition 3.4, we see that $((S_0)^{-1}, (S_1)^{-1})_\Sigma = (S_\Sigma)^{-1}$. The inequalities of the norms follow from Theorem 3.8 (i). □

A trivial example such that the equivalent statements (a″)–(f″) of Theorem 6.4 are fulfilled is the operator $S : E_\Sigma \supseteq \mathrm{D}(S) \longrightarrow F_\Sigma$ with $\mathrm{D}(S) := F_\Sigma$ and $Sx := x$, $x \in F_\Sigma$, where (E_0, E_1), (F_0, F_1) are compatible couples such that $F_j \subseteq E_j$ with continuous inclusion for $j \in \{0, 1\}$. As for the Fredholm properties of the operators S_0, S_1, S_Δ and S_Σ, it is possible to obtain further results from Theorem 6.4 with Proposition 2.2.

In literature, the spectra of the operators S_0 and S_1 were often investigated in concrete examples, see for instance [Dav07, p. 49 Example 2.2.11] or [Jör82, p. 330, Exercise 12.11 a]. From the theory of this section, it is possible to obtain results on the resolvent sets of S_0, S_1, S_Δ and S_Σ in the abstract case. Some of these results are known. For instance, Y. Chen has shown that the resolvent sets corresponding to the statements (b″), (c″) and (e″) are equal, see [Che01, p. 257, Theorem 2.1]. Note that statement (e″) often appears in literature, see for instance [Gri66, p. 168, Section 4.1], [HT56, p. 286, Theorem] or [Zaf73, p. 365, Theorem 3.1].

7. Example – differential operators

In this section, we consider differential operators as an example for the theory in the previous sections. Similarly as the well-known differential operators on L^p-spaces, we introduce differential operators on the intersection and the sum of two L^p-spaces and investigate their properties.

Assume $b, c \in \mathbb{R}$ and f is a scalar-valued function on $[b, c]$. Then f is said to be absolutely continuous on $[b, c]$ if and only if there exists a function $\varphi \in L^1([b, c])$ such that $f(x) = f(b) + \int_b^x \varphi(t)dt$ for all $x \in [b, c]$.

Let $I \subseteq \mathbb{R}$ be an open interval and $n \in \mathbb{N}$. We define the set $A_n(I)$ (A_n for short) by

$$A_n(I) := \{f : I \longrightarrow \mathbb{C} \ : \ f^{(n-1)} \text{ exists and is absolutely continuous on every compact subinterval of } I\}$$

and we define the differential expression τ (of order n) on I and its formal adjoint $\tau^\star$ on I by

$$(\tau f)(x) := \Sigma_{k=0}^n a_k (D^k f)(x), \quad x \in I,$$

$$(\tau^\star f)(x) := \Sigma_{k=0}^n (-1)^k (D^k (a_k f))(x), \quad x \in I,$$

where $f \in A_n$, $a_k \in C^k(\overline{I})$ (see p. 188) for $k \in \{0, 1, \ldots, n\}$ with $a_n(t) \neq 0$ for all $t \in I$ and D denotes the operator of differentiation.

If there is no possibility of confusion, we will write L^p instead of $L^p(I)$ for $1 \leq p \leq \infty$ to simplify the notation. The complicate notation of the maximal, restricted and minimal operator introduced in the next definition is necessary to state the theory precisely.

Definition 7.1. *Assume* $1 \leq p_0, p_1, q_0, q_1 \leq \infty$ *and* τ *is a differential expression on an open interval* I. *We define the maximal operators*

- $S^{\max}_{\tau,p_0,p_1,q_0,q_1,\Delta} : L^{p_0} \cap L^{p_1} \supseteq \mathrm{D}(S^{\max}_{\tau,p_0,p_1,q_0,q_1,\Delta}) \longrightarrow L^{q_0} \cap L^{q_1}$ ($S^{\max}_\Delta$ *for short*) *by*

$$\mathrm{D}(S^{\max}_{\tau,p_0,p_1,q_0,q_1,\Delta}) := \{f \in L^{p_0} \cap L^{p_1} : f \in A_n \text{ and } \tau f \in L^{q_0} \cap L^{q_1}\},$$
$$S^{\max}_{\tau,p_0,p_1,q_0,q_1,\Delta} f := \tau f \text{ for } f \in \mathrm{D}(S^{\max}_{\tau,p_0,p_1,q_0,q_1,\Delta}).$$

- $S^{\max}_{\tau,p_0,p_1,q_0,q_1,\Sigma} : L^{p_0} + L^{p_1} \supseteq \mathrm{D}(S^{\max}_{\tau,p_0,p_1,q_0,q_1,\Sigma}) \longrightarrow L^{q_0} + L^{q_1}$ ($S^{\max}_\Sigma$ *for short*) *by*

$$\mathrm{D}(S^{\max}_{\tau,p_0,p_1,q_0,q_1,\Sigma}) := \{f \in L^{p_0} + L^{p_1} : f \in A_n \text{ and } \tau f \in L^{q_0} + L^{q_1}\},$$
$$S^{\max}_{\tau,p_0,p_1,q_0,q_1,\Sigma} f := \tau f \text{ for } f \in \mathrm{D}(S^{\max}_{\tau,p_0,p_1,q_0,q_1,\Sigma}).$$

Let the restricted operator

- $S^R_{\tau,p_0,p_1,q_0,q_1,\Delta}$ (S^R_Δ *for short*) *be the restriction of* $S^{\max}_{\tau,p_0,p_1,q_0,q_1,\Delta}$,
- $S^R_{\tau,p_0,p_1,q_0,q_1,\Sigma}$ (S^R_Σ *for short*) *be the restriction of* $S^{\max}_{\tau,p_0,p_1,q_0,q_1,\Sigma}$

to those elements in the domain, which have compact support in the interior of I.

For $1 \le p, q \le \infty$, *define*

$$S^{\max}_{\tau,p,q} := S^{\max}_{\tau,p,p,q,q,\Delta} = S^{\max}_{\tau,p,p,q,q,\Sigma} \quad (S^{\max}_{p,q} \text{ for short}),$$
$$S^{R}_{\tau,p,q} := S^{R}_{\tau,p,p,q,q,\Delta} = S^{R}_{\tau,p,p,q,q,\Sigma} \quad (S^{R}_{p,q} \text{ for short}).$$

Moreover, we define the minimal operators

- $S^{\min}_{\tau,p_0,p_1,q_0,q_1,\Delta} : L^{p_0} \cap L^{p_1} \supseteq \mathrm{D}(S^{\min}_{\tau,p_0,p_1,q_0,q_1,\Delta}) \longrightarrow L^{q_0} \cap L^{q_1}$ ($S^{\min}_{\Delta}$ *for short*) *by*

$$S^{\min}_{\tau,p_0,p_1,q_0,q_1,\Delta} := \begin{cases} \overline{S^R_{\Delta}} & \text{if } 1 \le p_0,p_1,q_0,q_1 < \infty, \\ (S^{\max}_{\tau^*,q_0',q_1',p_0',p_1',\Sigma})' & \text{if } 1 < p_0,p_1,q_0,q_1 \le \infty. \end{cases}$$

- $S^{\min}_{\tau,p_0,p_1,q_0,q_1,\Sigma} : L^{p_0} + L^{p_1} \supseteq \mathrm{D}(S^{\min}_{\tau,p_0,p_1,q_0,q_1,\Sigma}) \longrightarrow L^{q_0} + L^{q_1}$ ($S^{\min}_{\Sigma}$ *for short*) *by*

$$S^{\min}_{\tau,p_0,p_1,q_0,q_1,\Sigma} := \begin{cases} \overline{S^R_{\Sigma}} & \text{if } 1 \le p_0,p_1,q_0,q_1 < \infty, \\ (S^{\max}_{\tau^*,q_0',q_1',p_0',p_1',\Delta})' & \text{if } 1 < p_0,p_1,q_0,q_1 \le \infty. \end{cases}$$

For $1 \le p,q < \infty$ *or* $1 < p,q \le \infty$, *define*

$$S^{\min}_{\tau,p,q} := S^{\min}_{\tau,p,p,q,q,\Delta} = S^{\min}_{\tau,p,p,q,q,\Sigma} \quad (S^{\min}_{p,q} \text{ for short}).$$

The differential operators $S^{\max}_{\tau,p,q}$, $S^{R}_{\tau,p,q}$ and $S^{\min}_{\tau,p,q}$ are the well-known maximal, restricted and minimal operators examined in [Gol66, Chapter VI].

It is possible to prove corresponding properties for the operators introduced in the previous definition as known for classical differential operators, see for instance [Gol66, Chapter VI]. In the following, we describe some of these properties and discuss further properties, which we need to show that the equivalent statements of Theorem 5.5 are fulfilled for the maximal and the minimal operator.

Let E, F be Banach spaces and $S : F' \supseteq \mathrm{D}(S) \longrightarrow E'$ be linear such that for all $0 \neq y \in F$, there exists $y' \in \mathrm{D}(S)$ with $\langle y, y'\rangle \neq 0$. Then the preconjugate ${}'S : E \supseteq \mathrm{D}({}'S) \longrightarrow F$ of S is defined by

$$\mathrm{D}({}'S) := \{x \in E : \text{ there exists } y \in F \text{ with } \langle x, Sy'\rangle = \langle y, y'\rangle \text{ for all } y' \in \mathrm{D}(S)\},$$

$${}'Sx := y \text{ for } x \in \mathrm{D}({}'S) \quad \text{if and only if} \quad \langle x, Sy'\rangle = \langle y, y'\rangle, \text{ for } y' \in \mathrm{D}(S).$$

Let $1 \le p \le \infty$. We denote by p' the number satisfying

$$1 = \frac{1}{p} + \frac{1}{p'},$$

where '$\frac{1}{\infty}$' is defined to be '0'. Obviously, $1 \le p' \le \infty$.

Theorem 7.2. *Assume* τ *is a differential expression on an open interval* I. *Then*

(i) $S^{\max}_{\tau^*,q_0',q_1',p_0',p_1',\Sigma} = (S^R_{\Delta})' = (\overline{S^R_{\Delta}})' = (S^{\min}_{\Delta})'$ *for* $1 \le p_0,p_1,q_0,q_1 < \infty$,
$S^{\max}_{\tau^*,q_0',q_1',p_0',p_1',\Sigma} = {}'(S^R_{\Delta})$ *for* $1 < p_0,p_1,q_0,q_1 \le \infty$,

(ii) $S^{\max}_{\tau^*,q_0',q_1',p_0',p_1',\Delta} = (S^R_{\Sigma})' = (\overline{S^R_{\Sigma}})' = (S^{\min}_{\Sigma})'$ *for* $1 \le p_0,p_1,q_0,q_1 < \infty$,
$S^{\max}_{\tau^*,q_0',q_1',p_0',p_1',\Delta} = {}'(S^R_{\Sigma})$ *for* $1 < p_0,p_1,q_0,q_1 \le \infty$.

Proof. Note that $L^{q_0'}+L^{q_1'}=(L^{q_0}\cap L^{q_1})'$ and $L^{q_0'}\cap L^{q_1'}=(L^{q_0}+L^{q_1})'$ by [BL76, p. 32, Theorem 2.7.1]. Since for $1\le p_0,p_1,q_0,q_1<\infty$, the set of all infinitely differentiable functions f on I with compact support is in the domain of the restricted operators, these operators are densely defined. Then we conclude similarly as in the proof of [Gol66, p. 130, Theorem VI.1.9]. □

In Theorem 7.3 and Theorem 7.5, we discuss further connections between the maximal and the minimal operators.

Let $1\le p_0,p_1\le\infty$ and I be an open interval. Assume $f\in L^{p_0}\cap L^{p_1}$ and $g\in L^{p_0'}+L^{p_1'}$. From the classical Hölder inequality, we obtain the extended Hölder inequality

$$\int_I |f(x)g(x)|dx \le \|f\|_{L^{p_0}\cap L^{p_1}}\|g\|_{L^{p_0'}+L^{p_1'}}. \tag{3}$$

Now, it is possible to show the 'if-implication' of the following theorem with similar arguments as in the proof of [Gol66, p. 139, Lemma VI.2.9]

Theorem 7.3. *Let $1\le p_0,p_1,q_0,q_1<\infty$ or $1<p_0,p_1,q_0,q_1\le\infty$. Assume τ is a differential expression of order n on $I=(b,c)$, where $-\infty\le b<c\le\infty$. Then $f\in \mathrm{D}(S_\Delta^{\min})$ ($f\in \mathrm{D}(S_\Sigma^{\min})$) if and only if $f\in \mathrm{D}(S_\Delta^{\max})$ ($f\in \mathrm{D}(S_\Sigma^{\max})$) and*

$$\lim_{\tilde b\downarrow b,\ \tilde c\uparrow c}\left[\sum_{k=1}^{n}\sum_{j=0}^{k-1}(-1)^j(a_kg)^{(j)}f^{(k-1-j)}\right]_{\tilde b}^{\tilde c}=0 \tag{4}$$

for all $g\in \mathrm{D}(S^{\max}_{\tau^\star,q_0',q_1',p_0',p_1',\Sigma})$ ($g\in \mathrm{D}(S^{\max}_{\tau^\star,q_0',q_1',p_0',p_1',\Delta})$).

Proof. We only prove the 'only-if-implication'. Assume $f\in \mathrm{D}(S_\Delta^{\max})$ and (4) holds. Set

$$S^{\max}_{\star,\Sigma}:=S^{\max}_{\tau^\star,q_0',q_1',p_0',p_1',\Sigma}.$$

The Lagrange formula, see [Gol66, p. 130, Lemma VI.1.8], and the extended Hölder inequality, see (3), imply that

$$\langle\tau f,g\rangle=\int_b^c(\tau f)g\,dx=\int_b^c f(\tau^\star g)dx=\langle f,\tau^\star g\rangle \tag{5}$$

for all $g\in \mathrm{D}(S^{\max}_{\star,\Sigma})$.

Let $1\le p_0,p_1,q_0,q_1<\infty$. Then $(S_\Delta^{\min})'=S^{\max}_{\star,\Sigma}$, see Theorem 7.2 (i). From (5), it follows that $f\in \mathrm{D}('((S_\Delta^{\min})'))$. Since $S_\Delta^{\min}$ is closed and densely defined, we obtain from [Gol66, p. 127, Lemma VI.1.4] that $f\in \mathrm{D}(S_\Delta^{\min})$.

Let $1<p_0,p_1,q_0,q_1\le\infty$. Then (5) implies that $f\in \mathrm{D}((S^{\max}_{\star,\Sigma})')=\mathrm{D}(S_\Delta^{\min})$. We conclude as above for $S_\Sigma^{\min}$. □

Similar conclusions as in the proof of [Wei03, p. 39, Satz 13.5] (cf. the proof of [Gol66, p. 140, Theorem VI.3.1]) yield the next theorem.

Theorem 7.4. *Let $1\le p_0,p_1,q_0,q_1\le\infty$ and τ be a differential expression of order n on $I=(b,c)$ such that $b,c\in\mathbb{R}$ and $a_n(b)\ne 0$, $a_n(c)\ne 0$. For $f\in \mathrm{D}(S_\Sigma^{\max})$ and*

$k \in \{0, 1, \ldots, n-1\}$, *the limits*

$$\lim_{\tilde{b}\downarrow b} f^{(k)}(\tilde{b}), \quad \lim_{\tilde{c}\uparrow c} f^{(k)}(\tilde{c})$$

exist. The same holds for $f \in \mathrm{D}(S_{\Delta}^{\max})$.

Moreover, suppose $1 \le p_0, p_1, p, q_0, q_1, q \le \infty$ and τ is a differential expression of order n on $I = (b, c)$ with $b, c \in \mathbb{R}$ and $a_n(b) \neq 0$, $a_n(c) \neq 0$. Then Theorem 7.4 implies

$$\begin{aligned}&\{f \in A_n : \lim_{\tilde{b}\downarrow b} f(\tilde{b}) \text{ and } \lim_{\tilde{c}\uparrow c} f(\tilde{c}) \text{ exist and } \tau f = 0\}\\ &= \mathrm{N}(S_{\Delta}^{\max}) = \mathrm{N}(S_{p,q}^{\max}) = \mathrm{N}(S_{\Sigma}^{\max}).\end{aligned} \tag{6}$$

Theorem 7.5. *Let* $1 \le p_0, p_1, q_0, q_1 < \infty$ *or* $1 < p_0, p_1, q_0, q_1 \le \infty$ *and* τ *be a differential expression of order* n *on* $I = (b, c)$ *such that* $b, c \in \mathbb{R}$ *and* $a_n(b) \neq 0$, $a_n(c) \neq 0$. *Then* $f \in \mathrm{D}(S_{\Delta}^{\min})$ *(*$f \in \mathrm{D}(S_{\Sigma}^{\min})$*) if and only if* $f \in \mathrm{D}(S_{\Delta}^{\max})$ *(*$f \in \mathrm{D}(S_{\Sigma}^{\max})$*) and*

$$\lim_{\tilde{b}\downarrow b} f^{(k)}(\tilde{b}) = 0 \text{ and } \lim_{\tilde{c}\uparrow c} f^{(k)}(\tilde{c}) = 0 \text{ for } k \in \{0, 1, \ldots, n-1\}. \tag{7}$$

Proof. The 'if-implication' follows similarly as in the proof of [Gol66, p. 139, Lemma VI.2.9].

Conversely, assume $f \in \mathrm{D}(S_{\Delta}^{\max})$ and (7) hold. Theorem 7.4 implies that

$$\lim_{\tilde{b}\downarrow b,\ \tilde{c}\uparrow c} \left[\sum_{k=1}^{n} \sum_{j=0}^{k-1} (-1)^j (a_k g)^{(j)} f^{(k-1-j)}\right]_{\tilde{b}}^{\tilde{c}} = 0$$

for all $g \in \mathrm{D}(S_{\tau^*, q_0', q_1', p_0', p_1', \Sigma}^{\max})$. Thus $f \in \mathrm{D}(S_{\Delta}^{\min})$ by Theorem 7.3.

It is possible to conclude similarly for $S_{\Sigma}^{\min}$. □

In Theorem 7.6 and Corollary 7.7, we describe the nullities and the deficiencies of the maximal and the minimal operators.

Theorem 7.6. *Let* $1 \le p_0, p_1, q_0, q_1 \le \infty$ *and* τ *be a differential expression of order* n *on* $I = (b, c)$ *with* $b, c \in \mathbb{R}$ *and* $a_n(b) \neq 0$, $a_n(c) \neq 0$. *For* $j \in \{\Delta, \Sigma\}$, *the operator* $S_j^{\max}$ *is surjective with* $\mathrm{nul}(S_j^{\max}) = n$.

Proof. Since $a_n(t) > 0$, it is possible to assume without loss of generality that $a_n(t) = 1$ for all $t \in [b, c]$. Let $g_{\Delta} \in L^{q_0} \cap L^{q_1}$, $g_{\Sigma} \in L^{q_0} + L^{q_1}$ and $j \in \{\Delta, \Sigma\}$. Then $g_j \in L^1$. We know from the proof of (i) of [Gol66, p. 140, Theorem VI.3.1] that there exists $f_j \in A_n$ such that f_j can be extended to a continuous function on $[b, c]$ and

$$\tau f_j = g_j.$$

Hence $f_j \in \mathrm{D}(S_j^{\max})$. Therefore $S_{\Delta}^{\max}$ and $S_{\Sigma}^{\max}$ are surjective.

It holds $\mathrm{nul}(S_{p_0,q_0}^{\max}) = n$, see [Gol66, p. 140, Theorem VI.3.1]. Then the theorem follows from (6). □

From Theorem 7.2 and Theorem 7.6, we obtain the next corollary.

Corollary 7.7. *Let $1 \le p_0, p_1, q_0, q_1 < \infty$ or $1 < p_0, p_1, q_0, q_1 \le \infty$. Assume τ is a differential expression of order n on $I = (b,c)$ with $b, c \in \mathbb{R}$ and $a_n(b) \neq 0$, $a_n(c) \neq 0$. For $j \in \{\Delta, \Sigma\}$, the operator $S_j^{\min}$ is injective with closed range and $\operatorname{def}(S_j^{\min}) = n$.*

The next theorem shows that the restricted, maximal and minimal operators on different L^p-spaces corresponding to the same differential expression are indeed an example for the abstract theory studied in this paper.

Theorem 7.8. *Suppose $1 \le p_0 \le p \le p_1 < \infty$ and $1 \le q_0 \le q \le q_1 < \infty$. Let τ be a differential expression on an open interval I. Then we have*

(i)
$$\check{(S_\Sigma^R)}_{L^p, L^q} = S_{p,q}^R,$$
$$\check{(S_{p,q}^R)}_{L^{p_0} \cap L^{p_1}, L^{q_0} \cap L^{q_1}} = S_\Delta^R,$$
$$\check{(S_\Sigma^R)}_{L^{p_0} \cap L^{p_1}, L^{q_0} \cap L^{q_1}} = S_\Delta^R,$$

(ii)
$$\check{(S_\Sigma^{\max})}_{L^p, L^q} = S_{p,q}^{\max},$$
$$\check{(S_{p,q}^{\max})}_{L^{p_0} \cap L^{p_1}, L^{q_0} \cap L^{q_1}} = S_\Delta^{\max},$$
$$\check{(S_\Sigma^{\max})}_{L^{p_0} \cap L^{p_1}, L^{q_0} \cap L^{q_1}} = S_\Delta^{\max}.$$

(iii) *If $I = (b,c)$ with $b, c \in \mathbb{R}$ and $a_n(b) \neq 0$, $a_n(c) \neq 0$, then*
$$\check{(S_\Sigma^{\min})}_{L^p, L^q} = S_{p,q}^{\min},$$
$$\check{(S_{p,q}^{\min})}_{L^{p_0} \cap L^{p_1}, L^{q_0} \cap L^{q_1}} = S_\Delta^{\min},$$
$$\check{(S_\Sigma^{\min})}_{L^{p_0} \cap L^{p_1}, L^{q_0} \cap L^{q_1}} = S_\Delta^{\min}.$$

Proof. Let $j \in \{0,1\}$. Obviously, $L^{p_0} \cap L^{p_1} \subseteq L^{p_j} \subseteq L^{p_0} + L^{p_1}$ with continuous inclusions. If $p_0 < p < p_1$, then we know that $L^{p_0} \cap L^{p_1} \subseteq L^p \subseteq L^{p_0} + L^{p_1}$ with continuous inclusions from [BL76, p. 106, Theorem 5.1.1]. Corresponding results hold for L^{q_j} and L^q. Therefore the induced operators are well defined. The proof of the equalities in (i) and (ii) is straightforward. From (ii) and Theorem 7.5, we obtain (iii). □

Now, it follows that the equivalent statements of Theorem 5.5 are fulfilled for the maximal and the minimal operator. Let $1 < p_0, p_1, q_0, q_1 < \infty$ and τ be a differential expression of order n on $I = (b,c)$ such that $b, c \in \mathbb{R}$ and $a_n(b) \neq 0$, $a_n(c) \neq 0$. Theorem 7.2 and Theorem 7.8 imply

$$(S_{p_j,q_j}^{\max})' = S_{\tau^\star, q_j', p_j'}^{\min} = \check{(S_{\tau^\star, q_0', q_1', p_0', p_1', \Sigma}^{\min})}_{L^{q_j'}, L^{p_j'}} = \check{((S_\Delta^{\max})')}_{L^{q_j'}, L^{p_j'}},$$
$$(S_{p_j,q_j}^{\min})' = S_{\tau^\star, q_j', p_j'}^{\max} = \check{(S_{\tau^\star, q_0', q_1', p_0', p_1', \Sigma}^{\max})}_{L^{q_j'}, L^{p_j'}} = \check{((S_\Delta^{\min})')}_{L^{q_j'}, L^{p_j'}}.$$

Hence Theorem 2.6 yields that $\mathrm{D}(S_\Delta^{\max})$ is a core of $S_{p_0,q_0}^{\max}$ and $S_{p_1,q_1}^{\max}$ and that $\mathrm{D}(S_\Delta^{\min})$ is a core of $S_{p_0,q_0}^{\min}$ and $S_{p_1,q_1}^{\min}$. From Theorem 7.6, Corollary 7.7 and Theorem 7.8, we see that the equivalent statements of Theorem 5.5 are fulfilled for the maximal and the minimal operator.

The next corollary shows that the 'sum' of two maximal (minimal) operators is equal to the maximal (minimal) operator on the sum of two L^p-spaces.

Corollary 7.9. *Let τ be a differential expression of order n on $I = (b,c)$ with $b, c \in \mathbb{R}$ and $a_n(b) \neq 0$, $a_n(c) \neq 0$. Then*

$$(S_{p_0,q_0}^{\max}, S_{p_1,q_1}^{\max})_\Sigma = S_\Sigma^{\max} \quad \textit{for } 1 \le p_0, p_1, q_0, q_1 < \infty,$$

$$(S_{p_0,q_0}^{\min}, S_{p_1,q_1}^{\min})_\Sigma = S_\Sigma^{\min} \quad \textit{for } 1 < p_0, p_1, q_0, q_1 < \infty.$$

Proof. Proposition 3.3 (i) and Theorem 7.8 (ii) imply that

$$(S_{p_0,q_0}^{\max}, S_{p_1,q_1}^{\max})_\Sigma \subseteq S_\Sigma^{\max}.$$

From Proposition 3.4 and Theorem 7.6, we obtain that the kernels and the ranges of these operators are equal. Then $(S_{p_0,q_0}^{\max}, S_{p_1,q_1}^{\max})_\Sigma = S_\Sigma^{\max}$ follows from Remark 3.7.

Since the equivalent statements of Theorem 5.5 are fulfilled for the minimal operator, we know that there exists a finite-dimensional subspace $H \subseteq F_\Delta$ such that

$$\mathrm{R}(S_{p_0,q_0}^{\min}) \oplus H = F_0, \quad \mathrm{R}((S_{p_0,q_0}^{\min}, S_{p_1,q_1}^{\min})_\Sigma) \oplus H = F_\Sigma,$$

see Theorem 5.5. Then Corollary 7.7 yields $\mathrm{R}(S_\Sigma^{\min}) \oplus H = F_\Sigma$. Thus

$$\mathrm{R}((S_{p_0,q_0}^{\min}, S_{p_1,q_1}^{\min})_\Sigma) = \mathrm{R}(S_\Sigma^{\min}).$$

Then the corollary follows from Remark 3.7 and Corollary 7.7. □

Acknowledgment

I would like to thank Prof. Dr. K.-H. Förster for many valuable suggestions and discussions.

References

[BL76] Bergh, J.; Löfström, J.: *Interpolation spaces.* Springer, 1976.

[Che01] Chen, Y.: The spectra of closed interpolation operators. *Integr. equ. oper. theory* 41 (2001), pp. 255–263.

[Dav07] Davies, E.B.: *Linear operators and their spectra.* Cambridge Univ. Pr., 2007.

[DS67] Dunford, N.; Schwartz, J.T.: *Linear operators, Part I: General theory.* 4. print. Wiley, 1967.

[EN00] Engel, K.-J.; Nagel, R.: *One-parameter semigroups for linear evolution equations.* Springer, 2000.

[Gol66] Goldberg, S.: *Unbounded linear operators.* McGRAW-HILL, 1966.

[Gri66] Grisvard, P.: Commutativité de deux foncteurs d'interpolation et applications. *J. Math. pures et appl.* 45 (1966), pp. 143–206.

[HT56] Halberg, C.J.A.; Taylor, A.E.: On the spectra of linked operators. *Pacific J. Math.* 6 (1956), pp. 283–290.

[Jör82] Jörgens, K.: *Linear integral operators.* Pitman, 1982.

[Kat66] Kato, T.: *Perturbation theory for linear operators.* Springer, 1966.

[Kra96] Krause, M.: *Fredholm theory in paraalgebras.* Tectum, 1996.

[Sch87] Schindler, K.: *Spektrale Eigenschaften linearer Operatoren auf Interpolationsräumen*, Universität des Saarlandes, Diss., 1987.

[Wei03] Weidmann, J.: *Lineare Operatoren in Hilberträumen, Teil II: Anwendungen.* Teubner, 2003.

[Zaf73] Zafran, M.: Multipliers, spectral theory, and the interpolation of closed operators. *Michigan Math. J.* 20 (1973), pp. 361–372.

Kerstin Günther
Technische Universität Berlin
MA 6-4
Str. des 17. Juni 136
D-10623 Berlin, Germany
e-mail: guenther@math.tu-berlin.de

Operator Theory:
Advances and Applications, Vol. 198, 211–227

Domain and Range Descriptions for Adjoint Relations, and Parallel Sums and Differences of Forms

Seppo Hassi, Zoltán Sebestyén and Henk de Snoo

To the memory of Peter Jonas

Abstract. The adjoint of a linear operator or relation from a Hilbert space $\mathfrak{H}$ to a Hilbert space $\mathfrak{K}$ is a closed linear relation. The domain and the range of the adjoint are characterized in terms of certain mappings defined on $\mathfrak{K}$ and $\mathfrak{H}$, respectively. These characterizations are applied to contractions between Hilbert spaces and to the form domains and ranges of the Friedrichs and Kreĭn-von Neumann extensions of a nonnegative operator or relation. Furthermore these characterizations are used to introduce and derive properties of the parallel sum and the parallel difference of a pair of forms on a linear space.

Mathematics Subject Classification (2000). 47A05, 47A06, 47A07, 47A64.

Keywords. Range, domain, nonnegative sesquilinear form, parallel sum, parallel difference.

1. Introduction

Let T be a linear operator or a linear relation (multivalued operator) from a Hilbert space $\mathfrak{H}$ to a Hilbert space $\mathfrak{K}$, which is not necessarily closed or densely defined. The adjoint T^* of T is a closed linear relation from $\mathfrak{K}$ to $\mathfrak{H}$. The present paper is concerned with the characterization of the elements in $\operatorname{dom} T^*$ and $\operatorname{ran} T^*$. In fact it will be shown that an element $h \in \mathfrak{K}$ belongs to $\operatorname{dom} T^*$ if and only if

$$\sup \{ (h,g) + (g,h) - (f,f) : \{f,g\} \in T \} < \infty, \tag{1.1}$$

The support by a bilateral agreement between the Eötvös Loránd University in Budapest and the University of Groningen is gratefully acknowledged. Part of the work was done while the third author was a Mercator visiting professor at the TU Berlin.

and that an element $k \in \mathfrak{H}$ belongs to $\operatorname{ran} T^*$ if and only if

$$\sup \{ (f,k) + (k,f) - (g,g) : \{f,g\} \in T \} < \infty. \tag{1.2}$$

The formulas (1.1) and (1.2) show the advantage of the language of relations: the formula (1.2) is in fact the same as the formula (1.1) when the relation T is replaced by its formal inverse T^{-1}. Moreover, it will be shown that (1.1) is equivalent to

$$\sup \{ |(g,h)|^2 : \{f,g\} \in T, (f,f) \leq 1 \} < \infty, \tag{1.3}$$

and that (1.2) is equivalent to

$$\sup \{ |(f,k)|^2 : \{f,g\} \in T, (g,g) \leq 1 \} < \infty. \tag{1.4}$$

Again the relation between (1.3) and (1.4) via the formal inverse of T is evident. The last two characterizations are versions of results which in their original form go back to Shmulyan [18]; cf. [9], [10], [14], [17].

Apart from their own intrinsic simplicity there are also some interesting applications of the above characterizations. First of all (1.1) and (1.2) give the range characterization of contractions between Hilbert spaces in terms of their defect operators (and conversely) as shown by P.A. Fillmore and J.T. Williams [9], motivated by similar results of L. de Branges and J. Rovnyak [7]. Another application is in the characterization of the form domains and ranges of the Friedrichs and Kreĭn-von Neumann extensions of a nonnegative operator or relation in a Hilbert space, cf., e.g., [5], [10], [11], [14], [17].

A further application of the above characterizations is in the theory of parallel sums and parallel differences of forms on a linear space, cf. [13]. This theory was developed parallel to the potential theoretic treatment for operators as developed by Eriksson and Leutwiler [8]. The original introduction of parallel sums for nonnegative operators in finite-dimensional spaces is due to [1], [2], [3]; see [9] for the infinite-dimensional case. Further developments for parallel sums and parallel differences in infinite-dimensional spaces were given by E.L. Pekarev and Yu.L. Shmulyan [15], [16]. In the present paper there is another approach to parallel sums and differences of forms. The underlying idea behind the present approach is that the pair of forms induces a linear relation between the Hilbert spaces generated by the forms, which makes it possible to apply the range characterization appearing in the first part of the paper. As was shown in [13] the parallel sum and difference of forms play an essential role in the Lebesgue type decomposition of one form with respect to another form.

The paper is organized as follows. In Section 2 there is a description of the domain and the range of T^* when T is a linear relation between Hilbert spaces $\mathfrak{H}$ and $\mathfrak{K}$. In Section 3 it is shown how these descriptions imply the range descriptions of contractions in terms of their defect operators (and conversely). The implications towards the Friedrichs and the Kreĭn-von Neumann extensions of a nonnegative operator or relation can be found in Section 4. The introduction of parallel sums and differences for forms and implications in that setting are given in Section 5.

2. Domain and range characterizations of adjoint relations

Let T be a, not necessarily closed, linear relation from a Hilbert space $\mathfrak{H}$ to a Hilbert space $\mathfrak{K}$. Define the mappings $\mathfrak{l} : \mathfrak{K} \to \mathbb{R} \cup \{\infty\}$ and $\mathfrak{m} : \mathfrak{K} \to \mathbb{R} \cup \{\infty\}$ by

$$\mathfrak{l}[h] = \sup\{\, (h,g) + (g,h) - (f,f) : \{f,g\} \in T \,\}, \quad h \in \mathfrak{K},$$

and

$$\mathfrak{m}[h] = \sup\{\, |(g,h)|^2 : \{f,g\} \in T, \, (f,f) \le 1 \,\}, \quad h \in \mathfrak{K}.$$

Define the mappings $\mathfrak{l}_* : \mathfrak{H} \to \mathbb{R} \cup \{\infty\}$ and $\mathfrak{m}_* : \mathfrak{H} \to \mathbb{R} \cup \{\infty\}$ by

$$\mathfrak{l}_*[k] = \sup\{\, (f,k) + (k,f) - (g,g) : \{f,g\} \in T \,\}, \quad k \in \mathfrak{H},$$

and

$$\mathfrak{m}_*[k] = \sup\{\, |(f,k)|^2 : \{f,g\} \in T, \, (g,g) \le 1 \,\}, \quad k \in \mathfrak{H}.$$

The mappings $\mathfrak{l}$ and $\mathfrak{m}$ will be used to characterize the elements in the domain of the adjoint relation T^*, and the mappings $\mathfrak{l}_*$ and $\mathfrak{m}_*$ will be used to characterize the elements in the range of the adjoint relation T^*.

Lemma 2.1. *Let T be a linear relation from a Hilbert space $\mathfrak{H}$ to a Hilbert space $\mathfrak{K}$. The mappings $\mathfrak{l}$ and $\mathfrak{m}$ are nonnegative and satisfy*

$$\mathfrak{l}[h] \ge \mathfrak{m}[h] \ge 0, \quad h \in \mathfrak{K}. \tag{2.1}$$

Similarly, the mappings $\mathfrak{l}_*$ and $\mathfrak{m}_*$ are nonnegative and satisfy

$$\mathfrak{l}_*[k] \ge \mathfrak{m}_*[k] \ge 0, \quad k \in \mathfrak{H}. \tag{2.2}$$

Proof. By definition, the mappings $\mathfrak{m}$ and $\mathfrak{m}_*$ are nonnegative. To see that the mapping $\mathfrak{l}$ and $\mathfrak{l}_*$ are nonnegative, consider the element $\{f,g\} = \{0,0\} \in T$ in their definitions.

Now let $h \in \mathfrak{K}$. Then, for any $\{f,g\} \in T$ and all $\lambda \in \mathbb{C}$, one has by definition:

$$\mathfrak{l}[h] \ge \bar{\lambda}(h,g) + \lambda(g,h) - |\lambda|^2(f,f).$$

Now $(g,h) = e^{i\varphi}|(g,h)|$ for some $\varphi \in \mathbb{R}$ and thus $(h,g) = e^{-i\varphi}|(g,h)|$. In particular, let $\lambda = e^{-i\varphi}t$ with $t \in \mathbb{R}$. Then it follows for all $t \in \mathbb{R}$ that:

$$\mathfrak{l}[h] \ge 2t|(g,h)| - t^2(f,f).$$

As a consequence, one concludes that if $\mathfrak{l}[h] < \infty$, then

$$|(g,h)|^2 - \mathfrak{l}[h](f,f) \le 0,$$

or, equivalently, $\mathfrak{m}[h] \le \mathfrak{l}[h] < \infty$. This proves the inequalities (2.1).

To get the inequalities in (2.2) apply the first part of the lemma to the inverse T^{-1}. Then the formulas and inequalities (2.2) for $\mathfrak{l}_*[k]$ and $\mathfrak{m}_*[k]$ are obtained by evaluating $\mathfrak{l}[h]$ and $\mathfrak{m}[h]$ with $h = k$ and making use of the fact that $\{f,g\} \in T$ if and only if $\{g,f\} \in T^{-1}$. □

The next lemma gives a subspace of $\mathfrak{K}$ on which the mappings $\mathfrak{l}$ and $\mathfrak{m}$ are finite and equal, and a subspace of $\mathfrak{H}$ on which the mappings $\mathfrak{l}_*$ and $\mathfrak{m}_*$ are finite and equal.

Lemma 2.2. *Let T be a linear relation from a Hilbert space $\mathfrak{H}$ to a Hilbert space $\mathfrak{K}$, let P_d be the orthogonal projection from $\mathfrak{H}$ onto $\overline{\operatorname{dom}}\, T$, and let P_r be the orthogonal projection from $\mathfrak{K}$ onto $\overline{\operatorname{ran}}\, T$. Then for $\{h,k\} \in T^*$ the following quantities are equal:*
(i) $(P_d k, P_d k)$;
(ii) $\mathfrak{l}[h]$;
(iii) $\mathfrak{m}[h]$.
Moreover, for $\{h,k\} \in T^*$ the following quantities are equal:
(iv) $(P_r h, P_r h)$;
(v) $\mathfrak{l}_*[k]$;
(vi) $\mathfrak{m}_*[k]$.

Proof. Note that $\{h,k\} \in T^*$ means that $(k,f) = (h,g)$ holds for all $\{f,g\} \in T$. The equality of the first three quantities can now be obtained as follows.

Equality of (i) and (ii): If $\{h,k\} \in T^*$, then

$$\begin{aligned}((I-P_d)k,(I-P_d)k) &= \inf\{\,(k-f,k-f) : f \in \operatorname{dom} T\,\}\\ &= (k,k) + \inf\{\,-(k,f)-(f,k)+(f,f) : f \in \operatorname{dom} T\,\}\\ &= (k,k) - \sup\{\,(k,f)+(f,k)-(f,f) : f \in \operatorname{dom} T\,\}\\ &= (k,k) - \sup\{\,(h,g)+(g,h)-(f,f) : \{f,g\} \in T\,\}.\end{aligned}$$

Equality of (i) and (iii): If $\{h,k\} \in T^*$, then

$$\begin{aligned}(P_d k, P_d k) &= \sup\{\,|(f,P_d k)|^2 : f \in \operatorname{dom} T,\ (f,f) \le 1\,\}\\ &= \sup\{\,|(f,k)|^2 : f \in \operatorname{dom} T,\ (f,f) \le 1\,\}\\ &= \sup\{\,|(g,h)|^2 : \{f,g\} \in T,\ (f,f) \le 1\,\}.\end{aligned}$$

This proves the first part of the lemma.

As to the second part, observe that $(T^{-1})^* = (T^*)^{-1}$ and $\overline{\operatorname{ran}}\, T = \overline{\operatorname{dom}}\, T^{-1}$. Now, applying the arguments in the first part of the proof to the inverse T^{-1}, one obtains the equality of the quantities (iv)–(vi) in the second part of the lemma. □

The following theorem provides a description of $\operatorname{dom} T^*$ in terms of the mappings $\mathfrak{l}$ and $\mathfrak{m}$, and a description of $\operatorname{ran} T^*$ in terms of the mappings $\mathfrak{l}_*$ and $\mathfrak{m}_*$.

Theorem 2.3. *Let T be a linear relation from a Hilbert space $\mathfrak{H}$ to a Hilbert space $\mathfrak{K}$. For an element $h \in \mathfrak{K}$ the following statements are equivalent:*
(i) $h \in \operatorname{dom} T^*$;
(ii) $\mathfrak{l}[h] < \infty$;
(iii) $\mathfrak{m}[h] < \infty$.
If any of the conditions (i), (ii), *or* (iii) *hold, then there exists a unique $k \in \overline{\operatorname{dom}}\, T$ such that $\{h,k\} \in T^*$ and $\mathfrak{l}[h] = \mathfrak{m}[h] = (k,k)$. Furthermore, for an element $k \in \mathfrak{H}$ the following statements are equivalent:*
(iv) $k \in \operatorname{ran} T^*$;
(v) $\mathfrak{l}_*[k] < \infty$;
(vi) $\mathfrak{m}_*[k] < \infty$.

If any of the conditions (iv), (v), *or* (vi) *hold, then there exists a unique* $h \in \overline{\operatorname{ran}} T$ *such that* $\{h, k\} \in T^*$ *and* $\mathfrak{l}_*[k] = \mathfrak{m}_*[k] = (h, h)$.

Proof. Assume that $h \in \operatorname{dom} T^*$. Then there exists an element $\tilde{k} \in \operatorname{ran} T^*$, such that $\{h, \tilde{k}\} \in T^*$. Decompose $\tilde{k} = k + k_\infty$ according to the decomposition $\mathfrak{H} = \overline{\operatorname{dom}} T \oplus \operatorname{mul} T^*$, then $\{h, k\} \in T^*$ and $k \in \overline{\operatorname{dom}} T$. Clearly, the element k with these properties is uniquely determined.

Now the equivalence of (i)–(iii) is shown.

(i) $\Rightarrow$ (ii), (iii) This follows from Lemma 2.2.

(ii) $\Rightarrow$ (iii) This follows from Lemma 2.1.

(iii) $\Rightarrow$ (i) The linear relation $M_h : \mathfrak{H} \to \mathbb{C}$ defined by

$$M_h = \{ \{f, (g, h)\} : \{f, g\} \in T \}$$

is single-valued, since $f = 0$ and the assumption $\mathfrak{m}[h] < \infty$ imply that $(g, h) = 0$. It follows that M_h is a continuous linear functional from $\operatorname{dom} T$ to $\mathbb{C}$ and uniquely extends to a continuous linear functional from $\overline{\operatorname{dom}} T$ to $\mathbb{C}$ with norm $\sqrt{\mathfrak{m}[h]}$. Hence there exists a unique element $k \in \overline{\operatorname{dom}} T$, such that

$$(g, h) = (f, k), \quad \{f, g\} \in T,$$

which implies that $\{h, k\} \in T^*$. Thus (i) is obtained.

Assume that any of the conditions (i), (ii), or (iii) hold, i.e., assume that $h \in \operatorname{dom} T^*$ and let $\{h, k\} \in T^*$ with $k \in \overline{\operatorname{dom}} T$. Then it follows from Lemma 2.2 that $\mathfrak{l}[h] = \mathfrak{m}[h] = (k, k)$. This proves the first part of the theorem.

The second part of the theorem is again obtained by applying the first part to the inverse; cf. the proofs of Lemmas 2.1, 2.2. □

Remark 2.4. The unique element $k \in \overline{\operatorname{dom}} T$ such that $\{h, k\} \in T^*$ can be written as $k = (T^*)_s h$, where $(T^*)_s = P_s T^*$ stands for the operator part of the closed relation T^*; here P_s stands for the orthogonal projection onto $\overline{\operatorname{dom}} T = (\operatorname{mul} T^*)^\perp$. Therefore, one can write

$$\sqrt{\mathfrak{m}[h]} = \sqrt{\mathfrak{l}[h]} = \|k\| = \|(T^*)_s h\|. \tag{2.3}$$

Using the definition of $\mathfrak{m}[h]$ one obtains from Theorem 2.3 the following statement: $h \in \operatorname{dom} T^*$ precisely when

$$|(g, h)|^2 \le \mathfrak{m}[h] \, \|f\|^2, \quad \{f, g\} \in T, \tag{2.4}$$

holds with $\mathfrak{m}[h] < \infty$. In particular, if T is an operator (2.4) takes the form

$$|(Tf, h)|^2 \le \mathfrak{m}[h] \, \|f\|^2, \quad f \in \operatorname{dom} T. \tag{2.5}$$

Likewise, the unique element $h \in \overline{\operatorname{ran}} T$ such that $\{h, k\} \in T^*$ can be written as $h = Q_s u$, where $\{u, k\} \in T^*$ and Q_s stands for the orthogonal projection onto $\overline{\operatorname{ran}} T = (\ker T^*)^\perp$. Now one gets from Theorem 2.3 the following statement: $k \in \operatorname{ran} T^*$ precisely when

$$|(f, k)|^2 \le \mathfrak{m}_*[k] \, \|g\|^2, \quad \{f, g\} \in T, \tag{2.6}$$

holds with $\mathfrak{m}_*[k] < \infty$. In particular, if T is an operator, then (2.6) takes the form

$$|(f,k)|^2 \leq \mathfrak{m}_*[k]\, \|Tf\|^2, \quad f \in \operatorname{dom} T. \tag{2.7}$$

If T is densely defined, then T^* is an operator and one can write $k = T^*h$. This clarifies the equalities $\mathfrak{m}[h] = \|k\|^2$ and $\mathfrak{m}_*[k] = \|h\|^2$ stated in Theorem 2.3.

In this form the characterization of $\operatorname{ran} T^*$ for a bounded operator T goes back to Shmul'yan [18]. For densely defined operators the range characterization was given in [17] and for nondensely operators and linear relations in [10].

Corollary 2.5. *With the notations as in Theorem* 2.3 *the following statements and equivalent:*

(i) $h \in \ker T^*$;
(ii) $\mathfrak{l}[h] = 0$;
(iii) $\mathfrak{m}[h] = 0$.

Likewise, the following statements are equivalent:

(iv) $k \in \operatorname{mul} T^*$;
(v) $\mathfrak{l}_*[k] = 0$;
(vi) $\mathfrak{m}_*[k] = 0$.

Proof. (i) $\Rightarrow$ (ii), (iii) Assume that $h \in \ker T^*$, so that $\{h, 0\} \in T^*$. Now Theorem 2.3 with $k = 0$ implies that $\mathfrak{l}[h] = \mathfrak{m}[h] = (k,k) = 0$.

(ii) $\Rightarrow$ (iii) This follows from Lemma 2.1.

(iii) $\Rightarrow$ (i) Assume that $\mathfrak{m}[h] = 0$. Then the definition of $\mathfrak{m}[h]$ shows that $h \in (\operatorname{ran} T)^\perp = \ker T^*$.

This proves the first part of the theorem. The proof of the second part is completely analogous. □

3. Range characterizations for contractions and their defect operators

This section shows how the domain and range descriptions of the adjoint of a linear relation in a Hilbert space yield descriptions of ranges of contractions and of the associated defect operators appearing in functional models for contractions.

Let $\mathfrak{H}$ and $\mathfrak{K}$ be Hilbert spaces and denote the bounded linear operators from all of $\mathfrak{H}$ to $\mathfrak{K}$ by $\mathbf{B}(\mathfrak{H}, \mathfrak{K})$. Let $T \in \mathbf{B}(\mathfrak{H}, \mathfrak{K})$ be a contraction, in which case also $T^* \in \mathbf{B}(\mathfrak{K}, \mathfrak{H})$ is a contraction. The corresponding defect operators $D \in \mathbf{B}(\mathfrak{H})$ and $D_* \in \mathbf{B}(\mathfrak{K})$, defined by

$$D = (I - T^*T)^{\frac{1}{2}}, \quad D_* = (I - TT^*)^{\frac{1}{2}},$$

are nonnegative contractions. The defect operators satisfy

$$TD = D_*T, \quad DT^* = T^*D_*, \tag{3.1}$$

cf. [20]. It is straightforward to see that the commutation relations in (3.1) lead to invariance results for $\operatorname{ran} T$ and $\operatorname{ran} T^*$:

$$\psi \in \operatorname{ran} T \Leftrightarrow D_* \psi \in \operatorname{ran} T \quad \text{and} \quad \psi \in \operatorname{ran} T^* \Leftrightarrow D\psi \in \operatorname{ran} T^*, \tag{3.2}$$

and to invariance results for $\operatorname{ran} D$ and $\operatorname{ran} D_*$:

$$\psi \in \operatorname{ran} D \Leftrightarrow T\psi \in \operatorname{ran} D_* \quad \text{and} \quad \psi \in \operatorname{ran} D_* \Leftrightarrow T^*\psi \in \operatorname{ran} D. \tag{3.3}$$

The following lemmas are immediate consequences of Theorem 2.3.

Lemma 3.1. *Let $T \in \mathbf{B}(\mathfrak{H}, \mathfrak{K})$ be a contraction. Then*

$$D\psi \in \operatorname{ran} T^* \quad \Leftrightarrow \quad \sup \left\{ \|\psi + Df\|^2 - \|f\|^2 : f \in \mathfrak{H} \right\} < \infty. \tag{3.4}$$

Moreover, if $D\psi = T^*h$, $h \in \overline{\operatorname{ran}}\, T$, then

$$\|h\|^2 + \|\psi\|^2 = \sup \left\{ \|\psi + Df\|^2 - \|f\|^2 : f \in \mathfrak{H} \right\}. \tag{3.5}$$

Proof. It follows from Theorem 2.3 that $D\psi \in \operatorname{ran} T^*$ if and only if $\mathfrak{l}_*[D\psi] < \infty$. Observe that in this case, where $T \in \mathbf{B}(\mathfrak{H}, \mathfrak{K})$, one has

$$\mathfrak{l}_*[D\psi] = \sup \{ (f, D\psi) + (D\psi, f) - (Tf, Tf) : f \in \mathfrak{H} \}.$$

Completing squares shows that

$$(f, D\psi) + (D\psi, f) - (Tf, Tf) = -\|\psi\|^2 + \|\psi + Df\|^2 - \|f\|^2,$$

so that

$$\mathfrak{l}_*[D\psi] = -\|\psi\|^2 + \sup \{ \|\psi + Df\|^2 - \|f\|^2 : f \in \mathfrak{H} \}.$$

This leads to the equivalence in (3.4). Moreover, if $D\psi = T^*h$ and $h \in \overline{\operatorname{ran}}\, T$, then by Theorem 2.3 one has $\|h\| = \mathfrak{l}_*[D\psi]$, so that

$$\|h\|^2 = -\|\psi\|^2 + \sup \{ \|\psi + Df\|^2 - \|f\|^2 : f \in \mathfrak{H} \}.$$

This completes the proof. □

Lemma 3.2. *Let $T \in \mathbf{B}(\mathfrak{H}, \mathfrak{K})$ be a contraction. Then*

$$T^*\psi \in \operatorname{ran} D \quad \Leftrightarrow \quad \sup \{ \|\psi + Tf\|^2 - \|f\|^2 : f \in \mathfrak{H} \} < \infty. \tag{3.6}$$

Moreover, if $T^*\psi = Dh$, $h \in \overline{\operatorname{ran}}\, D$, then

$$\|h\|^2 + \|\psi\|^2 = \sup \{ \|\psi + Tf\|^2 - \|f\|^2 : f \in \mathfrak{H} \}. \tag{3.7}$$

Proof. Observe that D is a selfadjoint operator. Hence, it follows from Theorem 2.3 that $T^*\psi \in \operatorname{ran} D$ if and only if $\mathfrak{l}_*[T^*\psi] < \infty$. Observe that in this case, where $D \in \mathbf{B}(\mathfrak{H})$, one has

$$\mathfrak{l}_*[T^*\psi] = \sup \{ (f, T^*\psi) + (T^*\psi, f) - (Df, Df) : f \in \mathfrak{H} \}.$$

Completing squares shows that

$$(f, T^*\psi) + (T^*\psi, f) - (Df, Df) = -\|\psi\|^2 + \|\psi + Tf\|^2 - \|f\|^2,$$

so that

$$\mathfrak{l}_*[T^*\psi] = -\|\psi\|^2 + \sup \{ \|\psi + Tf\|^2 - \|f\|^2 : f \in \mathfrak{H} \}.$$

This leads to the equivalence in (3.6). Moreover, if $T^*\psi = Dh$ and $h \in \overline{\operatorname{ran}} D$, then by Theorem 2.3 one has $\|h\| = \mathfrak{l}_*[T^*\psi]$, so that

$$\|h\|^2 = -\|\psi\|^2 + \sup\{\, \|\psi + Tf\|^2 - \|f\|^2 : f \in \mathfrak{H} \,\}.$$

This completes the proof. □

Theorem 3.3. *Let $T \in \mathbf{B}(\mathfrak{H}, \mathfrak{K})$ be a contraction. Then*

$$\psi \in \operatorname{ran} T^* \quad \Leftrightarrow \quad \sup\{\, \|\psi + Df\|^2 - \|f\|^2 : f \in \mathfrak{H} \,\} < \infty. \tag{3.8}$$

Moreover, if $\psi = T^*\varphi$, $\varphi \in \overline{\operatorname{ran}} T$, then

$$\|\varphi\|^2 = \sup\{\, \|\psi + Df\|^2 - \|f\|^2 : f \in \mathfrak{H} \,\}. \tag{3.9}$$

Proof. Recall that $\psi \in \operatorname{ran} T^*$ if and only if $D\psi \in \operatorname{ran} T^*$; see (3.2). Hence, the equivalence (3.8) follows from the equivalence (3.4) in Lemma 3.1. Now suppose that $\psi \in \operatorname{ran} T^*$, so that by (3.2) also $D\psi \in \operatorname{ran} T^*$. In this case there exist unique elements h and φ with

$$D\psi = T^*h, \quad h \in \overline{\operatorname{ran}} T \quad \text{and} \quad \psi = T^*\varphi, \quad \varphi \in \overline{\operatorname{ran}} T.$$

Observe that, via the commutation relations (3.1), one has

$$T^*h = D\psi = DT^*\varphi = T^*D_*\varphi. \tag{3.10}$$

It follows from (3.2) that $\varphi \in \overline{\operatorname{ran}} T$ implies $D_*\varphi \in \overline{\operatorname{ran}} T$. Since also $h \in \overline{\operatorname{ran}} T$, one has $h - D_*\varphi \in \overline{\operatorname{ran}} T = (\ker T^*)^\perp$, and it follows from (3.10) that $h = D_*\varphi$. Hence

$$\|h\|^2 + \|\psi\|^2 = \|D_*\varphi\|^2 + \|T^*\varphi\|^2 = \|\varphi\|^2,$$

so that (3.9) follows from (3.5). □

Theorem 3.4. *Let $T \in \mathbf{B}(\mathfrak{H}, \mathfrak{K})$ be a contraction. Then*

$$\psi \in \operatorname{ran} D_* \quad \Leftrightarrow \quad \sup\{\, \|\psi + Tf\|^2 - \|f\|^2 : f \in \mathfrak{H} \,\} < \infty. \tag{3.11}$$

Moreover, if $\psi = D_*\varphi$, $\varphi \in \overline{\operatorname{ran}} D_*$, then

$$\|\varphi\|^2 = \sup\{\, \|\psi + Tf\|^2 - \|f\|^2 : f \in \mathfrak{H} \,\}. \tag{3.12}$$

Proof. According to (3.3) $\psi \in \operatorname{ran} D_*$ if and only if $T^*\psi \in \operatorname{ran} D$. Hence, the equivalence (3.11) follows from the equivalence (3.6) in Lemma 3.2. Now suppose that $\psi \in \operatorname{ran} D_*$, so that by (3.3) also $T^*\psi \in \operatorname{ran} D$. In this case there exist unique elements h and φ with

$$T^*\psi = Dh, \quad h \in \overline{\operatorname{ran}} D \quad \text{and} \quad \psi = D_*\varphi, \quad \varphi \in \overline{\operatorname{ran}} D_*.$$

Now the commutation relations (3.1) yield

$$Dh = T^*\psi = T^*D_*\varphi = DT^*\varphi. \tag{3.13}$$

It follows from (3.3) that $\varphi \in \overline{\operatorname{ran}} D_*$ implies $T^*\varphi \in \overline{\operatorname{ran}} D$. Since also $h \in \overline{\operatorname{ran}} D$, one has $h - T^*\varphi \in \overline{\operatorname{ran}} D = (\ker D)^\perp$, and it follows from (3.13) that $h = T^*\varphi$. Hence

$$\|h\|^2 + \|\psi\|^2 = \|T^*\varphi\|^2 + \|D_*\varphi\|^2 = \|\varphi\|^2,$$

so that (3.12) follows from (3.7). □

Theorem 3.3 (with T^* and D replaced by T and D_*) goes back to P.A. Fillmore and J.P. Williams [9, Theorem 4.1], who refer to earlier work of L. de Branges and J. Rovnyak, [7, p. 23].

4. Descriptions involving the Friedrichs and Kreĭn-von Neumann extensions of nonnegative relations

This section contains some applications of the domain and range descriptions of the adjoint of a linear relation in a Hilbert space for nonnegative selfadjoint extensions of nonnegative operators and relations in a Hilbert space.

Let S be a nonnegative relation in a Hilbert space $\mathfrak{H}$. Provide the linear space $\operatorname{ran} S$ with a semi-inner product $\langle \cdot, \cdot \rangle$ by

$$\langle f', g' \rangle := (f', g) = (f, g'), \quad \{f, f'\}, \{g, g'\} \in S. \tag{4.1}$$

Note that if also $\{f_0, f'\}, \{g_0, g'\} \in S$, then the symmetry of S implies that

$$(f', g) = (f, g') = (f', g_0) = (f_0, g'), \tag{4.2}$$

which shows that the inner product (4.1) is well defined. Define the linear space $\mathfrak{R}_0$ by

$$\mathfrak{R}_0 = \{ f' : (f', f) = 0 \quad \text{for some } \{f, f'\} \in S \}. \tag{4.3}$$

Note that if $(f', f) = 0$ for $\{f, f'\} \in S$, then also $(f', f_0) = 0$ when $\{f_0, f'\} \in S$, cf. (4.2). In general, the space $\mathfrak{R}_0$ is nontrivial. Clearly the definition implies that $\operatorname{mul} S \subset \mathfrak{R}_0 \subset \operatorname{ran} S$. In fact,

$$\mathfrak{R}_0 = \operatorname{ran} S \cap \operatorname{mul} S^*. \tag{4.4}$$

cf. [12]. The quotient space $\operatorname{ran} S/\mathfrak{R}_0$ equipped with the inner product

$$\langle [f'], [g'] \rangle := (f', g) = (f, g'), \quad \{f, f'\}, \{g, g'\} \in S, \tag{4.5}$$

where $[f']$, $[g']$ denote the equivalence classes containing f' and g', is a pre-Hilbert space. The Hilbert space completion of $\operatorname{ran} S/\mathfrak{R}_0$ is denoted by $\mathfrak{H}_S$; its inner product is again denoted by $\langle \cdot, \cdot \rangle$. Define the linear relation Q from $\mathfrak{H}$ to $\mathfrak{H}_S$ by

$$Q = \{ \{f, [f']\} : \{f, f'\} \in S \}. \tag{4.6}$$

Moreover, define the linear relation J from $\mathfrak{H}_S$ to $\mathfrak{H}$ by

$$J = \{ \{[f'], f'\} : \{f, f'\} \in S \}. \tag{4.7}$$

Note that $\operatorname{dom} Q = \operatorname{dom} S$ and that $\operatorname{mul} Q = \{0\}$, i.e., Q is (the graph of) an operator, cf. [12]. Moreover, note that J is multivalued with $\operatorname{mul} J = \mathfrak{R}_0$ and that $\operatorname{ran} J = \operatorname{ran} S$. By construction the relation J is densely defined in $\mathfrak{H}_S$ and therefore J^* is an operator. The symmetry of S and the definitions (4.6) and (4.7) imply that

$$J \subset Q^*, \quad Q \subset J^*, \tag{4.8}$$

as follows from (4.5). In particular, $Q^{**} = \operatorname{clos} Q$ is a restriction of J^*, so that Q is closable, i.e., the closure Q^{**} of the operator Q, is also an operator.

Proposition 4.1. *Let S be a nonnegative relation in a Hilbert space $\mathfrak{H}$, let J be the linear relation from $\mathfrak{H}_S$ to $\mathfrak{H}$ defined by* (4.7), *and let Q be the operator from $\mathfrak{H}$ to $\mathfrak{H}_S$ defined by* (4.6). *Then*

$$\operatorname{dom} J^* = \{ h \in \mathfrak{H} : \sup \{ |(f',h)|^2 : \{f,f'\} \in S,\ (f',f) \leq 1 \} < \infty \},$$
$$\operatorname{ran} Q^* = \{ k \in \mathfrak{H} : \sup \{ |(f,k)|^2 : \{f,f'\} \in S,\ (f',f) \leq 1 \} < \infty \},$$

and

$$\ker J^* = \ker S^*, \quad \operatorname{mul} Q^* = \operatorname{mul} S^*.$$

Proof. Apply Theorem 2.3 with the relation J as in (4.7). Then the domain of the operator J^* is determined by the condition $h \in \operatorname{dom} J^*$ if and only if $\mathfrak{m}[h] < \infty$. Rewriting $\mathfrak{m}[h]$ by means of (4.5) and (4.7) one obtains

$$\operatorname{dom} J^* = \{ h \in \mathfrak{H} : \sup \{ |(f',h)|^2 : \{f,f'\} \in S,\ (f',f) \leq 1 \} < \infty \}.$$

From this formula it is clear that $h \in \ker J^*$, i.e., $\mathfrak{m}[h] = 0$ (see Corollary 2.5), if and only if $h \in (\operatorname{ran} S)^\perp = \ker S^*$.

Similarly, the range of Q^* is determined by the condition $k \in \operatorname{ran} Q^*$ if and only if $\mathfrak{m}_*[k] < \infty$, and now using (4.5) and (4.6) the description of $\operatorname{ran} Q^*$ follows. From the formula for $\operatorname{ran} Q^*$ it is clear that $k \in \operatorname{mul} Q^*$, i.e., $\mathfrak{m}_*[k] = 0$, if and only if $k \in (\operatorname{dom} S)^\perp = \operatorname{mul} S^*$. □

Since the defect numbers of S are equal, S has selfadjoint extensions in $\mathfrak{H}$. Furthermore, among the selfadjoint extensions of S there are also nonnegative selfadjoint extensions and among all nonnegative selfadjoint extensions there are two extreme nonnegative selfadjoint extensions, the Friedrichs extension S_F and the Kreĭn-von Neumann extension S_N. These extensions are related by $S_N = ((S^{-1})_F)^{-1}$. To motivate the descriptions given in Proposition 4.1 the following result is recalled: for completeness a short proof is given; cf. [6, Proposition 3.1], [12, Theorem 3.3], see also references therein.

Proposition 4.2. *Let S be a nonnegative relation in a Hilbert space $\mathfrak{H}$ and let J and Q be defined by* (4.7) *and* (4.6), *respectively. Then the Friedrichs extension of S is given by $S_F = Q^*Q^{**}$ and the Kreĭn-von Neumann is given by $S_N = J^{**}J^*$.*

Proof. If $\{f,f'\} \in S$ then $\{f,[f']\} \in Q \subset J^*$ and $\{[f'],f'\} \in J \subset J^{**}$, so that $\{f,f'\} \in J^{**}J^*$. Hence, $S \subset J^{**}J^*$ and therefore $J^{**}J^*$ is a nonnegative selfadjoint extension of S.

Now let $\{f,f'\} \in J^{**}J^*$. Then $\{f,J^*f\} \in J^*$ and $\{J^*f,f'\} \in J^{**}$. Since $J^{**} = \operatorname{clos} J$ there exists a sequence $(\{[f_n'],f_n'\}) \subset J$ with $\{f_n,f_n'\} \in S$, such that

$$[f_n'] \to J^*f \text{ in } \mathfrak{H}_S, \quad f_n' \to f' \text{ in } \mathfrak{H}.$$

In particular, $[f_n']$ is a Cauchy sequence in $\mathfrak{H}_S$, which in view of (4.5) means that

$$(f_n' - f_m', f_n - f_m) \to 0, \quad n,m \to \infty.$$

This together with $f'_n \to f'$ in $\mathfrak{H}$ shows that $f' \in \operatorname{ran} A_N^{1/2}$. Thus $\operatorname{ran} J^{**}J^* \subset \operatorname{ran} A_N^{1/2}$. Since A_N is the only selfadjoint extension of S, whose range belongs to $\operatorname{ran} A_N^{1/2}$, the equality $J^{**}J^* = S_N$ follows.

The proof of the equality $S_F = Q^*Q^{**}$ is completely analogous. □

As a consequence of Propositions 4.1 and 4.2 one gets the following equalities:

$$\operatorname{dom} S_N^{1/2} = \{\, h \in \mathfrak{H} : \sup\{\, |(f', h)|^2 : \{f, f'\} \in S,\ (f', f) \le 1 \,\} < \infty \,\},$$

$$\operatorname{ran} S_F^{1/2} = \{\, k \in \mathfrak{H} : \sup\{\, |(f, k)|^2 : \{f, f'\} \in S,\ (f', f) \le 1 \,\} < \infty \,\},$$

and

$$\ker S_N = \ker S^*, \quad \operatorname{mul} S_F = \operatorname{mul} S^*.$$

5. Parallel sums and parallel differences for forms

In this section the definitions of the mappings $\mathfrak{l}$ and $\mathfrak{m}$ are extended by replacing certain Hilbert space inner products by sesquilinear (not necessarily nonnegative) forms defined on a complex linear space $\mathfrak{D}$. It is shown how this leads to the notions of *parallel sum* and *parallel difference* in the general setting *of nonnegative forms*. This can be considered also as a further motivation for the mappings $\mathfrak{l}$ and $\mathfrak{m}$ as well as an application of the domain and range descriptions treated in Section 2.

Recall that a sesquilinear form on a complex linear space $\mathfrak{D}$ is a mapping from $\mathfrak{D} \times \mathfrak{D}$ to $\mathbb{C}$, which is linear in its first and anti-linear in its second argument. The sesquilinear form is said to be *nonnegative* if it maps $\mathfrak{D} \times \mathfrak{D}$ to $\mathbb{R}_+ = [0, \infty)$.

Lemma 5.1. *Let $\mathfrak{t}$ and $\mathfrak{v}$ be sesquilinear forms on a complex linear space $\mathfrak{D}$. Let the associated mappings $\mathfrak{l}, \mathfrak{m} : \mathfrak{D} \to \mathbb{R} \cup \{\infty\}$ be defined by*

$$\mathfrak{l}[\varphi] = \sup\{\, \mathfrak{t}(\varphi, g) + \mathfrak{t}(g, \varphi) - \mathfrak{v}[g] : g \in \mathfrak{D} \,\}, \quad \varphi \in \mathfrak{D},$$

and

$$\mathfrak{m}[\varphi] = \sup\left\{ |\mathfrak{t}(\varphi, g)|^2 : g \in \mathfrak{D},\ \mathfrak{v}[g] \le 1 \right\}, \quad \varphi \in \mathfrak{D}.$$

Then $\mathfrak{l}$ and $\mathfrak{m}$ are nonnegative and

$$\mathfrak{l}[\varphi] \ge \mathfrak{m}[\varphi], \quad \varphi \in \mathfrak{D}.$$

Furthermore, if $\mathfrak{l}[\varphi] < \infty$ at least for one $\varphi \in \mathfrak{D}$, then $\mathfrak{v}[g] \ge 0$ for every $g \in \mathfrak{D}$.

Proof. Let $\varphi \in \mathfrak{D}$. Clearly $\mathfrak{l}[\varphi] \ge 0$ and $\mathfrak{m}[\varphi] \ge 0$; take $g = 0$ in the above definitions. Moreover, for all $g \in \mathfrak{D}$ and all $\lambda \in \mathbb{C}$ it follows from the definition that

$$\mathfrak{l}[\varphi] \ge \bar{\lambda}\mathfrak{t}(\varphi, g) + \lambda\mathfrak{t}(g, \varphi) - |\lambda|^2\mathfrak{v}[g].$$

Now $\mathfrak{t}(\varphi, g) = e^{i\chi}|\mathfrak{t}(\varphi, g)|$ for some $\chi \in \mathbb{R}$ and thus $\mathfrak{t}(g, \varphi) = e^{-i\chi}|\mathfrak{t}(\varphi, g)|$. Choose $\lambda = e^{i\chi}t$ with $t \in \mathbb{R}$. Then for all $t \in \mathbb{R}$ one has

$$\mathfrak{l}[\varphi] \ge 2t|\mathfrak{t}(\varphi, g)| - t^2\mathfrak{v}[g]. \tag{5.1}$$

For $\mathfrak{l}[\varphi] < \infty$ this implies that

$$|\mathfrak{t}(\varphi, g)|^2 \leq \mathfrak{v}[g]\mathfrak{l}[\varphi], \quad \varphi, g \in \mathfrak{D}.$$

Since $\mathfrak{l}[\varphi] \geq 0$, one concludes that

$$\mathfrak{m}[\varphi] \leq \mathfrak{l}[\varphi], \quad \varphi \in \mathfrak{D}.$$

For $\mathfrak{l}[\varphi] = \infty$ the statement $\mathfrak{l}[\varphi] \geq \mathfrak{m}[\varphi]$ is clear.

Finally, it follows from (5.1) that if $\mathfrak{v}[g] < 0$ for some $g \in \mathfrak{D}$, then $\mathfrak{l}[\varphi] = \infty$ for every $\varphi \in \mathfrak{D}$. This completes the proof. □

By Lemma 5.1 the mappings $\mathfrak{l}$ and $\mathfrak{m}$ satisfy the inequality $\mathfrak{l} \geq \mathfrak{m}$. It turns out that if the mapping $\mathfrak{l}$ has values in $\mathbb{R}$, then equality $\mathfrak{l} = \mathfrak{m}$ holds. Recall that by Lemma 5.1 $\mathfrak{l}[\varphi] < \infty$ for some $\varphi \in \mathfrak{D}$ implies that the form $\mathfrak{v}$ is nonnegative. The proof of the next theorem is relying on the domain characterization for adjoints established in Theorem 2.3.

Theorem 5.2. *Let $\mathfrak{t}$ and $\mathfrak{v}$ be nonnegative forms on a complex linear space $\mathfrak{D}$. The following statements are equivalent:*

(i) $\mathfrak{l}[\varphi] = \sup\{\, \mathfrak{t}(\varphi, g) + \mathfrak{t}(g, \varphi) - \mathfrak{v}[g] : g \in \mathfrak{D} \,\} < \infty$ *for all* $\varphi \in \mathfrak{D}$;

(ii) $\mathfrak{m}[\varphi] = \sup\{\, |\mathfrak{t}(\varphi, g)|^2 : g \in \mathfrak{D},\, \mathfrak{v}[g] \leq 1 \,\} < \infty$ *for all* $\varphi \in \mathfrak{D}$.

If either (i) *or* (ii) *holds, then the mappings $\mathfrak{l}$ and $\mathfrak{m}$ are identical on $\mathfrak{D}$ and they are nonnegative forms on $\mathfrak{D}$.*

Proof. (i) ⇒ (ii) This implication follows from Lemma 5.1.

(ii) ⇒ (i) Assume that $\mathfrak{v}[g] = 0$ for some $g \in \mathfrak{D}$. Then for all $n \in \mathbb{N}$, $\mathfrak{v}[ng] = 0$ and hence in this case, for every $\varphi \in \mathfrak{D}$

$$\mathfrak{m}[\varphi] \geq |\mathfrak{t}(\varphi, ng)|^2 = n^2|\mathfrak{t}(\varphi, g)|,$$

and, in particular, $\mathfrak{m}[g] \geq n^2\mathfrak{t}[g]$. Since $\mathfrak{m}[g] < \infty$, this implies that $\mathfrak{t}[g] = 0$. Therefore, the inclusion

$$\ker \mathfrak{v} \subset \ker \mathfrak{t} \tag{5.2}$$

holds. Let $\mathfrak{H}_\mathfrak{v}$ and $\mathfrak{H}_\mathfrak{t}$ be the Hilbert spaces obtained by completing the quotient spaces

$$\mathfrak{D}/\ker \mathfrak{v} \quad \text{and} \quad \mathfrak{D}/\ker \mathfrak{t}$$

with respect to the induced inner products $\mathfrak{v}(\cdot,\cdot)$ and $\mathfrak{t}(\cdot,\cdot)$. Define the linear relation $\iota : \mathfrak{H}_\mathfrak{v} \to \mathfrak{H}_\mathfrak{t}$ by

$$\iota = \{\, \{\varphi + \ker \mathfrak{v}, \varphi + \ker \mathfrak{t}\} : \varphi \in \mathfrak{D} \,\},$$

so that $\operatorname{dom} \iota = \mathfrak{D}/\ker \mathfrak{v}$. Then in view of (5.2) ι is an operator (single-valued) and by construction it has a dense domain. Hence, its adjoint ι^* is an operator too. The mappings $\mathfrak{l}$ and $\mathfrak{m}$ in (i) and (ii) can be interpreted as mappings from $\mathfrak{H}_\mathfrak{t}$ to $\mathbb{R} \cup \{\infty\}$ associated to $T = \iota$ as in Section 2. Now, by Theorem 2.3 the assumption (ii) implies that $\mathfrak{D}/\ker \mathfrak{t} \subset \operatorname{dom} \iota^*$, so that $\operatorname{dom} \iota^*$ is dense in $\mathfrak{H}_{\mathfrak{t}}$. Moreover, by the same theorem, it follows that

$$\mathfrak{l}[\varphi] = \mathfrak{m}[\varphi] = (\iota^*(\varphi + \ker \mathfrak{t}), \iota^*(\varphi + \ker \mathfrak{t}))_{\mathfrak{v}}, \quad \varphi \in \mathfrak{D}.$$

This gives (i) and also implies that $\mathfrak{l}[\varphi]$ and $\mathfrak{m}[\varphi]$ are nonnegative forms on $\mathfrak{D}$. □

Let $\mathfrak{t}$ and $\mathfrak{w}$ be nonnegative forms on a complex linear space $\mathfrak{D}$. Define the *parallel sum* $\mathfrak{t} : \mathfrak{w}$ of $\mathfrak{t}$ and $\mathfrak{w}$ as a mapping from $\mathfrak{D}$ to $\mathbb{R} \cup \{\infty\}$ by

$$(\mathfrak{t} : \mathfrak{w})[\varphi] = \inf \{ \mathfrak{t}[\varphi - g] + \mathfrak{w}[g] : g \in \mathfrak{D} \}, \quad \varphi \in \mathfrak{D}. \tag{5.3}$$

The parallel sum $\mathfrak{t} : \mathfrak{w}$ of nonnegative forms $\mathfrak{t}$ and $\mathfrak{w}$ was introduced in [13], where the main properties of $\mathfrak{t} : \mathfrak{w}$ can be found. From Theorem 5.2 one obtains the following result for parallel sums.

Proposition 5.3. *Let $\mathfrak{t}$ and $\mathfrak{w}$ be nonnegative forms on a complex linear space $\mathfrak{D}$. Then:*

(i) $\mathfrak{l}[\varphi] = \sup \{ \mathfrak{t}(\varphi, g) + \mathfrak{t}(g, \varphi) - (\mathfrak{t} + \mathfrak{w})[g] : g \in \mathfrak{D} \} < \infty$ *for all* $\varphi \in \mathfrak{D}$;

(ii) $\mathfrak{m}[\varphi] = \sup \{ |\mathfrak{t}(\varphi, g)|^2 : g \in \mathfrak{D}, (\mathfrak{t} + \mathfrak{w})[g] \leq 1 \} < \infty$ *for all* $\varphi \in \mathfrak{D}$,

and the mappings $\mathfrak{l}$ and $\mathfrak{m}$ are equal on $\mathfrak{D}$ and they are nonnegative forms on $\mathfrak{D}$. The parallel sum $\mathfrak{t} : \mathfrak{w}$ is a nonnegative form on $\mathfrak{D}$ which satisfies the identities

$$\mathfrak{t} = \mathfrak{m} + (\mathfrak{t} : \mathfrak{w}) = \mathfrak{l} + (\mathfrak{t} : \mathfrak{w}). \tag{5.4}$$

Proof. (ii) Observe that $(\mathfrak{t} + \mathfrak{w})[g] \leq 1$ implies $\mathfrak{t}[g] \leq 1$, so that $\mathfrak{m}[\varphi] \leq \mathfrak{t}[\varphi]$ for all $\varphi \in \mathfrak{D}$. This shows (ii).

(i) Apply Theorem 5.2 with $\mathfrak{v} = \mathfrak{t} + \mathfrak{w} \geq 0$. Then, by (ii), $\mathfrak{l}[\varphi] = \mathfrak{m}[\varphi] < \infty$ for all $\varphi \in \mathfrak{D}$, and (i) has been shown.

To complete the proof, it suffices to show (5.4). Completing squares gives:

$$\mathfrak{t}(\varphi, g) + \mathfrak{t}(g, \varphi) - (\mathfrak{t} + \mathfrak{w})[g] = -\mathfrak{t}[\varphi - g] + \mathfrak{t}[\varphi] - \mathfrak{w}[g].$$

Therefore, it follows that

$$\begin{aligned} \mathfrak{m}[\varphi] = \mathfrak{l}[\varphi] &= \sup \{ -\mathfrak{t}[\varphi - g] + \mathfrak{t}[\varphi] - \mathfrak{w}[g] : g \in \mathfrak{D} \} \\ &= \mathfrak{t}[\varphi] - \inf \{ \mathfrak{t}[\varphi - g] + \mathfrak{w}[g] : g \in \mathfrak{D} \} \\ &= \mathfrak{t}[\varphi] - (\mathfrak{t} : \mathfrak{w})[\varphi]. \end{aligned}$$

By Theorem 5.2 $\mathfrak{m}$ is a nonnegative form on $\mathfrak{D}$, and since $(\mathfrak{t} : \mathfrak{w})[\varphi] = \mathfrak{t}[\varphi] - \mathfrak{m}[\varphi] \geq 0$, also $\mathfrak{t} : \mathfrak{w}$ is a nonnegative form on $\mathfrak{D}$. □

Let $\mathfrak{t}$ and $\mathfrak{s}$ be sesquilinear forms on a complex linear space $\mathfrak{D}$. Define the *parallel difference* $\mathfrak{t} \div \mathfrak{s}$ of $\mathfrak{t}$ and $\mathfrak{s}$ as a mapping from $\mathfrak{D}$ to $\mathbb{R} \cup \{\infty\}$ by

$$(\mathfrak{t} \div \mathfrak{s})[\varphi] = \sup \{ \mathfrak{t}[\varphi + g] - \mathfrak{s}[g] : g \in \mathfrak{D} \}. \tag{5.5}$$

Clearly, $(\mathfrak{t} \div \mathfrak{s})[\varphi] \geq \mathfrak{t}[\varphi]$ for all $\varphi \in \mathfrak{D}$. In particular, if $\mathfrak{t}[\varphi] \geq 0$ for some $\varphi \in \mathfrak{D}$, then $(\mathfrak{t} \div \mathfrak{s})[\varphi] \geq 0$.

Lemma 5.4. *Let $\mathfrak{t}$ and $\mathfrak{s}$ be sesquilinear forms on a complex linear space $\mathfrak{D}$. Let the mappings $\mathfrak{l}, \mathfrak{m} : \mathfrak{D} \to \mathbb{R} \cup \{\infty\}$ be defined by*

$$\mathfrak{l}[\varphi] = \sup \{ \mathfrak{t}(\varphi, g) + \mathfrak{t}(g, \varphi) - (\mathfrak{s} - \mathfrak{t})[g] : g \in \mathfrak{D} \}, \quad \varphi \in \mathfrak{D},$$

and

$$\mathfrak{m}[\varphi] = \sup\left\{ |\mathfrak{t}(\varphi, g)|^2 : g \in \mathfrak{D},\, (\mathfrak{s} - \mathfrak{t})[g] \leq 1 \right\}, \quad \varphi \in \mathfrak{D}.$$

Then

$$(\mathfrak{t} \div \mathfrak{s})[\varphi] = \mathfrak{l}[\varphi] + \mathfrak{t}[\varphi] \geq \mathfrak{m}[\varphi] + \mathfrak{t}[\varphi], \quad \varphi \in \mathfrak{D}, \tag{5.6}$$

and, in particular,

$$(\mathfrak{t} \div \mathfrak{s})[\varphi] < \infty \quad \textit{if and only if} \quad \mathfrak{l}[\varphi] < \infty.$$

Furthermore, if $(\mathfrak{t} \div \mathfrak{s})[\varphi] < \infty$ *or, equivalently, if* $\mathfrak{l}[\varphi] < \infty$ *for at least one* $\varphi \in \mathfrak{D}$, *then* $\mathfrak{s}[g] \geq \mathfrak{t}[g]$ *for every* $g \in \mathfrak{D}$.

Proof. Consider arbitrary $\varphi, g \in \mathfrak{D}$. Completing squares gives:

$$\mathfrak{t}(\varphi, g) + \mathfrak{t}(g, \varphi) - (\mathfrak{s} - \mathfrak{t})[g] = \mathfrak{t}[\varphi + g] - \mathfrak{t}[\varphi] - \mathfrak{s}[g].$$

Hence for every $\varphi \in \mathfrak{D}$,

$$\begin{aligned} \mathfrak{l}[\varphi] &= \sup\{ \mathfrak{t}[\varphi + g] - \mathfrak{t}[\varphi] - \mathfrak{s}[g] : g \in \mathfrak{D} \} \\ &= \sup\{ \mathfrak{t}[\varphi + g] - \mathfrak{s}[g] : g \in \mathfrak{D} \} - \mathfrak{t}[\varphi] = (\mathfrak{t} \div \mathfrak{s})[\varphi] - \mathfrak{t}[\varphi]. \end{aligned}$$

Moreover, according to Lemma 5.1, one has $\mathfrak{l}[\varphi] \geq \mathfrak{m}[\varphi]$ and this yields (5.6). Clearly, $(\mathfrak{t} \div \mathfrak{s})[\varphi] < \infty$ if and only if $\mathfrak{l}[\varphi] < \infty$, and in this case the inequality $\mathfrak{s}[g] \geq \mathfrak{t}[g]$ for all $g \in \mathfrak{D}$ is obtained from Lemma 5.1. □

Remark 5.5. The identity in (5.6) is obtained by completing squares. Completing squares in a different way shows that for all $\varphi \in \mathfrak{D}$,

$$\begin{aligned} \mathfrak{l}[\varphi] &= \sup\{ (2\mathfrak{t} - \mathfrak{s})[g] - \mathfrak{t}[\varphi - g] + \mathfrak{t}[\varphi] : g \in \mathfrak{D} \} \\ &= ((2\mathfrak{t} - \mathfrak{s}) \div \mathfrak{t})[\varphi] + \mathfrak{t}[\varphi], \end{aligned}$$

which gives a representation for $\mathfrak{l} - \mathfrak{t}$. In particular,

$$((2\mathfrak{t} - \mathfrak{s}) \div \mathfrak{t})[\varphi] < \infty \quad \text{if and only if} \quad \mathfrak{l}[\varphi] < \infty,$$

in which case $\mathfrak{s} \geq \mathfrak{t}$. Note that if $2\mathfrak{t} \geq \mathfrak{s}$, then

$$\mathfrak{l} - \mathfrak{t} = (2\mathfrak{t} - \mathfrak{s}) \div \mathfrak{t} \geq 0.$$

Theorem 5.2 yields the following result for parallel differences.

Proposition 5.6. *Let* $\mathfrak{t}$ *and* $\mathfrak{s}$ *be nonnegative forms on a complex linear space* $\mathfrak{D}$. *The following statements are equivalent:*

(i) $\mathfrak{l}[\varphi] = \sup\{ \mathfrak{t}(\varphi, g) + \mathfrak{t}(g, \varphi) - (\mathfrak{s} - \mathfrak{t})[g] : g \in \mathfrak{D} \} < \infty$ *for all* $\varphi \in \mathfrak{D}$;

(ii) $\mathfrak{m}[\varphi] = \sup\left\{ |\mathfrak{t}(\varphi, g)|^2 : g \in \mathfrak{D},\, (\mathfrak{s} - \mathfrak{t})[g] \leq 1 \right\} < \infty$ *for all* $\varphi \in \mathfrak{D}$.

If either (i) *or* (ii) *holds, then the mappings* $\mathfrak{l}$ *and* $\mathfrak{m}$ *are equal on* $\mathfrak{D}$, *they are nonnegative forms on* $\mathfrak{D}$, *and the parallel difference* $\mathfrak{t} \div \mathfrak{s}$ *is a nonnegative form on* $\mathfrak{D}$, *which satisfies the identities*

$$\mathfrak{t} \div \mathfrak{s} = \mathfrak{l} + \mathfrak{t} = \mathfrak{m} + \mathfrak{t}. \tag{5.7}$$

Proof. To apply Theorem 5.2 it is first shown that the condition (ii) implies that $(\mathfrak{s}-\mathfrak{t})[g] \geq 0$. Therefore, assume that $(\mathfrak{s}-\mathfrak{t})[g] \leq 0$ for some $g \in \mathfrak{D}$. Then $(\mathfrak{s}-\mathfrak{t})[ng] = n^2(\mathfrak{s}-\mathfrak{t})[g] \leq 0$ for all $n \in \mathbb{N}$. Consequently, for every $\varphi \in \mathfrak{D}$

$$\mathfrak{m}[\varphi] \geq |\mathfrak{t}(\varphi, ng)|^2 = n^2|\mathfrak{t}(\varphi, g)|,$$

and, in particular, $\mathfrak{m}[g] \geq n^2\mathfrak{t}[g]$. This implies that $\mathfrak{t}[g] = 0$ and, since $\mathfrak{s}$ is nonnegative, one concludes that $\mathfrak{s}[g] = 0$. Thus, it follows that $\mathfrak{s}-\mathfrak{t} \geq 0$ and, moreover,

$$\ker\,(\mathfrak{s}-\mathfrak{t}) \subset \ker\,\mathfrak{s} \cap \ker\,\mathfrak{t} \subset \ker\,\mathfrak{t}.$$

Now by Theorem 5.2 the statements (i) and (ii) are equivalent and the equality $\mathfrak{l}[\varphi] = \mathfrak{m}[\varphi]$ holds for all $\varphi \in \mathfrak{D}$. The equalities (5.7) follow from (5.6). In particular, $\mathfrak{t} \div \mathfrak{s}$ is a nonnegative form on $\mathfrak{D}$. □

Let $\mathfrak{t}$ and $\mathfrak{s}$ be nonnegative forms on $\mathfrak{D}$ such that $\mathfrak{s} \geq 2\mathfrak{t}$. Then $\mathfrak{w} = \mathfrak{s} - 2\mathfrak{t} \geq 0$. Use the identity (5.4) from Proposition 5.3:

$$\mathfrak{l}[\varphi] = \mathfrak{t}[\varphi] - (\mathfrak{t} : \mathfrak{w})[\varphi],$$

with $\mathfrak{w} = \mathfrak{s} - 2\mathfrak{t}$. Hence, in this case $\mathfrak{s} - \mathfrak{t} = \mathfrak{t} + \mathfrak{w}$ and the conditions (i) and (ii) in Proposition 5.6 are satisfied. It follows from (5.7) in Proposition 5.6 that $\mathfrak{t} \div \mathfrak{s}$ is a nonnegative form on $\mathfrak{D}$ and that

$$(\mathfrak{t} \div \mathfrak{s})[\varphi] - \mathfrak{t}[\varphi] = \mathfrak{t}[\varphi] - (\mathfrak{t} : (\mathfrak{s} - 2\mathfrak{t}))[\varphi].$$

Hence, if $\mathfrak{s} \geq 2\mathfrak{t}$, then one obtains the following decomposition of the form $\mathfrak{t}$ into a sum of two nonnegative forms:

$$\mathfrak{t} \div \mathfrak{s} + \mathfrak{t} : (\mathfrak{s} - 2\mathfrak{t}) = 2\mathfrak{t}. \tag{5.8}$$

This decomposition can be stated also in a more symmetric form as follows.

Corollary 5.7. *Let $\mathfrak{t}$ and $\mathfrak{w}$ be nonnegative forms on a complex linear space $\mathfrak{D}$ such that $\mathfrak{t} \leq \mathfrak{w}$. Then $\mathfrak{t} \div (\mathfrak{t} + \mathfrak{w})$ and $\mathfrak{t} : (\mathfrak{w} - \mathfrak{t})$ are nonnegative forms on $\mathfrak{D}$ and*

$$\mathfrak{t} \div (\mathfrak{t} + \mathfrak{w}) + (\mathfrak{t} : (\mathfrak{w} - \mathfrak{t})) = 2\mathfrak{t}.$$

Proof. Consider $\mathfrak{s} = \mathfrak{t} + \mathfrak{w} \geq 0$. Then $\mathfrak{s} - 2\mathfrak{t} = \mathfrak{w} - \mathfrak{t} \geq 0$, which implies that $\mathfrak{t} \div (\mathfrak{t} + \mathfrak{w})$ and $\mathfrak{t} : (\mathfrak{w} - \mathfrak{t})$ are nonnegative forms on $\mathfrak{D}$. It remains to apply (5.8) with $\mathfrak{s} = \mathfrak{t} + \mathfrak{w}$. □

Lemma 5.8. *Let $\mathfrak{t}$ and $\mathfrak{w}$ be nonnegative forms on a complex linear space $\mathfrak{D}$. Then*

$$((\mathfrak{t} : \mathfrak{w}) \div \mathfrak{w})[\varphi] = \sup_{h \in \mathfrak{D}} \left\{ \inf_{g \in \mathfrak{D}} \{\, \mathfrak{t}[h-g] + \mathfrak{w}[g] \,\} - \mathfrak{w}[\varphi + h] \right\}, \quad \varphi \in \mathfrak{D}, \tag{5.9}$$

is a nonnegative form on $\mathfrak{D}$ which satisfies the inequalities

$$0 \leq (\mathfrak{t} : \mathfrak{w})[\varphi] \leq ((\mathfrak{t} : \mathfrak{w}) \div \mathfrak{w})[\varphi] \leq \mathfrak{t}[\varphi], \quad \varphi \in \mathfrak{D}. \tag{5.10}$$

Proof. By Proposition 5.3 the parallel sum $\mathfrak{t} : \mathfrak{w}$ is a nonnegative form on $\mathfrak{D}$, and

$$0 \leq (\mathfrak{t} : \mathfrak{w})[\varphi] \leq ((\mathfrak{t} : \mathfrak{w}) \div \mathfrak{w})[\varphi], \quad \varphi \in \mathfrak{D}.$$

From the definition of parallel sum in (5.3) one obtains the estimate

$$(\mathfrak{t} : \mathfrak{w})[h] - \mathfrak{w}[\varphi + h] \leq \mathfrak{t}[h - g] + \mathfrak{w}[g] - \mathfrak{w}[\varphi + h], \quad h, g, \varphi \in \mathfrak{D},$$

which with $g = \varphi + h$ yields

$$(\mathfrak{t} : \mathfrak{w})[h] - \mathfrak{w}[\varphi + h] \leq \mathfrak{t}[-\varphi] = \mathfrak{t}[\varphi], \quad h, \varphi \in \mathfrak{D}.$$

Taking the supremum over $h \in \mathfrak{D}$ yields the estimate $((\mathfrak{t} : \mathfrak{w}) \dot{-} \mathfrak{w})[\varphi] \leq \mathfrak{t}[\varphi] < \infty$, $\varphi \in \mathfrak{D}$. In particular, the conditions (i) and (ii) in Proposition 5.6 are satisfied (see Lemma 5.4), and thus $(\mathfrak{t} : \mathfrak{w}) \dot{-} \mathfrak{w}$ is a nonnegative form on $\mathfrak{D}$ satisfying (5.10). □

Remark 5.9. Let $\mathfrak{t}$ and $\mathfrak{w}$ be nonnegative forms on a complex linear space $\mathfrak{D}$. The nonnegative form

$$\mathfrak{t}_{\rm reg}[\varphi] := ((\mathfrak{t} : \mathfrak{w}) \dot{-} \mathfrak{w})[\varphi], \quad \varphi \in \mathfrak{D}, \tag{5.11}$$

introduced in Lemma 5.8 is called in [13] the *regular part* of the form $\mathfrak{t}$ with respect to the form $\mathfrak{w}$. By Lemma 5.8 the form

$$\mathfrak{t}_{\rm sing}[\varphi] = \mathfrak{t}[\varphi] - \mathfrak{t}_{\rm reg}[\varphi]$$

is also nonnegative on $\mathfrak{D}$ and in [13] it is called the *singular part* of the form $\mathfrak{t}$ with respect to the form $\mathfrak{w}$. This gives rise to the following decomposition of the form $\mathfrak{t}$ with respect to the form $\mathfrak{w}$:

$$\mathfrak{t}[\varphi] = \mathfrak{t}_{\rm reg}[\varphi] + \mathfrak{t}_{\rm sing}[\varphi], \quad \varphi \in \mathfrak{D}, \tag{5.12}$$

which can be viewed as an analog and extension of the *Lebesgue decomposition* of a pair of finite positive measures: for further details concerning this decomposition, see [13]; for some special cases see also [4], [19].

References

[1] W.N. Anderson, Jr., "Shorted operators", SIAM J. Appl. Math., 20 (1971), 520–525.

[2] W.N. Anderson and R.J. Duffin, "Series and parallel addition of matrices", J. Math. Anal. Appl., 26 (1969), 576–594.

[3] W.N. Anderson and G.E. Trapp, "Shorted operators. II", SIAM J. Appl. Math., 28 (1975), 60–71.

[4] T. Ando, "Lebesgue-type decomposition of positive operators", Acta Sci. Math. (Szeged), 38 (1976), 253–260.

[5] T. Ando and K. Nishio, "Positive selfadjoint extensions of positive symmetric operators", Tôhoku Math. J., 22 (1970), 65–75.

[6] Yu.M. Arlinskiĭ, S. Hassi, Z. Sebestyén, and H.S.V. de Snoo, "On the class of extremal extensions of a nonnegative operator", Oper. Theory Adv. Appl. (B. Sz.-Nagy memorial volume), 127 (2001), 41–81.

[7] L. de Branges and J. Rovnyak, *Square summable power series*, Holt, Rinehart and Winston, New York, 1966.

[8] S.L. Eriksson and H. Leutwiler, "A potential theoretic approach to parallel addition", Math. Ann., 274 (1986), 301–317.

[9] P. Fillmore and J. Williams, "On operator ranges", Adv. in Math., 7 (1971), 254–281.

[10] S. Hassi, "On the Friedrichs and the Kreĭn-von Neumann extension of nonnegative relations", Acta Wasaensia, 122 (2004), 37–54.

[11] S. Hassi, M.M. Malamud, and H.S.V. de Snoo, "On Kreĭn's extension theory of nonnegative operators", Math. Nachr., 274/275 (2004), 40–73.

[12] S. Hassi, A. Sandovici, H.S.V. de Snoo, and H. Winkler, "A general factorization approach to the extension theory of nonnegative operators and relations", J. Operator Theory, 58 (2007), 351–386.

[13] S. Hassi, Z. Sebestyén, and H.S.V. de Snoo, "Lebesgue type decompositions for nonnegative forms", to appear in J. Functional Analysis.

[14] E. Jaimes, "On ranges of adjoint operators", Ann. Univ. Sci. Budapest, 33 (1990), 235–237.

[15] E.L. Pekarev, "Shorts of operators and some extremal problems", Acta Sci. Math. (Szeged), 56 (1992), 147–163.

[16] E.L. Pekarev and Yu.L. Shmulyan, "Parallel addition and parallel subtraction of operators", Izv. Akad. Nauk SSSR, Ser Mat., 40 (1976), 366–387.

[17] Z. Sebestyén, "On ranges of adjoint operators in Hilbert space", Acta Sci. Math. (Szeged), 46 (1983), 295–298.

[18] Yu.L. Shmulyan, "Two-sided division in a ring of operators", Math. Notes 1 (1967), 400–403.

[19] B. Simon, "A canonical decomposition for quadratic forms with applications to monotone convergence theorems", J. Functional Analysis, 28 (1978), 377–385.

[20] B. Sz.-Nagy and C. Foiaş, *Harmonic analysis of operators on Hilbert space*, North-Holland, New York, 1970.

Seppo Hassi
Department of Mathematics and Statistics
University of Vaasa
P.O. Box 700
65101 Vaasa, Finland
e-mail: `sha@uwasa.fi`

Zoltán Sebestyén
Department of Applied Analysis
Eötvös Loránd University
Pázmány Péter sétány 1/C
1117 Budapest, Hungary
e-mail: `sebesty@cs.elte.hu`

Henk de Snoo
Department of Mathematics and Computing Science
University of Groningen
P.O. Box 407
9700 AK Groningen, Nederland
e-mail: `desnoo@math.rug.nl`

Operator Theory:
Advances and Applications, Vol. 198, 229–237

The Königs Problem and Extreme Fixed Points

V.A. Khatskevich and V.A. Senderov

Dedicated to Peter Jonas

Abstract. We consider a linear-fractional mapping $\mathcal{F}_A$ of the unit operator ball, which is generated by a triangular operator. Under the assumption that $\mathcal{F}_A$ has an extreme fixed point C and under some natural restrictions on one of the diagonal elements of the operator block-matrix A, we prove the KE-property of $\mathcal{F}_A$. In this case, the structure of the other diagonal element is studied completely.

We consider specific cases in which for C one can take any arbitrary point of the unit sphere.

Mathematics Subject Classification (2000). Primary 47B50, 47A52.

Keywords. Bounded linear operator, Hilbert space, indefinite metric, Königs embedding property, linear fractional map, operator ball.

1. Introduction

The Königs embedding problem (the KE-problem), which was first used by G. Königs, P. Lévy, and J. Hadamard to solve various applied problems has more than a century long history. The general statement of this problem is as follows. Let D be a domain in a complex Banach space, $f \in \mathrm{Hol}(D)$. The question is: Does there exist a family $\{F(t)\}_{t\geq 0} \subset \mathrm{Hol}(D)$ continuously (in the topology of locally uniform convergence over D) depending on t and satisfying the conditions $F(0) = I$, $F(1) = f$, and $F(s+t) = F(s) \circ F(t)$ for all $s, t \geq 0$?

If the family $\{F(t)\}_{t\geq 0}$ exists, then it is said that f has the KE-property.

In the course of more than a century, this problem was studied by numerous mathematicians (see, e.g., [12–17]). In particular, T.E. Harris considered this problem in the process of constructing models for branching Markov processes. Another important field of analysis in which the problem of embedding of a discrete semigroup of a given self-mapping iterates into a continuous semigroup of self-mappings naturally arises, is the theory of composition operators on functional spaces (see, e.g., [18–22]).

In 1884, G. Königs showed [12] how the Schröder equation

$$\varphi \circ f = \lambda \varphi \qquad (*)$$

can be used to solve the problem locally. Namely, if the mapping f has a fixed point z_0 in the open unit disk D of the plane $\mathbb{C}$ and $|f'(z_0)| \neq 0, 1$, then for $\lambda = f'(z_0)$, equation $(*)$ has a unique solution $\varphi \in \mathrm{Hol}(D, \mathbb{C})$ normalized by the condition $\varphi'(z_0) = 1$. To construct a continuous semigroup, it suffices to set $f^t = \varphi^{-1} \circ (\lambda^t \varphi)$, where $t > 0$ and λ^t is one of the branches of the function $\psi(t) = \lambda^t$.

In 1981, C.C. Cowen showed [18] that to solve equation $(*)$, it suffices to solve it in the specific case $f = F_A$, where F_A is a linear-fractional mapping (l.f.m.) of the form

$$F_A(z) = (a_{21} + a_{22}z)(a_{11} + a_{12}z)^{-1},$$

where $a_{ij} \in \mathbb{C}$, $i, j = 1, 2$ and $A = \|a_{ij}\|_{i,j=1}^2$.

In recent years, new works concerning the KE-problem and its applications have appeared. So, the case in which D is the unit open ball of the space $\mathcal{L}(\mathfrak{H}_1, \mathfrak{H}_2)$, where $\mathfrak{H}_1$ and $\mathfrak{H}_2$ are Hilbert spaces and f is the transformation of D generated by the (plus-) operator A by formula (1), was considered in [1–3]. In the present paper, we deal precisely with this case. In [23, 24], the situation is, in fact, equivalent: continuous semigroups $\{A^t\}_{t>0}$ of plus-operators in a Krein space generating linear-fractional mappings (l.f.m.) of the unit operator ball (for the definition, see below) are considered.

Here we develop a new approach based on the properties of fixed points of mappings. It turns out that a wide class of mappings with a fixed point either in a sufficiently small neighborhood of zero or on the boundary of the unit ball has the KE-property. The results concerning the first of these cases were published in [3–5], the second case is studied in the present paper.

In the classical Königs theorem, a holomorphic transformation f of the unit circle D in the complex plane has an (interior) fixed point z_0 at which the derivative is nonzero: $f'(z_0) \neq 0$. Because the transformation $\varphi(z) = -\dfrac{z - z_0}{1 - z\bar{z}_0}$ is a biholomorphic automorphism of D taking 0 into z_0 and conversely, without loss of generality, we can assume that $z_0 = 0$. A similar argument shows that the latter statement also holds in the general case of an interior fixed point of a holomorphic transformation of an operator ball. In [1–2], deep methods of infinite-dimensional complex analysis are used to show that an l.f.m. $\mathcal{F}_A$ with an upper triangular block-matrix A satisfying the condition $A_{11} = I$ (i.e., of the form $\mathcal{F}_A(K) = A_{22}K(I + A_{12}K)^{-1}$) and such that the operator A_{22} is uniformly positive has the KE-property. In this case, it is clear that $\mathcal{F}_A$ has the fixed point $K_0 = 0$ and $\mathcal{F}'_A(0) = A_{22} \gg 0$. In [3–5], the theorem stating that such an l.f.m. has the KE-property is generalized to the cases dual with respect to the operator conjugation and inversion, including the cases $A_{22} = I$ and $A_{11} \gg 0$ and the affine transformations. Thus, the Königs theorem is proved for any (upper or lower) triangular plus-operator one of whose principal diagonal elements is equal to I and the other is uniformly positive.

On the other hand, it is easy to see that the inequality $a(= f'(z)) > 0$ holds for a nondegenerate affine ($f(z) = az + b$) mapping of the unit circle of the complex plane into itself with a boundary fixed point. This fact leads to the idea to generalize the latter inequality to the general case of affine $\mathcal{F}_A$ and then, using the above methods, to prove that any $\mathcal{F}_A$ with a triangular A and at least one identical element on the principal diagonal and with a boundary fixed point has the KE-property. But it became clear that, in many statements, one must consider only extreme boundary fixed points rather than any arbitrary boundary fixed points (see Example 8 below).

Moreover, it was shown that, in the case under study, the second diagonal element of the operator block-matrix A is not only uniformly positive but also scalar, i.e., it is equal to αI, where $\alpha > 0$. But in the general case (i.e., without the assumption about the identity of at least one of the diagonal elements), some simple formulas relating the diagonal elements of the operator block matrix A were obtained.

Thus, on one hand, the present paper continues the line of investigations opened by the classical papers of G. Königs, J. Hadamard, and other authors. On the other hand, we present new conditions for the embedding of a discrete semigroup of iterates of a plus-operator A acting in a Krein space into a continuous semigroup of plus-operators $\{A^t\}_{t>0}$.

The results were reported at the Workshop on Krein Spaces (Berlin 2007), at IWOTA 2008 (USA), and at KROMSH 2007 (Ukraine); a short version will be published in the Russian journal "Funktsionalnyi analiz i ego prilojenia" (Functional Anal. Appl.).

2. Statement of the problem and main results

Let $\mathfrak{H} = \mathfrak{H}_1 \oplus \mathfrak{H}_2$ be a complex indefinite ($\min\{\dim \mathfrak{H}_1, \dim \mathfrak{H}_2\} > 0$) Krein space, and let $\mathcal{K}$ be the open unit ball of the space $\mathcal{L}(\mathfrak{H}_1, \mathfrak{H}_2)$. The formula

$$K'_+ = \mathcal{F}_A(K_+) = (A_{21} + A_{22}K_+)(A_{11} + A_{12}K_+)^{-1}, \tag{1}$$

where K_+, $K'_+ \in \mathcal{L}(\mathfrak{H}_1, \mathfrak{H}_2)$ and $A_{ij} \in \mathcal{L}(\mathfrak{H}_j, \mathfrak{H}_i)$ for $i, j = 1, 2$, determines the l.f.m. $\mathcal{F} = \mathcal{F}_A = \{K_+, K'_+\}$ in $\mathcal{L}(\mathfrak{H}_1, \mathfrak{H}_2)$.

As a rule, we also preserve the notation $\mathcal{F}_A$ for the natural restrictions of the mapping under study.

In what follows, unless otherwise specified, we use definitions from functional analysis and the theory of spaces equipped with an indefinite metric (see the fundamental monographs [6] and [7]). In particular, a bounded linear operator T defined everywhere in $\mathfrak{H}$ is called a plus-operator if the inequality $[x, x] \geq 0$ implies the inequality $[Tx, Tx] \geq 0$. A plus-operator T is said to be strict if $\inf\limits_{[x,x]=1} [Tx, Tx] > 0$

and is said to be bistrict if, in addition, T^* is also a strict plus-operator. Further, according to the terminology introduced in [8–10], an operator A with the block matrix $\begin{pmatrix} A_{11} & A_{12} \\ A_{21} & A_{22} \end{pmatrix}$ is said to be upper triangular if $A_{21} = 0$ and lower triangular if $A_{12} = 0$.

We recall that the extreme points of the unit ball of the space $\mathcal{L}(\mathfrak{H}_1, \mathfrak{H}_2)$ are the maximal partial isometries, i.e., the isometries $(C\colon C^*C = I)$ and the coisometries $(C\colon CC^* = I)$ [11, Chapter XIII, Sec. 99].

We assume that a plus-operator A determines the affine mapping $\mathcal{F}_A$,

$$\mathcal{F}_A(K) = (A_{21} + A_{22}K)A_{11}^{-1}, \tag{2}$$

of the open ball $\mathcal{K}$ into itself. Obviously, $\mathcal{F}_A(\bar{\mathcal{K}}) \subseteq \bar{\mathcal{K}}$; in addition, the l.f.m. $\mathcal{F}_A\colon \bar{\mathcal{K}} \to \bar{\mathcal{K}}$ is continuous in the weak operator topology. Because, in addition, the nonempty convex set $\bar{\mathcal{K}}$ is compact in this topology, the mapping $\mathcal{F}_A$ has at least one fixed point in $\bar{\mathcal{K}}$ by the Schauder–Tikhonov–Glicksberg theorem (see, e.g., [7, Chapter. III, Theorem 3.6]).

Let C be a fixed point of an l.f.m. $\mathcal{F}_A$ of the form (2). Then $A_{21} + A_{22}C = CA_{11}$, or $A_{21} = CA_{11} - A_{22}C$. Substituting the last relation in (2), we obtain

$$\mathcal{F}_A(K) = A_{22}(K - C)A_{11}^{-1} + C, \tag{3}$$

where $K \in \mathcal{K}$.

Definition 1 (cf. [11, Chapter XIII, Sec. 98). The set of vectors x from the domain a linear operator C which satisfy the condition $\|Cx\| = \|x\|$ is called the *initial set* $\mathcal{J}(C)$ of the linear operator C The set $C(\mathcal{J}(C))$ is called the *final set* $R(C)$ of the linear operator C.

Theorem 2. *Let $\mathcal{F}_A\colon \mathcal{K} \to \mathcal{K}$ be an affine l.f.m. of the form*

$$\mathcal{F}_A(K) = A_{22}(K - C) + C,$$

where $K \in \mathcal{K}$ and $\|C\| = 1$. If $A_{22}x = \alpha x$, where $x \in R(C)\backslash\{0\}$, then $\alpha \in (0, 1]$.

Proof. Let $\|x\| = 1$, $\beta \in \mathbb{C}$, $|\beta| < 1$. Then $\mathcal{F}_A(\beta C) \in \mathcal{K}$, which implies that $\gamma = \|(A_{22}(\beta C - C) + C)y\| < 1$ for any vector y: $\|y\| = 1$. Since $x \in R(C)$, there exists y such that $\|y\| = 1$ and $Cy = x$. We obtain

$$\gamma = \|(A_{22}(\beta - 1)I + I)x\| = |\alpha(\beta - 1) + 1| < 1.$$

Thus, the affine mapping $w(\beta) = \alpha(\beta - 1) + 1$ of the complex plane takes the disk $\{\beta\colon |\beta| < 1\}$ into itself. The mapping $w(\beta)$ is a composition of the similarity transformation centered at point 1 with the coefficient $|\alpha|$ and the rotation about point 1 by an angle $\arg \alpha$ anticlockwise. Hence $\arg \alpha = 0$, $\alpha \in (0, 1]$. The proof of the theorem is complete. □

Corollary 3. *Let $\mathcal{F}_A\colon \mathcal{K} \to \mathcal{K}$ be an affine l.f.m. of the form*

$$\mathcal{F}_A(K) = A_{22}(K - C) + C,$$

where $K \in \mathcal{K}$ and C is a coisometry. Then $\sigma_p(A_{22}) \subseteq (0, 1]$.

Proof. We have $R(C) = J(C^*)$. Since C^* is an isometry, we also have $J(C^*) = \mathfrak{H}_2$. Now the rest of the proof follows from Theorem 2. □

Lemma 4. *Under the conditions of Theorem 2, $(A_{22}x, x) \neq 0$ for $x \in R(C)\backslash\{0\}$.*

Proof. Let $(A_{22}x, x) = 0$, where $A_{22}x \neq 0$, and let $x = Cy$, where $\|x\| = \|y\|$. Then $\|A_{22}(0 - C)y + Cy\| = \| - A_{22}x + x\| > \|x\|$, and we obtain $\|\mathcal{F}_A(0)\| > 1$. This is a contradiction. □

Corollary 5. *Under the conditions of Corollary 3, $(A_{22}x, x) \neq 0$ for $x \neq 0$.*

Theorem 6. *Under the conditions of Corollary 3, $A_{22} = \alpha I$, where $\alpha \in (0, 1]$.*

Proof. Let $\mathcal{P}\colon \mathfrak{H}_2 \to \mathfrak{H}_2'$ be an arbitrary two-dimensional orthoprojection. It suffices to prove that $\mathcal{P}A_{22}\mathcal{P} = \alpha I$. We assume that this is not the case for some $B = \mathcal{P}A_{22}\mathcal{P}$. In $\mathfrak{H}_2'$, we choose an orthonormal basis $\{e_1, e_2\}$ whose vectors are not eigenvectors of B.

It follows from Corollary 5 that all the entries of the matrix $\begin{pmatrix} \alpha & \gamma \\ \beta & \tau \end{pmatrix}$ of the operator B are nonzero in this basis. Corollary 5 also implies $\alpha\tau - \gamma\beta \neq 0$.

We define the operator $\widehat{K} \in \mathcal{L}(\mathfrak{H}_2)$ by the relations $\widehat{K}e_1 = xe_1 + ze_2$, where $1 > x > \dfrac{|\alpha|^2 - |\gamma|^2}{|\alpha|^2 + |\gamma|^2}$, $z = \dfrac{\alpha(1-x)}{\gamma}$, and $\widehat{K}(e_1^\perp) = \{0\}$. Since $|x|^2 + |z|^2 < 1$, it follows that $\|\widehat{K}\| < 1$. We have

$$\begin{pmatrix} \alpha & \gamma \\ \beta & \tau \end{pmatrix}\begin{pmatrix} x & 0 \\ z & 0 \end{pmatrix}\begin{pmatrix} 1 \\ 0 \end{pmatrix} = \begin{pmatrix} \alpha \\ \beta x + \tau z \end{pmatrix}.$$

Consider the vector $a = (A_{22}(\widehat{K} - I) + I)e_1$. We have $\mathcal{P}a = e_1 + (\beta x + \tau z - \beta)e_2$. Since $\alpha\tau \neq \gamma\beta$, we see that $\beta x + \tau z - \beta \neq 0$.

Hence $\|\mathcal{P}a\| > 1$, which implies that $\|a\| > 1$ and $\|\mathcal{F}_A(\widehat{K}C\| > 1$. This is a contradiction. □

Theorem 7. *Let $\mathcal{F}_A\colon \mathcal{K} \to \mathcal{K}$ be an affine l.f.m. of the form*

$$\mathcal{F}_A(K) = (K - C)A_{11}^{-1} + C,$$

where $K \in \mathcal{K}$ and C is an isometry. Then $A_{11}^{-1} = \alpha I$, where $\alpha \in (0, 1]$.

Proof. The proof of this theorem is similar to that of Theorem 6. □

The following argument shows that the coisometry in Theorem 6 or the isometry in Theorem 7 cannot be replaced by an arbitrary linear operator with norm 1.

Example 8 (V. Shulman). Let $\{e_i\}_{i=1}^3$ be an orthonormal basis in the space $\mathfrak{H}$, $\mathfrak{H}_1 = \operatorname{Lin}\{e_1\}$, $\mathfrak{H}_2 = \operatorname{Lin}\{e_2, e_3\}$, $Ae_1 = e_1$, $Ae_2 = e_2$, $Ae_3 = 0$. Obviously, $\mathcal{F}_A(\mathcal{K}) \subseteq \mathcal{K}$ and $\mathcal{F}_A(C) = C$ for $Ce_1 = e_2$. But $A_{22} \neq \alpha I$.

We note that since $\operatorname{Ker} A \neq \{0\}$, the operator A cannot be included in any continuous semigroup of operators $\{A(t)\}_{t \geq 0}$ satisfying the conditions $A(0) = I$ and $A(s + t) = A(s)A(t)$ for all $s, t \geq 0$. Indeed, in the case of such an inclusion, we would have $0 \in \sigma(A(\frac{s}{n}))$, where $A(s) = A$, for all $n \in N$, which contradicts the fact that $A(t)$ is continuous at the point $t = 0$.

But we note that if the operator $A_{22}(A_{11})$ is a priori collinear to I, then a more precise statement about it can be obtained for any linear operator C with the norm 1.

Proposition 9. *Let $\mathcal{F}_A \colon \mathcal{K} \to \mathcal{K}$ be an affine l.f.m. of the form*

$$\mathcal{F}_A(K) = \alpha(K - C) + C, \tag{4}$$

where $K \in \mathcal{K}$, $\alpha \in \mathbb{C}$, and $\|C\| = 1$.

Then $\alpha \in (0, 1]$.

Proof. The proof of this proposition is similar to that of Theorem 2. □

Remark 10. It is easy to see that the statement of Remark 9 is invertible: for any $C \in \mathcal{L}(\mathfrak{H}_1, \mathfrak{H}_2)$, $\|C\| = 1$, and $\alpha \in (0, 1]$, formula (4) determines an affine l.f.m. $\mathcal{F}_A \colon \mathcal{K} \to \mathcal{K}$.

Theorem 11. *Let $\mathcal{F}_A \colon \mathcal{K} \to \mathcal{K}$ be an affine l.f.m. of the form (3), where $K \in \mathcal{K}$ and A_{11} and C are unitary operators.*

Then $A_{22} = \alpha C A_{11} C^{-1}$, where $\alpha \in (0, 1]$.

Proof. We set

$$\begin{aligned}\psi_1(K) &= KA_{11},\\ \psi_2(K) &= (A_{21} + A_{22}K)A_{11}^{-1},\\ \psi_3(K) &= CA_{11}^{-1}C^{-1}K.\end{aligned}$$

It follows from the conditions of the theorem that all these mappings take the open ball $\mathcal{K}$ into itself. Hence the composition

$$\psi(K) = \psi_3(\psi_2(\psi_1(K))) = CA_{11}^{-1}C^{-1}(A_{21}A_{11}^{-1} + A_{22}K)$$

has the same property.

Next, we have

$$\begin{aligned}\psi_2(C) &= C,\\ \psi_2(\psi_1(CA_{11}^{-1})) &= C,\\ \psi_3(\psi_2(\psi_1(CA_{11}^{-1}))) &= CA_{11}^{-1}C^{-1}\psi_2(\psi_1(CA_{11}^{-1})) = (CA_{11}^{-1}C^{-1})C = CA_{11}^{-1}.\end{aligned}$$

Thus, $\psi(CA_{11}^{-1}) = CA_{11}^{-1}$, and the mapping $\psi \colon \mathcal{K} \to \mathcal{K}$ satisfies the conditions of Theorem 6. Applying this theorem, we complete the proof. □

Theorem 12. *Let $\mathcal{F}_A \colon \mathcal{K} \to \mathcal{K}$ be an affine l.f.m. of the form* (3), *where $K \in \mathcal{K}$, C and A_{22} are unitary operators. Then the statement of Theorem* 11 *holds.*

Proof. We consider the affine mapping ψ, which implies that

$$\psi(K) = (KA_{11}^{-1} + C - A_{22}CA_{11}^{-1})C^{-1}A_{22}C.$$

Just as in Theorem 11, we prove that the mapping ψ takes the open ball $\mathcal{K}$ into itself. Then we obviously have $\psi(A_{22}C) = A_{22}C$, and hence we are under the conditions of Theorem 7. To complete the proof, it suffices to use this theorem. □

Proposition 13. *Suppose that, on some subset of the space $\mathcal{L}(\mathfrak{H}_1, \mathfrak{H}_2)$, an upper triangular plus-operator A such that $0 \in \rho(A_{11})$ determines an l.f.m. $\mathcal{F}_A$; suppose also that $\mathcal{F}_A(C) = C$, where $C \neq 0$. Then*

(a) *A is a bistrict plus-operator;*
(b) *the l.f.m. $\mathcal{F}_A$ and $\mathcal{F}_{A^*}$ are determined at any point of the closed ball $\bar{\mathcal{K}}$;*
(c) *$\mathcal{F}_A(\mathcal{K}) \subseteq \mathcal{K}$, $\mathcal{F}_{A^*}(\mathcal{K}) \subseteq \mathcal{K}$, $\mathcal{F}_A(\bar{\mathcal{K}}) \subseteq \bar{\mathcal{K}}$, and $\mathcal{F}_{A^*}(\bar{\mathcal{K}}) \subseteq \bar{\mathcal{K}}$.*

Proof. (a) First, we prove that the plus-operator A is strict. Assume the contrary. Then it follows from Theorem 4.3 [7] that the lineal $A\mathfrak{H}$ is nonnegative, and hence the fact that $A\mathfrak{H}_1 = \mathfrak{H}_1$ implies that $A\mathfrak{H} = \mathfrak{H}_1$, which contradict the fact that $\mathcal{F}_A$ has a fixed point $C \neq 0$.

Now we prove that the strict plus-operator A is bistrict. This readily follows from Theorem 4.17 [7].

(b) This statement follows from (a) and Theorem 4.17 [7].

(c) This statement readily follows from (b) and Corollary 4.13 [7]. □

Lemma 14. *Let C be a unitary operator under the conditions of Proposition* 13. *Then $\mathcal{F}_{A^*}(-C) = -C$.*

Proof. We have

$$\begin{aligned} A_{22}C &= C(A_{11} + A_{12}C), \\ C^{-1}A_{22} &= A_{11}C^{-1} + A_{12}, \\ A_{12} + (-C^{-1})A_{22} &= A_{11}(-C^{-1}), \\ A_{12}^* + A_{22}^*(-C^{-1})^* &= (-C^{-1})^*A_{11}^*, \\ (A_{12}^* + A_{22}^*(-C^{-1})^*)(A_{11}^*)^{-1} &= (-C^{-1})^*. \end{aligned}$$

It remains to use the fact that $(-C^{-1})^* = -C$. The proof of the lemma is complete. □

Theorem 15. *Suppose that at least one of the operators A_{11} and A_{22} is unitary under the conditions of Lemma* 14. *Then the statement of Theorem* 11 *holds.*

Proof. This theorem is proved by using Proposition 13, Lemma 14, Theorem 11, and Theorem 12. □

From Theorems 6, 7, and 15, using Proposition 13 in the present paper and Theorem 3 in [3], we derive the main statements of the present paper.

Theorem 16. *Let $\mathcal{F}_A \colon \mathcal{K} \to \mathcal{K}$ be an affine l.f.m. of the form* (3), *where $K \in \mathcal{K}$ and*

either $A_{11} = I$ and C is a coisometry
or $A_{22} = I$ and C is an isometry.

Then the mapping $\mathcal{F}_A$ has the KE-property.

Proof. Because of Theorem 6 of the present paper in the first case and because of Theorem 7 of the present paper in the second case, we are under the conditions of Lemmas 1–3 from [3]. The rest of the proof can be readily obtained by using these lemmas. □

Theorem 17. *Each mapping $\mathcal{F}_A\colon \mathcal{K} \to \mathcal{K}$ satisfying the conditions of Proposition* 9 *has the KE-property.*

Proof. Because of Proposition 9, the proof follows from Lemmas 1–3 in [3]. □

Theorem 18. *Assume that, on some subset of the space $\mathcal{L}(\mathfrak{H}_1, \mathfrak{H}_2)$, the upper triangular plus-operator A such that $0 \in \rho(A_{11})$ determines an l.f.m. $\mathcal{F}_A$ such that $\mathcal{F}_A(C) = C$, where C is a unitary operator and at least one of the operators A_{11} and A_{22} is the unit operator. Then the restriction $\mathcal{F}_A\big|_{\mathcal{K}}$ has the KE-property.*

Proof. It follows from Proposition 13 that $\mathcal{F}_A\big|_{\mathcal{K}}$ maps the open ball $\mathcal{K}$ into itself. It also follows from Theorem 15 that this mapping satisfies the conditions of Lemmas 1–3 in [3]. Now the statement of the theorem follows from these lemmas. □

References

[1] V. Khatskevich, S. Reich, and D. Shoikhet, *Schröder's Functional Equation and the Königs Embedding Property.* Nonlin. Anal. **47** (2001), 3977–3988.

[2] V. Khatskevich, S. Reich, and D. Shoikhet, *Abel–Schröder Equations for Linear Fractional Mappings and the Königs Embedding Problem.* Acta Sci. Math. (Szeged) **69** (2003), 67–98.

[3] V. Khatskevich and V. Senderov, *The Königs Problem for Linear-Fractional Mappings.* Dokl. RAN **403** (2005), no. 5, 607–609 (in Russian).

[4] M. Elin and V. Khatskevich, *The Königs Embedding Problem for Operator Affine Mappings.* Contemporary Math. **382** (2005), 113–120.

[5] M. Elin and V. Khatskevich, *Triangular Plus-Operators in Banach Spaces: Applications to the Königs Embedding Problem.* J. Nonlinear and Convex Analysis **6** (2005), no. 1, 173–185.

[6] N. Danford and J. Schwartz, *Linear Operators.* Pt. 1: *General Theory.* New York, London: Intersci. Publ., 1958.

[7] T.Ya. Azizov and I.S. Iokhvidov, *Linear Operators in Spaces with Indefinite Metric.* Moscow: Nauka, 1986.

[8] V. Khatskevich and V. Senderov, *The Abel–Schröder Equations for Linear Fractional Maps of Operator Balls.* Dokl. Ross. Akad. Nauk **379** (4), 455–458 (2001).

[9] V. Khatskevich and V. Senderov, *Abel–Schröder Type Equations for Maps of Operator Balls.* Funct. Different. Equats. **10** (1/2), 239–258 (2003).

[10] V. Khatskevich and V. Senderov, *Basic Properties of Linear Fractional Mappings of Operator Balls: Scröder's Equation.* Fields Inst. Communs. **25**, 331–344 (2000).

[11] Paul R. Halmos, *A Hilbert Space Problem Book.* Toronto: London, 1967.

[12] G. Königs, *Recherches sur les intégrales de certaines équations fonctionnelles.* Annales Sci. École Normale Sup. (Sér 3), **1**, supplément, 3–41 (1884).

[13] P. Lévy, *Fonctions à croissance régulière et itération d'ordre fractionnaire.* Ann. Mat. Pura Appl. **5**, 269–298 (1928).

[14] J. Hadamard, *Two Works on Iteration and Related Questions.* Bull. Amer. Math. Soc. **50**, 67–75 (1944).

[15] T.E. Harris, *Some Mathematical Models for Branching Processes.* in *2nd Berkeley Symposium* (1951), pp. 305–328.

[16] I.N. Baker, *Fractional Iteration near a Fixed Point of Multiplier* 1. J. Australian Math. Soc. **4**, 143–148 (1964).

[17] S. Karlin and J. McGregor, *Embedding Iterates of Analytic Functions with Two Fixed Points into Continuous Groups.* Trans. Amer. Math. Soc. **132**, 137–145 (1968).

[18] C.C. Cowen, *Iteration and the Solution of Functional Equations for Functions Analytic in the Unit Disk.* Trans Amer. Math. Soc. **265**, 69–95 (1981).

[19] A.G. Siskakis, *Weighted composition semigroups on Hardy spaces.* Linear Algebra Appl. **84**, 359–371 (1986).

[20] C.C. Cowen and B.D. MacCluer, *Linear Fractional Maps of the Ball and Their Composition Operators.* Acta Sci. Math. (Szeged) **66**, 351–376 (2000).

[21] D. Alpay and V. Khatskevich, *Linear Fractional Transformations: Basic Properties, Applications to Spaces of Analytic Functions and Schröder's Equation.* Internat. J. Appl. Math. **2**, 459–476 (2000).

[22] Maria J. Martin, *Composition Operators with Linear Fractional Symbols and Their Adjoints.* in *First Advanced Course in Operator Theory and Complex Analysis.* University of Seville, June 2004.

[23] T.Ya. Azizov, A.I. Barsukov, and A. Dijksma, *The Cauchy Problem Associated with an* (ω, W)*-Dissipative Operator.* Methods of Funct. Anal. and Topology **10** (3), 1–6 (2004).

[24] T.Ya. Azizov, A.I. Barsukov, and A. Dijksma, *Decompositions of a Krein Space in Regular Subspaces Invariant under a Uniformly Bounded* C_0*-Semigroup of Bicontractions.* J. of Funct. Anal. **211** (2), 324–354 (2004).

V.A. Khatskevich
ORT Braude Academic College
College Campus
P.O. Box 78
Karmiel 21982, Israel
e-mail: `victor_kh@hotmail.com`

V.A. Senderov
Pyatnitskoe highway 23-2-156
Moscow, 125430, Russia
e-mail: `senderov@mccme.ru`

Operator Theory:
Advances and Applications, Vol. 198, 239–258

Bisectors, Isometries and Connected Components in Hilbert Spaces

Jean-Philippe Labrousse

Abstract. Let $\mathcal{H}$ be a separable Hilbert space over $\mathbb{C}$ and let $F(\mathcal{H})$ be the set of all closed linear subspaces of $\mathcal{H}$. For M, $N \in F(\mathcal{H})$ set $g(M,N) = \|P_M - P_N\|$ (known as the *gap metric*) where P_M, P_N denote respectively the orthogonal projections in $\mathcal{H}$ on M and on N. For M, $N \in F(\mathcal{H})$ such that ker $(P_M + P_N - I) = \{0\}$ the *bisector* $\beta(M,N)$ of M and N is a uniquely determined element of $F(\mathcal{H})$ such that (setting $\beta(M,N) = W$ and $U_W = 2P_W - I$)

(i) $P_W U_W = U_W P_N$ and

(ii) $(P_M + P_N)U_W = U_W(P_M + P_N)$ is positive definite.

A mapping Φ of $F(\mathcal{H})$ into itself is called an *isometry* if

$$\forall M,\ N \in F(\mathcal{H}),\ g(M,N) = g(\Phi(M),\Phi(N)).$$

This paper may be considered as a sequel to [4]) since it relies heavily on the notion of bisector defined therein, as well as the notation and several results proved in that earlier work, in order to determine the arcwise connected components of $F(\mathcal{H})$ and the properties of isometry on that space. This leads to a number of applications to linear relations.

Mathematics Subject Classification (2000). 47A06, 47A05.

Keywords. Linear relations, isometries, connected components.

1. The space of projections on a Hilbert space

Notation. In this paper $\mathcal{H}$ will denote a separable Hilbert space over $\mathbb{C}$, with norm $\|\,.\|$ and scalar product $\langle\,.\,,.\rangle$ and $L(\mathcal{H})$ will denote the Banach algebra of all bounded linear operators on $\mathcal{H}$. If $T \in L(\mathcal{H})$ then ker (T) and ran (T) will denote respectively the kernel and the range of T. Let $P(\mathcal{H})$ be the set of all orthogonal projections on closed subspaces of $\mathcal{H}$ and let $R(\mathcal{H})$ denote the set of all reflections (i.e., the set of all symmetric unitary operators) on $\mathcal{H}$. Both $P(\mathcal{H})$ and $R(\mathcal{H})$ are closed subspaces of $L(\mathcal{H})$ and are closely related to each other via the homeomorphism which maps every $P \in P(\mathcal{H})$ on $Q = 2P - \mathcal{I} \in R(\mathcal{H})$ where $\mathcal{I}$ is the identity mapping of $\mathcal{H}$ onto itself.

Finally let $F(\mathcal{H})$ denote the set of all closed linear subspaces of $\mathcal{H}$ and if $M \in F(\mathcal{H})$ let P_M denote the orthogonal projection of $\mathcal{H}$ onto M (equivalently $M = \operatorname{ran}(P_M)$). Then $F(\mathcal{H})$ is also closely related to $P(\mathcal{H})$ via the mapping $M \mapsto P_M$ which is an homeomorphism if we take as a metric on $F(\mathcal{H})$ the so-called gap metric defined by $g(M,N) = \|P_M - P_N\|$ where $M, N \in F(\mathcal{H})$.

Remark 1.1. *Since $F(\mathcal{H})$, $P(\mathcal{H})$ and $R(\mathcal{H})$ are so closely related most results obtained in this paper concerning one of them can readily be transposed to the other two but we shall not do so explicitly every time.*

Definition 1.2. *Let $M, N \in F(\mathcal{H})$. Set $\alpha = \dim M \cap N^\perp$ and $\beta = \dim M^\perp \cap N$*

$$\operatorname{Ind}_N(M) = \begin{cases} \alpha - \beta & \text{if} \quad \max\{\alpha,\beta\} < \infty, \\ \infty & \text{if} \quad \alpha = \infty \text{ and } \beta < \infty, \\ -\infty & \text{if} \quad \alpha < \infty \text{ and } \beta = \infty, \\ 0 & \text{if} \quad \alpha = \infty \text{ and } \beta = \infty. \end{cases} \tag{1.1}$$

Then

$$\operatorname{Ind}_N(M) = \operatorname{Ind}_{M^\perp}(N^\perp) = -\operatorname{Ind}_M(N) = -\operatorname{Ind}_{N^\perp}(M^\perp). \tag{1.2}$$

If $P, Q \in P(\mathcal{H})$ set $\operatorname{Ind}_Q(P) = \operatorname{Ind}_N(M)$ where $M = \operatorname{ran}(P)$ and $N = \operatorname{ran}(Q)$. If $U, V \in R(\mathcal{H})$ set $\operatorname{Ind}_V(U) = \operatorname{Ind}_Q(P)$ where $P = (\mathcal{I}+U)/2$, $Q = (\mathcal{I}+V)/2$

Definition 1.3. *Let X be a metric space with metric g and let a, $b \in X$. Set*

$$\operatorname{equid}(a,b) = \{c \in X \mid g(a,c) = g(c,b)\},$$

$$\operatorname{mid}(a,b) = \{c \in \operatorname{equid}(a,b) \mid \forall d \in \operatorname{equid}(a,b),\ g(a,c) \leq g(a,d)\}.$$

Definition 1.4. *Let $P, Q \in P(\mathcal{H})$. Then*

$$\operatorname{cent}(P,Q) = \{R \in P(\mathcal{H}) \mid R \in \operatorname{mid}(P,Q) \text{ and } (2R - \mathcal{I})P = Q(2R - \mathcal{I})\}.$$

Let $U, V \in R(\mathcal{H})$. Then

$$\operatorname{cent}(U,V) = \{W \in \operatorname{mid}(U,V) \mid WU = VW\}.$$

Proposition 1.5. *Let $M, N \in F(\mathcal{H})$ with $M + N^\perp$ closed. Then there exists a projection $P \in P(\mathcal{H})$ such that $\operatorname{ran}(P) = N^\perp \cap (M \cap N^\perp)^\perp$ and $\ker(P) = M + M^\perp \cap N$.*

Proof. The proposition is equivalent to stating that if $M + N^\perp$ is closed then

(i) $N^\perp \cap (M \cap N^\perp)^\perp \cap (M + M^\perp \cap N) = \{0\}$,

(ii) $N^\perp \cap (M \cap N^\perp)^\perp + M + M^\perp \cap N = \mathcal{H}$.

To prove (i) let $u \in N^\perp \cap (M \cap N^\perp)^\perp \cap (M + M^\perp \cap N) = \{0\}$. Then $u = s + t$ with $s \in M$ and $t \in M^\perp \cap N$ and clearly $\langle t, u\rangle = \langle t, s\rangle = 0$ so that $\|t\|^2 = \langle t, u\rangle - \langle t, s\rangle = 0$. Hence $t = 0$ so that $u = s \in M \cap N^\perp \cap (M \cap N^\perp)^\perp = 0$.

In order to establish (ii) note that if $M + N^\perp$ is closed then

$$\mathcal{H} = M + N^\perp + (M + N^\perp)^\perp = M + N^\perp + M^\perp \cap N.$$

Also $N^{\perp} = N^{\perp} \cap (M \cap N^{\perp})^{\perp} + M \cap N^{\perp}$ so that

$$M + N^{\perp} = M + N^{\perp} \cap (M \cap N^{\perp})^{\perp} + M \cap N^{\perp} = M + N^{\perp} \cap (M \cap N^{\perp})^{\perp}.$$

Hence

$$\mathcal{H} = M + N^{\perp} \cap (M \cap N^{\perp})^{\perp} + M^{\perp} \cap N = N^{\perp} \cap (M \cap N^{\perp})^{\perp} + M + M^{\perp} \cap N$$

and the proposition is proved. □

Proposition 1.6. *Let* $M, N \in F(\mathcal{H})$. *Then*

$$\dim\, [M^{\perp} \cap (M^{\perp} \cap N)^{\perp}] = \dim\, [N^{\perp} \cap (N^{\perp} \cap M)^{\perp}]. \tag{1.3}$$

Proof. Assume first that $\dim\, [N^{\perp} \cap (N^{\perp} \cap M)^{\perp}] < \infty$. Then $M + N^{\perp} = M + N^{\perp} \cap (N^{\perp} \cap M)^{\perp}$ is closed. Using the notation of the previous proposition, consider $P[M^{\perp} \cap (M^{\perp} \cap N)^{\perp}] \subseteq N^{\perp} \cap (N^{\perp} \cap M)^{\perp}$. Since

$$[M^{\perp} \cap (M^{\perp} \cap N)^{\perp}] \cap \ker\, (P) \subseteq (M^{\perp} \cap N)^{\perp} \cap (M^{\perp} \cap N) = \{0\}$$

it follows that $\dim\, [M^{\perp} \cap (M^{\perp} \cap N)^{\perp}] \leq \dim\, [N^{\perp} \cap (N^{\perp} \cap M)^{\perp}] < \infty$. Then

$$N + M^{\perp} = N + M^{\perp} \cap (M^{\perp} \cap N)^{\perp}$$

is closed and as above (interchanging M with N) this implies that

$$\dim\, [N^{\perp} \cap (N^{\perp} \cap M)^{\perp}] \leq \dim\, [M^{\perp} \cap (M^{\perp} \cap N)^{\perp}]$$

and the proposition is proved in that case. Clearly the proposition is also true if $\dim\, [M^{\perp} \cap (M^{\perp} \cap N)^{\perp}] < \infty$ and the only case left is when both dimensions are infinite an therefore also equal. □

Corollary 1.7. *Let* $M, N \in F(\mathcal{H})$. *Then*

(i) $\mathrm{Ind}_N(M) = 0 \Rightarrow \dim\, M^{\perp} = \dim\, N^{\perp}$.

(ii) *If* $\max\{\dim\, M^{\perp}, \dim\, N^{\perp}\} < \infty$, $\dim\, M^{\perp} = \dim\, N^{\perp} \Rightarrow \mathrm{Ind}_N(M) = 0$.

Proof. (i) First observe that

$$M^{\perp} = M^{\perp} \cap (M^{\perp} \cap N)^{\perp} + M^{\perp} \cap N$$

and

$$N^{\perp} = N^{\perp} \cap (N^{\perp} \cap M)^{\perp} + N^{\perp} \cap M$$

so that

$$\dim\, M^{\perp} = \dim\, (M^{\perp} \cap (M^{\perp} \cap N)^{\perp}) + \dim\, (M^{\perp} \cap N)$$

and

$$\dim\, N^{\perp} = \dim\, (N^{\perp} \cap (N^{\perp} \cap M)^{\perp}) + \dim\, (N^{\perp} \cap M).$$

Since $\mathrm{Ind}_N(M) = 0 \Rightarrow \dim\, (M^{\perp} \cap N) = \dim\, (N^{\perp} \cap M)$ the corollary follows from Proposition 1.6.

(ii) $\max\{\dim\, M^{\perp}, \dim\, N^{\perp}\} < \infty$ implies that

$$\dim\, (M^{\perp} \cap (M^{\perp} \cap N)^{\perp}) < \infty.$$

Hence, in view of Proposition 1.6, $\dim\, M^{\perp} = \dim\, N^{\perp} \Rightarrow \dim\, (M^{\perp} \cap N) = \dim\, (N^{\perp} \cap M)$ that is $\mathrm{Ind}_N(M) = 0$. □

Proposition 1.8. *Let $M, N \in F(\mathcal{H})$ be such that*

$$\dim N = \dim M = \dim N^\perp = \dim M^\perp = \infty.$$

Then there exists $N' \in F(\mathcal{H})$ such that

$$\operatorname{Ind}_M(N') = \operatorname{Ind}_N(N') = 0. \tag{1.4}$$

Proof. We must distinguish two cases.

(i) $\max\{\dim (M \cap N), \dim (M^\perp \cap N^\perp)\} = \infty$.

Without loss of generality we can assume that $\dim (M^\perp \cap N^\perp) = \infty$ since in view of (1.2) we can interchange M and N with $M^\perp$ and $N^\perp$. Then taking $N' = M^\perp \cap N^\perp$, $\dim N' = \infty$ and we see that

$$N' \cap M^\perp = N' \quad \text{and} \quad (N')^\perp \cap M = (\overline{M+N}) \cap M = M$$

so that $\operatorname{Ind}_M(N') = 0$. Also

$$N' \cap N^\perp = N' \quad \text{and} \quad (N')^\perp \cap N = (\overline{M+N}) \cap N = N$$

so that $\operatorname{Ind}_N(N') = 0$.

(ii) $\max\{\dim (M \cap N), \dim (M^\perp \cap N^\perp)\} < \infty$.

Set $N' = [N^\perp + M \cap N] \cap (M^\perp \cap N^\perp)^\perp$. Clearly

$$N' \in F(\mathcal{H}) \quad \text{and} \quad (N')^\perp = N \cap (M \cap N)^\perp + M^\perp \cap N^\perp \in F(\mathcal{H}).$$

Furthermore $M \cap (N')^\perp = M^\perp \cap N' = \{0\}$. Indeed, let $u \in M \cap (N')^\perp$. Then $u = x + y \in M$ with $x \in N \cap (M \cap M)^\perp$ and $y \in M^\perp \cap N^\perp$. Thus

$$\|y\|^2 = \langle y, u\rangle - \langle y, x\rangle = 0$$

so that

$$u = x \in M \cap N \cap (M \cap N)^\perp = \{0\}.$$

Similarly, if $v \in M^\perp \cap N'$ then $v = s + t \in M^\perp$ with $s \in N^\perp$ and $t \in M \cap N$. Then

$$\|t\|^2 = \langle t, v\rangle - \langle t, s\rangle = 0$$

so that $t = 0$ so that

$$v = s \in M^\perp \cap N^\perp \cap (M^\perp \cap M^\perp)^\perp = \{0\}$$

and hence $M^\perp \cap N' = \{0\}$. Therefore $\operatorname{Ind}_M(N') = 0$. Similarly it is easy to see that

$$\dim (N^\perp \cap N') = \dim [N^\perp \cap (M^\perp \cap N^\perp)^\perp] = \infty$$

and

$$\dim (N \cap (N')^\perp) = \dim N \cap (M \cap N)^\perp = \infty$$

so that $\operatorname{Ind}_N(N') = 0$. □

Proposition 1.9. *Let $f(x,y) = x\sqrt{1-y^2} + y\sqrt{1-x^2}$ and let $X, Y, U \in R(\mathcal{H})$ be such that $\frac{1}{4}\|X-U\|^2 + \frac{1}{4}\|Y-U\|^2 \le 1$. Then*

$$\frac{1}{2}\|X-Y\| \le f\left(\frac{1}{2}\|X-U\|, \frac{1}{2}\|Y-U\|\right).$$

Proof. This is Corollary 1.15 of [4] □

2. Bisectors

Proposition 2.1. *Let* $U, V \in R(\mathcal{H})$ *be such that* $\ker\,(U+V) = \{0\}$. *Then there exists a unique* $W \in R(\mathcal{H})$ *such that*

$$UW = WV \tag{2.1}$$

$$(U+V)W = W(U+V) \text{ is positive definite.} \tag{2.2}$$

W will be called the bisector of U and V and we shall write $W = \beta(U, V)$.

Proof. (2.1) and (2.2) follow from Proposition 1.1.6 of [4] □

Remark 2.2. *If* $\|U - V\| < 2$ *then* $\ker\,(U+V) = \{0\}$.

Proposition 2.3. *Let* $U, V \in R(\mathcal{H})$. *The following conditions are equivalent.*

(i) $\mathrm{Ind}_U(V) = 0$.

(ii) *There exists* $W \in \mathrm{cent}(U, V)$.

(iii) *There exists* $W \in R(\mathcal{H})$ *such that*

$$UW = WV, \tag{2.3}$$

$$(U+V)W = W(U+V) \text{ is positive,} \tag{2.4}$$

$$\|U - W\| = \sqrt{2 - \sqrt{4 - \|U-V\|^2}} \le \|U - WUW\| / \sqrt{2}. \tag{2.5}$$

Proof. (i) ⇒ (iii) Set $P = (\mathcal{I} + U)/2$ and $Q = (\mathcal{I} + V)/2$. Then

$$\ker\,(U+V) = \mathrm{ran}\,(P) \cap \mathrm{ran}\,(\mathcal{I} - Q) \oplus \mathrm{ran}\,(\mathcal{I} - P) \cap \mathrm{ran}\,(Q) \tag{2.6}$$

as was proved in Proposition 1.8 of [4] and it is easy to verify that $\ker\,(U+V)$ (and therefore also $(\ker\,(U+V))^{\perp}$) is invariant under U and V. If $\mathrm{Ind}_U(V) = 0$ then

$$\dim\,(\mathrm{ran}\,(P) \cap \mathrm{ran}\,(I - Q)) = \dim\,(\mathrm{ran}\,(\mathcal{I} - P) \cap \mathrm{ran}\,(Q)).$$

Let Φ be an isometry of $\mathrm{ran}\,(P) \cap \mathrm{ran}\,(\mathcal{I} - Q)$ onto $\mathrm{ran}\,(\mathcal{I} - P) \cap \mathrm{ran}\,(Q)$ (such an isometry exists since $\mathcal{H}$ is separable). Then $\Phi^{-1} = \Phi^*$ maps $\mathrm{ran}\,(\mathcal{I} - P) \cap \mathrm{ran}\,(Q)$ onto $\mathrm{ran}\,(P) \cap \mathrm{ran}\,(\mathcal{I} - Q)$ so that $W_0 = \Phi \oplus \Phi^{-1}$ is a reflection which maps $\ker\,(U+V)$ onto itself. Denote by U_0 (resp. V_0) the restriction of U (resp. of V) to $\ker\,(U+V)$ and let $u \in \ker\,(U+V)$ and $u = s + t$ with $s \in \mathrm{ran}\,(P) \cap \mathrm{ran}\,(\mathcal{I} - Q)$ and $v \in \mathrm{ran}\,(\mathcal{I} - P) \cap \mathrm{ran}\,(Q)$. Then $U_0 s = s$, $U_0 t = -t$ and $V_0 s = -s$, $V_0 t = t$ so that

$$W_0 U_0 u = W_0(U_0 s + U_0 t) = W_0 s - W_0 t = V_0 W_0 s + V_0 W_0 t = V_0 W_0 u.$$

Therefore $W_0 U_0 = V_0 W_0$ and $W_0(U_0 + V_0) = 0$ and the restrictions U_1 of U and V_1 of V to $\ker\,(U+V)^{\perp}$ satisfy the conditions of Proposition 2.1 so that there exists a $W_1 \in R(\ker\,(U+V)^{\perp})$ satisfying conditions (2.1) and (2.2). Hence $W = W_0 \oplus W_1 \in R(\mathcal{H})$ satisfies conditions (2.3) and (2.4). (2.5) follows from (2.3), (2.4) and Proposition 1.17 of [4].

(iii) ⇒ (ii) (2.3) and (2.5) imply that $\|U - W\| = \|V - W\|$, i.e., that $W \in \mathrm{equid}(U, V)$ and also that

$$\|U - V\|^2 = \|U - W\|^2 (4 - \|U - W\|^2) \tag{2.7}$$

Let $Z \in \mathrm{mid}(U,V)$ be such that $\|U-W\| > \|U-Z\|$. Then

$$\left\|\frac{U-Z}{2}\right\|^2 + \left\|\frac{V-Z}{2}\right\|^2 < 2\left\|\frac{U-W}{2}\right\|^2 \leq 1$$

and from Proposition 1.9 and from (2.7) it follows that

$$\left\|\frac{U-V}{2}\right\| \leq f\left(\left\|\frac{U-Z}{2}\right\|, \left\|\frac{V-Z}{2}\right\|\right) < f\left(\left\|\frac{U-W}{2}\right\|, \left\|\frac{U-W}{2}\right\|\right) = \left\|\frac{U-V}{2}\right\|$$

which is a contradiction. Hence $\|U-W\| \leq \|U-Z\|$ and consequently $W \in \mathrm{mid}(U,V)$ and in view of (2.3) it follows that $W \in \mathrm{cent}(U,V)$.

(ii) $\Rightarrow$ (i) Let $u \in \mathrm{ran}\,(P) \cap \mathrm{ran}\,(\mathcal{I}-Q)$. Then $\exists x \in \mathcal{H}$ such that $u = P2x = (\mathcal{I}+U)x$ and $\exists y \in \mathcal{H}$ such that $u = (\mathcal{I}-Q)2v = (\mathcal{I}-V)y$. Then

$$Wu = W(\mathcal{I}+U)x = (\mathcal{I}+V)Wx \in R(P)$$

and

$$Wu = W(\mathcal{I}-V)y = (\mathcal{I}-U)Wy \in R(\mathcal{I}-Q).$$

Similarly, W maps $\mathrm{ran}\,(\mathcal{I}-P)\cap\mathrm{ran}\,(Q)$ into $\mathrm{ran}\,(P)\cap\mathrm{ran}\,(\mathcal{I}-Q)$ and since $W^2 = \mathcal{I}$ it follows that W is an isomorphism of $\mathrm{ran}\,(\mathcal{I}-P)\cap\mathrm{ran}\,(Q)$ onto $\mathrm{ran}\,(P)\cap\mathrm{ran}\,(\mathcal{I}-Q)$ so that these two spaces have the same dimension. □

Proposition 2.4. *Let $U, W \in R(\mathcal{H})$ be such that $\|U-W\| \leq \sqrt{2}$. Then*

$$\|U-W\| \leq \|U-WUW\|/\sqrt{2}.$$

Proof. $W(WUW) = UW$.
Also for all $u \in \mathcal{H}$ we have

$$\begin{aligned}\langle (UW+WU)u,u\rangle &= \langle (2I-(U-W)^2)u,u\rangle \\ &= 2\|u\|^2 - \|(U-W)u\|^2 \\ &\geq (2-\|U-W\|^2)\|u\|^2 \geq 0.\end{aligned}$$

Hence $W(U+WUW) = (U+WUW)W = UW+WU$ is positive. Then, from Proposition 2.3 it follows that

$$\|U-W\|^2 = 2-\sqrt{4-\|U-WUW\|^2} \leq \|U-WUW\|^2/2 \tag{2.8}$$

and the proposition is proved. □

Let $U, V \in R(\mathcal{H})$ be such that $\|U-V\| \leq \sqrt{2}$ and define inductively the sequence of sets $\{\Delta_n\}$, $n = 0,1,2,\dots$ as follows

$$\Delta_0 = \left\{X \in R(\mathcal{H}) \mid \|X-U\| = \|X-V\| \leq \sqrt{2-\sqrt{2}}\right\}, \tag{2.9}$$

$$\text{for } n \in \mathbb{N},\ \Delta_n = \{Y \in \Delta_{n-1} \mid \forall X \in \Delta_{n-1},\ \|Y-X\| \leq \sqrt{2}^{\,(1-n)}\}. \tag{2.10}$$

Note that it is quite possible that there exists an n_0 such that the sequence will consist of empty sets for $n \geq n_0$.

Proposition 2.5. *Let $\delta(\Delta_n)$ denote the diameter of Δ_n. Then*

$$\lim_{n\to\infty} \delta(\Delta_n) = 0.$$

Proof. If $X, Y \in \Delta_0$ then since $\frac{1}{4}\|X-U\|^2 + \frac{1}{4}\|Y-U\|^2 \leq 2\,\frac{2-\sqrt{2}}{4} < 1$, we can use Proposition 1.9 to show that

$$\|X-Y\| \leq 2\,f(\|X-U\|/2, \|Y-U\|/2) \leq 2\,f\left(\frac{\sqrt{2-\sqrt{2}}}{2}, \frac{\sqrt{2-\sqrt{2}}}{2}\right) = \sqrt{2}$$

so that $\delta(\Delta_0) \leq \sqrt{2}$.
If $n \geq 1$ then since $\Delta_n \subseteq \Delta_{n-1}$ it follows from (2.10) that

$$\forall X, Y \in \Delta_n,\ \|X-Y\| \leq \sqrt{2}^{(1-n)} \tag{2.11}$$

so that $\delta(\Delta_n) \leq \sqrt{2}^{(1-n)}$ and the proposition follows at once. □

Corollary 2.6.

$$\Delta = \bigcap_{n\geq 0} \Delta_n \text{ is either empty or reduced to one element.}$$

Proof. Obvious. □

Proposition 2.7. *Let $U, V \in R(\mathcal{H})$ be such that $\|U-V\| \leq \sqrt{2}$. Then, if $W = \beta(U,V)$, $W \in \Delta$.*

Proof. This proof is largely inspired by the proof of a similar result in [6]. First we proceed to show by induction that if $Y \in \Delta_n$ then $WYW \in \Delta_n$. If $Y \in \Delta_0$ then $WYW \in \Delta_0$ since

$$\|U - WXW\| = \|WUW - X\| = \|V - X\|$$

and

$$\|V - WXW\| = \|WVW - X\| = \|X - U\|.$$

Also if $Y \in \Delta_n$ then $\forall X \in \Delta_{n-1}$, $\|X-Y\| \leq \sqrt{2}^{(1-n)}$ so that

$$\forall X \in \Delta_{n-1},\ \|WXW - WYW\| \leq \sqrt{2}^{(1-n)}$$

which implies that $WYW \in \Delta_n$ since, using the induction hypothesis, the mapping of Δ_{n-1} into itself given by $X \mapsto WXW$ is surjective. Now we proceed to show by induction that $\forall n \in \mathbb{N}$, $W \in \Delta_n$. The fact that $W \in \Delta_0$ is a consequence of (2.2) and of the fact that $\|U-V\| \leq \sqrt{2}$. Assume now that $W \in \Delta_{n-1}$ with $n \geq 1$. Then from (2.8) we see that $\forall Y \in \Delta_{n-1}$, $\|Y-W\| \leq \sqrt{2}^{(2-n)} \leq \sqrt{2}$. Hence, using (2.5),

$$\forall Y \in \Delta_{n-1},\ \|Y-W\| \leq \|Y - WYW\|/\sqrt{2} \leq \sqrt{2}^{(2-n)}/\sqrt{2} = \sqrt{2}^{(1-n)}$$

and consequently $W \in \Delta_n$. Therefore $\forall n \in \mathbb{N}$, $W \in \Delta_n$ that is, $\Delta = \{W\}$. □

Lemma 2.8. *Let $U, V, W \in R(\mathcal{H})$ be such that $\|U-W\| < \sqrt{2}$ and $\|W-V\| < \sqrt{2}$. Set $W_1 = \beta(U,W)$, $W_2 = \beta(W,V)$, $W = \beta(W_1, W_2)$. Then $W = \beta(U,V)$.*

Proof. Under the conditions of the lemma, using (2.5) we see that

$$\|W_1 - W\| < \sqrt{2-\sqrt{2}} \quad \text{and} \quad \|W - W_2\| < \sqrt{2-\sqrt{2}}$$

so that, using Proposition 1.9, $\|W_1 - W_2\| < \sqrt{2}$. Hence

$$\forall u \in \mathcal{H},\ \langle [W_1W_2 + W_2W_1)]u, u\rangle = 2\|u\|^2 - \|(W_1 - W_2)u\|^2 > 0.$$

Also

$$WUW = WW_1W_1UW_1W_1W = WW_1WWWW_1W = W_2WW_2 = V$$

so that $UW = WV$. Finally

$$W(U+V) = WW_1WW_1 + WW_2WW_2 = W_2W_1 + W_1W_2$$

is self-adjoint and positive definite, as we have just seen. □

Lemma 2.9. *Let $U, V \in R(\mathcal{H})$ be such that* $\mathrm{Ind}_U(V) = 0$ *and let $W \in R(\mathcal{H})$ satisfy (2.3) and (2.4). Set $W_1 = \beta(U,W)$, $W_2 = \beta(W,V)$. Then $W = \beta(W_1, W_2)$.*

Proof. Let $M = WW_1W$. Then

$$MWM = WW_1WWWW_1W = WW_1WW_1W = WUW = V$$

so that $MW = VM$. Also

$$M(V+W) = WW_1UW + WW_1WW = W[W_1(U+W)]W$$

is self-adjoint and positive definite since so is $W_1(U+W)$. Therefore, $M = \beta(W,V) = W_2$ and consequently $W_1W = WW_2$. But then

$$W(W_1 + W_2) = WW_1 + W_1W = W_1(U+W)$$

is positive definite so that finally $W = \beta(W_1, W_2)$. □

Definition 2.10. *Let $U, V \in R(\mathcal{H})$ be such that* $\mathrm{Ind}_U(V) = 0$ *and let $W \in R(\mathcal{H})$ satisfy* (2.3) *and* (2.4). *For all $m, n \in \mathbb{N}$ let $W_{m,n}$ be defined inductively as follows*

(i) $W_{0,0} = U$, $W_{0,1} = V$, $W_{1,1} = W$, $W_{m,n} = 0$ *if* $n < 0$ *or* $n > 2^m$,

(ii) $W_{m+1,2n} = W_{m,n}$ *where* $0 \le n \le 2^m$,

(iii) $W_{m+1,2n+1} = \Psi(W_{m,n}, W_{m,n+1})$, *where* $0 < n < 2^m$.

Remark 2.11. *This definition is given in [4] under slightly more restricted conditions (namely* ker $(U+V) = \{0\}$ *instead of* $\mathrm{Ind}_U(V) = 0$*). It is easy to check that this restriction is, in fact, unnecessary and that the results of §2 and §3 of [4] are still valid under the more general condition.*

Lemma 2.12. *Let $U, V \in R(\mathcal{H})$ be such that* $\mathrm{Ind}_U(V) = 0$ *and let $\|U - V\| = 2\sin\varphi$ with $0 \le \pi/2$. Then*

$$\|W_{m,q} - W_{m,0}\| = 2\sin(q2^{-m}\varphi).$$

Proof. Let us first prove by induction that

$$\|W_{m,q} - W_{m,0}\| \leq 2\sin(q2^{-m}\varphi),\ q = 1, 2, \ldots, 2^m \tag{2.12}$$

For $q = 1$ this follows from Proposition 2.7 of [4]. Suppose that the lemma has been proved for $q = n < 2^m$. Then from Proposition 1.9 we see that

$$\begin{aligned}
\|W_{m,q+1} - W_{m,0}\| &\leq \|W_{m,q} - W_{m,0}\|\sqrt{1 - \tfrac{1}{4}\|W_{m,1} - W_{m,0}\|^2} \\
&\quad + \|W_{m,1} - W_{m,0}\|\sqrt{1 - \tfrac{1}{4}\|W_{m,q} - W_{m,0}\|^2} \\
&\leq 2\sin(q2^{-m}\varphi)\cos(2^{-m}\varphi) + 2\sin(2^{-m}\varphi)\cos(q2^{-m}\varphi) \\
&\leq 2\sin((q+1)2^{-m}\varphi)
\end{aligned}$$

which establishes (2.12). Suppose now that $\|W_{m,q} - W_{m,0}\| < 2\sin(q2^{-m}\varphi)$. Then using again Proposition 1.9 we get, setting $r = 2^m - q$,

$$\begin{aligned}
\|U - V\| &\leq \|W_{m,q} - W_{m,0}\|\sqrt{1 - \tfrac{1}{4}\|W_{m,r} - W_{m,0}\|^2} \\
&\quad + \|W_{m,r} - W_{m,0}\|\sqrt{1 - \tfrac{1}{4}\|W_{m,q} - W_{m,0}\|^2} \\
&< 2\sin(q2^{-m}\varphi)\cos(r2^{-m}\varphi) + 2\sin(r2^{-m}\varphi)\cos(q2^{-m}\varphi) \\
&= 2\sin(\varphi) = \|W_{m,q} - W_{m,0}\|,
\end{aligned}$$

a contradiction. Hence the lemma is proved. □

Lemma 2.13. *Let $U, V \in R(\mathcal{H})$ be such that $\mathrm{Ind}_U(V) = 0$. Then there exist two sequence $\{U_m\} \subseteq R(\mathcal{H})$ and $\{V_m\} \subseteq R(\mathcal{H})$ satisfying the following conditions*

$$\|U - U_m\| = \|V_m - V\| \ \text{and}\ \ \|U_m - V_m\| < \|U - V\| \tag{2.13}$$

$$\lim_{m\to\infty} \|U - U_m\| = \lim_{m\to\infty} \|V - V_m\| = 0 \tag{2.14}$$

$$\forall m, \beta(U_m, V_m) \in \mathrm{cent}\,(U, V). \tag{2.15}$$

Proof. With the notation of [4] take $U_m = W_{m,1}$ and $V_m = W_{m,2^m-1}$. Then setting $\|U - V\| = 2\sin\varphi$ it follows from Proposition 2.7 of [4] that

$$\|U - U_m\| = \|V - V_m\| = 2\sin(2^{-m}\varphi)$$

and from (2.12) that

$$\|U_m - V_m\| = 2\sin(\varphi - 2^{-m+1}\varphi) < 2\sin\varphi = \|U - V\|$$

so that (2.13) and (2.14) are proved. (2.15) follows from Proposition 2.8 of [4] taking $n = 2^m$ and $p = 2^{m-1}$. □

Lemma 2.14. *Let $U, V \in R(\mathcal{H})$.*

$$\|UV + VU\| = \max\{\|U + V\|^2, \|U - V\|^2\} - 2. \tag{2.16}$$

Proof. Let $u \in \mathcal{H}$. Then

$$\|(U + V)u\|^2 = \langle(2\mathcal{I} + UV + VU)u, u\rangle \tag{2.17}$$

$$\|(U - V)u\|^2 = \langle(2\mathcal{I} - UV - VU)u, u\rangle \tag{2.18}$$

Hence

$$\begin{aligned}|\langle (UV+VU)u,u\rangle| &\le \max\{\|(U+V)u\|^2,\ \|(U-V)u\|^2\}-2\|u\|^2\\ &\le (\max\{\|(U+V)\|^2,\ \|(U-V)\|^2\}-2)\|u\|^2\end{aligned}$$

and therefore $\|UV+VU\| \le \max\{\|U+V\|^2, \|U-V\|^2\}-2$.

From (2.17) it follows that

$$\|(U+V)u\|^2 \le 2\|u\|^2 + \|(UV+VU)u\|\|u\|$$

so that

$$\|U+V\|^2 \le 2+\|UV+VU\|$$

and from (2.18) it follows that

$$\|(U-V)u\|^2 \le 2\|u\|^2 + \|(UV+VU)u\|\|u\|$$

so that $\|U-V\|^2 \le 2+\|UV+VU\|$. Hence

$$\max\{\|U+V\|^2, \|U-V\|^2\}-2 \le \|UV+VU\|$$

and this concludes the proof of the lemma. □

3. Arcwise connected components of $P(\mathcal{H})$

Definition 3.1. *Let X be a metric space with metric g and let a,b be two elements of X. A continuous injection $h:[0,1]\to X$ will be called a path connecting a to b in X if $h(0)=a$ and $h(1)=b$. A subset $\Pi=\{t_0,t_1,\dots,t_n\}$ of $\mathbb{R}$ such that $0=t_0<t_1<\dots<t_n=1$ will be called a partition of $[0,1]$. Given a partition Π of $[0,1]$ set*

$$L(a,b;h,\Pi)=\sum_{i=1}^{n} g(h(t_{i-1}),h(t_i))$$

and let $L(a,b;h)=\sup L(a,b;h,\Pi)$ where the supremum is taken over all partitions Π of $[0,1]$. $L(a,b;h)$ will be called the length of the path h.

For a discussion of the properties of $L(a,b;h)$ we refer to the introduction of [4].

Proposition 3.2. *Let $P,Q\in P(\mathcal{H})$ be such that $\mathrm{Ind}_Q(P)=0$. Then there is a path connecting P to Q such that $L(P,Q;h)=\varphi$ where $0\le\varphi\le\pi/2$ and $\sin\varphi=\|P-Q\|$. Furthermore, for every path h' connecting P to Q, $\varphi=L(P,Q;h)\le L(P,Q;h')$.*

Proof. An immediate consequence of Proposition 2.3 and of Theorem 3.1 of [4] where the path is defined as the limit of the $W_{m,n}$ of Definition 2.10 when $m\to\infty$. □

Lemma 3.3. *Let $P,Q\in P(\mathcal{H})$ be connected by a path of length $\le\pi/2$. Then there exists $R\in P(\mathcal{H})$ such that $\|P-R\|\le 1/\sqrt{2}$, $\|R-Q\|\le 1/\sqrt{2}$.*

Proof. P and Q are connected by a path h of length $\leq \pi/2$ and let $t_0 \in [0,1]$ be such that $L(P, h(t_0); h) = L(h(t_0), Q; h)$ (such a t_0 exists in view of the continuity in t of $L(P, h(t); h)$ as shown in Proposition 0.4 of [4]). Then setting $R = h(t_0)$, $\|P - R\| \leq L(P, R; h) \leq \pi/4 < 1$, $\|R - Q\| \leq L(R, Q; h) \leq \pi/4 < 1$. Then setting $\|P - R\| = \sin\varphi$, $0 \leq \varphi < \pi/2$ it follows from Proposition 3.2 that there exists a geodesic path connecting P to R of length φ so that $\varphi \leq \pi/4$ and hence $\|P - R\| = \sin\varphi \leq 1/\sqrt{2}$. $\|R - Q\| \leq 1/\sqrt{2}$ is established in the same way. □

Proposition 3.4. *Let $P, Q \in P(\mathcal{H})$ be connected by a path of length $\leq \pi/2$. Then* $\mathrm{Ind}_Q(P) = 0$.

Proof. Set $U = 2P - \mathcal{I}$ and $V = 2Q - \mathcal{I}$. Let R be the projection described in the previous lemma and set $X = 2R - \mathcal{I}$. Then

$$\|U - X\| \leq \sqrt{2} \quad \text{and} \quad \|V - X\| \leq \sqrt{2}.$$

Set $X_1 = \beta(U, X)$, $X_2 = \beta(V, X)$. Then, using Proposition 1.9, $\|X_1 - X_2\| \leq \sqrt{2}$. If $S = X_1X_2 + X_2X_1$, then S is self-adjoint. Let $u \in \mathrm{ran}\,(\mathcal{I} + U) \cap \mathrm{ran}\,(\mathcal{I} - V)$. Then

$$\begin{aligned} Su &= X_2X_1u + X_1X_2u \\ &= X_2X_1[(\mathcal{I} + U)/2]u + X_1X_2[(\mathcal{I} - V)/2]u \\ &= X_2[(\mathcal{I} + X)/2]X_1u + X_1[(\mathcal{I} - X)/2]X_2u \\ &= [(\mathcal{I} + V)/2]X_2X_1u + [(\mathcal{I} - U)/2]X_1X_2u \in \mathrm{ran}\,(\mathcal{I} + V) + \mathrm{ran}\,(\mathcal{I} - U). \end{aligned}$$

Hence $Su \perp u$. Similarly if $u \in \mathrm{ran}\,(\mathcal{I} + U) \cap \mathrm{ran}\,(\mathcal{I} - V)$ we see that $Su \perp u$. But $S = 2 - (X_1 - X_2)^2$ so that since $\|X_1 - X_2\| \leq \sqrt{2}$, S is positive and consequently using the generalized Schwarz inequality, (cf. [7], p. 262) for all $u \in H$,

$$\|Su\|^2 \leq \langle Su, u\rangle \langle S^2u, Su\rangle$$

so that $u \in \ker\,(U + V) \Rightarrow Su = 0$. In other words $\ker\,(U + V) \subseteq \ker\,(S)$. Finally, as we have seen above, if $u \in \mathrm{ran}\,(\mathcal{I} + U) \cap \mathrm{ran}\,(\mathcal{I} - V)$, $X_2X_1u \in \mathrm{ran}\,(\mathcal{I} + V)$ and since $X_2X_1u = -X_1X_2u \in \mathrm{ran}\,(\mathcal{I} - U)$ we see that X_2X_1 is a one-to-one mapping of $\mathrm{ran}\,(\mathcal{I}+U) \cap \mathrm{ran}\,(\mathcal{I}-V)$ onto $\mathrm{ran}\,(\mathcal{I}+V) \cap \mathrm{ran}\,(\mathcal{I}-U)$ so that $\mathrm{Ind}_U(V) = 0$. □

Definition 3.5. *Let $P, Q \in P(\mathcal{H})$. Set*

$$d(P, Q) = \begin{cases} \inf_h L(P, Q; h) & \text{if } P \text{ and } Q \text{ are connected,} \\ \pi & \text{otherwise,} \end{cases} \tag{3.1}$$

where the infimum is taken over all paths h connecting P to Q.

Proposition 3.6. *d is a metric on $P(\mathcal{H})$, equivalent to g since*

$$\forall P, Q \in P(\mathcal{H})\ ,\ g(P, Q) \leq d(P, Q) \leq \pi g(P, Q). \tag{3.2}$$

From now on it is the metric we shall use on $P(\mathcal{H})$ and on $F(\mathcal{H})$.

Proof. If P and Q are in the same connected component then $g(P, Q) = \|P - Q\| = \sin\varphi$ for some $0 \leq \varphi \leq \pi/2$ and $d(P, Q) = \varphi$. So $\sin\varphi \leq \varphi \leq \pi \sin\varphi$. If P and Q are not in the same connected component then $g(P, Q) = 1$ and $d(P, Q) = \pi$. □

Corollary 3.7. *Let $P, Q \in P(\mathcal{H})$. Then the following conditions are equivalent.*

(i) *P and Q are connected by a path of length $\leq \pi/2$.*

(ii) $\mathrm{Ind}_Q(P) = 0$.

In that case $d(P, Q) \leq 1$.

Proof. (ii) $\Rightarrow$ (i) Follows at once from Proposition 2.3 and from Proposition 3.2.

(i) $\Rightarrow$ (ii) Consequence of Proposition 3.4. □

Proposition 3.8. *Let $M \in F(\mathcal{H})$ be such that* $\dim M^\perp < \infty$. *Then the arcwise connected component of P_M is the set of all $Q \in P(\mathcal{H})$ such that* $\dim \ker (I - Q) = \dim M^\perp$ *and any two elements of that connected component can be connected by a path of length less or equal to $\pi/2$. Furthermore $\pi/2$ is the lowest possible such bound.*

Proof. Suppose that $\dim \ker (\mathcal{I} - Q) = \dim M^\perp = d$. Set $N = \mathrm{ran}\,(Q)$. Then in view of Corollary 1.7, $\mathrm{Ind}_N(M) = 0$ so that it follows from Proposition 2.3 that there exists a path of length at most equal to $\pi/2$ connecting them.

Conversely, if P_M and Q are in the same connected component there exists a continuous mapping $P : [0, 1] \to P(\mathcal{H})$ such that $P(0) = P_M$ and $P(1) = Q$. Because h is uniformly continuous there exists $\delta > 0$ such that if $|t - t'| < \delta$ then $g(P(t), P(t')) < 1$ so that $\mathrm{Ind}_{P(t')}(P(t)) = 0 - 0 = 0$ and using again Cor 1.7 this implies that $\dim \mathrm{ran}\ (\mathcal{I} - P(t)) = \dim \mathrm{ran}\ (\mathcal{I} - P(t'))$. Hence $\dim \mathrm{ran}\ (\mathcal{I} - P(t))$ is locally constant and therefore $\dim \ker\ (\mathcal{I} - Q) = \dim M^\perp = d$. Furthermore if $M, N \in P(\mathcal{H})$ are such that $M \cap N^\perp \neq \{0\}$ then $\|P_M - P_N\| = 1$ and the geodesic path connecting them has length $\pi/2$. □

Corollary 3.9. *Let $M \in F(\mathcal{H})$ be such that* $\dim M < \infty$. *Then the arcwise connected component of P_M is the set of all $Q \in P(\mathcal{H})$ such that* $\dim \mathrm{ran}\,(Q) = \dim M$ *and any two elements of that connected component can be connected by a path of length less or equal to $\pi/2$. Furthermore $\pi/2$ is the lowest possible such bound.*

Proof. Obvious. □

Proposition 3.10. *Let $M, N \in F(\mathcal{H})$. Assume that* $\dim M = \dim M^\perp = \dim N = \dim N^\perp = \infty$. *Then P_M and P_N are in the same connected component and there is a path connecting P_M to P_N of length less or equal to π.*

Proof. Let $N' \in F(\mathcal{H})$ be as in Proposition 1.8. Then using Proposition 2.3, we see that there is a path connecting P_M to $P_{N'}$ of length less or equal to $\pi/2$ and a path connecting $P_{N'}$ to P_N of length less or equal to $\pi/2$ so that the proposition is proved. □

Definition 3.11. *Let $m, n \in \mathbb{N} \cup \{0\}$). Set*

$$C_{m,n} = \{P \in P(\mathcal{H}) \mid \dim \ker\ (P) = m\,;\ \dim \mathrm{ran}\,(P) = n\}$$

Proposition 3.12. *Let* $\dim \mathcal{H} = \infty$. *Then*

(i) *$\forall m \in \mathbb{N} \cup \{0\}$, $C_{m,\infty}$ and $C_{\infty,m}$ are connected components of $P(\mathcal{H})$*

(ii) *$C_{\infty,\infty}$ is a connected component of $P(\mathcal{H})$.*

Proof. An immediate consequence of the previous results. □

Proposition 3.13. *For every* $X \subseteq P(\mathcal{H})$ *let* $\delta(X)$ *denote its diameter. Then*

(i) $\delta(C_{\infty,0}) = \delta(C_{0,\infty}) = 0$,
(ii) $\forall m \in \mathbb{N}$, $\delta(C_{m,\infty}) = \delta(C_{\infty,m}) = \pi/2$,
(iii) $\pi/2 < \delta(C_{\infty,\infty}) \leq \pi$.

Proof. (i) $C_{\infty,0} = \{\mathcal{H}\}$, a singleton. Similarly $C_{0,\infty} = \{0\}$, a singleton.

(ii) follows from Proposition 3.8 and Corollary 3.9.

(iii) Let $M \in F(\mathcal{H})$ be such that $\dim M = \operatorname{codim} M = \infty$ and let $N \in F(\mathcal{H})$ be such that $M \subseteq N$ and $\dim N \cap M^{\perp} = 1$. Then $\dim M \cap N^{\perp} = 0$ and therefore $\operatorname{Ind}_M(N) \neq 0$. Set $P = P_M$ and $Q = P_N$. Then P, $Q \in C_{\infty,\infty}$ and in view of Corollary 3.7,

$$d(P, Q) > \pi/2 \tag{3.3}$$

which implies the first inequality. The second is an immediate consequence of Proposition 3.10. □

Conjecture $\delta(C_{\infty,\infty}) = \pi$

4. Isometries

Definition 4.1. *Let* Φ *be a mapping of* $R(\mathcal{H})$ *into itself such that*

$$\forall U, V \in R(\mathcal{H}),\ \|\Phi(U) - \Phi(V)\| = \|U - V\|.$$

Then Φ *will be called and isometry on* $R(\mathcal{H})$. *We shall also denote by* Φ *the corresponding isometries on* $P(\mathcal{H})$ *and* $F(\mathcal{H})$. *Note that an isometry is injective.*

Proposition 4.2. *Let* Φ *be an isometric mapping of* $R(\mathcal{H})$ *into itself and let* $U, V \in R(\mathcal{H})$ *be such that* $\|U - V\| \leq \sqrt{2}$. *Then for* $W = \beta(U, V)$ *we have*

$$\Phi(W) = \beta(\Phi(U), \Phi(V)).$$

Proof. Since Φ is an isometry $\|\Phi(U) - \Phi(V)\| \leq \sqrt{2}$. Let $\{\Delta_n^{'}\}$ be the sequence constructed as in the previous section starting with $\Phi(U)$ and $\Phi(V)$. It is easy to see that $\Phi(\Delta_0) \subseteq \Delta_0^{'}$. Then by induction it follows that $\forall n \in \mathbb{N}$, $\Phi(\Delta_n) \subseteq \Delta_n^{'}$ so that $\Phi(\Delta) \subseteq \Delta^{'}$ and since both Δ and $\Delta^{'}$ consist of exactly one element, namely $W = \beta(U, V)$ and $W' = \beta(\Phi(U), \Phi(V))$, we must have $W' = \Phi(W)$. □

Theorem 4.3. *Let* Φ *be an isometry of* $R(\mathcal{H})$ *into itself and let* $U, V \in R(\mathcal{H})$ *be such that* $\operatorname{Ind}_U(V) = 0$ *and let* $W \in R(\mathcal{H})$ *satisfy* (2.3) *and* (2.4). *Then*

(i) $\Phi(W)\Phi(U) = \Phi(V)\Phi(W)$.
(ii) $\Phi(W)(\Phi(U) + \Phi(V)) = (\Phi(U) + \Phi(V))\Phi(W)$ *is positive.*

Proof. Let $W_1 = \beta(U, W)$ and $W_2 = \beta(W, V)$. Then from Lemma 2.9 it follows that $W = \beta(W_1, W_2)$. Using Proposition 2.1 and Proposition 1.9 we see that

$$\|U - W\| \leq \sqrt{2},\ \|W - V\| \leq \sqrt{2},\ \|W_1 - W_2\| \leq \sqrt{2}$$

and hence from the previous proposition it follows that

$$\Phi(W_1) = \beta(\Phi(U), \Phi(W)), \quad \Phi(W_2) = \beta(\Phi(W), \Phi(V))$$

and

$$\Phi(W) = \beta(\Phi(W_1), \Phi(W_2)).$$

Now

$$\begin{aligned}\Phi(W)\Phi(U)\Phi(W) &= \Phi(W)\Phi(W_1)\Phi(W_1)\Phi(U)\Phi(W_1)\Phi(W_1)\Phi(W)\\ &= \Phi(W)\Phi(W_1)\Phi(W)\Phi(W_1)\Phi(W)\\ &= \Phi(W_2)\Phi(W)\Phi(W_2) = \Phi(V)\end{aligned}$$

so that

$$\Phi(U)\Phi(W) = \Phi(W)\Phi(V).$$

And

$$\|\Phi(W_1) - \Phi(W_2)\| \le \sqrt{2}$$

so that

$\forall u \in \mathcal{H}$, $([\Phi(W_1)\Phi(W_2) + \Phi(W_2)\Phi(W_1)]u, u) = 2\|u\|^2 - \|(\Phi(W_1) - \Phi(W_2))u\|^2 \ge 0$

Hence

$$\begin{aligned}\Phi(W)(\Phi(U) + \Phi(V)) &= \Phi(W)\Phi(W_1)\Phi(W)\Phi(W_1) + \Phi(W)\Phi(W_2)\Phi(W)\Phi(W_2)\\ &= \Phi(W_2)\Phi(W_1) + \Phi(W_1)\Phi(W_2)\end{aligned}$$

is self-adjoint and positive. □

Corollary 4.4. *Let Φ be an isometry of $R(\mathcal{H})$ into itself and let $U, V \in R(\mathcal{H})$ be such that*

$$\ker\,(U + V) = \{0\} \quad \textit{and} \quad \ker\,(\Phi(U) + \Phi(V)) = \{0\}.$$

Then

$$\Phi(\beta(U, V)) = \beta(\Phi(U), \Phi(V)).$$

Proof. This follows at once from the previous proposition and the fact that under the conditions of the corollary both (i) and (ii) imply that $W = \beta(U, V)$ and $\Phi(W) = \beta(\Phi(U), \Phi(V))$. □

Proposition 4.5. *Let Φ be an isometry on $R(\mathcal{H})$ and let $U, V \in R(\mathcal{H})$ be such that $\|U - V\| < \sqrt{2}$. Then*

$$\Phi(UVU) = \Phi(U)\Phi(V)\Phi(U).$$

Proof. From Proposition 2.4 it follows that $U = \beta(V, UVU)$. Hence from Corollary 4.4 it follows that $\beta(\Phi(V), \Phi(UVU)) = \Phi(U)$ which implies that $\Phi(UVU) = \Phi(U)\Phi(V)\Phi(U)$. □

Corollary 4.6. *Let Φ be an isometry on $R(\mathcal{H})$ and let $U, V \in R(\mathcal{H})$ be such that $\|U - V\| \le \sqrt{2}$. Then*

$$\Phi(UVU) = \Phi(U)\Phi(V)\Phi(U).$$

Proof. Immediate consequence of the previous proposition and of Lemma 2.13. □

Proposition 4.7. *Let Φ be an isometry of $R(\mathcal{H})$ into itself and let $U, V \in R(\mathcal{H})$ be such that* $\operatorname{Ind}_U(V) = 0$. *Then if $W \in \operatorname{cent}(U, V)$ then*

$$\Phi(W) \in \operatorname{cent}(\Phi(U), \Phi(V)).$$

Proof. Since $W \in \operatorname{cent}(U, V)$ then $\|U - W\| \leq \sqrt{2}$. Then $V = WUW$ so that

$$\Phi(V) = \Phi(WUW) = \Phi(W)\Phi(U)\Phi(W)$$

and therefore $\Phi(W) \in \operatorname{cent}(\Phi(U), \Phi(V))$ □

Proposition 4.8. *Let Φ be an isometry of $R(\mathcal{H})$ into itself and let $U, V \in R(\mathcal{H})$. Then* $\operatorname{Ind}_U(V) = 0$ *if and only if* $\operatorname{Ind}_{\Phi(U)}(\Phi(V)) = 0$.

Proof. It is clear that an isometry preserves the length of the geodesics paths connecting two projections. Hence, setting $P = (\mathcal{I} + U)/2$ and $Q = (\mathcal{I} + V)/2$,

$$\operatorname{Ind}_U(V) = 0 \Leftrightarrow d(P, Q) \leq \pi/2 \Leftrightarrow d(\Phi(P), \Phi(Q)) \leq \pi/2 \Leftrightarrow \operatorname{Ind}_{\Phi(U)}(\Phi(V)) = 0.$$

□

Proposition 4.9. *Let Φ be an isometry of $F(\mathcal{H})$ into itself. Then Φ maps each connected component onto a connected component with the same diameter and hence, in particular, $C_{\infty,\infty}$ into itself.*

Proof. If $M, N \in F(\mathcal{H})$ are in the same connected component then there is a path h connecting them and it is clear that consequently $\Phi \circ h$ is a path connecting $\Phi(M)$ to $\Phi(N)$. Hence the diameter of the image by Φ of a connected component must be equal to the diameter of the connected component. The Proposition now follows from Proposition 3.13. □

Corollary 4.10. *Let Φ be an isometry on $R(\mathcal{H})$. Then*

$$\Phi(\mathcal{I}) = \mathcal{I} \text{ or } \Phi(\mathcal{I}) = -\mathcal{I}.$$

Proof. Consequence of the previous proposition and of Proposition 3.13. □

5. Applications to linear relations

Notation. Let $\mathbf{H} = \mathcal{H} \oplus \mathcal{H}$ denote the direct sum of $\mathcal{H}$ with itself, equipped with its natural Hilbert space structure and I the identity operator on $\mathbf{H}$. The space $LR(\mathcal{H})$ of all closed linear relations on $\mathcal{H}$ can be identified with the space $F(\mathbf{H})$ of their graphs. If E is a linear relation on $\mathcal{H}$ we shall denote its graph by $G(E)$ and by P_E the orthogonal projection in $\mathbf{H}$ onto $G(E)$ and set $U_E = 2P_E - I$. Then $U_E \in R(\mathbf{H})$. Conversely if $U \in R(\mathbf{H})$, $\operatorname{rel}(U) \in LR(\mathcal{H})$ will be the linear relation E such that $U_E = U$. $LR(\mathcal{H})$ is a complete metric space under the equivalent metrics g, defined by $g(E, F) = g(G(E), G(F))$ and d defined by $d(E, F) = d(G(E), G(F))$. Clearly $LR(\mathcal{H})$ contains $C(\mathcal{H})$, the set of all closed densely defined linear operators on $\mathcal{H}$ which in turn contains $L(\mathcal{H})$. If $E \in LR(\mathcal{H})$ then $\operatorname{dom}(E)$ and $\operatorname{ran}(E)$ will denote respectively the domain and the range of E while $\ker(E)$ and $\operatorname{mul}(E)$ will denote respectively the kernel and the multiple part of E (i.e., $\operatorname{mul}(E) = \{v \in \mathcal{H} \mid \{0, v\} \in G(E)\}$.

Definition 5.1. *If $E,\ F \in LR(\mathcal{H})$ let mid(E,F) (respectively* $\operatorname{cent}(E,F)$*) be the set of $E' \in LR(\mathcal{H})$ such that $G(E') \in \operatorname{mid}(G(E),G(F))$ (respectively $G(E') \in \operatorname{cent}(G(E),G(F))$). Also, if $E,\ F \in LR(\mathcal{H})$ let*

$$\operatorname{Ind}_E(F) = \operatorname{Ind}_{G(E)}(G(F)).$$

In particular,

$$\operatorname{Ind}_1(E) = \operatorname{Ind}_{G(0)}(G(E)) = \dim \operatorname{mul}(E) - \dim (\operatorname{dom}(E))^{\perp}$$

and

$$\operatorname{Ind}_2(E) = \operatorname{Ind}_1(E^{-1}) = \dim \ker (E) - \dim (\operatorname{ran}(E))^{\perp},$$

the Fredholm index. Let $E,F \in LR(\mathcal{H})$. Set $U_W = \beta(U_E,U_F)$. Then $\beta(E,F)$ will be the linear relation whose graph is the range of $(I+U_W)/2$ in $\mathbf{H}$.

Definition 5.2. *Let J, K and L be reflections on $\mathbf{H}$ defined as follows.*

$$J = \begin{pmatrix} 0 & I \\ I & 0 \end{pmatrix}; \quad K = \begin{pmatrix} 0 & -iI \\ iI & 0 \end{pmatrix}; \quad L = \begin{pmatrix} I & 0 \\ 0 & -I \end{pmatrix}.$$

Note that this is not how J was defined in ([1]).

Definition 5.3. *Let $E \in LR(\mathcal{H})$. Then*

$$E \text{ is self-adjoint if and only if } U_E + KU_EK = 0, \tag{5.1}$$

$$E \text{ is unitary if and only if } U_E + LU_EL = 0, \tag{5.2}$$

$$E \text{ is self-inverse if and only if } U_E - JU_EJ = 0. \tag{5.3}$$

Naturally (as shown in [1]) these definitions coincide with the usual definitions when E is an operator.

For further references on linear relations see for instance [2], [1], and [5].

Notation. Let $SA(\mathcal{H}) = \{F \in LR(\mathcal{H}) \mid F \text{ is self-adjoint }\}$.

Proposition 5.4. *We have*

(i) $SA(\mathcal{H}) = \operatorname{mid}(i\mathcal{I}, -i\mathcal{I})$,
(ii) $\operatorname{cent}(i\mathcal{I}, -i\mathcal{I}) = \operatorname{mid}(i\mathcal{I}, -i\mathcal{I})$.

Proof. (i) Let $F \in LR(\mathcal{H})$ be self-adjoint. Then $U_FK + KU_F = 0$ and from (2.16) it follows that $\|U_F + K\| \le \sqrt{2}$ and $\|U_F - K\| \le \sqrt{2}$. Suppose that one at least of the inequalities is strict. Then, using Proposition 1.9

$$\begin{aligned} 2 &= \|U_F + U_F\| \\ &\le \|U_F + K\|\sqrt{1 - \frac{1}{4}\|U_F - K\|^2} + \|U_F - K\|\sqrt{1 - \frac{1}{4}\|U_F + K\|^2} \\ &< \sqrt{2}/\sqrt{2} + \sqrt{2}/\sqrt{2} = 2, \end{aligned}$$

a contradiction. Hence $\|U_F + K\| = \|U_F - K\| = \sqrt{2}$ which implies, since $P_{i\mathcal{I}} = (I+K)/2$ and $P_{-i\mathcal{I}} = (I-K)/2$, that

$$g(i\mathcal{I}, F) = g(-i\mathcal{I}, F) = 1/\sqrt{2}. \tag{5.4}$$

The converse is obvious using Lemma 2.14.

(ii) Let $F \in \text{mid}(i\mathcal{I}, -i\mathcal{I})$. Then F is self-adjoint so that $U_F K + K U_F = 0$ and therefore $U_F P_{-i\mathcal{I}} = P_{i\mathcal{I}} U_F$ which shows that $F \in \text{cent}(i\mathcal{I}, -i\mathcal{I})$. □

Proposition 5.5. *Let $E \in LR(\mathcal{H})$ be such that*

$$G(E) \cap (G(E^*))^{\perp} = G(E^*) \cap G(E)^{\perp} = \{0\}. \tag{5.5}$$

Then $\beta(E, E^*)$ exists and is self-adjoint.

Proof. (5.5) implies that ker $(U_E + U_{E^*}) = \{0\}$. Note that the mapping $E \mapsto E^*$ is an involutive isometry on $LR(\mathcal{H})$. Hence it follows from Theorem 4.3 that

$$(\beta(E, E^*))^* = \beta(E^*, E) = \beta(E, E^*)$$

and the proposition is proved. □

Proposition 5.6. *Let $U, V, X \in R(\mathbf{H})$ be such that* ker $(U + V) = \{0\}$. *Then*

$$\beta(-XUX, -XVX) = -X\beta(U, V)X.$$

Proof. The mapping of $R(\mathbf{H})$ onto itself given by $T \mapsto -XTX$ is an isometry so that the proposition follows at once from Theorem 4.3. □

Corollary 5.7. *Let $E, F \in LR(\mathcal{H})$ be unitary such that* ker $(U_E + U_F) = \{0\}$. *Then $\beta(E, F)$ is unitary.*

Proof. Take $U = U_E$ and $V = U_F$ and $X = J$. Then

$$\beta(U_E, U_F) = \beta(-JU_E J, -JU_F J) = -J\beta(U_E, U_F)J$$

which establishes the corollary. □

Proposition 5.8. *Let $E \in LR(\mathcal{H})$. Then*

(i) $G(E) \cap G(E^*)^{\perp} = G(E) \cap G(i\mathcal{I}) + G(E) \cap G(-i\mathcal{I})$.

(ii) $G(E^*) \cap G(E)^{\perp} = G(E)^{\perp} \cap G(i\mathcal{I}) + G(E)^{\perp} \cap G(-i\mathcal{I})$.

Proof. (i) We have

$$G(E \cap G(E^*)^{\perp} \supset G(E) \cap G(i\mathcal{I}) + G(E) \cap G(-i\mathcal{I}).$$

Let $\{u, v\} \in G(E) \cap G(i\mathcal{I})$. Then $\{u, iu\} \in G(E)$ so that

$$\{u, iu\} = K\{u, iu\} \in G(E^*)^{\perp}.$$

Similarly, let $\{u, v\} \in G(E) \cap G(-i\mathcal{I})$. Then $\{u, -iu\} \in G(E)$ so that

$$\{u, -iu\} = -K\{u, iu\} \in G(E^*)^{\perp}.$$

(ii) We have

$$G(E^*)^{\perp} \cap G(E) \subset G(E) \cap G(i\mathcal{I}) + G(E) \cap G(-i\mathcal{I}).$$

Let $\{u, v\} \in G(E) \cap G(E^*)^{\perp}$. Then $\{-iv, iu\} = K\{u, v\} \in G(E)$. Hence

$$\{u - iv, i(u - iv\} = \{u - iv, iu + v\} \in G(E) \cap G(i\mathcal{I})$$

and

$$\{u + iv, -i(u + iv\} = \{u + iv, -iu + v\} \in G(E) \cap G(-i\mathcal{I})$$

so that

$$2\{u,v\} = \{u-iv, iu+v\} + \{u+iv, -iu+v\} \in G(E)\cap G(i\mathcal{I}) + G(E)\cap G(-i\mathcal{I})$$

and (i) is proved. (ii) is proved in a quite similar way. □

Proposition 5.9. *Let $E \in LR(\mathcal{H})$. Then the following conditions are equivalent.*

(i) *There exists $F \in LR(\mathcal{H})$ such that*

$$F \text{ is self-adjoint with } g(E,F) \le 1/\sqrt{2}. \tag{5.6}$$

(ii) $\mathrm{Ind}_E\,(-i\mathcal{I}) = \mathrm{Ind}_E\,(i\mathcal{I}) = 0.$ (5.7)

Proof. (ii) ⇒ (i) Let Φ_1 be an isometry of $G(E)\cap G(i\mathcal{I})$ onto $G(E^\perp)\cap G(-i\mathcal{I})$ and Φ_2 be an isometry of $G(E)\cap G(-i\mathcal{I})$ onto $G(E)^\perp\cap G(i\mathcal{I})$. Then $\Phi = \Phi_1\oplus\Phi_2$ maps $G(E)\cap G(E^*)^\perp$ onto $G(E^*)\cap G(E)^\perp$. Furthermore, if $u \in G(E)\cap G(E^*)^\perp$ is such that $u = s+t$ with $s \in G(E)\cap (G(iI)$ and $t \in G(E)\cap G(-i\mathcal{I})$ then $Ku = Ks + Kt = s - t$ so that

$$\Phi K u = \Phi(s-t) = \Phi_1 s - \Phi_2 t = -K\Phi_1 s - K\Phi_2 t = -K\Phi u.$$

Consequently $\Phi K = K\Phi$ and therefore $\Phi^{-1}K = K\Phi^{-1}$ so that $W_1 = \Phi\oplus\Phi^{-1}$ maps $\ker\,(U_E+U_{E^*})$ onto itself and $KW_1 = -W_1K$ It is clear that, as in Proposition 2.1, $U_EW_1 = W_1U_{E^*}$. From Proposition 5.5 we see that there exists $W_2 \in R(\ker\,(U_E + U_{E^*}))^\perp$ such that $KW_2 = -W_2K$ and $U_EW_1 = W_1U_{E^*}$. Hence setting $W = W_1\oplus W_2$ we see, using Proposition 2.4 that $WK + KW = 0$ and $\|U_E - W\| = \|U_E - W\| \le \sqrt{2}$.

(i) ⇒ (ii) From (5.4) it follows that $g(F, iI) \le 1/\sqrt{2}$ and therefore that E is arcwise connected to iI by a path of length $\le \pi/2$. Then from Proposition 3.4 we see that $\mathrm{Ind}_E\,(i\mathcal{I}) = 0$. The same reasoning shows that $\mathrm{Ind}_E\,(-i\mathcal{I}) = 0$. □

Remark 5.10. $\mathrm{Ind}_E\,(-i\mathcal{I})$ *and* $\mathrm{Ind}_E\,(i\mathcal{I})$ *extend to arbitrary elements of $LR(\mathcal{H})$ the notion of defect numbers of symmetric relation (see* [1], *Definition* 2.12*). Thus, if $E \in LR(\mathcal{H})$ is symmetric it satisfies* (5.7) *if and only if it is self-adjoint. Also, using Proposition* 3.9 *of* [1] *it follows that if* $\mathrm{Ind}_E\,(-i\mathcal{I}) = \mathrm{Ind}_E\,(i\mathcal{I}) > 0$ *there exists an extension E' of E (i.e., $G(E') \supseteq G(E)$) such that E' satisfies* (5.6) *and if* $\mathrm{Ind}_E\,(-i\mathcal{I}) = \mathrm{Ind}_E\,(i\mathcal{I}) < 0$ *there exists a restriction E'' of E, (i.e., $G(E'') \subseteq G(E)$) such that E'' satisfies* (5.6).

Proposition 5.11. *Let $E \in LR(\mathcal{H})$. The following conditions are equivalent.*

(i) *There exists $F \in LR(\mathcal{H})$ such that F is unitary and $g(E,F) \le 1/\sqrt{2}$.* (5.8)

(ii) $\mathrm{Ind}_1\,(E) = \mathrm{Ind}_2\,(E) = 0.$ (5.9)

Proof. Using the Cayley transform (cf. Corollary 3.8 in [1]) and Proposition 5.9. □

Proposition 5.12. *Let $E \in LR(\mathcal{H})$ be such that*

$$G(E)\cap G(E^{-1})^\perp = G(E^{-1})\cap G(E)^\perp = \{0\}. \tag{5.10}$$

Then $\beta(E, E^{-1})$ exists and is self-inverse.

Proof. (5.10) implies that ker $(U_E + U_{E^{-1}}) = \{0\}$. Note that the mapping $E \mapsto E^{-1}$ is an involutive isometry on $LR(\mathcal{H})$. Hence it follows from Theorem 4.3 that $(\beta(E, E^{-1}))^{-1} = \beta(E^{-1}, E) = \beta(E, E^{-1})$ and the proposition is proved. □

Proposition 5.13. *(Fixed point theorem) Let Φ be an isometry on $R(\mathbf{H})$ and let $U \in R(\mathbf{H})$ be such that* ker $(U + \Phi(U)) = \{0\}$. *Then*

$$\Phi\beta(U, \Phi(U)) = \beta(U, \Phi(U)).$$

Proof. Obvious. □

Proposition 5.14. *Let V be a unitary operator on $\mathcal{H}$ such that* ker $(\mathcal{I} + V) = \{0\}$. *Then there exists a unique W such that W is unitary, $W + W^*$ is positive definite and $W^2 = V$.*

Proof. Under the hypothesis ker $(U_V + L) = \{0\}$. Let

$$\beta(U_V, L) = \begin{pmatrix} 0 & W^* \\ W & 0 \end{pmatrix}.$$

Then

$$U_V = \begin{pmatrix} 0 & V^* \\ V & 0 \end{pmatrix} = \begin{pmatrix} 0 & W^* \\ W & 0 \end{pmatrix} \begin{pmatrix} 0 & \mathcal{I} \\ \mathcal{I} & 0 \end{pmatrix} \begin{pmatrix} 0 & W^* \\ W & 0 \end{pmatrix}$$

and the rest of the proof is a simple verification using Corollary 5.7. □

Proposition 5.15. *Let V be a unitary relation (and hence an operator). Then there exists a path $h(t)$ connecting V to $\mathcal{I}$ such that $\forall t$, $h(t)$ is unitary.*

Proof. Since V is unitary then

$$U_V = \begin{pmatrix} 0 & V^* \\ V & 0 \end{pmatrix}.$$

So that

$$\text{ker } (U_V + L) = \text{ker } (V + I) \oplus \text{ker } (V^* + \mathcal{I})$$

and since ker $(V + \mathcal{I}) =$ ker $(V^* + \mathcal{I})$ the mapping of ker $(U_V + L)$ onto itself given by $W_0 : \{u, v\} \mapsto \{v, u\}$ satisfies the conditions of Proposition 2.3 so that there exists $W = U_E \in R(\mathbf{H})$ with E unitary and such that U_V, L and W satisfy (2.3) and (2.4). Then using Corollary 5.7 we see that the path $h(t)$ connecting U_V to L via W is such that, setting $h(t) = U_{f(t)}$, $f(t)$ is unitary and is a path connecting V to $\mathcal{I}$. □

Remark 5.16. *This yields another proof of the well-known fact that the set of all unitary relations is arcwise connected.*

References

[1] Z. Boulmaarouf and J.-Ph. Labrousse, *The Cayley transform of linear relations*, J. of the Egyptian Math. Soc. **2**, (1994) 53–65.

[2] R. Cross *Multivalued Linear Operators*, Marcel Dekker Inc., New York (1998).

[3] M. Fernandez-Miranda and J.-Ph. Labrousse, *The Cayley transform of linear relations*, Proceedings of the Amer. Math. Soc. **133**(2), (2005) 493–498.

[4] J.-Ph. Labrousse, *Geodesics in the space of linear relations on a Hilbert space*, Proc. of the 18th OT Conference, The Theta Foundation, Bucharest, (2000) 213–234.

[5] J.-Ph. Labrousse, *Idempotent linear relations*, Spectral Theory and its Applications, The Theta Foundation, Bucharest, (2003) 121–141.

[6] S. Mazur and S. Ulam, *Sur les transformations isométriques d'espaces vectoriels normés*, C R Acad. des Sci, Paris **194**, (1932) 946–948.

[7] F. Riesz and B. Sz.-Nagy, *Functional Analysis*, F. Ungar, New York, 1955.

Jean-Philippe Labrousse
Laboratoire J.A. Dieudonné
Université de Nice-Sophia Antipolis
Parc Valrose
Nice Cedex 02, France

Operator Theory:
Advances and Applications, Vol. 198, 259–264

On Norms in Indefinite Inner Product Spaces

Matthias Langer and Annemarie Luger

In memory of Peter Jonas, who knew Krein spaces so well

Abstract. In a Krein space various norms can be defined by choosing different underlying fundamental decompositions. In this note we consider this dependence explicitly and draw the conclusion that – even in a Pontryagin space – there does not exist a natural choice motivated by minimizing properties.

Mathematics Subject Classification (2000). 46C05,46C20.

Keywords. Krein space, Pontryagin space, indefinite inner product.

1. Introduction

Let $(\mathcal{K}, [\cdot\,,\cdot\,])$ be a Krein space with fundamental decomposition

$$\mathcal{K} = \mathcal{K}_+ \,[\dot{+}]\, \mathcal{K}_-, \tag{1.1}$$

that is, $\mathcal{K}_+$ and $\mathcal{K}_-$ denote positive and negative subspaces, respectively, which are orthogonal with respect to the indefinite inner product $[\cdot\,,\cdot\,]$ and $\mathcal{K}$ is the direct sum of $\mathcal{K}_+$ and $\mathcal{K}_-$. Denote by P_+ and P_- the corresponding $[\cdot\,,\cdot\,]$-self-adjoint projections onto $\mathcal{K}_+$ and $\mathcal{K}_-$, respectively. Hence every element $x \in \mathcal{K}$ can be written in a unique way as

$$x = x_+ + x_- \qquad \text{with } x_\pm \in \mathcal{K}_\pm.$$

Then the corresponding fundamental symmetry $J := P_+ - P_-$

$$Jx = x_+ - x_-$$

gives rise to the positive definite inner product

$$(\cdot\,,\cdot\,)_J := [J\cdot\,,\cdot\,].$$

We denote the corresponding norm by

$$\|x\|_J := (x,x)_J^{\frac{1}{2}} = [Jx,x]^{\frac{1}{2}} \qquad x \in \mathcal{K}.$$

The authors gratefully acknowledge the support of the "Engineering and Physical Sciences Research Council" (EPSRC, United Kingdom), grant no. EP/E037844/1 (first author) and the "Fond zur Förderung der wissenschaftlichen Forschung" (FWF, Austria), grant number J2540-N13 (second author).

Conversely, every mapping J in $\mathcal{K}$ with $J = J^+ = J^{-1}$ can be seen as a fundamental symmetry giving rise to a fundamental decomposition $\mathcal{K} = \mathcal{K}_{J,+} [\dot{+}] \mathcal{K}_{J,-}$ where $\mathcal{K}_{J,+} = \operatorname{ran}(I+J)$ and $\mathcal{K}_{J,-} = \operatorname{ran}(I-J)$. See [1] for basics on Krein spaces.

Obviously, the norm $\|\cdot\|_J$ depends on the fundamental decomposition (1.1) of $\mathcal{K}$. Choosing a different one, say $\mathcal{K} = \mathcal{K}'_+ [\dot{+}] \mathcal{K}'_-$, leads to a norm which for certain vectors will be larger than the original one. Roughly speaking, if the spaces $\mathcal{K}'_+$ and $\mathcal{K}'_-$ "come closer" to the neutral set $\mathcal{K}_0$, these norms will in general be unbounded.

However, our investigation is motivated by a different question, namely when does the norm become small. This note deals with the elementary but enlightening observation how the norm (of a single element) actually depends on the choice of the fundamental decomposition.

2. The result

We give here an explicit description of all values that the norm of an element x can attain.

Theorem 1. *Assume that $\mathcal{K}$ is a Krein space such that $[\cdot,\cdot]$ is indefinite and let $x \in \mathcal{K}$, $x \neq 0$. Then the following holds.*

(i) *If $[x,x] \neq 0$, then*

$$\{\|x\|_J : J \text{ is a fundamental symmetry}\} = \left[\,\big|[x,x]\big|^{\frac{1}{2}}, \infty\right). \tag{2.1}$$

Moreover,

$$\|x\|_J = \big|[x,x]\big|^{\frac{1}{2}} \quad \text{if and only if} \quad x \in \mathcal{K}_{J,+} \cup \mathcal{K}_{J,-},$$

where $\mathcal{K} = \mathcal{K}_{J,+} [\dot{+}] \mathcal{K}_{J,-}$ is the fundamental decomposition connected with J.

(ii) *If $[x,x] = 0$, then*

$$\{\|x\|_J : J \text{ is a fundamental symmetry}\} = (0, \infty). \tag{2.2}$$

Proof. (i) Without loss of generality assume that $[x,x] < 0$. Let J be any fundamental symmetry, $\mathcal{K} = \mathcal{K}_{J,+} [\dot{+}] \mathcal{K}_{J,-}$ the corresponding fundamental decomposition and $P_{J,\pm}$ the orthogonal projections onto $\mathcal{K}_{J,\pm}$. We can write the J-inner product as

$$(x,x)_J = [Jx,x] = [(2P_{J,+} - I)x, x] = 2[P_{J,+}x, P_{J,+}x] - [x,x]. \tag{2.3}$$

It immediately follows that $(x,x)_J \geq -[x,x] = \big|[x,x]\big|$. Moreover, $(x,x)_J = \big|[x,x]\big|$ if and only if $P_{J,+}x = 0$, i.e., $x \in \mathcal{K}_{J,-}$.

Next we show that every value in the interval on the right-hand side of (2.1) is attained. Let $\mathcal{K} = \mathcal{M}_+ [\dot{+}] \mathcal{M}_-$ be a fundamental decomposition of $\mathcal{K}$ such that $x \in \mathcal{M}_-$. Choose an arbitrary element $a \in \mathcal{M}_+$, $a \neq 0$, and let $\mathcal{L}_+$ and $\mathcal{L}_-$ be such that

$$\mathcal{M}_+ = \mathcal{L}_+ [\dot{+}] \operatorname{span}\{a\}, \qquad \mathcal{M}_- = \mathcal{L}_- [\dot{+}] \operatorname{span}\{x\}.$$

Consider the elements

$$u(s) = sa + (1-s)x, \qquad s \in [0,1].$$

Since $[u(0), u(0)] < 0$ and $[u(1), u(1)] > 0$ and the inner product is continuous, there exists an $s_0 \in (0,1)$ such that $[u(s_0), u(s_0)] = 0$. Set $b := u(s_0)$; then

$$[b,b] = 0, \quad [a,b] = s_0[a,a] > 0 \quad \text{and} \quad [b,x] = (1-s_0)[x,x] < 0. \tag{2.4}$$

Now let

$$v(t) := ta + (1-t)b, \qquad t \in (0,1],$$

which is a positive element for $t \in (0,1]$. Moreover, set $\mathcal{K}_{t,+} := \mathcal{L}_+ + \operatorname{span}\{v(t)\}$, which is an orthogonal sum since both a and b are orthogonal to $\mathcal{L}_+$. Hence the orthogonal projection $P_{t,+}$ onto $\mathcal{K}_{t,+}$ can be written as

$$P_{t,+} = P_{\mathcal{L}_+} + P_{v(t)},$$

where $P_{\mathcal{L}_+}$ is the orthogonal projection onto $\mathcal{L}_+$ and $P_{v(t)}$ is the orthogonal projection onto $\operatorname{span}\{v(t)\}$, which has the form

$$P_{v(t)}y = \frac{[y, v(t)]}{[v(t), v(t)]} v(t).$$

Furthermore, let $w(t)$ be a non-zero element in $\operatorname{span}\{a, x\}$ which is orthogonal to $v(t)$ and hence negative. With $\mathcal{K}_{t,-} := \mathcal{L}_- + \operatorname{span}\{w(t)\}$ we have a fundamental decomposition $\mathcal{K} = \mathcal{K}_{t,+} [\dot{+}] \mathcal{K}_{t,-}$ and a corresponding fundamental symmetry $J_t = 2P_{t,+} - I$. Now let us calculate $(x,x)_{J_t}$. Using the relation $P_{t,+}x = P_{v(t)}x$ and (2.4), we obtain

$$\begin{aligned}
[P_{t,+}x, P_{t,+}x] = [P_{t,+}x, x] &= \frac{\big|[x, v(t)]\big|^2}{[v(t), v(t)]} \\
&= \frac{\big|[x, ta + (1-t)b]\big|^2}{[ta + (1-t)b, ta + (1-t)b]} \\
&= \frac{(1-t)^2 \big|[x,b]\big|^2}{t^2[a,a] + 2t(1-t)[a,b]}.
\end{aligned} \tag{2.5}$$

If t varies in $(0,1]$, then the expression in (2.5) takes all values in $[0,\infty)$ since $[a,a] > 0$, $[x,b] \neq 0$ and the denominator is always positive. Now using (2.3) we see that $(x,x)_{J_t}$ takes all values in $[-[x,x], \infty) = [\,|[x,x]|, \infty)$.

(ii) It is clear that $\|x\|_J > 0$ for every J since $x \neq 0$. We have to show that all values in $(0,\infty)$ are attained. Let y be another neutral element that satisfies $[x,y] = 1$, i.e., y is skewly linked to x, and define $a := \frac{1}{\sqrt{2}}(x+y)$, $b := \frac{1}{\sqrt{2}}(x-y)$. Then

$$x = \frac{1}{\sqrt{2}}(a+b), \qquad [a,a] = 1, \qquad [b,b] = -1, \qquad [a,b] = 0.$$

Now let $\mathcal{K} = \mathcal{M}_+ [\dot{+}] \mathcal{M}_-$ be a fundamental decomposition such that $\mathcal{M}_+ = \mathcal{L}_+ [\dot{+}] \operatorname{span}\{a\}$ and $\mathcal{M}_- = \mathcal{L}_- [\dot{+}] \operatorname{span}\{b\}$ with some subspaces $\mathcal{L}_\pm$. Set

$$v(t) := a + tb, \qquad t \in (-1,1);$$

then

$$[v(t), v(t)] = [a, a] + 2t\,\mathrm{Re}[a, b] + t^2[b, b] = 1 - t^2 > 0, \qquad t \in (-1, 1).$$

Hence $\mathcal{K}_{t,+} := \mathcal{L}_+ \,[\dot{+}]\, \mathrm{span}\{v(t)\}$ is a maximal uniformly positive subspace. As in part (i) we can write the projection $P_{t,+}$ onto $\mathcal{K}_{t,+}$ as $P_{t,+} = P_{\mathcal{L}_+} + P_{v(t)}$ and hence

$$\begin{aligned}[P_{t,+}x, P_{t,+}x] = [P_{t,+}x, x] &= \frac{\big|[x, v(t)]\big|^2}{[v(t), v(t)]} \\ &= \frac{\big|[a+b, a+tb]\big|^2}{2(1-t^2)} = \frac{|1-t|^2}{2(1-t^2)} = \frac{1-t}{2(1+t)}\end{aligned}$$

which takes all values in $(0, \infty)$ if t varies in $(-1, 1)$. Since $[x, x] = 0$, we obtain (2.2) by (2.3). □

3. A simple example

Let us have an even closer look at the dependence of the norm on the fundamental decomposition in the simplest possible situation, a Pontryagin space of dimension 2 with one negative square: $\mathcal{K} = \mathbb{C}^2$ with inner product

$$\left[\begin{pmatrix} x \\ y \end{pmatrix}, \begin{pmatrix} x' \\ y' \end{pmatrix}\right] = x\overline{x'} - y\overline{y'}.$$

There is no natural fundamental decomposition a priori, but we choose $\mathcal{K}_+ = \mathrm{span}\{\vec{e}_+\}$, $\mathcal{K}_- = \mathrm{span}\{\vec{e}_-\}$ where $\vec{e}_+ = \binom{1}{0}$, $\vec{e}_- = \binom{0}{1}$ and compare it with other fundamental decompositions.

For every $t \in \mathbb{C}$ with $|t| < 1$ the vectors

$$\vec{f}_+(t) := \frac{1}{\sqrt{1-|t|^2}}\begin{pmatrix} 1 \\ t \end{pmatrix}, \qquad \vec{f}_-(t) := \frac{1}{\sqrt{1-|t|^2}}\begin{pmatrix} \overline{t} \\ 1 \end{pmatrix},$$

give also rise to a fundamental decomposition $\mathcal{K} = \mathcal{K}_{t,+} \,[\dot{+}]\, \mathcal{K}_{t,-}$ with $\mathcal{K}_{t,\pm} = \mathrm{span}\{\vec{f}_\pm(t)\}$. The corresponding norm becomes

$$\left\|\begin{pmatrix} x \\ y \end{pmatrix}\right\|_t^2 = \frac{1}{1-|t|^2}\Big((1+|t|^2)\big(|x|^2+|y|^2\big) - 2\big(tx\overline{y} + \overline{t}\overline{x}y\big)\Big) \quad \text{for } \begin{pmatrix} x \\ y \end{pmatrix} \in \mathbb{C}^2. \tag{3.1}$$

For $t = 0$ we obtain the fundamental decomposition given by $\vec{e}_\pm$. To compare the norms, we calculate the difference of their squares for $\binom{x}{y} \in \mathbb{C}^2$,

$$\begin{aligned}\left\|\begin{pmatrix} x \\ y \end{pmatrix}\right\|_t^2 - \left\|\begin{pmatrix} x \\ y \end{pmatrix}\right\|_0^2 &= \frac{2}{1-|t|^2}\Big(|t|^2\big(|x|^2 + |y|^2\big) - \big(tx\overline{y} + \overline{t}\overline{x}y\big)\Big) \\ &= \frac{2|t|^2|x|^2}{1-|t|^2}\left(1 + |\alpha|^2 - \frac{\overline{\alpha}}{\overline{t}} - \frac{\alpha}{t}\right)\end{aligned} \tag{3.2}$$

with $\alpha := \frac{y}{x}$ if $x \neq 0$ and $t \neq 0$ (the case $x = 0$ can be treated separately and fits into what is said below). For simplicity of the presentation we continue with real vectors and real t only, that is, $x, y, t \in \mathbb{R}$. Then (3.2) can be written as

$$\left\| \begin{pmatrix} x \\ y \end{pmatrix} \right\|_t^2 - \left\| \begin{pmatrix} x \\ y \end{pmatrix} \right\|_0^2 = \frac{2|t|^2|x|^2}{1-|t|^2} \Big(\alpha - g_+(t)\Big)\Big(\alpha - g_-(t)\Big)$$

with

$$g_\pm(t) = \frac{1 \mp \sqrt{1-t^2}}{t}.$$

Hence, for a given t, there are two slopes α, i.e., two directions of vectors $\binom{x}{y}$, for which the two norms $\|\binom{x}{y}\|_t$ and $\|\binom{x}{y}\|_0$ coincide. Moreover,

$$\left\| \begin{pmatrix} x \\ y \end{pmatrix} \right\|_0 < \left\| \begin{pmatrix} x \\ y \end{pmatrix} \right\|_t \iff \begin{cases} \alpha < g_+(t) \quad \vee \quad \alpha > g_-(t) & \text{if } t > 0, \\ \alpha < g_-(t) \quad \vee \quad \alpha > g_+(t) & \text{if } t < 0. \end{cases}$$

The following pictures illustrate this.

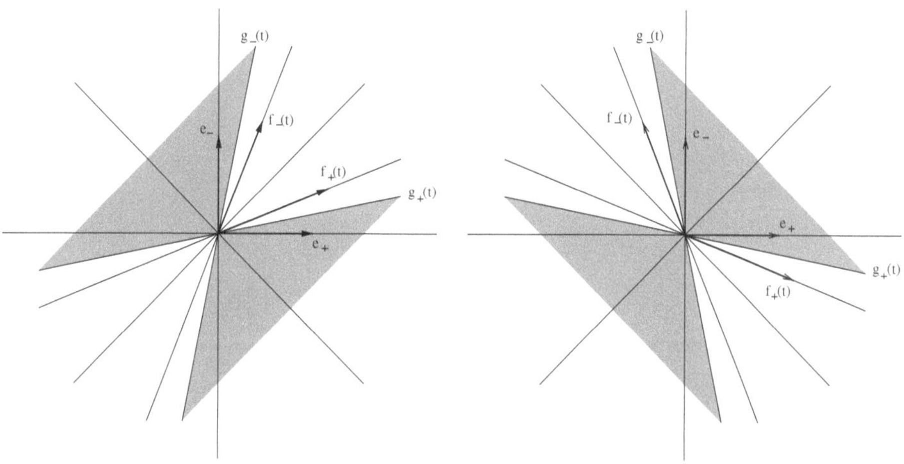

$t > 0$ $t < 0$

In both cases the shaded region depicts those vectors $\vec{x}$ for which $\|\vec{x}\|_t > \|\vec{x}\|_0$ for fixed t. Note that for the slopes the following relations are valid:

$$g_+(t) < t < 1 < \frac{1}{t} < g_-(t) \qquad \text{if } t > 0$$

and

$$g_-(t) < \frac{1}{t} < 1 < t < g_+(t) \qquad \text{if } t < 0.$$

Note that t and $\frac{1}{t}$ are the slopes of the directions of $\vec{f}_+(t)$ and $\vec{f}_-(t)$, respectively.

From (3.1) one can also see that no norm is uniformly large. If t tends non-tangentially to a point t_0 on the unit circle, then for non-neutral vectors $\vec{x}$ the norm $\|\vec{x}\|_t$ tends to infinity. The same is true for some neutral vectors, but for the neutral vectors $\vec{x} = c\binom{1}{t_0}$, $c \in \mathbb{C}$, the norm remains bounded.

4. Concluding remarks

One of our main observations is that there exists no norm which is uniformly small, i.e., one which is minimal for every vector. We want to mention that this fact was already mentioned by Rolf Nevanlinna in [2]; however, he does not describe all possible values of different norms for a given element.

In the case of a "proper" Krein space (i.e., with infinite-dimensional positive and negative subspaces) this observation is probably not so striking, since the choice of the fundamental decomposition is included in the definition and any other choice can seem unnatural. In a Pontryagin space, however, it is well known that all these norms are equivalent and hence the topology is unique. So the question for a particular norm comes into play only if one is interested in actual lengths. This plays a significant role for instance in measuring lengths of curves on manifolds. Our considerations here show that, in particular, on a symplectic manifold, there is no "natural" choice of a norm in the tangent space by uniformly minimizing properties.

References

[1] J. Bognár, *Indefinite inner product spaces*, Ergebnisse der Mathematik und ihrer Grenzgebiete, Springer, 1974.

[2] R. Nevanlinna, *Erweiterung der Theorie des Hilbertschen Raumes*, Comm. Sém. Math. Univ. Lund [Medd. Lunds Univ. Mat. Sem.] (1952), 160–168.

Matthias Langer
Department of Mathematics and Statistics
University of Strathclyde
26 Richmond Street
Glasgow G1 1XH, United Kingdom
e-mail: ml@maths.strath.ac.uk

Annemarie Luger
Department of Mathematics
Lund Institute of Technology
Box 118
SE-221 00 Lund, Sweden
e-mail: luger@maths.lth.se

Operator Theory:
Advances and Applications, Vol. 198, 265–290

Interpolation of Sobolev Spaces and Indefinite Elliptic Spectral Problems

S.G. Pyatkov

Abstract. We consider the question on conditions ensuring the equality

$$(W_p^r(\Omega), L_{p,g}(\Omega))_{1-s,q} = (\overset{\circ}{W}{}_p^r(\Omega), L_{p,g}(\Omega))_{1-s,q},$$

where $L_{p,g}(\Omega)$ is the Lebesgue space with weight. The results are applied to examining the Riesz basis property for elliptic eigenvalue problems of the form $Lu = \lambda g(x)u$ ($x \in G \subset R^n$), where the function $g(x)$ can change its sign in G.

Mathematics Subject Classification (2000). Primary 46B70; Secondary 35P10; 46C20.

Keywords. Interpolation of Banach spaces, elliptic eigenvalue problem, indefinite spectral problem, indefinite inner product, Krein space, Sobolev space.

1. Introduction

Let Ω be a bounded domain with a Lipschitz boundary Γ and let the symbol $W_p^r(\Omega)$ stand for the conventional Sobolev space (see the definition in [1]). By $\overset{\circ}{W}{}_p^r(\Omega)$ we mean the closure of the class $C_0^\infty(\Omega)$ in the norm of $W_p^r(\Omega)$. The symbol $(H_1, H_2)_{1-s,q}$ (H_1, H_2 are Banach spaces) denotes the space constructed by the real interpolation method (see [1]). Our main results are connected with the following property:

$$\exists s \in (0,1): \quad (W_p^r(\Omega), L_{p,g}(\Omega))_{1-s,q} = (\overset{\circ}{W}{}_p^r(\Omega), L_{p,g}(\Omega))_{1-s,q}, \tag{1.1}$$

where $r > 0$, $1 < p, q < \infty$, and $L_{p,g}(\Omega)$ is the space of measurable (in Ω) functions $u(x)$, with $\|u\|_{L_{p,g}(\Omega)} = \|u(x)|g(x)|^{1/p}\|_{L_p(\Omega)} < \infty$. We also present applications to elliptic eigenvalue problems with an indefinite weight function of the form

$$Lu = \lambda Bu, \quad x \in G \subset R^n, \tag{1.2}$$

$$B_j u|_\Gamma = 0, \quad j = \overline{1,m}, \tag{1.3}$$

where L is an elliptic differential operator of order $2m$ defined in a domain $G \subset R^n$ with boundary Γ, the B_j's are differential operators defined on Γ, and $Bu =$

$g(x)u$ with $g(x)$ a real-valued measurable function changing sign in G. We assume that there exist open subsets G^+ and G^- of G such that $\mu(\overline{G^\pm} \setminus G^\pm) = 0$ (μ is the Lebesgue measure), $g(x) > 0$ almost everywhere in G^+, $g(x) < 0$ almost everywhere in G^-, and $g(x) = 0$ almost everywhere in $G^0 = G \setminus (\overline{G^+} \cup \overline{G^-})$. For example, it is possible that $G^0 = \emptyset$.

We describe some problems, where the condition (1.1) arises. Let A be a positive operator in the Hilbert space $L_{2,g}(\Omega) = H$ such that $\overset{\circ}{W}{}_2^r(\Omega) \subset D(A), D(A^*) \subset W_2^r(\Omega)$. Denote its domain by H_1 and put $H_{-1} = (D(A^*))^*$. This space can be identified with a negative space constructed on the pair $D(A^*)$ and H. In this case the condition (1.1) serves as a good sufficient condition ensuring the interpolation equality

$$(H_1, H_{-1})_{1/2,2} = H, \tag{1.4}$$

which arises in many fields. An abstract analog of (1.1) is the following condition: $\exists s \in (0,1) : (D(A), H)_{1-s,2} = (D(A^*), H)_{1-s,2}$ and it was used (probably, for the first time) in [4]. Consider an abstract situation in which A is an arbitrary positive operator in a Hilbert space H. In this case the condition (1.4) is a sufficient and sometimes necessary condition ensuring the fact that a Riesz projection of A corresponding to an unbounded component of the spectrum is bounded ([2, 3, 4]). Let S be a sector of the complex plane with the vertex at the origin such that $\sigma(A) \subset S$ and let $\varphi(\lambda)$ be an analytic bounded function on $\overline{S}$. The question is when the operator $\varphi(A)$ is bounded. If this is true for every φ then we say that the operator A possesses the bounded H^∞-calculus. It is possible to prove that A possesses the H^∞-calculus if and only if (1.4) holds (see [7]). The condition (1.4) is also equivalent to the condition that the imaginary powers $A^{i\xi}$ of A are bounded operators [8, 7]. The corresponding theory in the Banach case can be found in [6]. Let H be a Krein space and let A be a J-dissipative operator in H. Under some simple conditions the equality (1.4) guarantees the existence of maximal semidefinite invariant subspaces (see Section 4 of Chapter 1 in [2] or [3, 5]). The equalities of the type (1.4) arise when studying boundary value problems for forward-backward parabolic equations (see, for instance, [2, 9] and elliptic eigenvalue problem of the form (1.2), (1.3).

The spectral problems (1.2), (1.3) were studied in many papers. Primarily, we should mention the early articles by Hilbert who proved the existence of infinitely many positive and negative eigenvalues in the case when $m = 1$, $n = 1$, with L a positive operator, and who considered the corresponding eigenfunction expansion. Similar questions were also discussed in the articles by R.G.D. Richardson, O. Haupt, H. Hilb (1905–1915). The first results in the multidimensional case are due to E. Holmgren (1907). The asymptotic distribution of the eigenvalues of the problem (1.2), (1.3) was established in the articles by M.Sh. Birman and M.Z. Solomyak (1977–1979), J. Fleckinger and M.L. Lapidus, E.I.M. Veling, A.B. Mingarelli, and some other authors (1980–1990). Studying the completeness questions for the root functions of the problem (1.2), (1.3) is a comparatively recent matter. Many articles are devoted to the model problems arising in mathematical physics. Attention is focused on the completeness questions for the eigenfunctions in $L_{2,g}(G^+ \cup G^-)$ and for the eigenfunctions in $L_{2,g}(G^+)$ ($L_{2,g}(G^-)$) corresponding to positive (negative) eigenvalues. The most general results in this direction were obtained by Faierman M. (1989–1990). He even considered the case of a nonselfadjoint operator L. The first articles devoted to the unconditional basis property appeared only recently (R. Beals, S. Pyatkov, 1984, 1985). Later, this question has been extensively studied up to this time, but mainly in the one-dimensional case. We should mention the articles Najman B., Curgus B., P. Binding, R. Hryniv, H. Langer, H. Volkmer, A. Fleige, A. Parfenov. The multi-dimensional case was studied in the articles by B. Curgus and B. Najman, M. Faerman, G.F. Roach, S.G. Pyatkov, A. Parfenov. Sufficiently complete bibliography can be found in [2, 3, 15]. Among the recent article we mention [9]–[12], [16]–[24].

In the present article, we present sufficient conditions ensuring (1.1). They look like the celebrated A_p-conditions (see, [26], [25], [13]) and are stated in Theorems 3.7–3.11. Next, the results obtained are applied to the study of the problems (1.2), (1.3). In Theorems 4.4, 4.5, we exhibit some conditions that ensure the Riesz basis property of eigenfunctions and associated functions of problem (1.2), (1.3) in the space $L_{2,g}(G \setminus G^0)$ and the Riesz basis property of "halves" of eigenfunctions and associated functions in $L_{2,g}(G^+)$ and $L_{2,g}(G^-)$, respectively. If $L > 0$ then these halves comprise eigenfunctions corresponding to positive and negative eigenvalues. The sufficient conditions obtained are much weaker than those in [2, 3].

2. Preliminaries

Here, we present our conditions on the operators L and the function g and recall some known results and facts. We consider eigenvalue problem (1.2), (1.3), where the operator L is defined by the differential expression

$$Lu = \sum_{|\alpha|,|\beta| \le m} D^\alpha a_{\alpha,\beta} D^\beta u(x)$$

and the boundary conditions

$$B_j u|_\Gamma = 0 \quad (j = \overline{1,m}),$$

with B_j differential operators defined on $\Gamma = \partial G$, i.e.,

$$B_j u = \sum_{|\alpha| \le m_j} b_{\alpha,j} D^\alpha u(x) \ (j = \overline{1,m}, m_j < 2m).$$

Let G be a bounded (for simplicity) domain and let $\Gamma \in C^{2m}$. The conditions on the coefficients of L and B_j are described as follows:

(I) $\quad a_{\alpha,\beta}(x) = (-1)^{|\alpha|+|\beta|}\overline{a_{\beta,\alpha}}, \quad a_{\alpha,\beta} \in C^{\max(|\alpha|,|\beta|)}(\bar{G}), \quad b_{\alpha,j} \in C^{2m-m_j}(\Gamma)$

for each α, β, and j;

(II) $$a(x,\xi) = \sum_{|\alpha|,|\beta|=m} a_{\alpha,\beta}\xi^{\alpha+\beta} \ne 0 \text{ for } 0 \ne \xi \in \mathbb{R}^n \text{ and } x \in \overline{G};$$

for every pair of linear independent vectors $\xi,\eta \in \mathbb{R}^n$ and every $x \in \overline{G}$, the polynomial $a(x,\xi+\tau\eta)$ in τ has exactly m roots with positive imaginary part (and, thus, m roots with negative imaginary part); the system of boundary operators $\{B_j\}$ is normal on Γ and covers the operator L on Γ (see the definitions in [1]).

Next, we put

$$D(L) = \{u \in W_2^{2m}(G) : B_j u|_\Gamma = 0,\ j = 1,2,\dots,m\},$$
$$H_1 = \{u \in W_2^m(G) : B_j u|_\Gamma = 0,\ m_j < m\}$$

and assume that

(III) L with the domain $D(L)$ is a selfadjoint operator in $L_2(G)$ and there exist constants $c_1, c_2 > 0$, and c_3 such that

$$c_1\|u\|^2_{W_2^m(G)} \geq (Lu,u) \geq c_2\|u\|^2_{W_2^m(G)} - c_3\|u\|^2_{L_2(G)} \quad \forall u \in D(L);$$

(IV) $g(x) \in L_p(G)$, $p > \dfrac{n}{2m}$ for $m < \dfrac{n}{2}$, $p = 1$ for $m > \dfrac{n}{2}$, $p > 1$ for $m = \dfrac{n}{2}$.

The operators $L, B : L_2(G) \to L_2(G)$ are selfadjoint and the last our condition is as follows

$$\text{(V)} \quad \ker B \cap \ker L = \{0\}.$$

We suppose that the norm in H_1 coincides with the conventional norm of the space $W_2^m(G)$. Denote by $\chi_A(x)$ the characteristic function of the set A. Put $F_0 = L_{2,g}(G \setminus G^0)$. By $(\cdot,\cdot)$ we mean the inner product in $L_2(G)$, i.e.,

$$(u,v) = \int_G u(x)\overline{v(x)}\,dx.$$

The inner product $(\cdot,\cdot)_0$ in F_0 is defined as $(u,v)_0 = (|g(x)|u,v)$. The space F_0 becomes a Krein space if we introduce an indefinite metric by the equality $[u,v]_0 = (Ju(x),v(x))_0$, with J the operator of multiplication by the function $\chi_{G^+} - \chi_{G^-}$. Given Banach spaces H_1, H_2, the symbol by $L(H_1,H_2)$ stands for the space of bounded linear operators defined on H_1 with values in H_2. The inner product in $L_2(G)$ admits an extension to the duality relation between H_1 and the negative space H_1' (see the definition in Section 1 of Chapter 1 in [29]) which coincides with the dual space of H_1 and we have $H_1 \subset L_2(G) \subset H_1'$; moreover, the operator L and the operator $B|_{H_1\cap D(B)}$ (under the conditions (I)–(IV)) are extendable to operators of the class $L(H_1,H_1')$ (see the proof in [15]). We now explain what we call eigenfunctions and associated functions for problem (1.2), (1.3).

By definition, $\{u_i\}_{i=0}^n$ ($u_i \in H_1$) is a *chain of e.a.e (eigenfunctions and associated functions) of problem* (1.2), (1.3) *corresponding to* $\lambda \in \mathbb{C}$ whenever

$$Lu_i - \lambda Bu_i - Bu_{i-1} = 0,\ i = \overline{0,n},\ u_{-1} = 0.$$

Put $C_1 = \{u \in H_1 : (Lu,v) = 0\ \forall v \in \ker B \cap H_1\}$. We define the space F_1 as

$$F_1 = \{u \in F_0 :\ \exists v \in C_1 :\ \ v(x) = u(x)\ a.e.\ in\ G^+ \cup G^-\}$$

and endow this space with the norm

$$\|f\|_{F_1} = \inf_{\psi\in\ker B\cap H_1} \|f+\psi\|_{H_1}.$$

As is known (see, for instance, [15]), under the conditions (I)–(IV) the eigenfunctions and associated functions of the problem (1.2), (1.3) constitute a Riesz basis for F_1 and are complete in $L_2(G \setminus G^0)$. Put $F_s = (F_1, F_0)_{1-s,2}$. Define the space F_{-1} as the completion of F_0 with respect to the norm

$$\|u\|_{-1} = \sup_{v \in F_1} |[u, v]_0| / \|v\|_{F_1}.$$

We use the following interpolation condition ([2, 3, 15]) under which the eigenfunctions and associated functions of the problem (1.2), (1.3) after the corresponding normalization form a Riesz basis for $L_2(G \setminus G^0)$:

$$(F_1, F_{-1})_{1/2,2} = F_0. \tag{2.1}$$

A sufficient condition for (2.1) is described in the following lemma (see [2]).

Lemma 2.1. *Let*

$$\exists s_0 \in (0,1) : \quad J \in L(F_{s_0}, F_{s_0}). \tag{2.2}$$

Then the equality (2.1) *is valid.*

Present also the following important criterion by A. Parfenov [9, 10].

Theorem 2.2. *Let the Banach space Z, X be Banach spaces, $Z \subset X$, and let a Banach space Z_0 be a closed subspace of Z. Then, $\exists s > 0 : (Z, X)_{1-s,q} = (Z_0, X)_{1-s,q}$ $(q > 1)$ if and only if there exist a map $T : X \to X$, $N_1 \in (0, 1)$, $N_2 > 1$ such that, for all $u \in Z$, $u - Tu \in Z_0$, $\|Tu\|_Z \leq N_2 \|u\|_Z$, and $\|Tu\|_X \leq N_1 \|u\|_X$.*

In the articles [9, 10] this criterion is stated for the case of $q = 2$. In the general case the same arguments as those in [9, 10] can be applied. Moreover, the statement is a simple consequence of the criterion for $q = 2$ and the reiteration theorem [1, Section 1.10.2].

3. Interpolation

Definition 3.1. We say that Ω is a Lipschitz domain (a domain with a Lipschitz boundary) if, for any $x_0 \in \Gamma = \partial\Omega$, there exists a neighborhood U about x_0 and a local coordinate system y obtained by rotation and translation of the origin from the initial one in which

$$\begin{aligned} \overline{U} \cap \Omega &= \{y \in \mathbb{R}^n : y' \in \overline{B_r}, \omega(y') < y_n \leq \omega(y') + \delta\}, \\ \overline{U} \cap (\mathbb{R}^n \setminus \overline{\Omega}) &= \{y \in \mathbb{R}^n : \omega(y') - \delta \leq y_n < \omega(y')\}, \\ \Gamma \cap \overline{U} &= \{y \in \mathbb{R}^n : y' \in \overline{B_r}, \ y_n = \omega(y')\}, \end{aligned}$$

where $y' = (y_1, y_2, \ldots, y_{n-1})$, $B_r = \{y' : \ |y'| < r\}$, $\delta > 0$ is a constant, and the function ω satisfies the Lipschitz condition in B_r.

By $B_r(y)$ we mean the ball of radius r centered at y. We say that nonnegative functions ω_1, ω_2 are equivalent in Ω ($\omega_1 \sim \omega_2$) whenever $\exists c_1, c_2 > 0 : \ \omega_1(x) c_1 \leq$

$\omega_2(x) \leq c_2\omega_1(x)$ for almost all $x \in \Omega$. By $\rho(x)$ we mean a regularized distance in Ω, i.e., a function in $C^\infty(\Omega)$ such that

$$\rho(x) \sim \operatorname{dist}(x, \Gamma) = \rho_0(x), \quad \exists c(\alpha) > 0 : |D^\alpha \rho(x)| \leq c(\alpha)\rho^{1-|\alpha|}(x) \quad \forall x \in \Omega. \tag{3.1}$$

Such a function exists and a method of construction of this function and more general functions of this kind is presented in [1, Section 3.2.3]. Let Ω be a Lipschitz domain and let U be a domain from Definition 3.1. For convenience, we always assume that any such domain $U = \{y \in \mathbb{R}^n : y' \in B_r, \omega(y') - \delta < y_n < \omega(y') + \delta\}$ is contained in a bigger domain $U^* = \{y \in \mathbb{R}^n : y' \in B_{r_1}, \omega(y') - \delta < y_n < \omega(y') + \delta\}$ $(r_1 > r)$ with the same properties (those in Definition 3.1). In this case, it is easy to establish that

$$\exists\, \alpha, \beta > 0 : \ \alpha|y_n - \omega(y')| \leq \rho(x(y)) \leq \beta|y_n - \omega(y')| \quad \forall y \in U, \tag{3.2}$$

where $x(y)$ is the point corresponding to y in the initial coordinate system. Let a function $\varphi(\eta)$ be an averaging kernel, i.e., $\varphi \in C_0^\infty(\mathbb{R}^n)$ and $\int_{\mathbb{R}^n} \varphi(\eta)\, d\eta = 1$. We often consider averaging kernels with the properties

$$\int_{\mathbb{R}^n} \varphi(\eta)\eta^\alpha\, d\eta = 0 \quad \forall 1 \leq |\alpha| \leq r, \tag{3.3}$$

where r is a positive integer. The averaging kernels with these properties exist (see [27, Section 7]). For a Lipschitz domain Ω we can find finitely many open domains $U_i = \{y \in \mathbb{R}^n : y' \in B_{r_i}, \omega(y') - \delta < y_n < \omega(y') + \delta\}$ $(i = 1, 2, \ldots, N)$ covering Γ and possessing the properties from Definition 3.1. By assumption, some domains $U_i^* = \{y \in \mathbb{R}^n : y' \in B_{r_i^*}, \omega(y') - \delta < y_n < \omega(y') + \delta\}$ $(i = 1, 2, \ldots, N)$ with $r_i^* > r_i$ also possess the same properties. In what follows, we consider coverings possessing this property. Given $\varepsilon_0 \in (0, 1)$, for any such system we can construct a system of functions $\rho_i(y) \in C^\infty(\overline{U_i} \setminus \Gamma)$ (y is a local coordinate system) satisfying (3.1) and such that

$$(1 - \varepsilon_0)\, |y_n - \omega(y')| \leq \rho_i \leq (1 + \varepsilon_0)\, |y_n - \omega(y')|\,, \quad 1 - \varepsilon_0 \leq \frac{\partial \rho_i(y)}{\partial y_n} \leq 1 + \varepsilon_0 \tag{3.4}$$

for all $y \in U_i$ (the functions ω change in dependence of i). The functions ρ_i can be constructed as follows. Choose a nonnegative averaging kernel $\varphi(\xi)$, with $\operatorname{supp} \varphi \subset B_{q_0}(0)$. Let $C = \operatorname{ess\,sup}_{x \in \Omega} |\nabla \rho(x)|$ and let $M = \operatorname{ess\,sup}_{y' \in B_{r_i^*}} |\nabla \omega(y')|$. There exists a constant $c_0 > 0$ such that $\rho(x) \leq c_0\rho_0(x)$ $\forall x \in \Omega$. Choose q_0 so that $q_0 < \min(r_i^* - r_i, \varepsilon_0/MC, 1/c_0)$. Then the function

$$\rho_i(y) = y_n - \frac{1}{\rho^n(x(y))} \int_{\mathbb{R}^n} \varphi((\eta - y)/\rho(x(y)))\omega(\eta')\, d\eta$$

meets our conditions. A nonnegative measurable function $g(x)$ is called a slowly varying function in Ω if

$$\exists q < 1, \ M > 1 : \quad \operatorname*{ess\,sup}_{y \in B_{q\rho_0(x)}(x)} g(y) \leq M \operatorname*{ess\,inf}_{y \in B_{q\rho_0(x)}(x)} g(y)$$

for almost all $x \in \Omega$.

We will use the straightening of the boundary, i.e., the transformation $z' = y'$, $z_n = y_n - \omega(y')$. This transformation takes U from Definition 3.1 onto a cylinder $z(U)$. The inverse transformation $y = y(z)$ can be written as $y' = z'$, $y_n = z_n + \omega(z')$. Given $y \in U \cap \Omega$, put

$$K_y(a) = \{\eta \in \Omega : \ |\eta' - y'| < a(y_n - \eta_n)\}, \quad a > 0. \tag{3.5}$$

As it follows from the definitions, we have the following statement.

Lemma 3.2. *Let U be a domain from Definition 3.1 and let $z = z(y)$ be the straightening of the boundary. Assume that $K_y(a) \subset U$ and $a < 1/(2M)$, with M the Lipschitz constant of the function ω in B_r. Then $L_z(a/(1+aM)) \subset z(K_y(a)) \subset L_z(a/(1-aM))$, where $L_z(a_1) = \{\xi : \ |\xi' - z'| < a_1(z_n - \xi_n), \ 0 < \xi_n < z_n\}$.*

In this section, we assume that $g(x)$ is a nonnegative function which is positive a.e. in some neighborhood about Γ, and $g \in L_1(\Omega)$, unless otherwise stated. We will use the following condition

(A) there exist a finite covering U_i $(i = 1, 2, \dots, N)$ of Γ (the domains U_i possess the properties from Definition 3.1) and the corresponding local coordinate systems such that for some $a > 0, c > 0$ and almost all $y \in U_i \cap \Omega$

$$\int_{K_y(a) \cap U_i} g(\eta)\, d\eta \le c\mu(K_y(a))g(y), \quad i = 1, 2, \dots, N, \tag{3.6}$$

(here the function $g(y)$ is written in the local coordinate system y).

We can always assume that a, M, δ are independent of i and $a < 1/(2M)$ (recall that M is the Lipschitz constant of the corresponding function ω in U_i). The parameter r in the definition of U_i and the function ω can change in dependence on i. For brevity, we preserve the same notation ω for all functions ω. The condition (A) is an analog of the A_1-condition [26].

Let $K_y(a) \subset U_i$. Straightening the boundary and using (A) and Lemma 3.2, we arrive at the inequality

$$\int_{L_z(a_1)} g_0(\xi)\, d\xi \le c^* \mu(L_z(a_1)) g_0(z), \quad a_1 \le a/(1 + Ma), \ z = z(y), \tag{3.7}$$

where $g_0(\xi) = g(\xi', \xi_n + \omega(\xi'))$. Put $L_{qz}(a) = \{\xi \in L_z(a) : \ 0 < \xi_n < qz_n\}$, $q \in (0, 1)$.

Lemma 3.3. *Assume that Ω is a Lipschitz domain and* (A) *holds. Fix $a_2 < a_1 \le a/(1 + Ma)$. Then*

a) *for any $\varepsilon > 0$ there exists a constant $q \in (0, 1)$ such that*

$$\int_{L_{qz}(a_2)} g_0(\xi)\, d\xi \le \varepsilon \int_{L_z(a_2)} g_0(\xi)\, d\xi \tag{3.8}$$

for all $z \in A_i = z(U_i \cap \Omega)$ such that $K_{y(z)}(a) \subset U_i$ $(i = 1, 2, \dots, N)$;

b) *for any* $\varepsilon > 0$ *there exists a constant* $q \in (0,1)$ *such that*

$$\int_{L_{qz}(a_2)} g_0(\xi)\, d\xi \le \varepsilon \int_{L_z(a_1)} g_0(\xi)\, d\xi, \tag{3.9}$$

for all $z \in A_i$ *such that* $K_{y(z)}(a) \subset U_i$ $(i = 1, 2, \ldots, N)$;

c) *for any* $\varepsilon > 0$ *there exists a constant* $q \in (0,1)$ *such that*

$$\int_{L_{qz}(a_2)} g_0(\xi)\, d\xi \le \varepsilon \int_{L_z(a_1)\setminus L_{qz}(a_1)} g_0(\xi)\, d\xi \tag{3.10}$$

for all $z \in A_i$ *such that* $K_{y(z)}(a) \subset U_i$ $(i = 1, 2, \ldots, N)$.

Proof. First, we prove a). Fix i, divide (3.7) by $\mu(L_z(a_1))$, and integrate the result over $L_\eta(a_2)$ $(K_{y(\eta)}(a) \subset U_i)$. The inequality obtained can be written as

$$\int_{A_i}\int_{A_i} g_0(\xi)\chi_{L_\eta(a_2)}(z)\frac{\chi_{L_z(a_1)}(\xi)}{\mu(L_z(a_1))}\, dz d\xi \le c\int_{L_\eta(a_2)} g_0(\xi)\, d\xi. \tag{3.11}$$

The left-hand side is estimated from below by the quantity

$$c_0\int_{L_{q\eta}(a_2)} g_0(\xi)\int_{A_i}\chi_{L_\eta(a_2)}(z)\frac{\chi_{L_z(a_1)}(\xi)}{z_n^n}\, dz d\xi, \quad q \in (0,1), \tag{3.12}$$

where $c_0 = 1/\mu(L_e(a_1))$ and $e = (0, \ldots, 0, 1)$. We estimate the function

$$a(\eta,\xi) = \int_{A_i}\chi_{L_\eta(a_2)}(z)\frac{\chi_{L_z(a_1)}(\xi)}{z_n^n}\, dz, \quad \xi \in L_{q\eta}(a_2),$$

from below. Let $q < 1/4$. We have

$$a(\eta,\xi) \ge \int_{\xi_n}^{\eta_n/2}\frac{1}{z_n^n}\mu_{n-1}(B_{z_n})\, dz_n, \tag{3.13}$$

where $\mu_{n-1}(\cdot)$ is the $n-1$-dimensional Lebesgue measure and $B_{z_n} = \{h' : |h' - \eta'| < a_2(\eta_n - z_n),\ |\xi' - h'| < a_1(z_n - \xi_n)\}$, $0 < z_n < \eta_n/2$. Put $\tilde{h}' = h'/\eta_n$, $\tilde{z} = z/\eta_n$, $\tilde{\xi} = \xi/\eta_n$, $\tilde{B}_{\tilde{z}_n} = \{\tilde{h}' : |\tilde{h}' - \tilde{\eta}'| < a_2(1 - \tilde{z}_n),\ |\tilde{\xi}' - \tilde{h}'| < a_1(\tilde{z}_n - \tilde{\xi}_n)\}$, $0 < \tilde{z}_n < 1/2$. Since $a_2 < a_1$ and $\tilde{\xi}_n < 1/4$, from geometric arguments (we just use similarity) it follows that there exists a constant $\delta_1 = \delta_1(a_2, a_1)$ such that $\mu_{n-1}(\tilde{B}_{\tilde{z}_n}) \ge \delta_1(\tilde{z}_n - \tilde{\xi}_n)^{n-1}$. Make the change of variables $z_n/\eta_n = \tau$ in (3.13). We arrive at the estimate

$$a(\eta,\xi) \ge \int_{\tilde{\xi}_n}^{1/2}\frac{1}{\tau^n}\mu_{n-1}(\tilde{B}_\tau)\, d\tau \ge \delta_1\int_{\tilde{\xi}_n}^{1/2}\frac{(\tau - \tilde{\xi}_n)^{n-1}}{\tau^n}\, d\tau.$$

Using the fact that $0 < \tilde{\xi}_n \le q$ and the change of variables $\tau = \tilde{\xi}_n r$, we infer

$$a(\eta,\xi) \ge \delta_1\int_1^{1/(2q)}\frac{(r-1)^{n-1}}{r^n}\, dr \ge \delta_2\ln(1/q), \tag{3.14}$$

for all $q \leq 1/4$, where $\delta_2 > 0$ is a constant. The relations (3.11) and (3.12) yield

$$\int_{L_{q\eta}(a_2)} g_0(\xi)\, d\xi \leq \frac{c}{\ln(1/q)} \int_{L_\eta(a_2)} g_0(\xi)\, d\xi,$$

where $c > 0$ is some constant independent of η. Choosing an appropriate q we obtain the claim. The statements b) and c) are consequences of a). □

Remark 3.4. Condition (3.9) is an analog of the A_∞-condition [25].

Remark 3.5. Consider the one-dimensional case and let $\Omega = (0, 1)$. In this case the boundary of Ω consists of two points and every of the conditions (3.8)–(3.10) must be divided into two inequalities. For instance, a) (for the function g) is stated as follows:

b_1) for every $\varepsilon > 0$ there exists a constant $q < 1$ such that

$$\int_0^{qz} g(\xi)\, d\xi \leq \varepsilon \int_0^z g(\xi)\, d\xi, \forall z \in (0, 1); \tag{3.15}$$

b_2) for every $\varepsilon > 0$ there exists a constant $q < 1$ such that

$$\int_{qz+1-q}^1 g(\xi)\, d\xi \leq \varepsilon \int_z^1 g(\xi)\, d\xi \quad \forall z \in (0, 1). \tag{3.16}$$

As it follows from the results in [9]–[11] (see, for instance, Theorem 3 in [9]), conditions b_1), b_2) are necessary and sufficient conditions for the property (1.1). They are weaker than (A). Analogs of conditions b_1), b_2) in the multidimensional case are the conditions a), b), c). In some cases (for small values of p or in the case of slowly varying function g) similar conditions were successfully used (see Sections 8, 9 in [9]).

Lemma 3.6. *Let $g \in L_1(\Omega)$ be a slowly varying function. Then the condition* (A) *is equivalent to the condition*

(A′) *there exist a finite covering U_i $(i = 1, 2, \ldots, N)$ of Γ (the domains U_i possess the properties from Definition* 3.1*) and the corresponding local coordinate systems such that for some $a > 0, c > 0, q \in (0, 1)$ and all $y \in U_i \cap \Omega$ $(i = 1, 2, \ldots, N)$*

$$\int_{K_y(a)\cap U_i} g(\eta)\, d\eta \leq c \int_{K_y(a)\cap U_i \cap B_{q\rho_0(y)}(y)} g(\eta)\, d\eta$$

(here the function $g(y)$ is written in the local coordinate system y).

Proof. The proof results from the definitions. □

Let us proceed with the our main results. For simplicity, we assume below that the number r is an integer. The results are valid for arbitrary r, but this assumption simplifies all arguments.

Theorem 3.7. *Under the condition* (A) *for $p \in (1, \infty)$*

$$\exists s \in (0, 1) \quad (W_p^r(\Omega), L_{p,g}(\Omega))_{1-s,q} = (\overset{\circ}{W}{}_p^r(\Omega), L_{p,g}(\Omega))_{1-s,q} \tag{3.17}$$

for every $q \in (1, \infty)$.

Proof. To prove the claim, we use Theorem 2.2. So we need to prove the existence of a map $T : L_{p,g}(\Omega) \to L_{p,g}(\Omega)$ such that, for some constants $N_1 \in (0,1)$ and $N_2 > 1$ and for all $u \in W_p^r(\Omega)$, $u - Tu \in \overset{\circ}{W}{}_p^r(\Omega)$, $\|Tu\|_{W_p^r(\Omega)} \le N_2 \|u\|_{W_p^r(\Omega)}$, and $\|Tu\|_{L_{2,g}(\Omega)} \le N_1 \|u\|_{L_{2,g}(\Omega)}$. Consider the covering $\{U_i\}_{i=1}^N$ from (A). Fix $\varepsilon_0 \in (0,1)$ and find the corresponding functions $\rho_i(y)$ $(i = 1, 2, \dots, N)$. Fix a parameter i. We have an initial coordinate system x and a local coordinate system $y = y(x)$ $(x = x(y))$ and $U_i = \{y \in \mathbb{R}^n : |y'| < r_i, \ \omega(y') - \delta < y_n < \omega(y') + \delta\}$. After the straightening the boundary $z = z(y)$ we have $z(U_i) = \{z \in \mathbb{R}^n : |z'| < r_i, \ -\delta < z_n < \delta\}$. Put $U_{i0} = \{y \in U_i : |y'| < r_{1i}\}$. For every i, choose $r_{1i} < r_i$ so that the collection $\{U_{i0}\}$ is a covering of Γ again. Construct a partition of unity $\{\varphi_i(x)\}_{i=1}^N$ on Γ subordinate to this covering such that $0 \le \varphi_i(x) \le 1$ for all x and i. Let $\Omega_{\delta_1} = \{x \in \Omega : \rho(x) < \delta_1\}$. By construction, there exists $\delta_1 > 0$ such that, for every i and $y \in U_{i0} \cap \Omega_{\delta_1}$, we have $K_y(a) \subset U_i$. Construct a function $\psi(x) \in C_0^\infty(\mathbb{R}^n)$ such that $\psi(x) = 1$ for $x \in \Omega_{\alpha\delta_1/2\beta}$, $\psi(x) = 0$ for $x \in \Omega \setminus \Omega_{3\alpha\delta_1/4\beta}$, and $0 \le \psi(x) \le 1$ for all x. Given $a_1 = a/(1 + Ma)$, take $a_2 < a_1$, and assign $a_3 = a_2/(1 + Ma_2)$. We can assume that the constant a_2 is sufficiently small and we have

$$a_3 \sup_{y \in U_i \cap \Omega} |\nabla_{y'} \rho_i(y)| \le (1 - \varepsilon_0)/2 \quad (i = 1, 2, \dots, N). \tag{3.18}$$

Let $q_0 \in (0,1)$ and let $K_\sigma = \{\xi \in \mathbb{R}^n : |\xi'| < a_3 \xi_n, \ q_0\sigma < \xi_n < \sigma\}$, where the parameter $\sigma > 0$ will be chosen later. Construct an averaging kernel ψ_0 satisfying (3.3) and such that $\operatorname{supp} \psi_0 \in K_1$. Then the function $\varphi(\eta) = \frac{1}{\sigma^n} \psi_0(\frac{\eta}{\sigma})$ is also an averaging kernel, $\operatorname{supp} \varphi \subset K_\sigma$, and $\int_{\mathbb{R}^n} |\varphi(\eta)|\, d\eta = \int_{\mathbb{R}^n} |\psi_0(\eta)|\, d\eta = c_1$, $|\varphi(\eta)| \le c_0/\sigma^n$, with $c_0 = \sup_{x \in \mathbb{R}^n} |\psi_0(x)|$. Therefore, we can assume that the constants c_0, c_1 are independent of σ. Assign

$$T_i u = u_i(x) = \frac{1}{\rho_i^n(y)} \int_{U_i \cap \Omega} v(\eta) \varphi\left(\frac{\eta - y}{\rho_i(y)}\right) d\eta \Big|_{y = y(x)}, \quad v(\eta) = \varphi_i \psi u(x(\eta)).$$

This function is defined for $y \in U_i \cap \Omega$. Let us find $\operatorname{supp} T_i u$ for these y. Let $\eta \in B = \operatorname{supp} v(\eta)$ and let $z = z(y)$ as well as $\xi = \xi(\eta)$ be the straightening the boundary. The condition $\rho(x(\eta)) < 3\alpha\delta_1/4\beta$ and (3.2) yield $\xi_n < 3\delta_1/4\beta$. Since on the support of the integrand $\xi_n > z_n$, this inequality implies that if $z_n \ge 3\delta_1/4\beta$ then $u_i(x(y(z))) = 0$. Thus, if $A = \{y \in U_i \cap \Omega : u_i(x(y)) \ne 0\}$ then $z(A) \subset \{z \in \mathbb{R}^n : |z'| \le r_i, \ 0 < z_n < 3\delta_1/4\beta\}$. In view of the inequality $\rho(x(y(z))) \le \beta z_n$, we have that $A \subset \Omega_{\delta_1}$. If $\frac{\eta - y}{\rho_i} \in K_\sigma$ then $|\eta' - y'| < a_3(\eta_n - y_n)$ and $q_0\sigma\rho_i(y) < \eta_n - y_n < \sigma\rho_i(y)$. Using the equalities $\xi_n = \eta_n - \omega(\eta')$ and $z_n = y_n - \omega(y')$ we infer

$$|\xi' - z'| < a_3(\xi_n - z_n)/(1 - Ma_3) = a_2(\xi_n - z_n), \tag{3.19}$$

$$\sigma(1 - \varepsilon_0) z_n q_0 (1 - Ma_3) < \xi_n - z_n < \sigma(1 + \varepsilon_0) z_n (1 - Ma_3)/(1 - 2Ma_3). \tag{3.20}$$

As a consequence, we have that

$$\frac{\xi_n(1 - 2Ma_3)}{1 - 2Ma_3 + \sigma(1 + \varepsilon_0)(1 - Ma_3)} < z_n < \frac{\xi_n}{1 + \sigma(1 - \varepsilon_0) q_0 (1 - Ma_3)}. \tag{3.21}$$

The inequality (3.19) implies that $|\xi' - z'| < a_2\xi_n < a_2\delta_1/\alpha$. Denote by δ_2 the distance from $z(\operatorname{supp}\varphi_i)$ to $\partial z(U_{i0})$. Choose δ_1 so that $\delta_1 < \delta\alpha$ and $\delta_1 < \delta_2\alpha/a_2$. Let $U_{\delta_1} = U_{i0}\cap\Omega_{\delta_1}$. The last inequalities for δ_1 ensure that $\operatorname{dist}(A, \partial(U_{i0}\cap\Omega)\setminus\Gamma) > 0$, $A \subset U_{\delta_1}$, and $z(U_{\delta_1}) \subset \{z \in \mathbb{R}^n : \ |z'| < r_{1i}, 0 < z_n < \delta_1/\alpha\}$. Finally, we want to guarantee

$$\{y + \rho_i(y)\eta : \ \eta \in K_\sigma, \ y \in U_{\delta_1}\} \subset U_i \cap \Omega. \tag{3.22}$$

Decreasing δ_1 if necessary we can assume that $\delta_1 < \delta\alpha/(1 + (1+\varepsilon_0)\sigma)$ and $\delta_1 < (r_i - r_{1i})\alpha/(\sigma a_3(1+\varepsilon_0))$. In this case, let $\xi = y + \rho_i(y)\eta$ with $y \in U_{\delta_1}$ and $\eta \in K_\sigma$. Then $\xi_n - \omega(y') = y_n - \omega(y') + \rho_i(y)\eta_n < z_n(1+(1+\varepsilon_0)\sigma) < \delta_1(1+(1+\varepsilon_0)\sigma)/\alpha < \delta$ and $|y' - \xi'| = \rho_i(y)|\eta'| < a_3\sigma(1+\varepsilon_0)z_n < a_3(1+\varepsilon_0)\sigma\delta_1/\alpha < r_i - r_{1i}$. These inequalities ensure (3.22). Now we can fix δ_1. Thus, we have that δ_1 is a sufficiently small parameter and $\delta_1 = \delta_1(\sigma) \approx 1/\sigma$. Extend the function $u_i(x)$ on the whole Ω by zero. We look for a function Tu in the form

$$Tu = \sum_{i=1}^{N} T_i u(x) = \sum_{i=1}^{N} u_i(x).$$

Demonstrate that, for $u \in W_p^r(\Omega)$, $u - Tu \in \overset{\circ}{W}{}_p^r(\Omega)$ and there exists a constant N_2 such that $\|Tu\|_{W_p^r(\Omega)} \leq N_2\|u\|_{W_p^r(\Omega)}$. In view of (3.22), for $y \in U_{\delta_1}$ we have

$$T_i u - v = \int_{K_\sigma} (v(y + \rho_i(y)\xi) - v(y))\varphi(\xi)\, d\xi. \tag{3.23}$$

We have (see, for instance, Section 1.1.7 in [28]) that

$$D_y^\alpha(v(y + \rho_i(y)\xi)) = \sum_{1\leq|\beta|\leq|\alpha|} \varphi_\beta^\alpha(y,\xi) D_\eta^\beta v(y + \rho_i(y)\xi) \quad (\eta = y + \rho_i(y)\xi, \ |\alpha| \leq r),$$

where

$$\varphi_\beta^\alpha(y,\xi) = \sum_s c_s \prod_{k=1}^{n} \prod_j D_y^{s_{kj}}(y_k + \rho_i(y)\xi_k).$$

Here c_s are some constants and the summation is performed over all collections of multi-indices $s = \{s_{kj}\}$, satisfying the conditions

$$\sum_{k,j} s_{kj} = \alpha, \ |s_{kj}| \geq 1, \ \sum_{k,j}(|s_{kj}| - 1) = |\alpha| - |\beta|.$$

If $|s_{kj}| = 1$ then $|D_y^{s_{kj}}(y_k + \rho_i(y)\xi_k)| \leq (1 + c_1|\xi|) \leq c_0$. If $|s_{kj}| > 1$ then $|D_y^{s_{kj}}(y_k + \rho_i(y)\xi_k)| = |D_y^{s_{kj}}\rho_i(y)\xi_k| \leq c_1\rho_i(y)^{1-|s_{kj}|}$. These inequalities imply that

$$|\varphi_\beta^\alpha| \leq c_2\rho_i(y)^{\sum_{k,j}(1-|s_{kj}|)} = c_2\rho_i(y)^{|\beta|-|\alpha|}. \tag{3.24}$$

Next, we use the following Taylor formula:

$$\begin{aligned} V(y + \rho_i(y)\xi) - V(y) &= \textstyle\sum_{1\leq|\alpha|\leq s-1} \frac{D^\alpha V(y)\rho_i^{|\alpha|}(y)\xi^\alpha}{\alpha!} \\ &\quad + \textstyle\int_0^1 (1-\tau)^{s-1} s \sum_{|\beta|=s} \frac{D^\beta V(y + \tau\rho_i(y)\xi)}{\beta!}\rho_i^s(y)\xi^\beta\, d\tau. \end{aligned} \tag{3.25}$$

From (3.23) and the properties of the averaging kernel we infer

$$D^\alpha(T_iu-v)=\int_{K_\sigma} D^\alpha_y(v(y+\rho_i(y)\xi)-v(y))\varphi(\xi)\,d\xi$$

$$=\int_{K_\sigma}\sum_{1\le|\beta|\le|\alpha|}\varphi^\alpha_\beta(y,\xi)(D^\beta_\eta v(y+\rho_i(y)\xi)-D^\beta_y v(y))\varphi(\xi)\,d\xi\ \ (\eta=y+\rho_i(y)\xi).$$

In view of (3.25) we conclude that

$$D^\beta_\eta v(y+\rho_i(y)\xi)-D^\beta_y v(y)=P_\beta(y,\xi)$$

$$+\int_0^1(1-\tau)^{r-|\beta|-1}(r-|\beta|)\sum_{|\gamma|=r-|\beta|}D^{\gamma+\beta}v(y+\tau\rho_i(y)\xi)\rho_i^{r-|\beta|}(y)\xi^\gamma\,d\tau,$$

where P_β is a polynomial in ξ of degree $r-|\beta|-1$ without a free term. Inserting this expression into the previous equality, we arrive at the representation

$$D^\alpha(T_iu-v)=\int_{K_\sigma}\sum_{1\le|\beta|\le|\alpha|}\varphi^\alpha_\beta(y,\xi)\varphi(\xi)$$

$$\times\int_0^1(1-\tau)^{r-|\beta|-1}(r-|\beta|)\sum_{|\gamma|=r-|\beta|}D^{\gamma+\beta}v(y+\tau\rho_i(y)\xi)\rho_i^{r-|\beta|}(y)\xi^\gamma\,d\tau d\xi.$$

Using (3.24) we obtain the estimate

$$|D^\alpha(T_iu-v)|\le\frac{c}{\sigma^n}\int_{K_\sigma}\rho_i^{r-|\alpha|}(y)\int_0^1\sum_{|\gamma|=r}|D^\gamma v(y+\tau\rho_i(y)\xi)|\,d\tau d\xi,\qquad(3.26)$$

with c a constant independent of v and σ. Let $|\alpha|=r$. Applying the Minkowski inequality, we derive

$$\begin{aligned}&\|D^\alpha(T_iu-v)\|_{L_p(U_i\cap\Omega)}=\|D^\alpha(T_iu-v)\|_{L_p(U_{\delta_1})}\\&\le\tfrac{c}{\sigma^n}\textstyle\int_{K_\sigma}\int_0^1\sum_{|\gamma|=r}\|D^\gamma v(y+\tau\rho_i(y)\xi)\|_{L_p(U_{\delta_1})}\,d\tau d\xi.\end{aligned}\qquad(3.27)$$

Next, we make the change of variables $\eta=\eta(y)=y+\tau\rho_i(y)\xi$. Let us show that the Jacobian of this transformation is bounded from below by some constant independent of $\tau\in(0,1)$ and $\xi\in K_\sigma$. Using induction it is not difficult to obtain that

$$\frac{\partial(\eta_1,\eta_2,\dots,\eta_n)}{\partial(y_1,y_2,\dots,y_n)}=1+\xi_n\rho_{iy_n}\tau+\sum_{k=1}^{n-1}\tau\xi_k\rho_{iy_k},$$

where $\rho_{iy_k}=\frac{\partial\rho_i}{\partial y_k}$. Next, we use (3.18) and derive

$$\frac{\partial(\eta_1,\eta_2,\dots,\eta_n)}{\partial(y_1,y_2,\dots,y_n)}\ge1+(1-\varepsilon_0)\xi_n\tau-\tau|\xi'||\nabla_{y'}\rho_i|\ge1+\tau\xi_n(1-\varepsilon_0)/2\ge1.$$

If we use (3.27) then we conclude that

$$\|D^\alpha(T_iu-v)\|_{L_p(U_i\cap\Omega)}=\|D^\alpha(T_iu-v)\|_{L_p(U_{\delta_1})}\le c_1\|v\|_{W^r_p(U_{\delta_1})}.\qquad(3.28)$$

where c_1 is some constant independent of v. So we have estimated the derivatives $D^\alpha(T_iu-v)$. A similar estimate is valid for the function T_iu-v. The arguments are much simpler in this case. We just use the Minkowski inequality. Therefore, we can conclude that

$$\exists N_2 > 0: \quad \|Tu\|_{W_p^r(\Omega)} \leq N_2 \|u\|_{W_p^r(\Omega)} \quad \forall u \in W_p^r(\Omega).$$

Given $u \in W_p^r(\Omega)$, demonstrate that $u - Tu \in \overset{\circ}{W}{}_p^r(\Omega)$. Using (3.26) and the arguments after this inequality we easily obtain the estimates

$$\sum_{|\alpha| \leq r} \|\rho_i^{|\alpha|-r}(y) D^\alpha (T_i u - v)\|_{L_p(U_i \cap \Omega)} \leq c \|v\|_{W_p^r(\Omega)}, \ i = 1, 2, \ldots, N.$$

The definition of the operator T yields the estimate

$$\sum_{|\alpha| \leq r} \|\rho_0^{|\alpha|-r}(x) D^\alpha (Tu - \psi u)\|_{L_p(\Omega)} \leq c \|u\|_{W_p^r(\Omega)} < \infty.$$

The results of Section 3.2.4 in [1] (we use the statement of Theorem 1 about the density of the class $C_0^\infty(\Omega)$ and Theorem 3 about equivalent norms) imply that there exists a sequence $v_n \in C_0^\infty(\Omega)$ such that

$$\|Tu - \psi u - v_n\|_{W_p^r(\Omega)} \leq \sum_{|\alpha| \leq r} \|\rho_0^{|\alpha|-r}(x) D^\alpha (Tu - \psi u - v_n)\|_{L_p(\Omega)} \to 0$$

as $n \to \infty$. Hence, $\psi u - Tu \in \overset{\circ}{W}{}_p^r(\Omega)$. The inclusion $(1-\psi)u \in \overset{\circ}{W}{}_p^r(\Omega)$ is obvious since this function has a compact support in Ω.

Now we estimate $\|Tu\|_{L_{p,g}(\Omega)}$. We have that

$$\|Tu\|_{L_{p,g}(\Omega)} \leq \sum_{i=1}^N \|T_i u\|_{L_p(U_i \cap \Omega)}, \quad T_i u = \frac{1}{\rho_i^n(y)} \int_{U_{\delta_1}} v(\eta) \varphi\left(\frac{\eta - y}{\rho_i(y)}\right) d\eta. \quad (3.29)$$

The Hölder inequality yields ($y \in U_{\delta_1}$)

$$|T_i u|^p \leq \frac{1}{\rho_i^n(y)} \int_{U_{\delta_1}} |v(\eta)|^p \left|\varphi\left(\frac{\eta - y}{\rho_i(y)}\right)\right| d\eta \left(\frac{1}{\rho_i^n(y)} \int_{U_{\delta_1}} \left|\varphi\left(\frac{\eta - y}{\rho_i(y)}\right)\right| d\eta\right)^{p/q}. \quad (3.30)$$

The last factor is estimated by

$$c_1 = \left(\int_{K_\sigma} |\varphi(\xi)| \, d\xi\right)^{p/q}, \quad 1/p + 1/q = 1.$$

Recall that this constant is independent of σ by construction. Multiplying (3.30) by $g(y)$ ($g(y) \equiv g(x(y))$) and integrating (3.30) over U_{δ_1} we arrive at the estimate

$$I = \int_{U_{\delta_1}} g(y) |T_i u|^p \, dy \leq c_1 \int_{U_{\delta_1}} |v(\eta)|^p \int_{U_{\delta_1}} g(y) \frac{1}{\rho_i^n(y)} \left|\varphi\left(\frac{\eta - y}{\rho_i(y)}\right)\right| dy d\eta.$$

Put $A_{\delta_1} = \{z : \ |z'| < r_{1i}, 0 < z_n < \delta_1/\alpha\}$. After the changes of variables $\xi = \xi(\eta)$, $z = z(y)$ (i.e., $z' = y'$ and $z_n = y_n - \omega(y')$), we arrive at the estimate

$$I \leq c_1 \int_{A_{\delta_1}} |v(\eta(\xi))|^p \int_{A_{\delta_1}} g(y(z)) \frac{1}{\rho_i^n(y(z))} \left|\varphi\left(\frac{\eta(\xi) - y(z)}{\rho_i(y(z))}\right)\right| dz d\xi = I_1.$$

In view of (3.4) $\rho_i(y(z)) \geq (1-\varepsilon_0)z_n$ and by (3.19), (3.21), and the properties of φ, we have that

$$\big|\varphi((\eta(\xi)-y(z))/\rho_i(y(z)))\big| \leq c_2\chi_{B_\xi}(z)/\sigma^n,$$

where

$$B_\xi = \{z:\ |\xi'-z'| < a_2(\xi_n - z_n),\ \alpha_1\xi_n < z_n < \alpha_2\xi_n\},$$

with $\alpha_1 = (1-2Ma_3)/(1-2Ma_3+\sigma(1+\varepsilon_0)(1-Ma_3))$ and $\alpha_2 = 1/(1+\sigma(1-\varepsilon_0)q_0(1-Ma_3))$. Thus, we infer

$$I_1 \leq \frac{c_3}{\sigma^n}\int_{A_{\delta_1}} |v(\eta(\xi))|^p \int_{A_{\delta_1}} g_0(z)\frac{1}{z_n^n}\chi_{B_\xi}(z)\,dzd\xi = I_2.$$

By definition, $B_\xi \subset L_{q\xi}(a_2)$ with $q = \alpha_2 \approx 1/\sigma$. As a corollary of this fact, we conclude that

$$I_2 \leq \frac{c_4}{\alpha_1^n\sigma^n}\int_{A_{\delta_1}} |v(\eta(\xi))|^p \frac{1}{\xi_n^n}\int_{L_{q\xi}(a_2)} g_0(z)\,dzd\xi = I_3.$$

Note that $1/\xi_n^n = c_5/\mu(L_\xi(a_1))$, with c_5 some constant independent of σ. We have that $q(\sigma) \to 0$ and the quantity $c_4/(\alpha_1^n\sigma^n)$ is bounded as $\sigma \to \infty$. Fix $\varepsilon > 0$. By Lemmas 3.3 and 3.2, there exists σ_0 such that for $\sigma \geq \sigma_0$

$$\begin{aligned} I_3 &\leq \frac{\varepsilon}{c^*}\int_{A_{\delta_1}} |v(\eta(\xi))|^p \frac{1}{\mu(L_\xi(a_1))}\int_{L_\xi(a_1)} g_0(z)\,dzd\xi \\ &\leq \varepsilon\int_{A_{\delta_1}} |v(\eta(\xi))|^p g_0(\xi)d\xi = \varepsilon\int_{U_{\delta_1}} |v(\eta)|^p g(\eta)d\eta. \end{aligned}$$

Final estimate looks as follows:

$$I = \int_{U_{\delta_1}} g(y)|T_iu(y)|^p\,dy \leq \varepsilon\int_{U_{\delta_1}} |v(\eta)|^p g(\eta)d\eta,\quad \sigma \geq \sigma_0(\varepsilon).$$

Without loss of generality, we may assume that the constant σ_0 is independent of $i = 1, 2, \ldots, N$. Using this estimates and (3.29), we derive the inequality

$$\|Tu\|_{L_{p,g}(\Omega)} \leq N\varepsilon\|u\|_{L_{p,g}(\Omega)}.$$

We take $\varepsilon = 1/(2N)$ and find the corresponding σ. In this case,

$$\|Tu\|_{L_{p,g}(\Omega)} \leq \frac{1}{2}\|u\|_{L_{p,g}(\Omega)}$$

and $N_1 = 1/2$. Theorem 2.2 proves the claim. □

In the following theorem we present necessary conditions. Consider an arbitrary neighborhood U from the definition of a Lipschitz domain and a cone $K_y(a) \subset U$ with $a < 1/(2M)$. Put $K_{qy} = \{\eta \in K_y(a) : \omega(\eta') < \eta_n < \omega(\eta') + q\rho(y)\}$, where $q < 1/\beta$ (see (3.2)).

Theorem 3.8. *Assume that*

$$\exists s \in (0,1) \ \ (W_p^r(\Omega), L_{p,g}(\Omega))_{1-s,q} = (\overset{\circ}{W}{}_p^r(\Omega), L_{p,g}(\Omega))_{1-s,q}, \quad p, q > 1,$$

and $\gamma = r - n/p > 0$. *Let* U *be a domain from Definition* 3.1 *and* y *is the corresponding coordinate system. Then*

(B) *there exists* $a_0 > 0$ *such that, for any given numbers* $a_1 < a \le a_0$, *the following statement holds: given* $\varepsilon > 0$, *there exists a constant* $q \in (0,1)$ *such that*

$$\int_{K_{qy}(a_1)} g(\eta)\, d\eta \le \varepsilon \int_{K_y(a)} g(\eta)\, d\eta \quad \forall y \in U : \ K_y(a) \subset U.$$

Proof. The arguments coincide with those of Theorem 4 in [9]. So we omit them. □

Next, we present some condition which is an analog of the classical A_p-condition.

Let U is a domain from Definition 3.1. As before, we consider a local coordinate system y. For $y \in U \cap \Omega$, we put

$$K_y^*(a,b) = \{\eta \in \mathbb{R}^n : \ |\eta' - y'| < a(\eta_n - y_n), \ y_n < \eta_n < y_n + b(y_n - \omega(y'))\}.$$

Consider a nonnegative function $g \in L_1(\Omega)$ such that $g^{-1/(p-1)} \in L_{1,loc}(\Omega)$. We assume that

(C) There exist a finite covering U_i $(i = 1, 2, \dots, N)$ of Γ (the domains U_i possess the properties from Definition 3.1) such that for some $a, b, R_0 > 0$ and all $y \in U_i \cap \Omega$ such that $K_y(a), K_y^*(a,b) \subset U_i \cap \Omega$ $(i = 1, 2, \dots, N)$

$$\frac{1}{\mu(K_y(a))} \int\limits_{K_y(a)} g(\eta)\, d\eta \Big(\frac{1}{\mu(K_y^*(a,b))} \int\limits_{K_y^*(a,b)} g(\eta)^{\frac{-1}{p-1}}\, d\eta \Big)^{p-1} \le R_0 \tag{3.31}$$

(here the function $g(y)$ is written in the local coordinate system y).

Theorem 3.9. *Assume that* Ω *is a bounded Lipschitz domain, the condition* (C) *holds,* $p \in (1, \infty)$, *and the function* g *satisfies* (B) *in every domain* U_i. *Then*

$$\exists s \in (0,1) \ \ (W_p^r(\Omega), L_{p,g}(\Omega))_{1-s,q} = (\overset{\circ}{W}{}_p^r(\Omega), L_{p,g}(\Omega))_{1-s,q} \ \ \forall q \in (1, \infty).$$

Proof. Without loss of generality we can assume that $a < 1/(2M)$ in (C) (as before, M is the Lipschitz constant of ω). Given a function $u \in W_p^r(\Omega)$, define the map T as in the proof of Theorem 3.7. Every function $u_i(x)$ $(i = 1, 2, \dots, N)$ is defined by the parameters $a_1 = a/(1 + Ma)$, $a_2 < a_1$, $a_3 = a_2/(1 + Ma_2)$, q_0, σ, and δ_1. We describe some additional requirements for the parameters involved. First, we choose and fix a sufficiently small parameter ε_0 in (3.4) and a parameter $q_0 < 1$ such that

$$(1 + \varepsilon_0)^2 / [(1 - \varepsilon_0) q_0]^2 < (1 + b(1 - Ma)). \tag{3.32}$$

Next, we choose the parameters a_2 and δ_1. The parameter δ_1 is chosen as in Theorem 3.7 and decreasing δ_1 if necessary we can assume that $K_y^*(a,b) \subset U_i \cap \Omega$

for all $y \in U_{\delta_1}$. Since the other parameters have been fixed, we can assume that δ_1 depends on a_2 and σ. The parameters a_2 and σ will be specified below. The first part of the theorem coincides with that of Theorem 3.7. All we need is to prove the estimate

$$\|Tu\|_{L_{p,g}(\Omega)} \leq N_1\|u\|_{L_{p,g}(\Omega)} \quad (N_1 < 1), \quad u \in W_p^r(\Omega).$$

First, we describe the consequences of (C). Let U_i be a domain from (C). Denote

$$L_z(a_1, b_1) = \{\xi \in \mathbb{R}^n : \ |\xi' - z'| < a_1(\xi_n - z_n), \ 0 < \xi_n - z_n < b_1 z_n\}, \quad b_1 > 0.$$

Consider the straightening of the boundary $z = z(y)$. We can state that

$$L_z(a_2, b(1 - Ma)) \subset z(K_y^*(a, b)), \quad a_2 \leq a_1 = a/(1 + Ma).$$

Indeed, put $\xi' = \eta'$, $\xi_n = \eta_n - \omega(\eta')$, $z' = y'$, $z_n = y_n - \omega(y')$. Let $\xi \in L_z(a_2, b(1 - Ma))$. Hence, $|\xi' - z'| < a_1(\xi_n - z_n)$ and $0 < \xi_n - z_n < b(1 - Ma)z_n$. The definitions of the transformations $\xi(\eta), z(y)$ imply that

$$|\eta' - y'| < a(\eta_n - y_n), \quad 0 < \eta_n - y_n < b(1 - Ma)(y_n - \omega(y')) + M|\eta' - y'| \leq$$
$$\leq b(1 - Ma)(y_n - \omega(y')) + Ma|\eta_n - y_n|.$$

As a consequence, we derive $0 < \eta_n - y_n < b(y_n - \omega(y'))$, i.e., $\eta = \eta(\xi) \subset K_y^*(a, b)$. This inclusion implies that $\xi \in z(K_y^*(a, b))$. By Lemma 3.2 we infer

$$L_z(a_2, b(1 - Ma)) \subset z(K_y^*(a, b)), \quad L_z(a_2) \subset z(K_y(a)) \subset L_z(a/(1 - aM)) \quad (3.33)$$

for every $y \in U_i \cap \Omega$ such that $K_y(a), K_y^*(a, b) \subset U_i \cap \Omega$ and $a_2 \leq a_1 = a/(1 + Ma)$. Hence, the condition (C) implies that ($\eta' = \xi'$, $\eta_n = \xi_n + \omega(\xi')$)

$$\frac{1}{z_n^n} \int\limits_{L_z(a_2)} g(\eta(\xi))\, d\xi \Big(\frac{1}{z_n^n} \int\limits_{L_z(a_2, b(1 - Ma))} g(\eta(\xi))^{\frac{-1}{p-1}}\, d\xi\Big)^{p-1} \leq R_1 \quad (3.34)$$

for all $z = z(y)$ such that $K_y(a), K_y^*(a, b) \subset U_i \cap \Omega$ $(i = 1, 2, \ldots, N)$. Here R_1 is a constant independent of z. Now we estimate $\|Tu\|_{L_{p,g}(\Omega)}$. We have that

$$\|Tu\|_{L_{p,g}(\Omega)} \leq \sum_{i=1}^N \|T_i u\|_{L_p(U_i \cap \Omega)}, \quad T_i u = \frac{1}{\rho^n(y)} \int_{U_{\delta_1}} v(\eta)\varphi\Big(\frac{\eta - y}{\rho_i(y)}\Big)\, d\eta. \quad (3.35)$$

The Hölder inequality yields $(y \in U_{\delta_1})$

$$|T_i u|^p \leq \frac{1}{\rho_i^n(y)} \int\limits_{U_{\delta_1}} |v(\eta)|^p g(\eta) \Big|\varphi\Big(\frac{\eta - y}{\rho_i(y)}\Big)\Big|\, d\eta \Big(\frac{1}{\rho_i^n(y)} \int\limits_{U_{\delta_1}} (g(\tau))^{\frac{-1}{p-1}} \Big|\varphi\Big(\frac{\tau - y}{\rho_i(y)}\Big)\Big|\, d\tau\Big)^{p-1}. \quad (3.36)$$

Multiplying (3.36) by $g(y)$ $(g(y) \equiv g(x(y)))$ and integrating the result over U_{δ_1} we arrive at the estimate

$$I = \int_{U_{\delta_1}} g(y)|T_i u|^p\, dy \leq \int_{U_{\delta_1}} |v(\eta)|^p g(\eta) R(\eta)\, d\eta, \quad (3.37)$$

where

$$R(\eta) = \int_{U_{\delta_1}} g(y) \frac{1}{\rho_i^n(y)} \left| \varphi\Big(\frac{\eta - y}{\rho_i(y)}\Big) \right| \Big(\frac{1}{\rho_i^n(y)} \int_{U_{\delta_1}} (g(\tau))^{\frac{-1}{p-1}} \left| \varphi\Big(\frac{\tau - y}{\rho_i(y)}\Big) \right| d\tau \Big)^{p-1} dy.$$

We need to estimate the function $R(\eta)$. Make the change of variables $z = z(y)$, $\zeta = \zeta(\tau)$, and put $\xi = \xi(\eta)$ (i.e., $z' = y'$, $z_n = y_n - \omega(y')$ and so on). As in the proof of Theorem 3.7, we obtain that

$$\big|\varphi\big((\eta(\xi) - y(z))/\rho_i(y(z))\big)\big| \le c_2 \chi_{B_\xi}(z)/\sigma^n,$$

where

$$B_\xi = \{z : \; |\xi' - z'| < a_2(\xi_n - z_n), \; \alpha_1 \xi_n < z_n < \alpha_2 \xi_n\},$$

$\alpha_1 = (1-2Ma_3)/(1-2Ma_3+\sigma(1+\varepsilon_0)(1-Ma_3))$, and $\alpha_2 = 1/(1+\sigma(1-\varepsilon_0)q_0(1-Ma_3))$. Similarly, we derive

$$\left|\varphi\Big(\frac{\tau - y}{\rho_i(y)}\Big)\right| \le c_3 \chi_{C_z}(\zeta)/\sigma^n,$$

where

$$C_z = \{\zeta : \; |\zeta' - z'| < a_2(\zeta_n - z_n), \; \alpha_1 \zeta_n < z_n < \alpha_2 \zeta_n\}.$$

In this case the function $R(\eta(\xi))$ is estimated as follows

$$R(\eta(\xi)) \le c \int_{A_{\delta_1}} g(y(z)) \frac{1}{\sigma^n z_n^n} \chi_{B_\xi}(z) \Big(\frac{1}{z_n^n \sigma^n} \int_{A_{\delta_1}} (g(\tau(\zeta)))^{\frac{-1}{p-1}} \chi_{C_z}(\zeta)\, d\zeta \Big)^{p-1} dz,$$

where $A_{\delta_1} = \{z : \; |z'| < r_{1i}, \; 0 < z_n < \delta_1/\alpha\}$ and c is some constant. Let $\zeta \in C_z$ and let $z \in B_\xi$. The definitions of these sets yield

$$\zeta_n < z_n/\alpha_1 < \alpha_2 \xi_n/\alpha_1, \quad \zeta_n > z_n/\alpha_2 > \alpha_1 \xi_n/\alpha_2. \tag{3.38}$$

Therefore, we have

$$\zeta_n < \alpha_2 \xi_n/\alpha_1, \quad \zeta_n > \alpha_1 \xi_n/\alpha_2. \tag{3.39}$$

Similarly, we have

$$|\zeta' - \xi'| \le |\zeta' - z'| + |z' - \xi'| < a_2[(\xi_n - z_n) + (\zeta_n - z_n)].$$

In view of (3.38) and (3.39),

$$\zeta_n - z_n < z_n(1 - \alpha_1)/\alpha_1 < \zeta_n(1 - \alpha_1)\alpha_2/\alpha_1$$

and

$$\xi_n - z_n \le z_n(1 - \alpha_1)/\alpha_1 < \zeta_n(1 - \alpha_1)\alpha_2/\alpha_1.$$

Therefore, we arrive at the estimate

$$|\zeta' - \xi'| < 2a_2 \zeta_n (1 - \alpha_1)\alpha_2/\alpha_1. \tag{3.40}$$

Thus, the conditions $z \in B_\xi$ and $\zeta \in C_z$ imply that

$$\zeta \in M_\xi = \{\zeta : \; |\zeta' - \xi'| < a_4 \zeta_n, \; \zeta_n < \alpha_2 \xi_n/\alpha_1, \quad \zeta_n > \alpha_1 \xi_n/\alpha_2\},$$

where $a_4 = 2a_2(1 - \alpha_1)\alpha_2/\alpha_1$, i.e., $C_z \subset M_\xi$. In view of (3.32) the limit of $\left(\frac{\alpha_1}{\alpha_2}\right)^2$ as $\sigma \to \infty$ and $a_2 \to 0$ (and as a consequence $a_3 \to 0$) is equal to $q_1/b_0 < 1$, with

$q_1 > 1$ and $b_0 = (1 + b(1 - Ma))$. Take $1 < q_2 < \min(3, q_1)$. There exists a_{02}, σ_0 such that $(\alpha_1/\alpha_2)^2 \geq q_2/b_0$ for all $0 < a_2 \leq a_{02}$ and $\sigma \geq \sigma_0$. In this case,

$$M_\xi \subset \left\{\zeta : |\zeta' - \xi'| < a_5\zeta_n \quad \left(a_5 = \frac{2a_2\sqrt{b_0}}{\sqrt{q_2}}\right), \quad \frac{\xi_n\sqrt{b_0}}{\sqrt{q_2}} > \zeta_n > \frac{\xi_n\sqrt{q_2}}{\sqrt{b_0}}\right\}.$$

Now we fix a_2 putting $a_2 = a_{02}$. Let $\gamma = 1/\sqrt{q_2 b_0}$. As is easily seen, we have $M_\xi \subset L_{\xi^0}((a_6, b(1 - Ma))$, with $a_6 = a_5/(1 - 1/q_2)$ and $\xi^0 = (\xi', \gamma\xi_n)$. Consider the set B_ξ. It is not difficult to establish that $B_\xi \subset L_{q\xi^0}(a_7)$ with $a_7 = a_2(1-\alpha_2)/(\gamma-\alpha_2)$ and $q = \alpha_2/\gamma$. Note that $\alpha_2 \to 0$ as $\sigma \to \infty$. It is not difficult to check that $a_2/\gamma < a_6$. So there exists $\sigma_1 \geq \sigma_0$ and the number α_{02} independent of σ such that $a_7 \leq a_8 = a_2(1 - \alpha_{02})(\gamma - \alpha_{02}) < a_6$ for all $\sigma \geq \sigma_1$. Finally, we arrive at the containments $B_\xi \subset L_{q\xi^0}(a_8)$ and $C_z \subset M_\xi \subset L_{\xi^0}(a_6, b(1 - Ma))$ for all $z \in B_\xi$. So we can estimate the function $R(\eta)$ as follows

$$R(\eta(\xi)) \leq \frac{c_1}{\xi_n^n} \int_{B_\xi} g(y(z))\, dz \Big(\frac{1}{\xi_n^n} \int\limits_{M_\xi} (g(\tau(\zeta)))^{\frac{-1}{p-1}}\, d\zeta\Big)^{p-1} \leq$$

$$\leq \frac{c_1}{\xi_n^n} \int_{L_{q\xi^0}(a_8)} g(y(z))\, dz \Big(\frac{1}{\xi_n^n} \int\limits_{L_{\xi^0}(a_6, b(1-Ma))} (g(\tau(\zeta)))^{\frac{-1}{p-1}}\, d\zeta\Big)^{p-1}.$$

Next, we use the condition (B) and (3.34). Given $\varepsilon > 0$, we can find $\sigma \geq \sigma_1$ such that

$$R(\eta(\xi)) \leq \varepsilon \frac{1}{\xi_n^n} \int_{L_{\xi^0}(a_8)} g(y(z))\, dz \Big(\frac{1}{\xi_n^n} \int\limits_{L_{\xi^0}(a_6, b(1-Ma))} (g(\tau(\zeta)))^{\frac{-1}{p-1}}\, d\zeta\Big)^{p-1} \leq \varepsilon R_1.$$

Repeating the arguments for all $i = 1, 2, \ldots, N$ and using (3.35), (3.36), and (3.37) we arrive at the estimate

$$\|Tu\|_{L_{p,g}(\Omega)} \leq N\varepsilon R_1 \|u\|_{L_p(\Omega)},$$

Choosing $\varepsilon < 1/(2NR_1)$ and referring to Theorem 2.2 we obtain the claim. □

We can simplify the conditions of Theorem 3.9 and assume that

(C′) there exist a finite covering U_i $(i = 1, 2, \ldots, N)$ of Γ (the domains U_i possess the properties from Definition 3.1) such that for some $a, b, \varepsilon_1, R_2 > 0$ and all $y \in U_i \cap \Omega$ such that $K_y(a), K_y^*(a, b) \subset U_i \cap \Omega$ $(i = 1, 2, \ldots, N)$

$$\frac{1}{\mu^*(K_y(a))} \int\limits_{K_y(a)\cap U_i} \frac{g(\eta)}{\rho_0^{\varepsilon_1}(\eta)}\, d\eta \Big(\frac{1}{\mu(K_y^*(a, b))} \int\limits_{K_y^*(a,b)\cap\Omega} g(\eta)^{\frac{-1}{p-1}}\, d\eta\Big)^{p-1} \leq R_2 \quad (3.41)$$

(here the function $g(y)$ is written in the local coordinate system y and $\mu^*(K_y(a)) = \int_{K_y} \rho_0^{-\varepsilon_1}(\eta)\, d\eta$).

Theorem 3.10. *Assume that the condition* (C′) *is fulfilled. Then the statement of Theorem* 3.9 *holds.*

Proof. The proof coincides with the proof of Theorem 3.9. The only distinction is that the estimate for the function $R(\eta)$ looks as follows

$$R(\eta) \leq \frac{c}{\sigma^{\varepsilon_1}} R_2,$$

where the constant c is independent of σ. So the norm of the operator T is sufficiently small for a sufficiently large σ and we do not use the condition (B) at all. □

Let Ω be a domain of the form

$$\Omega = \{y \in \mathbb{R}^n : \ |y'| \in B_{r_0}, \ \omega(y') < y_n < \delta\}, \quad y' = (y_1, \dots, y_{n-1}), \tag{3.42}$$

where $B_{r_0} = \{y' : \ |y'| < r_0\}$, $\delta > \sup_{y' \in B_{r_0}} \omega(y') = \delta_0$, and ω is a Lipschitz function. Put $\Gamma = \partial\Omega$ and $\Gamma_0 = \{y \in \partial\Omega : y_n = \omega(y'), \ |y'| \leq r_0\}$. Let $W^r_{p0}(\Omega) = \{u \in W^r_p(\Omega) : \ u|_{\partial\Omega \setminus \Gamma_0} = 0\}$. Rewrite our conditions (A)–(C) for this particular case. Given $y \in \Omega$, put

$$K_y(a) = \{\eta \in \Omega : \ |\eta' - y'| < a(y_n - \eta_n)\}, \quad a > 0. \tag{3.43}$$

Denote by $\rho(y)$ the function satisfying (3.4) for some $\varepsilon_0 \in (0,1)$. Let $\Omega_{\delta_1} = \{y \in \Omega : \ y_n - \omega(y') < \delta_1\}$. In what follows, we assume that the function $g(y) \in L_1(\Omega)$ possesses the following property: for every $\delta_1 > 0$, there exists a constant $\delta_2 > 0$ such that $g(y) \geq \delta_2$ almost everywhere in $\Omega \setminus \Omega_{\delta_1}$.

We will use the following conditions.

(D) For some $a > 0, c > 0$ and almost all $y \in \Omega$ $(i = 1, 2, \dots, N)$

$$\int_{K_y(a)} g(\eta)\, d\eta \leq c\mu(K_y(a)) g(y). \tag{3.44}$$

Let $K_{qy}(a) = \{\eta \in K_y(a) : \omega(y') < \eta_n < \omega(y') + q\rho(y)\}$.

(E) there exists $a_0 > 0$ such that, for any given numbers $a_1 < a \leq a_0$, the following statement holds: given $\varepsilon > 0$, there exists a constant $q \in (0,1)$ such that

$$\int_{K_{qy}(a_1)} g(\eta)\, d\eta \leq \varepsilon \int_{K_y(a)} g(\eta)\, d\eta \quad \forall y \in \Omega.$$

For $y \in \Omega$, we put

$$K^*_y(a,b) = \{\eta \in \Omega : \ |\eta' - y'| < a(\eta_n - y_n), \ y_n < \eta_n < y_n + b(y_n - \omega(y))\}$$

We also use the conditions

(F) For some $a, b, R_0 > 0$, $\delta_2 > 0$, and all $y \in \Omega_{\delta_2}$

$$\frac{1}{\mu(K_y(a))} \int\limits_{K_y(a)} g(\eta)\, d\eta \Big(\frac{1}{\mu(K^*_y(a,b))} \int\limits_{K^*_y(a,b)} g(\eta)^{\frac{-1}{p-1}}\, d\eta \Big)^{p-1} \leq R_0. \tag{3.45}$$

(G) For some $a, b, \delta_2, \varepsilon_1, R_1 > 0$ and all $y \in \Omega_{\delta_2}$

$$\frac{1}{\mu^*(K_y(a))} \int\limits_{K_y(a)} \frac{g(\eta)}{\rho_0^{\varepsilon_1}(\eta)}\, d\eta \Big(\frac{1}{\mu(K_y^*(a,b))} \int\limits_{K_y^*(a,b)\cap\Omega} g(\eta)^{\frac{-1}{p-1}}\, d\eta \Big)^{p-1} \le R_1 \qquad (3.46)$$

(here $\mu^*(K_y(a)) = \int_{K_y} \rho_0^{-\varepsilon_1}(\eta)\, d\eta$).

Theorem 3.11. *Assume that* Ω *is a domain of the form* (3.42), $p \in (1, \infty)$, *and either the condition* (D), *or the condition* (E) *and* (F), *or the condition* (G) *holds. Then*

$$\exists s \in (0,1): \ (W^r_{p0}(\Omega), L_{p,g}(\Omega))_{1-s,q} = (\overset{\circ}{W}{}^r_p(\Omega), L_{p,g}(\Omega))_{1-s,q} \qquad (3.47)$$

for every $q \in (1, \infty)$.

Proof. To obtain the claim, we repeat the arguments of the previous Theorems 3.7, 3.9, and 3.10. We look for a map $T : L_{2,g}(\Omega) \to L_{2,g}(\Omega)$ such that, for some constants $N_1 \in (0,1)$ and $N_2 > 1$ and for all $u \in W^r_{p0}(\Omega)$, $u - Tu \in \overset{\circ}{W}{}^r_p(\Omega)$,

$$\|Tu\|_{W^r_p(\Omega)} \le N_2 \|u\|_{W^r_p(\Omega)}, \quad \|Tu\|_{L_{2,g}(\Omega)} \le N_1 \|u\|_{L_{2,g}(\Omega)}. \qquad (3.48)$$

This operator T is defined as

$$Tu = \frac{1}{\rho^n(y)} \int_\Omega v(\eta) \varphi\left(\frac{\eta - y}{\rho(y)}\right) d\eta, \quad v(\eta) = \psi(\eta) u(\eta),$$

where the averaging kernel φ is as that of the proof of Theorem 3.7 and the function ψ is constructed as follows. We denote $\Omega_{\beta,\delta_1} = \{y \in \Omega_{\delta_1} : \ y_n - \omega(y') \le (r_0 - |y'|)/\beta\}$ $(\beta > 0)$. Let $\varphi_0(\eta)$ be a nonnegative averaging kernel with $\operatorname{supp} \varphi_0 \subset B_1 = \{x : |x| < 1\}$. We put

$$\psi(y) = \frac{1}{(\rho\varepsilon_1)^n} \int_\Omega \varphi_0\left(\frac{\eta - y}{\varepsilon_1 \rho(y)}\right) \chi_{\Omega_{1,\delta_1}}(\eta)\, d\eta$$

choosing a sufficiently small parameter $\varepsilon_1 = \varepsilon_1(\varepsilon_0, M)$ so that $\operatorname{supp} \psi(y) \subset \Omega_{1/2, 2\delta_1}$ and $\Omega_{2,2\delta_1/3} \subset \{y \in \Omega_{1,\delta_1} : \ \psi(y) = 1\}$. Here M is the Lipschitz constant of $\omega(y')$ and ε_0 is a constant in (3.4). It suffices to take $\varepsilon_1 = 1/(2(1+\varepsilon_0)(2+M))$. In this case the function $\psi(y)$ admits an estimate

$$|D^\alpha \psi(y)| \le c(\alpha)/(r_0 - |y'|)^\alpha \quad (\forall \alpha : \ |\alpha| \ge 1)$$

and $\operatorname{supp} \psi \subset \Omega_{2\delta_1}$. This choice of the function ψ ensures that for a suitable cone K_σ and sufficiently small δ_1 we have that $\{y + \rho(y)\eta : \ \eta \in K_\sigma, \ y \in \operatorname{supp} \psi\} \subset \Omega$. The estimates (3.48) are derived as the corresponding estimates in Theorems 3.7, 3.9, and 3.10. The only problem is to prove that the map $u \to \psi u$ belongs to the class $L(W^r_{p0}(\Omega))$. To justify this fact, we use the inequalities of the Hardy type

$$\int_{\Omega_{2\delta_1}} \frac{|v|^p}{(r_0 - |y'|)^{kp}}\, dy \le c \int\limits_{\Omega_{2\delta_1}} \frac{|\nabla_{y'} v|^p + |v|^p}{(r_0 - |y'|)^{(k-1)p}}\, dy, \quad v \in W^1_{p0}(\Omega), \ k = 1, 2, \ldots.$$

which easily follows from the estimates of Section 3.2.6. in [1] with the use of the change of variables $z' = y'$, $z_n = y_n - \omega(y')$. ☐

Example. Let Ω be a bounded Lipschitz domain and let

$$g(x) = \rho_0^\alpha \prod_{i=1}^{N} (\ln(\rho_i(x)\alpha_i))^{\beta_i}, \quad \alpha_i > 0, \ \beta_i \in \mathbb{R},$$

where $\rho_0 = \operatorname{dist}(x, \partial\Omega)$ and $\rho_i = \operatorname{dist}(x, S_i)$ ($S_i \subset \partial\Omega$ are some closed sets). In this case, if $\alpha > -1$ then the condition (A) holds.

Example. Let Ω be a bounded Lipschitz domain and $g = \rho_0^\alpha(x) \prod_{i=1}^{N} \rho_i^{\beta_i}(x)$ ($\alpha, \beta_i \in \mathbb{R}$), with $\rho_0 = \operatorname{dist}(x, \partial\Omega)$ and $\rho_i(x) = \operatorname{dist}(x, S_i)$ ($S_i \subset \partial\Omega$ are some closed subsets). In this case the function g is a slowly varying function. If $\alpha + \sum_{\beta_i < 0} \beta_i > -1$ then the condition (A) holds.

Example. Assume that Ω is a bounded Lipschitz domain, $g(x) = e^{-\rho_0^{-r}(x)} \prod_{i=0}^{N} \rho_i^{\beta_i}(x)$, where $\rho_0(x) = \operatorname{dist}(x, \partial\Omega)$, $\rho_i(x) = \operatorname{dist}(x, S_i)$ ($S_i \subset \partial\Omega$ are closed subsets), $r > 0$, and $\alpha, \beta_i \in \mathbb{R}$. Then the condition (A) is fulfilled.

Remark 3.12. Generally, $A_p \Rightarrow A_q$ with $p < q$. It is easy to construct an example such that (A) does not hold but (C) is valid.

4. Elliptic eigenvalue problems

In this section we present the simplest conditions ensuring the Riesz basis property in the problems (1.2), (1.3). By Γ we mean ∂G. Let Ω be a Lipschitz domain. A nonnegative function $g \in L_1(\Omega)$ is called regular in Ω whenever it satisfies the conditions of one of Theorems 3.7, 3.9, and 3.10, where $p = 2$.

(C_1) The set G^+ consists of finitely many connected components with Lipschitz boundaries and the function g is regular in every connected component of Ω.

Let $\{D_i\}_{i=1}^M$ be the collection of connected components of G^+ such that, for every i, $\partial D_i \cap (\Gamma \cup \partial G^-) \neq \emptyset$. Put

$$G_0^+ = \partial G^+ \cap (\Gamma \cup \partial G^-), \quad G_\delta^+ = \{x \in G : \ \operatorname{dist}(x, G_0^+) < \delta\}.$$

(C_2) The set $\{D_i\}_{i=1}^M$ is finite, consists of Lipschitz domains, $\bigcup_{i=1}^{M} \partial D_i \cap (\Gamma \cup \partial G^-) = G_0^+$, there exists $\delta_0 > 0$ such that $\operatorname{dist}(G^+ \setminus G_\delta^+, \Gamma\} > 0$ for every $\delta < \delta_0$, and the subspace $\ker B \cap H_1$ is nondegenerate in H_1, i.e., $K_0 = \{0\}$ (see the definition in Section 2). For every i there exists a function $g_0(x)$ which is regular in D_i and coincides with g on some set of the form $D_{i\delta} = \{x \in D_i : \ \operatorname{dist}(x, \partial D_i \cap (\Gamma \cup \partial G^-)) < \delta\}$ ($\delta > 0$).

Let $\{D_i\}_{i=1}^M$ be the set of connected components of G^+ such that $\partial D_i \cap \partial G^- \neq \emptyset$ for every i.

(C_3) The set $\{D_i\}_{i=1}^M$ consists of finitely many connected components with Lipschitz boundaries, $\bigcup_{i=1}^M \partial D_i \cap \partial G^- = \partial G^+ \cap \partial G^-$, the subspace $\ker B \cap H_1$ is nondegenerate in H_1, and either $H_1 = W_2^m(G)$ or H_1 coincides with $\overset{\circ}{W}{}_2^m(G)$. For every i, there exists a function $g_0(x)$ which is regular in D_i and coincides with g on some set of the form $D_{i\delta} = \{x \in D_i : \operatorname{dist}(x, \partial D_i \cap \partial G^-) < \delta\}$ ($\delta > 0$).

Remark 4.1. The sets G^+ and G^- are interchangeable in (C_1)–(C_3). In order to state any of these conditions for G^-, it suffices to alter the signs there. We need the corresponding condition to be valid for one of the sets G^+ or G^-. We have stated them for the set G^+.

Lemma 4.2. *Let one of the conditions* (C_1), (C_2), (C_3) hold. *Then*

$$(F_1, F_{-1})_{1/2,2} = F_0.$$

Proof. Assume, for instance, that the condition (C_1) holds. Let Ω be a connected component of G^+ and let $u \in F_1$. Consider a map $Pu = u|_\Omega$. Obviously, $P \in L(F_1, W_2^m(\Omega)) \cap L(L_{2,g}(G^+ \cup G^-), L_{2,g}(\Omega))$. Hence (see, the statement (a) of Theorem 1.3.3 in [1]), we infer $P \in L(F_s, (W_2^m(\Omega), L_{2,g}(\Omega))_{1-s,2})$ for every $s \in (0,1)$. Since g is regular in Ω, we have that

$$\exists s_0 \in (0,1): \quad (W_2^m(\Omega), L_{2,g}(\Omega))_{1-s_0,2} = (\overset{\circ}{W}{}_2^m(\Omega), L_{2,g}(\Omega))_{1-s_0,2}. \tag{4.1}$$

Given an interpolation couple (A_0, A_1), by Theorem 1.3.3 in [1] the space $(A_0, A_1)_{s_0,2}$ belongs to the class $J(s_0) \cap K(s_0)$ (see the definitions of this class in Section 1.10.1 of [1]). By the reiteration theorem (see Section 1.10.2 in [1]), we have that

$$(A_0, (A_0, A_1)_{s_0,2})_{\theta,2} = (A_0, A_1)_{s,2}, \quad \theta s_0 = s, \quad s \in (0, s_0).$$

This equality and (4.1) yield

$$\begin{aligned}(L_{2,g}(\Omega), W_2^m(\Omega))_{s,2} &= (L_{2,g}(\Omega), (L_{2,g}(\Omega), W_2^m(\Omega))_{s_0,2})_{\theta,2} = \\ = (L_{2,g}(\Omega), (L_{2,g}(\Omega), \overset{\circ}{W}{}_2^m(\Omega))_{s_0,2})_{\theta,2} &= (L_{2,g}(\Omega), \overset{\circ}{W}{}_2^m(\Omega))_{s,2} \ \ \forall s \in (0, s_0).\end{aligned}$$

The above properties of the operator P imply that

$$\forall s \in (0, s_0] \ \ P \in L(F_s, (\overset{\circ}{W}{}_2^m(\Omega), L_{2,g}(\Omega))_{1-s,2}).$$

The map

$$Su = \begin{cases} u(x), & x \in \Omega \\ 0, & x \in (G^+ \cup G^-) \setminus \Omega \end{cases}$$

belongs to the class $L(\overset{\circ}{W}{}_2^m(\Omega), F_1,) \cap L(L_{2,g}(\Omega), L_{2,g}(G^+ \cup G^-))$ and, a consequence, we have that $S \in L((\overset{\circ}{W}{}_2^m(\Omega), L_{2,g}(\Omega))_{1-s,2}, F_s)$ for all $s \in (0,1)$. Therefore, we have $SP \in L(F_s, F_s)$ for $s \in (0, s_0]$. Repeating the arguments for every connected component of G^+ and summing the corresponding operators of the form SP, we arrive at the following statement: there exists $s_0 > 0$ such that

$$E^+ \in L(F_s, F_s) \quad \forall s \in (0, s_0],$$

where E^+ is the operator of multiplication by the characteristic function of the set G^+. Hence, $J \in L(F_s, F_s)$ for all $s \in (0, s_0]$. By Lemma 2.1, we obtain the claim.

Assume that condition (C_2) holds. Choose a sufficiently small $\delta > 0$ ($\delta < \delta_0$) so that in every $D_{k\delta}$ the function $g(x)$ coincides with some regular (in D_k) function. Consider the sets $G^+ \setminus G^+_{\delta/2}$ and G^+_δ and construct the functions $\varphi, \psi \geq 0$ such that $\varphi, \psi \in C^\infty(\overline{G})$, $\operatorname{supp}\varphi \subset G$, $\varphi + \psi = 1$ in G^+, $\varphi = 1$ in $G^+ \setminus G^+_\delta$, and $\varphi = 0$ in $G^+_{\delta/2}$. The function φ is constructed with the use the averaging method applied to the characteristic function of some neighborhood about $G^+ \setminus G^+_\delta$. Obviously, the operator $P_\varphi : u \to \varphi(x)u(x)$ defines a continuous mapping of F_1 into F_1 and F_0 into F_0, i.e.,

$$P_\varphi \in L(F_s, F_s) \ \forall s \in [0, 1].$$

Fix an arbitrary number k in the set $\{1, 2, \ldots, M\}$ and consider the connected component $\Omega = D_k$. For $\Omega = D_k$, we construct the operators $P_k u = u|_\Omega \psi \in L(F_1, W_2^m(\Omega)) \cap L(L_{2,g}(G^+ \cup G^-), L_{2,g_0}(\Omega))$, where the function $g_0(x)$ is regular in D_k and coincides with $g(x)$ in $D_{k\delta}$. In view of the regularity of g_0, there exists $s_0 \in (0, 1)$ such that

$$(W_2^m(\Omega), L_{2,g_0}(\Omega))_{1-s,2} = (\overset{\circ}{W}{}_2^m(\Omega), L_{2,g_0}(\Omega))_{1-s,2} = F_s(\Omega) \quad \forall s \in (0, s_0].$$

As in the first case, we can prove that the operator

$$S_k u = \begin{cases} \psi u(x), & x \in \Omega \\ 0, & x \in (G^+ \cup G^-) \setminus \Omega \end{cases}.$$

possesses the property $S_k \in L(F_s(\Omega), F_s)$ ($s \in (0, s_0]$). Repeating the arguments for every connected component D_i and summing the operators obtained, we find $s_0 > 0$ such that the operator

$$P_\psi u = \begin{cases} \psi u, \ x \in \bigcup\limits_{j=1}^{N} D_k \\ 0, \ x \notin \bigcup\limits_{j=1}^{N} D_k \end{cases}.$$

is of the class $P_\psi \in L(F_s, F_s) \ \forall s \in (0, s_0]$. We have that $P_\psi + P_\varphi \in L(F_s, F_s)$ ($\forall s \in (0, s_0]$). By construction, we infer

$$E^+ u = (P_\psi + P_\varphi)u = \begin{cases} u, & x \in G^+ \\ 0, & x \in G^- \end{cases}, \ \forall u \in F_0.$$

Now, it suffices to apply Lemma 2.1.

Assume that condition (C_3) holds. If $H_1 = W_2^m(G)$ or H_1 coincides with $\overset{\circ}{W}{}_2^m(G)$ then the operator of multiplication by every function $\varphi(x) \in C^\infty(\overline{G})$ whose support is included into a sufficiently small neighborhood about $\partial G^+ \cap \partial G^-$ is continuous as an operator from F_1 into F_1. With this fact in hand, the proof is a simplified version of the proof of the claim when condition (C_2) holds and we drop it. $\square$

Remark 4.3. We describe the most typical conditions ensuring the claim. To some extent, they can be generalized if we use also Theorem 3.11.

The following theorems are the consequences of Lemma 4.2 and the corresponding theorems in [15].

Theorem 4.4. *Let conditions* (I)–(V) *and one of the conditions* (C_i) $(i = 1,2,3)$ *hold. Then, in the space* $L_{2,g}(G \setminus G^0)$, *we can construct a Riesz basis consisting of eigenfunctions and associated functions of problem* (1.2), (1.3). *Every function* $f(x) \in F_0 = L_{2,g}(G \setminus G^0)$ *is uniquely representable in the form*

$$f = \sum_{i=1}^{\infty} u_i^+ c_i^+ + \sum_{i=1}^{\infty} u_i^- c_i^- + \sum_{i=1}^{M} u_i c_i \quad (M < \infty),$$

where the series are convergent in the norm of F_0, u_i^+ (u_i^-) *are eigenfunctions corresponding to positive* (*negative*) *eigenvalues,*

$$[u_i^\pm, u_j^\pm]_0 = \pm\delta_{ij}, c_i^\pm = \pm[f, u_i^\pm]_0, [u_i^\pm, u_j]_0 = 0, [u_i^+, u_j^-]_0 = 0,$$

and $\{u_j\}_{j=1}^M$ *is basis for a finite-dimensional subspace* N *which is the linear span of finitely many eigenfunctions and associated functions of the problem. Moreover, the norm in* F_0 *is equivalent to the norm*

$$\|f\|_{F_0}^2 = \sum_{i=1}^{\infty}(|c_i^+|^2 + |c_i^-|^2) + \sum_{i=1}^{M}|c_i|^2.$$

If $f \in F_1$ *then the norm in* F_1 *is equivalent to the norm*

$$\|f\|_{F_1}^2 = \sum_{i=1}^{\infty}(|\lambda_i^+||c_i^+|^2 + |\lambda_i^-||c_i^-|^2) + \sum_{i=1}^{M}|c_i|^2,$$

where $\lambda_i^\pm$ *are the eigenvalues corresponding to eigenfunctions* $u_i^\pm$.

Theorem 4.5. *Under the conditions of Theorem* 4.4 *eigenfunctions to problem* (1.2), (1.3) *corresponding to positive* (*negative*) *eigenvalues together with some finite set of eigenfunctions and associated functions form a Riesz basis for* $L_{2,g}(G^+)$ $(L_{2,g}(G^-))$. *More exactly, if* $f \in L_{2,g}(G^+)$ $(f \in L_{2,g}(G^-))$ *then the function* f *is representable as*

$$f = \sum_{i=1}^{\infty} u_i^+ c_i^+ + \sum_{i=1}^{M^+} v_i^+ a_i^+ \quad (M^+ < \infty), \tag{4.2}$$

or as

$$f = \sum_{i=1}^{\infty} u_i^- c_i^- + \sum_{i=1}^{M^-} v_i^- a_i^- \quad (M^- < \infty), \tag{4.3}$$

where $u_i^\pm$ *are eigenfunctions of the problem* (1.2), (1.3) *corresponding to positive or negative eigenvalues, respectively, and* $\{v_i^\pm\}$ *are some finite sets of eigen- and associated functions. The functions* $v_i^\pm$ *are not defined uniquely but we can choose them so that the representation* (4.2) *or* (4.3) *is unique. In this case, the norm in* $L_{2,g}(G^+)$ *or in* $L_{2,g}(G^-)$ *is equivalent to the norm*

$$\|f\|^2_{L_{2,g}(G^\pm)} = \sum_{i=1}^{\infty} |c_i^\pm|^2 + \sum_{i=1}^{M^\pm} |a_i^\pm|^2.$$

References

[1] H. Triebel, *Interpolation Theory, Function spaces, Differential operators*, VEB Deucher Verlag Wiss., Berlin (1977).

[2] I.E. Egorov, S.G. Pyatkov, and S.V. Popov *Nonclassical Operator-Differential Equations*, Nauka, Novosibirsk (2000).

[3] S.G. Pyatkov, *Operator theory. Nonclassical problems*, VSP, Utrecht (2002).

[4] P. Grisvard, *An approach to the singular solutions of elliptic problems via the theory of differential equations in Banach Spaces*, Lect. Notes Math., **1223** (1986), 131–156.

[5] S.G. Pyatkov, *Maximal semidefinite invariant subspaces for some classes of operators // Conditionally Well-Posed Problems.* TVP/TSP, Utrecht (1993), 336–338.

[6] R. Denk, M. Hieber, and J. Prüss, R-boundedness, *Fourier multipliers, and problems of elliptic and parabolic type*, Memoirs of the AMS, **166** (2003).

[7] P. Ausher, A. McIntosh, and A. Nahmrod, *Holomorphic functional calculi of operators, quadratic estimates and interpolation*, Indiana Univ. Math. J., **46** (1997), 375–403.

[8] S.G. Pyatkov, *On some properties of imaginary powers of linear operators*, Matem. Zam. YaGU, **7** (2000), 137–145.

[9] A.I. Parfenov, *On existence of a contracting mapping preserving the boundary values*, Vectnik NGU, Ser. Matem., Mekhan., Inform., 2007. **7** (2007), 69–91 (in Russian).

[10] A.I. Parfenov, *A contracting operator and boundary values*, Preprint no. 155 of the Sobolev Institute of Mathematics, Novosibirsk, Russia (2005).

[11] A.I. Parfenov, *On an embedding criterion for interpolation spaces and application to indefinite spectral problems*, Siberian Math. J., **44** (2003), 638–644.

[12] A.I. Parfenov, *The Curgus condition in indefinite Sturm-Liouville problems*, Siber. Advan. in Math., **15** (2005), 68–103.

[13] Bui Huy Qui. *Weighted Besov and Triebel spaces: interpolation by the real method*, Hiroshima Math. J., **12** (1982), 581–606.

[14] S.G. Pyatkov, *Interpolation of weighted Sobolev spaces*, Siber. Adv. in Math., **10** (2000), 83–132.

[15] S.G. Pyatkov, *Elliptic eigenvalue problems involving an indefinite weight functions*, Siber. Adv. in Math., **10** (2000), 134–150.

[16] S.G. Pyatkov, *Some properties of eigenfunctions and associated functions of indefinite Sturm-Liouville problems*, Nonclassical Equations of Mathematical Physics, The Sobolev Institute of Mathematics, Novosibirsk (2005), 240–251.

[17] P. Binding and R. Hryniv, *Full- and partial-range completeness*, Operator Theory: Adv. Appl. 130 (2001), 121–133.

[18] P.A. Binding and B. Curgus, *A counterexample in Sturm-Liouville completeness theory*, Proc. Roy. Soc. Edinburgh Sect. A 134 (2004), 244–248.

[19] A. Fleige, *The Riesz basis property of an indefinite Sturm-Liouville problem with a non-odd weight function*, Integral Equations, Oper. Theory **60** (2008), 237–246.

[20] P. Binding, B. Curgus, *Riesz bases of root vectors of indefinite Sturm-Liouville problems with eigenparameter dependent boundary conditions. I.* Basel: Birkhäuser. Operator Theory: Advances and Applications **163** (2006), 75–95.

[21] P. Binding, B. Curgus, *Form domains and eigenfunction expansions for differential equations with eigenparameter dependent boundary conditions*, Can. J. Math. **54** (2002), 1142–1164.

[22] P. Binding, H. Langer, M. Moller, *Oscillation results for Sturm–Liouville problems with an indefinite weight function*, J. Comput. Appl. Math. **171** (2004), 93–101.

[23] M. Faierman, M. Moller, *Eigenvalue asymptotics for an elliptic boundary problem* Proc. R. Soc. Edinb., Sect. A, Math., **137** (2007), 281–302.

[24] M. Faierman, *An elliptic boundary problem involving a semi-definite weight*, Proc. R. Soc. Edinb., Sect. A, Math., **134** (2004), 109–136.

[25] R.R. Coifman, *Weighted norm inequalities for maximal function and singular integrals*, Studia Matem., **51** (1974), 241–250.

[26] B. Muckenhoupt, *Weighted norm inequalities for the Hardy maximal function*, Trans. of AMS, **165** (1972), 207–226.

[27] O.V. Besov, V.P. Il′in, and S.M. Nikol′skiĭ *Integral Representations of Functions, and Embedding Theorems*, Nauka, Moscow (1975).

[28] V.G. Maz'ya, *Sobolev Spaces*, Springer-Verlag Berlin–New York (1985).

[29] Yu.M. Beresanskii, *Eigenfunction Expansion of Selfajoint Operators*, Naukova dumka, Kiev (1965).

S.G. Pyatkov
Ugra State University,
Chekhov st. 16,
628007 Hanty-Mansiisk, Russia
e-mail: s_pyatkov@ugrasu.ru

Operator Theory:
Advances and Applications, Vol. 198, 291–304

On the Spectrum of a Quantum Dot with Impurity in the Lobachevsky Plane

P. Šťovíček and M. Tušek

Abstract. A model of a quantum dot with impurity in the Lobachevsky plane is considered. Relying on explicit formulae for the Green function and the Krein Q-function which have been derived in a previous work we focus on the numerical analysis of the spectrum. The analysis is complicated by the fact that the basic formulae are expressed in terms of spheroidal functions with general characteristic exponents. The effect of the curvature on eigenvalues and eigenfunctions is investigated. Moreover, there is given an asymptotic expansion of eigenvalues as the curvature radius tends to infinity (the flat case limit).

Mathematics Subject Classification (2000). 81Q10, 81Q05, 33E10.

Keywords. Quantum dot, Lobachevsky plane, point interaction, spectrum.

1. Introduction

The influence of the hyperbolic geometry on the properties of quantum mechanical systems is a subject of continual theoretical interest for at least two decades. Numerous models have been studied so far, let us mention just few of them [1, 2, 3, 4]. Naturally, the quantum harmonic oscillator is one of the analyzed examples [5, 6]. It should be stressed, however, that the choice of an appropriate potential on the hyperbolic plane is ambiguous in this case, and several possibilities have been proposed in the literature. In [7], we have modeled a quantum dot in the Lobachevsky plane by an unbounded potential which can be interpreted, too, as a harmonic oscillator potential for this nontrivial geometry. The studied examples also comprise point interactions [8] which are frequently used to model impurities.

A Hamiltonian describing a quantum dot with impurity has been introduced in [7]. The main result of this paper is derivation of explicit formulae for the Green function and the Krein Q-function. The formulae are expressed in terms of spheroidal functions which are used rather rarely in the framework of mathematical physics. Further analysis is complicated by the complexity of spheroidal functions. In particular, the Green function depends on the characteristic exponent of the spheroidal functions in question rather than directly on the spectral parameter. In fact, it seems to be possible to obtain a more detailed information on eigenvalues and eigenfunctions only by means of numerical methods. The particular case, when the Hamiltonian is restricted to the eigenspace of the angular momentum with eigenvalue 0, is worked out in [9]. In the current contribution we aim to extend the numerical analysis to the general case and to complete it with additional details.

The Hamiltonian describing a quantum dot with impurity in the Lobachevsky plane, as introduced in [7], is a selfadjoint extension of the following symmetric operator:

$$H = -\left(\frac{\partial^2}{\partial\varrho^2} + \frac{1}{a}\coth\left(\frac{\varrho}{a}\right)\frac{\partial}{\partial\varrho} + \frac{1}{a^2}\sinh^{-2}\left(\frac{\varrho}{a}\right)\frac{\partial^2}{\partial\phi^2} + \frac{1}{4a^2}\right) + \frac{1}{4}\,a^2\omega^2\sinh^2\left(\frac{\varrho}{a}\right),$$
$$\mathrm{Dom}(H) = C_0^\infty((0,\infty)\times S^1) \subset L^2((0,\infty)\times S^1, a\,\sinh(\varrho/a)\mathrm{d}\varrho\,\mathrm{d}\phi)\,,$$

where (ϱ,ϕ) are the geodesic polar coordinates on the Lobachevsky plane and a stands for the so-called curvature radius which is related to the scalar curvature by the formula $R = -2/a^2$. The deficiency indices of H are known to be $(1,1)$ and we denote each selfadjoint extension by $H(\chi)$ where the real parameter χ appears in the boundary conditions for the domain of definition: $f(\varrho,\phi)$ belongs to $\mathrm{Dom}(H(\chi))$ if there exist $f_0, f_1 \in \mathbb{C}$ so that $f_1 : f_0 = \chi : 1$ and

$$f(\varrho,\phi) = -\frac{1}{2\pi}\,f_0\log(\varrho) + f_1 + o(1) \quad \text{as } \varrho \to 0+$$

(the case $\chi = \infty$ means that $f_0 = 0$ and f_1 is arbitrary), see [7] for details. $H(\infty)$ is nothing but the Friedrichs extension of H. The Hamiltonian $H(\infty)$ is interpreted as corresponding to the unperturbed case and describing a quantum dot with no impurity.

After the substitution $\xi = \cosh(\varrho/a)$ and the scaling $H = a^{-2}\tilde{H}$, we make use of the rotational symmetry (which amounts to a Fourier transform in the variable ϕ) to decompose $\tilde{H}$ into a direct sum as follows

$$\tilde{H} = \bigoplus_{m=-\infty}^{\infty} \tilde{H}_m,$$
$$\tilde{H}_m = -\frac{\partial}{\partial\xi}(\xi^2-1)\frac{\partial}{\partial\xi} + \frac{m^2}{\xi^2-1} + \frac{a^4\omega^2}{4}(\xi^2-1) - \frac{1}{4},$$
$$\mathrm{Dom}(\tilde{H}_m) = C_0^\infty(1,\infty) \subset L^2((1,\infty),\mathrm{d}\xi).$$

Let us denote by H_m, $m \in \mathbb{Z}$, the restriction of $H(\infty)$ to the eigenspace of the angular momentum with eigenvalue m. This means that H_m is a self-adjoint extension of $a^{-2}\tilde{H}_m$. It is known (Proposition 2.1 in [7]) that $\tilde{H}_m$ is essentially selfadjoint for $m \neq 0$. Thus, in this case, H_m is the closure of $a^{-2}\tilde{H}_m$. Concerning the case $m = 0$, H_0 is the Friedrichs extension of $a^{-2}\tilde{H}_0$. For quite general reasons, the spectrum of H_m, for any m, is semibounded below, discrete and simple [10]. We denote the eigenvalues of H_m in ascending order by $E_{n,m}(a^2)$, $n \in \mathbb{N}_0$.

The spectrum of the total Hamiltonian $H(\chi)$, $\chi \neq \infty$, consists of two parts (in a full analogy with the Euclidean case [11]):

1. The first part is formed by those eigenvalues of $H(\chi)$ which belong, at the same time, to the spectrum of $H(\infty)$. More precisely, this part is exactly the union of eigenvalues of H_m for m running over $\mathbb{Z} \setminus \{0\}$. Their multiplicities are discussed below in Section 5.
2. The second part is formed by solutions to the equation
$$Q^H(z) = \chi \tag{1.1}$$
with respect to the variable z where Q^H stands for the Krein Q-function of $H(\infty)$. Let us denote the solutions in ascending order by $\epsilon_n(a^2, \chi)$, $n \in \mathbb{N}_0$. These eigenvalues are sometimes called the point levels and their multiplicities are at least one. In more detail, $\epsilon_n(a^2, \chi)$ is a simple eigenvalue of $H(\chi)$ if it does not lie in the spectrum of $H(\infty)$, and this happens if and only if $\epsilon_n(a^2, \chi)$ does not coincide with any eigenvalue $E_{\ell,m}(a^2)$ for $\ell \in \mathbb{N}_0$ and $m \in \mathbb{Z}$, $m \neq 0$.

Remark. The lowest point level, $\epsilon_0(a^2, \chi)$, lies below the lowest eigenvalue of $H(\infty)$ which is $E_{0,0}(a^2)$, and the point levels with higher indices satisfy the inequalities $E_{n-1,0}(a^2) < \epsilon_n(a^2, \chi) < E_{n,0}(a^2)$, $n = 1, 2, 3, \ldots$.

2. Spectrum of the unperturbed Hamiltonian $H(\infty)$

Our goal is to find the eigenvalues of the mth partial Hamiltonian H_m, i.e., to find square integrable solutions of the equation
$$H_m \psi(\xi) = z\psi(\xi),$$
or, equivalently,
$$\tilde{H}_m \psi(\xi) = a^2 z\psi(\xi).$$
This equation coincides with the equation of the spheroidal functions (A.1) provided we set $\mu = |m|$, $\theta = -a^4\omega^2/16$, and the characteristic exponent ν is chosen so that
$$\lambda_\nu^m\left(-\frac{a^4\omega^2}{16}\right) = -a^2 z - \frac{1}{4}.$$
The only solution (up to a multiplicative constant) that is square integrable near infinity is $S_\nu^{|m|(3)}(\xi, -a^4\omega^2/16)$.

Proposition A.3 describes the asymptotic expansion of this function at $\xi = 1$ for $m \in \mathbb{N}$. It follows that the condition on the square integrability is equivalent to the equality
$$\mathrm{e}^{i(3\nu+1/2)\pi} K_{-\nu-1}^m\left(-\frac{a^4\omega^2}{16}\right) + K_\nu^m\left(-\frac{a^4\omega^2}{16}\right) = 0. \tag{2.1}$$

Furthermore, in [7] we have derived that

$$S_\nu^{0(3)}(\xi,\theta) = \alpha\log(\xi-1) + \beta + O((\xi-1)\log(\xi-1)) \quad \text{as } \xi \to 1+,$$

where

$$\alpha = \frac{i\tan(\nu\pi)\,\mathrm{e}^{-i(2\nu+1/2)\pi}}{2\pi s_\nu^0(\theta)}\left(\mathrm{e}^{i(3\nu+1/2)\pi}K_{-\nu-1}^0(\theta) + K_\nu^0(\theta)\right).$$

Taking into account that the Friedrichs extension has continuous eigenfunctions we conclude that equation (2.1) guarantees square integrability in the case $m = 0$, too.

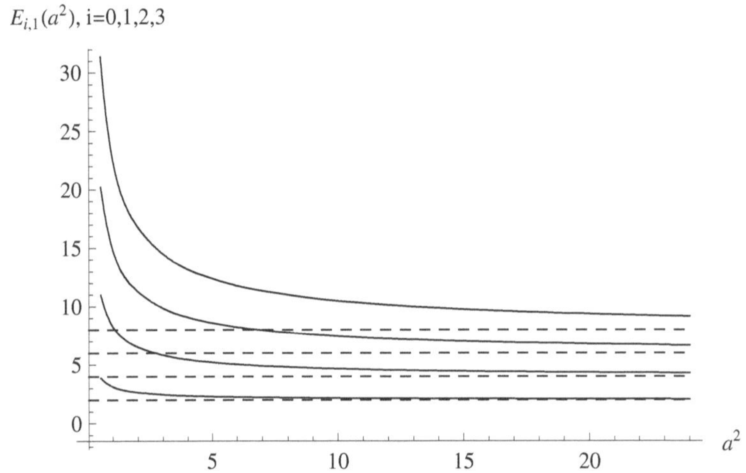

FIGURE 1. Eigenvalues of the partial Hamiltonian H_1

As far as we see it, equation (2.1) can be solved only by means of numerical methods. For this purpose we made use of the computer algebra system *Mathematica 6.0*. For the numerical computations we set $\omega = 1$. The particular case $m = 0$ has been examined in [9]. It turns out that an analogous procedure can be also applied for nonzero values of the angular momentum. As an illustration, Figure 1 depicts several first eigenvalues of the Hamiltonian H_1 as functions of the curvature radius a. The dashed asymptotic lines correspond to the flat limit ($a \to \infty$).

Denote the nth normalized eigenfunction of the mth partial Hamiltonian $\tilde{H}_m$ by $\tilde{\psi}_{n,m}(\xi)$. Obviously, the eigenfunctions for the values of the angular momentum m and $-m$ are the same and are proportional to $S_\nu^{|m|(3)}(\xi, -a^4\omega^2/16)$, with ν satisfying equation (2.1). Let us return to the original radial variable ϱ and, moreover, regard $\tilde{H}_m$ as an operator acting on $L^2(\mathbb{R}^+, \mathrm{d}\varrho)$. This amounts to an obvious isometry

$$L^2(\mathbb{R}^+, a^{-1}\sinh(\varrho/a)\mathrm{d}\varrho) \to L^2(\mathbb{R}^+, \mathrm{d}\varrho):\ f(\varrho) \mapsto a^{-1/2}\sinh^{1/2}(\varrho/a)f(\varrho).$$

The corresponding normalized eigenfunction of $\tilde{H}_m$, with an eigenvalue $a^2 z$, equals

$$\psi_{n,m}(\varrho) = \left(\frac{1}{a}\sinh\left(\frac{\varrho}{a}\right)\right)^{1/2} \tilde{\psi}_{n,m}\left(\cosh\left(\frac{\varrho}{a}\right)\right). \tag{2.2}$$

At the same time, relation (2.2) gives the normalized eigenfunction of H_m (considered on $L^2(\mathbb{R}^+, \mathrm{d}\varrho)$) with the eigenvalue z. The same Hilbert space may be used also in the limit Euclidean case ($a = \infty$). The eigenfunctions $\Phi_{n,m}$ in the flat case are well known and satisfy

$$\Phi_{n,m} \propto \varrho^{|m|+1/2} \mathrm{e}^{-\omega\varrho^2/4} {}_1F_1\left(-n, |m|+1, \frac{\omega\varrho^2}{2}\right). \tag{2.3}$$

The fact that we stick to the same Hilbert space in all cases facilitates the comparison of eigenfunctions for various values of the curvature radius a. We present plots of several first eigenfunctions of H_1 (Figures 2, 3, 4) for the values of the curvature radius $a = 1$ (the solid line), 10 (the dashed line), and ∞ (the dotted line). Again, see [9] for analogous plots in the case of the Hamiltonian H_0. Note that, in general, the smaller is the curvature radius a the more localized is the particle in the region near the origin.

3. The point levels

As has been stated, the point levels are solutions to equation (1.1) with respect to the spectral parameter z. Since, in general, $Q(\bar{z}) = \overline{Q(z)}$ the function $Q(z)$ takes real values on the real axis. Let $\tilde{H}(\infty) = a^2 H(\infty)$ be the Friedrichs extension of $\tilde{H}$. An explicit formula for the Krein Q-function $Q^{\tilde{H}}(z)$ of $\tilde{H}(\infty)$ has been derived

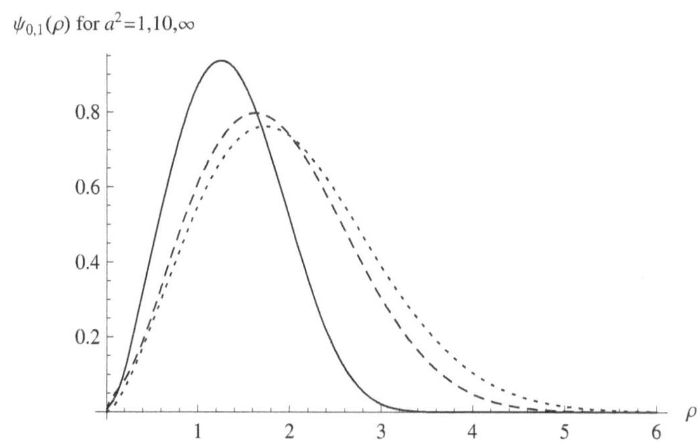

FIGURE 2. The first eigenfunction of the partial Hamiltonian H_1

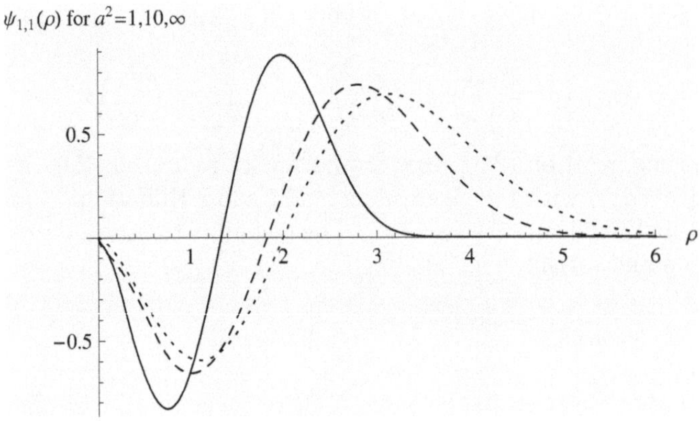

FIGURE 3. The second eigenfunction of the partial Hamiltonian H_1

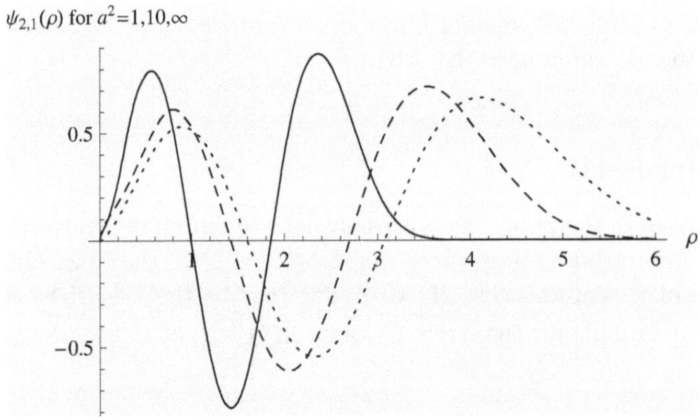

FIGURE 4. The third eigenfunction of the partial Hamiltonian H_1

in [7]:

$$Q^{\tilde{H}}(z) = -\frac{1}{4\pi a^2}\left(-\log(2) - 2\Psi(1) + 2\,\Psi s_\nu\left(-\frac{a^4\omega^2}{16}\right) s_\nu^0\left(-\frac{a^4\omega^2}{16}\right)\right)$$
$$+\frac{1}{2a^2\tan(\nu\pi)}\left(\mathrm{e}^{i\pi(3\nu+3/2)}\,\frac{K^0_{-\nu-1}(-\frac{a^4\omega^2}{16})}{K^0_\nu(-\frac{a^4\omega^2}{16})} - 1\right)^{-1} + \frac{\log(2a^2)}{4\pi a^2},$$

where ν is chosen so that

$$\lambda_\nu^0\left(-\frac{a^4\omega^2}{16}\right) = -z - \frac{1}{4}.$$

The symbol $K^0_\nu(\theta)$ stands for the so-called spheroidal joining factor,

$$\Psi s_\nu(\theta) := \sum_{r=-\infty}^{\infty} (-1)^r a^0_{\nu,r}(\theta)\, \Psi(\nu+1+2r),$$

where the coefficients $a^0_{\nu,r}(\theta)$, $r \in \mathbb{Z}$, come from the expansion of spheroidal functions in terms of Bessel functions (for details see [7, the Appendix])), and $s^0_\nu(\theta)$ is defined by formula (A.4). One can obtain the Krein Q-function of $H(\infty)$ simply by scaling $Q^H(z) = a^2\, Q^{\tilde{H}}(a^2 z)$.

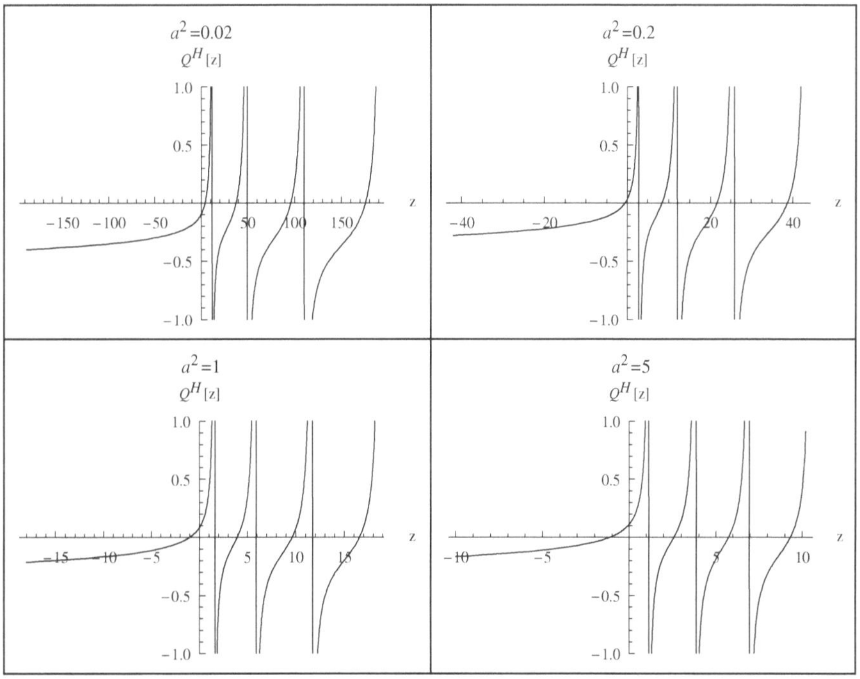

FIGURE 5. The Krein Q-function Q^H for $a^2 = 0.02,\ 0.2,\ 1,\ 5$

Since we know the explicit expression for the Krein Q-function as a function of the characteristic exponent ν rather than of the spectral parameter z itself it is of importance to know for which values of ν the spectral parameter z is real. Propositions A.1 and A.2 give the answer. For $\nu \in \mathbb{R}$ and for ν of the form $\nu = -1/2 + it$ where t is real, the spheroidal eigenvalue $\lambda^m_\nu(-a^4\omega^2/16)$ is real, and so the same is true for z. Moreover, these values of ν reproduce the whole real z axis. With this knowledge, one can plot the Krein Q-function $Q^H = Q^H(z)$ for an arbitrary value of the curvature radius a. Note that for $a = \infty$, the Krein Q-function is well known as a function of the spectral parameter z [12] and equals

(setting $\omega = 1$, Ψ is the logarithmic derivative of the gamma function)

$$Q(z) = \frac{1}{4\pi}\left(-\Psi\left(\frac{1-z}{2}\right) + \log(2) + 2\Psi(1)\right).$$

Next, in Figure 5, we present plots of the Krein Q-function for several distinct values of the curvature radius a. Moreover, in Figure 6 one can compare the behavior of the Krein Q-function for a comparatively large value of the curvature radius ($a^2 = 24$) and for the Euclidean case ($a = \infty$).

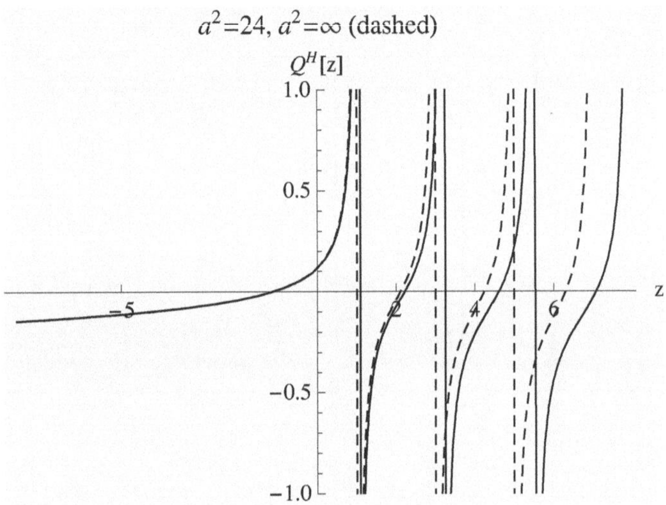

FIGURE 6. Comparison of the Krein Q-functions for $a^2 = 24$ and $a^2 = \infty$

Again, equation (1.1) can be solved only numerically. Fixing the parameter χ one may be interested in the behavior of the point levels as functions of the curvature radius a. See Figure 7 for the corresponding plots, with $\chi = 0$, where the dashed asymptotic lines again correspond to the flat case limit ($a = \infty$). Note that for the curvature radius a large enough, the lowest eigenvalue is negative provided χ is chosen smaller than $Q(0) \simeq 0.1195$.

4. Asymptotic behavior for large values of a

The mth partial Hamiltonian H_m, if considered on $L^2(\mathbb{R}^+, \mathrm{d}\varrho)$, acts like

$$H_m = -\frac{\partial^2}{\partial\varrho^2} + \frac{m^2 - \frac{1}{4}}{a^2 \sinh^2(\frac{\varrho}{a})} + \frac{1}{4}a^2\omega^2 \sinh^2\left(\frac{\varrho}{a}\right) =: -\frac{\partial^2}{\partial\varrho^2} + V_m(a, \varrho).$$

For a fixed $\varrho \neq 0$, one can easily derive that

$$V_m(a, \varrho) = \frac{m^2 - \frac{1}{4}}{\varrho^2} + \frac{1}{4}\omega^2\varrho^2 + \frac{\frac{1}{4} - m^2}{3a^2} + \frac{\omega^2\varrho^4}{12a^2} + O\left(\frac{1}{a^4}\right) \quad \text{as } a \to \infty.$$

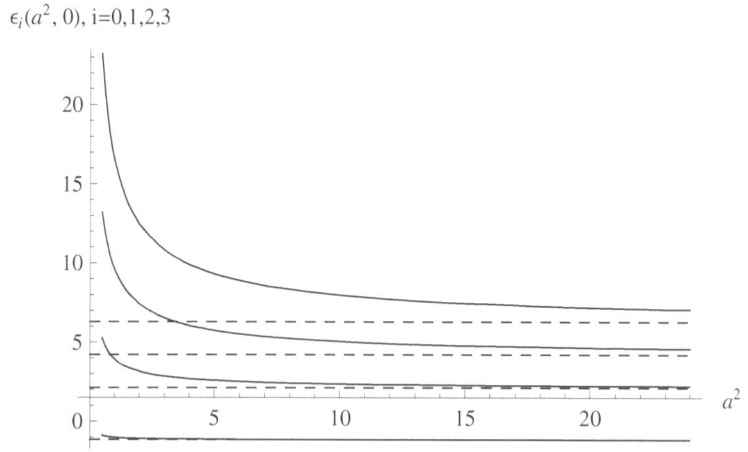

FIGURE 7. Point levels for H(0)

Recall that the mth partial Hamiltonian of the isotropic harmonic oscillator on the Euclidean plane, H_m^E, if considered on $L^2(\mathbb{R}^+, \mathrm{d}\varrho)$, has the form

$$H_m^E := -\frac{\partial^2}{\partial\varrho^2} + \frac{m^2 - \frac{1}{4}}{\varrho^2} + \frac{1}{4}\omega^2\varrho^2.$$

This suggests that it may be useful to view the Hamiltonian H_m, for large values of the curvature radius a, as a perturbation of H_m^E,

$$H_m \sim H_m^E + \frac{1}{12a^2}(1 - 4m^2 + \omega^2\varrho^4) =: H_m^E + \frac{1}{12a^2}U_m(\varrho).$$

The eigenvalues of the compared Hamiltonians have the same asymptotic expansions up to the order $1/a^2$ as $a \to \infty$.

Let us denote the nth eigenvalue of the Hamiltonian H_m^E by $E_{n,m}^E$, $n \in \mathbb{N}_0$. It is well known that

$$E_{n,m}^E = (2n + |m| + 1)\,\omega$$

and that the multiplicity of $E_{n,m}^E$ in the spectrum of H^E equals $2n + |m| + 1$. The asymptotic behavior of $E_{n,m}(a^2)$ may be deduced from the standard perturbation theory and is given by the formula

$$E_{n,m}(a^2) = E_{n,m}^E + \frac{1}{12a^2}\frac{\langle\Phi_{n,m}, U_m\Phi_{n,m}\rangle}{\langle\Phi_{n,m}, \Phi_{n,m}\rangle} + O\left(\frac{1}{a^4}\right) \quad \text{as } a \to \infty, \tag{4.1}$$

where $\Phi_{n,m}$ denotes a (not necessarily normalized) eigenfunction of H_m^E associated with the eigenvalue $E_{n,m}^E$ (see (2.3)). The scalar products occurring in formula (4.1) can be readily evaluated in $L^2(\mathbb{R}^+, \mathrm{d}\varrho)$ with the help of Proposition A.4. The

	$E_{0,0}$	$E_{1,0}$	$E_{2,0}$	$E_{0,1}$	$E_{1,1}$	$E_{2,1}$
numerical	1.0265	3.162	5.42	2.060	4.259	6.58
asymptotic	1.0268	3.169	5.46	2.058	4.258	6.59
error (%)	−0.03	−0.22	−0.74	0.10	0.02	−0.15

TABLE 1. Comparison of numerical and asymptotic results for the eigenvalues, $a^2 = 24$

resulting formula takes the form

$$E_{n,m}(a^2) = (2n + |m| + 1)\,\omega + \left(2n(n + |m| + 1) + |m| + \frac{3}{4}\right)\frac{1}{a^2} + O\left(\frac{1}{a^4}\right) \tag{4.2}$$

as $a \to \infty$. This asymptotic approximation of eigenvalues has been tested numerically for large values of the curvature radius a. The asymptotic eigenvalues for $a^2 = 24$ are compared with the precise numerical results in Table 1. It is of interest to note that the asymptotic coefficient in front of the a^{-2} term does not depend on the frequency ω.

5. The multiplicities

Since $H_{-m} = H_m$ the eigenvalues $E_{n,m}(a^2)$ of the total Hamiltonian $H(\infty)$ are at least twice degenerated if $m \neq 0$. From the asymptotic expansion (4.2) it follows, after some straightforward algebra, that no additional degeneracy occurs and thus theses eigenvalues are exactly twice degenerated at least for sufficiently large values of a.

Applying the methods developed in [11] one may complete the analysis of the spectrum of the total Hamiltonian $H(\chi)$ for $\chi \neq \infty$. Namely, the spectrum of $H(\chi)$ contains eigenvalues $E_{n,m}(a^2)$, $m > 0$, with multiplicity 2 if $Q^H(E_{n,m}(a^2)) \neq \chi$, and with multiplicity 3 if $Q^H(E_{n,m}(a^2)) = \chi$. The rest of the spectrum of $H(\chi)$ is formed by those solutions to equation (1.1) which do not belong to the spectrum of $H(\infty)$. The multiplicity of all these eigenvalues in the spectrum of $H(\chi)$ equals 1.

Appendix: Auxiliary results

In this appendix we summarize several auxiliary results. Firstly, for our purposes we need the following observations concerning spheroidal functions. The spheroidal functions are solutions to the equation

$$(1-\xi^2)\frac{\partial^2\psi}{\partial\xi^2} - 2\xi\frac{\partial\psi}{\partial\xi} + \left[\lambda_\nu^\mu(\theta) + 4\theta(1-\xi^2) - \mu^2(1-\xi^2)^{-1}\right]\psi = 0. \tag{A.1}$$

For the notation and properties of spheroidal functions see [13]. A detailed information on this subject can be found in [14], but be aware of somewhat different notation. A very brief overview of spheroidal functions is also given in the Appendix of [7].

In the last named source, the following proposition has been proved in the particular case $m = 0$. But, as one can verify by a direct inspection, the proof applies to the general case $m \in \mathbb{Z}$ as well.

Proposition A.1. *Let* $\nu, \theta \in \mathbb{R}$, $m \in \mathbb{Z}$. *Then* $\lambda_\nu^m(\theta) \in \mathbb{R}$.

The following claim is also of interest.

Proposition A.2. *Let* $\nu = -1/2 + it$ *where* $t \in \mathbb{R}$, *and* $\theta \in \mathbb{R}$, $m \in \mathbb{Z}$. *Then* $\lambda_\nu^m(\theta) \in \mathbb{R}$.

Proof. Let us recall that the coefficients $a_{\nu,r}^m(\theta)$ in the series expansion of spheroidal functions in terms of Bessel functions satisfy a three term recurrence relation (see [7, the Appendix]),

$$\beta_{\nu,r}^m(\theta)a_{\nu,r-1}^m(\theta) + \alpha_{\nu,r}^m(\theta)a_{\nu,r}^m(\theta) + \gamma_{\nu,r}^m(\theta)a_{\nu,r+1}^\mu(\theta) = \lambda_\nu^m(\theta)a_{\nu,r}^m(\theta). \qquad \text{(A.2)}$$

One may view the set of equations (A.2), with $r \in \mathbb{Z}$, as an eigenvalue equation for $\lambda_\nu^m(\theta)$ that is an analytic function of θ. A particular solution is fixed by the condition $\lambda_\nu^m(0) = \nu(\nu+1)$. Consider the set of complex conjugated equations. Since $\overline{\beta_{\nu,r}^m} = \beta_{\overline{\nu},r}^m(\theta) = \beta_{-\nu-1,r}^m(\theta)$, and the similar is true for $\alpha_{\nu,r}^m(\theta)$ and $\gamma_{\nu,r}^m(\theta)$, it holds true that

$$\beta_{-\nu-1,r}^m(\theta)\overline{a_{\nu,r-1}^m(\theta)} + \alpha_{-\nu-1,r}^m\overline{a_{\nu,r}^m(\theta)} + \gamma_{-\nu-1,r}^m(\theta)\overline{a_{\nu,r+1}^\mu(\theta)} = \overline{\lambda_\nu^m(\theta)}\,\overline{a_{\nu,r}^m(\theta)}.$$

Since for each ν of the considered form,

$$\lambda_{-\nu-1}^m(0) = (-\nu-1)(-\nu) = \nu(\nu+1) = \overline{\nu(\nu+1)} = \overline{\lambda_\nu^m(0)},$$

one has $\lambda_{-\nu-1}^m(\theta) = \overline{\lambda_\nu^m(\theta)}$. Moreover, $\lambda_{-\nu-1}^m(\theta) = \lambda_\nu^m(\theta)$ in general. We conclude that $\lambda_\nu^m(\theta) \in \mathbb{R}$. □

Another auxiliary result concerns the asymptotic expansion of the radial spheroidal function of the third kind.

Proposition A.3. *Let* $\nu \notin \{-1/2 + k |\ k \in \mathbb{Z}\}$, $m \in \mathbb{N}$. *Then*

$$S_\nu^{m(3)}(\xi,\theta) \sim \frac{(-1)^m 2^{m/2-1}\Gamma(m)\tan(\nu\pi)}{\pi s_\nu^m(\theta)\,\mathrm{e}^{-i(\nu+3/2)\pi}}\left(K_{-\nu-1}^m(\theta) + \frac{K_\nu^m(\theta)}{\mathrm{e}^{i(3\nu+1/2)\pi}}\right)(\xi-1)^{-m/2}$$

as $\xi \to 1+$. (A.3)

Proof. By the definition of the radial spheroidal function of the third kind,

$$S_\nu^{m(3)}(\xi,\theta) := \frac{1}{i\cos(\nu\pi)}\left(S_{-\nu-1}^{m(1)}(\xi,\theta) + i\mathrm{e}^{-i\nu\pi}S_\nu^{m(1)}(\xi,\theta)\right),$$

and by the relation between the radial and the angular spheroidal functions,

$$S_\nu^{m(1)}(\xi,\theta) = -\frac{\sin(\nu\pi)}{\pi}\,\mathrm{e}^{-i\nu\pi}K_\nu^m(\theta)\,Qs_{-\nu-1}^m(\xi,\theta),$$

one has

$$S_\nu^{m(3)}(\xi,\theta) = \frac{i\tan(\nu\pi)}{\pi e^{-i(\nu+1)\pi}}\left(K_{-\nu-1}^m(\theta)Qs_\nu^m(\xi,\theta) + \frac{K_\nu^m(\theta)Qs_{-\nu-1}^m(\xi,\theta)}{e^{i(3\nu+1/2)\pi}}\right).$$

Using the definition

$$Qs_\nu^m(\xi,\theta) = \sum_{r=-\infty}^{\infty}(-1)^r a_{\nu,r}^m(\theta)Q_{\nu+2r}^m(\xi)$$

and due to the well-known asymptotic expansion for the Legendre functions [13],

$$Q_\nu^m(\xi) \sim (-1)^m 2^{m/2-1}\Gamma(m)(\xi-1)^{-m/2} \quad \text{as } \xi \to 1+,$$

one derives that

$$Qs_\nu^m(\xi,\theta) \sim \frac{(-1)^m 2^{m/2-1}\Gamma(m)}{(\xi-1)^{m/2}s_\nu^m(\theta)} \quad \text{as } \xi \to 1+,$$

where

$$(s_\nu^m(\theta))^{-1} := \sum_{r=-\infty}^{\infty}(-1)^r a_{\nu,r}^m(\theta) = \sum_{r=-\infty}^{\infty}(-1)^r a_{-\nu-1,-r}^m(\theta) = (s_{-\nu-1}^m(\theta))^{-1}. \tag{A.4}$$

Hence $Qs_{-\nu-1}^m(\xi,\theta) \sim Qs_\nu^m(\xi,\theta)$ as $\xi \to 1+$, and one immediately obtains (A.3). □

Further some auxiliary computations follow that we need for evaluation of scalar products of eigenfunctions (see (4.1)).

Proposition A.4. *Let ${}_1F_1(a,b,t)$ stand for the Kummer confluent hypergeometric function, and $n,m,l \in \mathbb{N}_0$. Then*

$$\begin{aligned}&\int_0^\infty t^{m+l}e^{-t}\,{}_1F_1(-n,1+m,t)^2\,dt\\ &= (m!)^2\sum_{k=\max\{0,n-l\}}^{n}(-1)^{n+k}\binom{n}{k}\frac{(k+l)!}{(k+m)!}\binom{k+m+l}{n+m}.\end{aligned} \tag{A.5}$$

Proof. By definition,

$${}_1F_1(-n,1+m,t) := \sum_{k=0}^{n}\frac{(-n)_k\,t^k}{(1+m)_k\,k!} = m!\sum_{k=0}^{n}(-1)^k\binom{n}{k}\frac{t^k}{(m+k)!}.$$

Let us denote the LHS of (A.5) by I. Then the integral representation of the gamma function implies

$$I = (m!)^2\sum_{j,k=0}^{n}(-1)^{j+k}\binom{n}{j}\binom{n}{k}\frac{(j+k+m+l)!}{(m+j)!(m+k)!}. \tag{A.6}$$

Partial summation in (A.6) can be carried out,

$$\sum_{j=0}^{n}(-1)^j\binom{n}{j}\frac{(j+k+m+l)!}{(m+j)!} = \frac{d^{k+l}}{dx^{k+l}}\left(x^{k+m+l}(1-x)^n\right)\Big|_{x=1}. \tag{A.7}$$

Expression (A.7) vanishes for $k < n - l$ and equals

$$(-1)^n (k+l)! \binom{k+m+l}{n+m}$$

for $k \geq n - l$. The proposition follows immediately. □

Corollary A.5. *In the case $l = 0$, (A.5) takes a particularly simple form:*

$$\int_0^\infty t^m \mathrm{e}^{-t} {}_1F_1(-n, 1+m, t)^2 \, \mathrm{d}t = \frac{n!}{(m+n)!} \, .$$

References

[1] A. Comtet, *On the Landau Levels on the Hyperbolic Plane*, Ann. Physics **173** (1987), 185–209.

[2] M. Antoine, A. Comtet, and S. Ouvry, *Scattering on a Hyperbolic Torus in a Constant Magnetic Field*, J. Phys. A: Math. Gen. **23** (1990), 3699–3710.

[3] Yu.A. Kuperin, R.V. Romanov, and H.E. Rudin, *Scattering on the Hyperbolic Plane in the Aharonov-Bohm Gauge Field*, Lett. Math. Phys. **31** (1994), 271–278.

[4] O. Lisovyy, *Aharonov-Bohm Effect on the Poincaré Disk*, J. Math. Phys. **48** (2007), 052112.

[5] D.V. Bulaev, V.A. Geyler, and V.A. Margulis, *Effect of Surface Curvature on Magnetic Moment and Persistent Currents in the Two-Dimensional Quantum Ring and Dots*, Phys. Rev. B **69** (2004), 195313.

[6] J.F. Cariñena, M.F. Rañada, and M. Santander, *The Quantum Harmonic Oscillator on the Sphere and the Hyperbolic Plane*, Ann. Physics **322** (2007), 2249–2278.

[7] V. Geyler, P. Šťovíček, and M. Tušek, *A Quantum Dot with Impurity in the Lobachevsky Plane*, in Proceedings of the 6th Workshop on Operator Theory in Krein Spaces, Birkhäuser, 2008 (to appear); arXiv:0709.2790v3 (2007).

[8] J. Brüning and V. Geyler, *Gauge-Periodic Point Perturbations on the Lobachevsky Plane*, Theor. Math. Phys. **119** (1999), 687–697.

[9] P. Šťovíček and M. Tušek, *On the Harmonic Oscillator on the Lobachevsky Plane*, Russian J. Math. Phys. **14** (2007), 493–497.

[10] J. Weidmann, *Linear Operators in Hilbert Spaces.* Springer, 1980.

[11] J. Brüning, V. Geyler, and I. Lobanov, *Spectral Properties of a Short-Range Impurity in a Quantum Dot*, J. Math. Phys. **46** (2004), 1267–1290.

[12] V. Geyler and I. Popov, *Eigenvalues Imbedded in the Band Spectrum for a Periodic Array of Quantum Dots*, Rep. Math. Phys. **39** (1997), 275–281.

[13] H. Bateman and A. Erdélyi, *Higher Transcendental Functions III.* McGraw-Hill Book Company, 1955.

[14] J. Meixner and F.V. Schäfke, *Mathieusche Funktionen und Sphäroidfunktionen.* Springer-Verlag, 1954.

Acknowledgment

The authors wish to acknowledge gratefully partial support of the Ministry of Education of Czech Republic under the research plan MSM6840770039 (P.Š.) and from the grant No. LC06002 (M.T.). Moreover, one of the authors wish to acknowledge partial support from the grant No. 202/08/H072 of the Czech Science Foundation (M.T.).

P. Šťovíček and M. Tušek
Department of Mathematics
Faculty of Nuclear Sciences
Czech Technical University,
Prague, Czech Republic
e-mail: `stovicek@fjfi.cvut.cz`
`tusekmat@fjfi.cvut.cz`

Operator Theory:
Advances and Applications, Vol. 198, 305–307

Note on Interlacing for Hyperbolic Quadratic Pencils

Krešimir Veselić

Dedicated to the memory of Peter Jonas, dear colleague and friend

Abstract. We prove interlacing inequalities for the eigenvalues of the submatrices of (weakly) hyperbolic and gyroscopic quadratic pencils.

Mathematics Subject Classification (2000). 15-02, 47-02.

Keywords. Second order systems, overdamped linear systems, eigenvalue inclusions.

In [1], Theorem 4.3 an interlacing result for quadratic pencils was derived. It is the aim of this note to offer an elementary proof of this fact. Our main tool will be the Sylvester theorem of inertia. A Hermitian quadratic pencil

$$K(\lambda) = \lambda^2 M + \lambda C + K \tag{1}$$

of order n is called *hyperbolic*, if M is positive definite and $K(\mu)$ is negative definite for some real μ. It is well known, see [2], that the eigenvalues of $K(\cdot)$ can be written as

$$\lambda_n^- \leq \cdots \leq \lambda_1^- < \mu < \lambda_1^+ \leq \cdots \leq \lambda_n^+. \tag{2}$$

Let

$$K = \begin{bmatrix} K_{11} & K_{12} \\ K_{12}^* & K_{22} \end{bmatrix} \tag{3}$$

be any given partition, where K_{11} is of order m (and similarly for $M, C, K(\lambda)$). Then $K_{11}(\mu)$ is again negative definite and $K_{11}(\cdot)$ itself hyperbolic with the eigenvalues

$$\alpha_m^- \leq \cdots \leq \alpha_1^- < \mu < \alpha_1^+ \leq \cdots \leq \alpha_m^+. \tag{4}$$

This work was partly done while the author stayed at the University of Osijek, under the support of National Foundation for Science, Higher Education and Technological Development of the Republic of Croatia.

The interlacing inequalities are

$$\alpha_k^+ \geq \lambda_k^+, \quad \alpha_{k+m-n}^+ \leq \lambda_k^+ \tag{5}$$

$$\alpha_k^- \leq \lambda_k^-, \quad \alpha_{k+m-n}^- \geq \lambda_k^- \tag{6}$$

for all possible k. To prove these inequalities we will use the formula

$$\Pi_+(K(\lambda)) = k, \quad \text{whenever } \lambda_k^+ < \lambda < \lambda_{k+1}^+ \tag{7}$$

where

$$\Pi(H) = \{\Pi_+(H), \Pi_0(H), \Pi_-(H)\}$$

is the inertia of a Hermitian matrix H.

To prove this formula note that without loss of generality we may assume that C, K are positive definite. Otherwise make the spectral shift $\lambda = \lambda_0 + \nu$, $\lambda_0 > 0$ such that both $2\lambda_0 M + C$ and $K(\lambda_0)$ become positive definite and the new pencil

$$\hat{K}(\nu) = K(\lambda) = \nu^2 M + \nu(2\lambda_0 M + C) + K(\lambda_0)$$

has the eigenvalues shifted by $-\lambda_0$ so that all of them become negative. Set

$$A = \begin{bmatrix} 0 & K^{1/2}M^{-1/2} \\ M^{-1/2}K^{1/2} & M^{-1/2}CM^{-1/2} \end{bmatrix}, \quad J = \begin{bmatrix} 1 & 0 \\ 0 & -1 \end{bmatrix}. \tag{8}$$

Then

$$JA - \lambda J = Z \begin{bmatrix} -\lambda & 0 \\ 0 & \frac{K(\lambda)}{\lambda} \end{bmatrix} Z^*, \quad Z = \begin{bmatrix} 1 & 0 \\ -M^{-1/2}K^{1/2}/\lambda & M^{-1/2} \end{bmatrix}.$$

Thus, $\Pi_+(K(\lambda)) = \Pi_-(JA - \lambda J)$ and the latter is equal to k; this is best seen in diagonalising the J-Hermitian matrix A by a J-unitary similarity.

We proceed to prove the first inequality from (5). Supposing the contrary there would exist a λ, not an eigenvalue of $K_{11}(\cdot)$, with $\alpha_k^+ < \lambda < \lambda_k^+$. Now

$$K(\lambda) = W \begin{bmatrix} K_{11}(\lambda) & 0 \\ 0 & K_{22}(\lambda) - K_{12}(\lambda)^* K_{11}(\lambda)^{-1} K_{11}(\lambda) \end{bmatrix} W^* \tag{9}$$

with

$$W = \begin{bmatrix} 1 & 0 \\ K_{12}(\lambda)^* K_{11}(\lambda)^{-1} & 1 \end{bmatrix} \tag{10}$$

By the Sylvester inertia theorem

$$\Pi_+(K_{11}(\lambda)) \leq \Pi_+(K(\lambda)) \leq \Pi_+(K_{11}(\lambda)) + n - m. \tag{11}$$

Now, $\alpha_k^+ < \lambda < \lambda_k^+$ would imply

$$\Pi_+(K_{11}(\lambda)) \geq k, \quad \Pi_+(K(\lambda)) \leq k - 1$$

which contradicts the first inequality in (11). In a similar way would $\lambda_k^+ < \lambda < \alpha_{k+m-n}^+$ imply

$$\Pi_+(K(\lambda)) \geq k, \quad \Pi_+(K_{11}(\lambda)) \leq k + m - n - 1$$

which contradicts the second inequality in (11). This proves (5). The proof of (6) is completely analogous.

Weakly hyperbolic pencils. The pencil $K(\cdot)$ is *weakly hyperbolic*, if M is positive definite and $K(\mu)$ is negative semidefinite for some real μ. In this case the two strict inequalities appearing in (2) and (4) are correspondingly weakened. The proof of (5) and (6) will immediately be carried over, if the validity of (7) is extended to the weakly hyperbolic case. This can be done by the continuity argument. The pencil $K(\lambda, \epsilon) = \lambda^2 \epsilon M + \lambda C + K$ will be hyperbolic for every $\epsilon < 1$. In the limit $\epsilon \to 1$ the inequality (7) is established by the known continuity of the eigenvalues. Thus, the interlacings (5), (6) hold in the weakly hyperbolic case as well.

Gyroscopic pencils. The pencil $K(\cdot)$ is *gyroscopic*, if C is skew-Hermitian and M, K Hermitian positive definite. By substituting $\lambda = i\mu$ we obtain

$$\hat{K}(\mu) = K(i\mu) = -\mu^2 M + i\mu C + K$$

and $-\hat{K}(\cdot)$ is hyperbolic, so (5), (6) immediately apply.

References

[1] Azizov, T., Dijksma, A., Förster, K.-H., Jonas, P., Quadratic (weakly) Hyperbolic Matrix Polynomials: Direct and Inverse Spectral Problems, Operator Theory: Advances and Applications, Vol. 198, Birkhäuser Basel 2009 (this volume).

[2] Gohberg, I., Lancaster, P., Rodman, L., Matrices and indefinite scalar products, Birkhäuser Basel 1983.

Krešimir Veselić
Fernuniversität Hagen
Fakultät für Mathematik und Informatik
Postfach 940
D-58084 Hagen, Germany
e-mail: kresimir.veselic@fernuni-hagen.de.

Operator Theory: Advances and Applications (OT)

Edited by
Israel Gohberg, Tel Aviv University, Israel

BIRKHÄUSER

This series is devoted to the publication of current research in operator theory, with particular emphasis on applications to classical analysis and the theory of integral equations, as well as to numerical analysis, mathematical physics and mathematical methods in electrical engineering.

OT 198: Behrndt, J. / Förster, K.-H. / Trunk, C. (eds.), Recent Advances in Operator Theory in Hilbert and Krein Spaces (2009).
ISBN 978-3-0346-0179-5

OT 197: Alpay, D. / Vinnikov, V. (eds.), Characteristic Functions, Scattering Functions and Transfer Functions. The Moshe Livsic Memorial Volume (2009).
ISBN 978-3-0346-0182-5

OT 196: Maz'ya, V. / Soloviev, A., Boundary Integral Equations on Contours with Peaks (2009). ISBN 978-3-0346-0170-2

OT 195: Grobler, J.J. / Labuschagne, L.E. / Möller, M. (eds.), Operator Algebras, Operator Theory and Applications (2009).
ISBN 978-3-0346-0173-3

OT 194: González, M. / Martínez-Abejón, A,, Tauberian Operators (2009).
ISBN 978-3-7643-8997-0

OT 193: Cialdea, A. / Lanzara, F. / P.E. Ricci (eds.), Analysis, Partial Differential Equations and Applications. The Vladimir Maz'ya Anniversary Volume (2009). ISBN 978-3-7643-9897-2

OT 192: Gohberg, I. / Leiterer, J., Holomorphic Operator Functions of One Variable and Applications. Methods from Complex Analysis in Several Variables (2009).
ISBN 978-3-0346-0125-2

OT 190/191: Adamyan, V. / Berezansky, Yu.M. / Gohberg, I. / Gorbachuk, M.L. / Gorbachuk, V. / Kochubei, A.N. / Langer, H. / Popov, G. (eds.), Modern Analysis and Applications. The Mark Krein Centenary Conference (2009).
Volume 1 (OT 190): Operator Theory and Related Topics. ISBN 978-3-7643-9918-4
Volume 2 (OT 191): Differential Operators and Mechanics. ISBN 978-3-7643-9920-7
Set Volumes 1/2: ISBN 978-3-7643-9924-5

OT 189: Rodino, L. / Wong, M.W. (eds.), New Developments in Pseudo-Differential Operators (2008). ISBN 978-3-7643-8968-0

OT 188: Behrndt, J. / Förster, K.-H. / Langer, H. / Trunk, C. (eds.), Spectral Theory in Inner Product Spaces and Applications (2008).
ISBN 978-3-7643-8910-9

OT 187: Ando, T. / Curto, R.E. / Jung, I.B. / Lee, W.Y. (eds.), Recent Advances in Operator Theory and Applications (2008).
ISBN 978-3-7643-8892-8

OT 186: Janas, J. / Kurasov, P. / Laptev, A. / Naboko, S. / Stolz, G. (eds.), Methods of Spectral Analysis in Mathematical Physics (2008).
ISBN 978-3-7643-8754-9

OT 185: Vasilevski, N., Commutative Algebras of Toeplitz Operators on the Bergman Space (2008).
ISBN 978-3-7643-8725-9

OT 184: Qin, Y., Nonlinear Parabolic-Hyperbolic Coupled Systems and Their Attractors (2008). Subseries **A**dvances in **P**artial **D**ifferential **E**quations. ISBN 978-3-7643-8813-3

OT 183: Nazaikinskii, V.E.. / Savin, A.Yu. / Sternin, B.Yu., Elliptic Theory and Noncommutative Geometry (2008). Subseries **A**dvances in **P**artial **D**ifferential **E**quations.
ISBN 978-3-7643-8774-7

OT 182: van der Mee, C., Exponentially Dichotomous Operators and Applications (2008). Subseries **L**inear **O**perators and **L**inear **S**ystems. ISBN 978-3-7643-8732-7

OT 181: Bastos, A. / Gohberg, I. / Lebre, A.B. / Speck, F.-O. (eds.), Operator Algebras, Operator Theory and Applications (2008).
ISBN 978-3-7643-8683-2

MIX
Papier aus verantwortungsvollen Quellen
Paper from responsible sources
FSC® C105338

If you have any concerns about our products,
you can contact us on
ProductSafety@springernature.com

In case Publisher is established outside the EU,
the EU authorized representative is:
Springer Nature Customer Service Center GmbH
Europaplatz 3, 69115 Heidelberg, Germany

Printed by Libri Plureos GmbH
in Hamburg, Germany